NUMBER THEORY

WITH COMPUTER APPLICATIONS

NUMBER THEORY

WITH

COMPUTER APPLICATIONS

RAMANUJACHARY KUMANDURI

Department of Mathematics
Columbia University

CRISTINA ROMERO

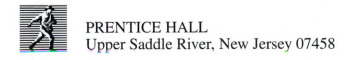

PRENTICE HALL
Upper Saddle River, New Jersey 07458

Library of Congress Cataloging-in-Publication Data

Kumanduri, Ramanujachary
    Number theory with computer applications / Ramanujachary
Kumanduri, Cristina Romero.
       p. cm.
    Includes index.
    ISBN 0-13-801812-X (alk. paper)
    1. Number theory.   2. Number theory—Computer-assisted
instruction.   I. Romero, Cristina.   II. Title
QA241.K85 1998   97-16756
512'.7—dc21   CIP

Acquisition Editor: George Lobell
Editorial Assistant: Gale Epps
Editorial Director: Tim Bozik
Editor-in-Chief: Jerome Grant
AVP, Production and Manufacturing: David W. Riccardi
Production Editor: Elaine W. Wetterau
Managing Editor: Linda Mihatov Behrens
Executive Managing Editor: Kathleen Schiaparelli
Manufacturing Buyer: Alan Fischer
Manufacturing Manager: Trudy Pisciotti
Director of Marketing: John Tweeddale
Marketing Manager: Melody Marcus
Creative Director: Paula Maylahn
Art Director: Jayne Conte
Cover Design: Jayne Conte
Cover Illustration: Moroccan Tiles from a Building in Tangiers, Photography by John Hedgecoe

©1998 by Prentice-Hall, Inc.
Upper Saddle River, New Jersey 07458

10 9 8 7 6 5 4 3 2

ISBN 0-13-801812-X

Prentice-Hall International (UK) Limited, *London*
Prentice-Hall of Australia Pty. Limited, *Sydney*
Prentice-Hall Canada Inc., *Toronto*
Prentice-Hall Hispanoamericana, S.A., *Mexico*
Prentice-Hall of India Private Limited, *New Delhi*
Prentice-Hall of Japan, Inc., *Tokyo*
Prentice-Hall Asia Pte. Ltd., *Singapore*
Editora Prentice-Hall do Brasil, Ltda., *Rio de Janeiro*

List of Photographs

The photograph of D. H. Lehmer's sieve machine is from The Computer Museum, Boston. The photograph of Plimpton 322 is from the Rare Book and Manuscript Library, Columbia University. All the other photographs are from the David Eugene Smith Collection, Rare Book and Manuscript Library, Columbia University.

Contents

Preface

This text evolved from a course taught by the first author at Columbia University. Our goal has been to write a flexible text that includes the basic theory along with numerous contemporary applications of number theory. The material is suitable for an introductory course in number theory for undergraduate students. The student is assumed to have completed a course in high-school algebra, but no knowledge of calculus is necessary for the first half of the book.

A novel feature of the book is the numerous computer experiments by which one can discover interesting properties of integers. Experiment and pattern recognition are an important part of number theory, the means by which new results are discovered. The purpose of the experiments is to gather sufficient numerical evidence to observe the underlying patterns and make a reasonable conjecture. The reader is invited to discover the important results and to prove them. While many results of the experiments are proved in a later chapter of the text when it is essential for the development of the subject, many more are left for the reader's enjoyment.

To the Student: You should do as many exercises as possible to internalize and reinforce the basic concepts. In addition, many interesting theorems are presented in the exercises and projects. There are also a number of exercises interspersed in the body of the text. Many of these are similar to the examples preceding them, and doing these exercises is a way to test your understanding of the examples. Many of the theoretical exercises reinforce the proof techniques. Careful study of the proofs given in the text is essential to solving them.

To the Instructor: There is plenty of material in the text for a one-year course in number theory. An introductory semester-long course with an emphasis on contemporary applications can be fashioned out of Chapters 1-9. If the emphasis is not on applications, then Chapters 1–4, 7, 9, 11, and 14 comprise an excellent introduction to the subject. The later chapters are suitable for independent study, or a second, more advanced course in number theory. Chapters 15 and 16 require some knowledge of complex analysis, so are suitable for advanced undergraduate students. Chapter 18 can be studied in conjunction with the theory of quadratic fields and ideal class groups.

The following diagram shows the prerequisites of various chapters. A more detailed organization of the individual sections is available from the web page listed below.

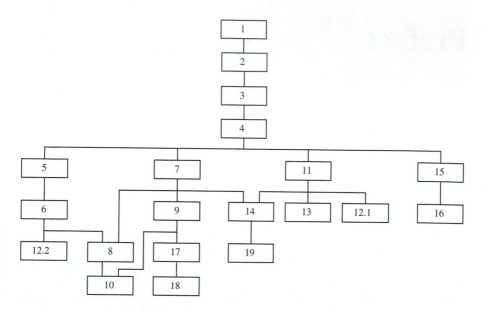

While it is possible to use the text without using any computers, it is best to use some type of software for doing the computer exercises. The students should be encouraged to write their own programs as much as possible. See the note below on electronic resources for more information.

There are several projects at the end of many chapters. A project might explore a challenging application, or delve deeper into more advanced topics. The projects can be assigned to groups of two or three students. Classroom experience has shown that projects are one of the most enjoyable aspects of the number theory course.

Electronic Resources: The authors have placed a variety of electronic resources for this book at their site on the world wide web. The reader can obtain instruction manuals for using *Mathematica, Maple,* C, or C++ with this book. The instruction manuals also contain many examples of useful programs implementing the algorithms presented in the book. Instructors can also obtain complete programs for all the algorithms. For information on software, additional exercises, projects, and other number theory related information, point your web browser to

        http://www.math.columbia.edu/~rama/book.html

Acknowledgments: Many people have played a part in the development of this book. The first author wishes to express his gratitude to the many students who read preliminary versions of the text, suggested changes, and found errors. These students include Michael Feldman, Ezra Freedman, Rachel Sendrovic, Chaya Wolman, Eric Jay Wolf, and Justin Zaglio. We thank Romuald Dabrowski and Susanna Fishel for their suggestions and encouragement.

The authors are also grateful to the following reviewers for their detailed comments and suggestions.

Tilak de Alwis, Southeastern Louisiana University

David Farmer, Bucknell University

Robert E. Kennedy, Central Missouri State University

Charles J. Parry, Virginia Tech

Dan Reich, Temple University

Jay W. Sweet, Jr., Florida International University

We are particularly grateful to David Farmer, who read the entire manuscript and made numerous suggestions that led to substantial changes. We wish to thank the staff at Prentice Hall for their assistance. We thank George Lobell, Executive Editor; Elaine Wetterau, Production Editor; and Gale Epps, Editorial Assistant, for their encouragement and support. Special kudos to Elaine Wetterau, who read the manuscript several times and made numerous suggestions that improved the style and quality of the presentation.

R. K.
C. R.

NUMBER THEORY

WITH COMPUTER APPLICATIONS

Chapter 1

Introduction

God made integers, all else is the work of man.

— L. KRONECKER

Number theory is the study of the divisibility properties of the integers. The natural numbers are one of the oldest and the most fundamental mathematical objects. Since ancient time, human beings have been fascinated by the magical, mystical properties of numbers. The numerous intriguing properties of numbers have led a great number of mathematicians and non-mathematicians to devote considerable energies to their study. The result has been the development of a beautiful and powerful theory, whose aim is to answer questions about the integers. Our goal is to study the elementary aspects of a wide array of techniques available to mathematicians. We begin with a description of some classical problems in number theory.

One of the oldest mathematical constructions is that of a right triangle with integer sides. If a, b, and c are the sides of the right triangle, then, as it is well known, $a^2 + b^2 = c^2$ (see Figure 1.1 for a proof). A triple of integers (a, b, c) satisfying $a^2 + b^2 = c^2$ is called a Pythagorean triple. Here are a few examples of Pythagorean triples.

a	3	5	7	8	9	11
b	4	12	24	15	40	60
c	5	13	25	17	41	61

One of the earliest results in number theory (due to Greek geometers) is a complete description of Pythagorean triples. In this classification, one sees that the hypotenuse is a multiple of a sum of two squares. For example, $5 = 2^2 + 1^2$, $13 = 3^2 + 2^2$, etc. We can show that 3 and 7 are not values for

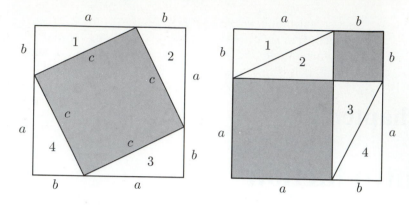

Figure 1.1 Proof of the Pythagorean Theorem

the hypotenuse of a right triangle because they are not a multiple of a sum of two integral squares.

One of the giants in the history of number theory is Pierre de Fermat. He determined which integers occur as the hypotenuse of a right triangle. Fermat showed that a prime number of the form $4k + 1$ occurs as the hypotenuse of a right triangle, but a prime of the form $4k + 3$ is never the hypotenuse of a right triangle. This result of Fermat follows from the study of representation of integers as a sum of two squares.

Fermat also studied the sums of four squares and conjectured that every natural number is the sum of four squares. For example,

$$7 = 2^2 + 1^2 + 1^2 + 1^2$$
$$14 = 3^2 + 2^2 + 1^2 + 0^2$$
$$37 = 6^2 + 1^2 + 0^2 + 0^2.$$

Fermat's conjecture was later proved by Joseph Louis Lagrange.

Mathematicians like Leonhard Euler, Adrien-Marie Legendre, and Carl Friedrich Gauss studied numbers of the form $x^2 + 2y^2$, $x^2 + 3y^2$, etc. This study leads to the beautiful theory of quadratic forms and the quadratic reciprocity law. An example of the type of theorems that are proved in this theory is the following: a prime number is represented in the form $x^2 + 5y^2$ if and only if it is of the form $20k + 1$ or $20k + 9$. For example, $29 = 3^2 + 5 \cdot 2^2$ and $41 = 6^2 + 5 \cdot 1^2$.

Another source of speculation, since ancient times, has been about numbers arising from geometric figures. The triangular numbers are

$$1, 3, 6, 10, 15, 21, 28, 36, \ldots$$

The nth triangular number counts the number of dots in the triangle with n dots in each side. These are shown in Figure 1.2. Similarly, we define square and pentagonal numbers to correspond to the number of dots in the corresponding geometric shape. An interesting problem is to find numbers that are simultaneously square and triangular. For example, 1 and 36 are square and triangular. It is possible to find many more. The reader is invited to find a few more numbers that are square and triangular and to look for a relationship among them.

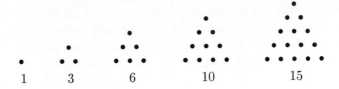

1   3   6   10   15

Figure 1.2 Triangular numbers

Fermat conjectured that every positive integer is a sum of at most three triangular numbers; every natural number is a sum of at most four square numbers; every natural number is a sum of at most five pentagonal numbers; and so on for hexagonal, heptagonal, and other polygonal numbers. For example, regarding the sum of triangular numbers, we have

$$32 = 28 + 3 + 1$$
$$33 = 15 + 15 + 3$$
$$34 = 28 + 6$$
$$35 = 28 + 6 + 1.$$

Gauss proved Fermat's conjecture for triangular numbers, as noted in his journal entry for July 10, 1796.

$$\text{EΥPHKA!} \quad \text{num} = \triangle + \triangle + \triangle.$$

In studying the divisibility properties of integers, we soon find that there are some numbers that are not divisible by any other. These are the prime numbers; all other numbers can be factored into a product of prime numbers. The prime numbers are the basic building blocks of the integers; hence their study is of utmost importance in number theory.

The fundamental result about prime factorization is also the source of one of the most difficult problems in mathematics. How do we tell if a number is prime? If it is not prime, how can we find its prime factors? It is easy to see that 17 is prime, and with some effort, we can see that $2^{17} - 1$ is prime, but it becomes very difficult to decide if $2^{127} - 1$ is prime. Recent advances have made it possible for us to check in a matter of seconds if a number with several hundred digits is prime.

The problem of computing the prime factors of large composite numbers is considered intractable, and this difficulty is the source of numerous contemporary applications of number theory to cryptography. The science of cryptography deals with the transmission of secure information. The problem of integer factorization can be used in clever ways to construct secure cryptosystems. All of these are based on the proposition, which is yet to be proved, that integer factorization is a difficult problem.

The study of the distribution of prime numbers is another important part of number theory. Euclid showed that there are infinitely many primes, and Peter Gustav Lejeune Dirichlet showed that there are infinitely many primes in every suitable arithmetic progression. But for sequences which are not arithmetic progressions, no results are available. For example, no one knows if there are infinitely many primes of the form $n^2 + 1$ or if there are infinitely many primes of the form $2^p - 1$. Most questions about the number of primes and their distribution require the use of complex function theory.

Number theory is characterized by the ease with which one can discover interesting facts, yet obtaining a rigorous proof of these facts can frustrate the greatest mathematicians. Numerous examples of facts discovered by computation, but whose proofs are maddeningly difficult, can be found in the text. Usually, a host of tools from different branches of mathematics must be employed in their solution. Many branches of mathematics have their origins in number theory. For example, the attempts to prove that the equation $x^n + y^n = z^n$ has no nontrivial solutions for $n \geq 3$ (Fermat's Last Theorem) led to the development of ideal theory of algebraic numbers. The problem spurred advances in arithmetic and algebraic geometry. Fermat's Last Theorem was finally settled by exploiting the connections between elliptic curves and modular forms. Even though Fermat's Last Theorem is without consequence, attempts to solve it have led to the development of important mathematics.

--- **Exercises** ---

1. Show that if there are integers r and s such that

$$a = r^2 - s^2, \quad b = 2rs, \quad c = r^2 + s^2,$$

   then (a, b, c) forms a Pythagorean triple.

2. Use the form of the Pythagorean triples given in the previous exercise to find an example [different from $(3, 4, 5)$] of a right triangle with consecutive legs.

3. Find a right triangle with smallest leg of length 13.

4. Are there infinitely many right triangles in which the hypotenuse and a leg are consecutive integers?

5. Show that the nth triangular number is $\dfrac{n(n+1)}{2}$.

6. The pentagonal numbers are shown in Figure 1.3. Determine a formula for the nth pentagonal number.

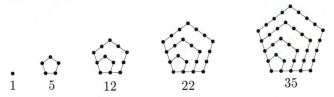

1   5   12   22   35

Figure 1.3 Pentagonal numbers

7. (Nicomanchus, 300 A.D.) Show that the sum of two consecutive triangular numbers is a square. Prove that the sum of the nth square and $(n-1)$st triangular numbers is the nth pentagonal number.

8. In how many ways can you express 84 as a sum of four squares?

9. Make a list of numbers that can be written as sums of three squares. Which numbers are not in your list? What can you say about the numbers that are not sums of three squares?

10. Write a computer program to verify Fermat's conjecture about the representation of natural numbers as sums of polygonal numbers.

11. Find all integer solutions to the equation $x^y = y^x$ when $x \neq y$.

12. Write a computer program to find integer solutions to the equation $x^y y^z = z^x$. Are there infinitely many solutions?

Chapter 2

Divisibility and Primes

Why add prime numbers?
Prime numbers are made to be multiplied,
not added.

— LEV LANDAU

2.1   Divisibility

Our object of study is the integers under multiplication. It is the operation of multiplication that creates a complicated and interesting structure, and is the focus of our study. The entire subject of number theory has its roots in trying to understand the multiplicative structure of the integers. Hence we start with the very basic notion of divisibility.

Definition 2.1.1. If a and b are integers, we say that a **divides** b (denoted as $a \mid b$) if there exists an integer c such that $b = ac$. If no such c exists, then a does not divide b (denoted by $a \nmid b$). If a divides b, we say that a is a **divisor** of b, and b is divisible by a.

When we are given the definition of a new concept, it is wise to consider several examples to see what the definition means (and what it doesn't).

Example 2.1.2.   1. Since $24 = 2 \cdot 12$, and $24 = 3 \cdot 8$, we have $2 \mid 24$, $3 \mid 24$, $8 \mid 24$, and $12 \mid 24$. Also $24 = (-2)(-12)$; therefore, $-2 \mid 24$ and $-12 \mid 24$.

2. No nonzero integer is divisible by 0. If $a \neq 0$, then there cannot exist a c such that $a = 0 \cdot c$. The expression $0 \mid 0$ makes sense because $0 = 0 \cdot c$

7

for any integer c. This is very different from $\dfrac{0}{0}$: the latter is supposed to represent a fraction and is undefined, whereas $0 \mid 0$ is a divisibility assertion and makes sense because of the definition.

3. Every number a divides 0 because $0 = a \cdot 0$ is always true for any a.

4. The only positive divisors of 5 are 1 and 5. The only positive divisors of 17 are 1 and 17.

Exercise. What are all the divisors of 24?

Exercise. What are the first seven positive integers n whose only positive divisors are 1 and n?

We now state the basic properties of divisibility.

Lemma 2.1.3. *Suppose $a, b, c, x,$ and y are integers.*

(a) *If $a \mid b$ and $x \mid y$, then $ax \mid by$.*

(b) *If $a \mid b$ and $b \mid c$, then $a \mid c$.*

(c) *If $a \mid b$ and $b \neq 0$, then $|a| \leq |b|$.*

(d) *If $a \mid b$ and $a \mid c$, then $a \mid bx + cy$.*

PROOF. We prove parts (a) and (c) and leave the other two as exercises.

(a) The definition of divisibility implies that there exist integers u and v such that $b = au$ and $y = xv$. Multiplying the two integers, we get $by = (au)(xv) = (ax)(uv)$, which implies that $ax \mid by$.

(c) There exists an integer $u \neq 0$ such that $b = au$. Now $|b| = |a||u|$. Since u is an integer, we must have $|u| \geq 1$; hence $|a| \leq |b|$. ∎

It is clear that if $a \mid b$, then $-a \mid b$. We consider only positive divisors in many cases, as the negative divisors are taken into account by multiplying by -1. A positive number a always has a and 1 as divisors. A special set of numbers has no other positive divisors.

Definition 2.1.4. A positive integer $p > 1$ is **prime** if the only positive divisors of p are 1 and p. If p is not prime, then it is said to be **composite**.

The number 1 is not considered prime, even though it has only 1 as a positive divisor. The properties of 1 and -1 are very different from those of the prime numbers. This exclusion will also make the statement of many theorems simpler.

A fundamental tool in our study is the division theorem. Most of you have encountered it on numerous occasions, but seldom with a proof. The division theorem states that if we divide a by a positive integer b, then the remainder can be taken to be positive and less than b. One way to prove this is to find the quotient q and compute $a - bq$ to obtain the remainder. If b is negative, we still want to take a positive remainder. The proof we give is based on taking differences $a - b, a - 2b, \ldots$, and is given to illustrate the use of the well-ordering principle. You are asked to complete the alternate proof, of first computing the quotient, in Exercise 17.

The Division Theorem. *Given two integers a and b with $b \neq 0$, there exist unique integers q and r such that*

$$a = bq + r \quad and \quad 0 \leq r < |b|.$$

Remark. The number q is the **quotient** of a divided by b, and r is the **remainder**.

PROOF. The main conclusion of the theorem is the inequalities on r. These inequalities show that the remainder is the smallest positive number of the form $a - bq$. To prove the inequalities, we consider all combinations $a - bm$ where m is an integer.

$$S = \{\ldots, a + 2b, a + b, a, a - b, a - 2b, a - 3b, \ldots\}.$$

We claim that S has some positive elements in it, hence a smallest positive number. First, there are positive numbers in S because if $a > b$, then $a - b$ is positive, and if $a < b < 0$, then $a - kb$ is positive for k large. (If $a < 0 < b$, then $a - kb$ is positive when k is sufficiently negative.) Hence there is a smallest nonnegative number r in S. Let q be such that $a - bq = r$; then we must have $0 \leq r < |b|$. Otherwise, if $r \geq |b|$ and b is positive, then $a - b(q + 1)$ is less than r, positive, and lies in S. This follows from $a - b(q + 1) = a - bq - b = r - b$. This contradicts the choice of r as the smallest nonnegative integer in S. This proves that $0 \leq r < b$ for $b > 0$. The proof is similar for $b < 0$.

For the uniqueness of q and r, suppose that q_1, r_1 and q_2, r_2 both satisfy the condition in the theorem, that is,

$$a = bq_1 + r_1 = bq_2 + r_2 \text{ with } 0 \leq r_1 < |b| \text{ and } 0 \leq r_2 < |b|.$$

Subtracting the equations gives $b(q_1 - q_2) = r_2 - r_1$. Now $|r_2 - r_1| < |b|$, (why?), which implies that $q_1 - q_2 = 0$ (as $|b||q_1 - q_2| < |b|$ and $q_1 - q_2$ is an integer); therefore, $r_2 - r_1 = 0$, that is, $r_1 = r_2$, and hence $q_1 = q_2$. This proves the uniqueness. ∎

Observe that the main step of the proof is the choice of the remainder. Its existence follows from the *well-ordering principle*, which states that *every nonempty set of positive integers contains a smallest element*. To use this property, we first show that there are positive numbers in the set we are looking at. This technique, of first restricting the study to positive integers so that a smallest positive integer with some property can be chosen, will be employed frequently in the text.

Here are some examples to illustrate the division theorem:

Example 2.1.5.   1. Apply the theorem to $a = 37$, $b = 15$. Since $37 = 2 \cdot 15 + 7$, the quotient is 2 and the remainder 7.

2. If $a = 37$ and $b = -15$, then $37 = -2(-15) + 7$ implies that the remainder is 7.

3. If $a = -37$ and $b = 15$, then $-37 = -3 \cdot 15 + 8$ implies that $r = 8$, as we must have a positive remainder.

4. There are two possible remainders, 0 or 1, upon division by 2. A number divisible by 2 is called **even**; otherwise it is called **odd**. Hence an even number can be written as $2k$ and an odd number as $2k + 1$ for some k.

5. Division by 4 can yield the remainders $0, 1, 2, 3$. Hence every integer lies in one of the sequences $\{4k\}$, $\{4k + 1\}$, $\{4k + 2\}$, or $\{4k + 3\}$.

Exercise. What are the quotient and the remainder when -25 is divided by 8?

Example 2.1.6. An important application of the division theorem is the representation of integers to a positive base. For example, every positive integer n can be uniquely represented in base 10 as

$$n = a_k 10^k + a_{k-1} 10^{k-1} + \cdots + a_1 10 + a_0, \quad 0 \le a_i \le 9, \quad 0 \le i \le k.$$

Given n, there is a unique a_0 such that $n = 10q + a_0, 0 \le a_0 \le 9$. Next, divide q by 10 to get a_1; so $q = 10q_1 + a_1$ with $0 \le a_1 \le 9$, and hence $n = 10^2 q_1 + 10a_1 + a_0$. Continuing, we obtain the decimal representation. At each step, the remainders are unique, proving that the decimal representation of an integer is unique.

The same technique shows that an integer n has a unique base b representation:

$$n = a_k b^k + a_{k-1} b^{k-1} + \cdots + a_1 b + a_0, \quad 0 \le a_i < b, \quad 0 \le i \le k.$$

Example 2.1.7. Show that $2 \mid n^2 - n$ for all n.

There are many ways to prove this assertion. Observe that $n^2 - n = n(n-1)$ is the product of two consecutive integers. By the division theorem, one of them is even, so $n^2 - n$ is a multiple of 2.

Another way to prove the assertion is by mathematical induction. (See Appendix A.) If $n = 1$, then the assertion is clear. If the assertion is true for $n = k$, then we have to show that it holds for $n = k + 1$. If $2 \mid k^2 - k$, then we want to show that $2 \mid (k+1)^2 - (k+1)$. Expanding the square yields $(k+1)^2 - (k+1) = (k^2 - k) + 2k$. The first term is a multiple of 2 by the induction hypothesis; hence $2 \mid (k+1)^2 - (k+1)$. The statement is then true for all natural numbers by induction.

The theorem is not really an algorithm in the sense of yielding a procedure to compute the quotient and the remainder. Clearly, it is infeasible to consider the differences $a - kb$ until a positive value less than $|b|$ is found. In practice the quotient is computed by long division. When $b > 0$, the reader will recall that the quotient q is simply the integral part of the fraction a/b; and the remainder is then $a - bq$. The integral part of a real number is an important quantity; hence we introduce a special notation for it.

Definition 2.1.8. If x is a real number, the **greatest integer function** or **floor function** $\lfloor x \rfloor$ is defined to be the largest integer less than or equal to x. Similarly, the smallest integer greater than or equal to x is denoted by $\lceil x \rceil$, called the **ceiling function**.

Example 2.1.9.   1. It is clear that $\lfloor 2 \rfloor = 2$, $\lfloor -3.5 \rfloor = -4$, $\lceil -3.5 \rceil = -3$.

2. For any x, $\lfloor x \rfloor \leq \lceil x \rceil$, and the equality occurs when x is an integer.

Exercise. What is $\lfloor -23/7 \rfloor$?

The quotient q in the division theorem can be expressed as $\lfloor a/b \rfloor$ for $b > 0$. For $b < 0$, the quotient is the smallest integer greater than or equal to a/b, that is, $\lceil a/b \rceil$. Once q is found, r is obtained by setting $r = a - bq$. So we have a simple formula for the quotient in the division theorem. The remainder is an important quantity, and so we introduce notation for it. We write

$$r = a \bmod b$$

for the remainder obtained when a is divided by b.

The floor and ceiling functions have many uses, one of which is in counting integers in an interval.

Example 2.1.10. Consider the closed interval $[x, y]$. If x and y are integers, then it is easy to see that the number of integers in the interval is $y - x + 1$. We want to give a similar formula when x and y are not integers. To avoid

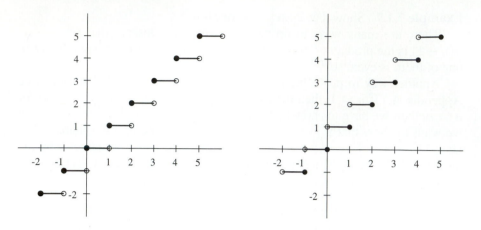

Figure 2.1.1 Graph of $\lfloor x \rfloor$ **Figure 2.1.2** Graph of $\lceil x \rceil$

considering many cases regarding the integrality of x or y, we use the floor and ceiling functions appropriately. The the number of integers in $[x, y]$ is the same as the number of integers k satisfying $\lceil x \rceil \leq k \leq \lfloor y \rfloor$. Hence there are $\lfloor y \rfloor - \lceil x \rceil + 1$ integers in $[x, y]$.

Lemma 2.1.11. *Let n and d be positive integers. The number of positive multiples of d that are less than or equal to n is $\lfloor n/d \rfloor$.*

PROOF. The positive multiples of d are $d, 2d, 3d, \ldots, kd, \ldots$. Consider any multiple kd such that $kd \leq n$. Then $k \leq n/d$, so $1 \leq k \leq \lfloor n/d \rfloor$. Then the previous example shows that the number of such k is $\lfloor n/d \rfloor$. ∎

Example 2.1.12. The number of multiples of 7 less than or equal to 200 is $\lfloor 200/7 \rfloor = 28$. The number of multiples of 7 less than or equal to 300 is $\lfloor 300/7 \rfloor = 42$. Hence the number of multiples of 7 in the interval $[200, 300]$ is $42 - 28 = 14$. Here we use the fact that 200 is not a multiple of 7.

Exercise. Determine the number of positive multiples of 11 less than 1000.

We state the properties of $\lfloor x \rfloor$ in the following lemma. These properties are used frequently in later sections.

Lemma 2.1.13 (Properties of $\lfloor x \rfloor$). *If x and y are real numbers and n is an integer, then*

(a) $x - 1 < \lfloor x \rfloor \leq x$.

(b) $\lfloor x + n \rfloor = \lfloor x \rfloor + n$.

(c) $\lfloor x + y \rfloor \geq \lfloor x \rfloor + \lfloor y \rfloor$.

(d) *If n is positive, then* $\left\lfloor \dfrac{x}{n} \right\rfloor = \left\lfloor \dfrac{\lfloor x \rfloor}{n} \right\rfloor$.

PROOF. We prove (b) and (d) and leave the proofs of the other parts to the reader. The proofs use the definition of $\lfloor x \rfloor$ as the largest integer $\leq x$.

(b) If $\lfloor x \rfloor = k$, then $k \leq x$ and $k + n \leq x + n$; hence $k + n \leq \lfloor x + n \rfloor$. On the other hand, $\lfloor x + n \rfloor \leq x + n$ implies that $\lfloor x + n \rfloor - n \leq x$. Since $\lfloor x + n \rfloor - n$ is an integer, it must be less than $\lfloor x \rfloor$.

(d) Let $\lfloor x \rfloor = nq + r$, with $0 \leq r < |n|$ and $x = \lfloor x \rfloor + \epsilon, 0 \leq \epsilon < 1$. Since n is positive, we have $q = \left\lfloor \dfrac{\lfloor x \rfloor}{n} \right\rfloor$ and $\lfloor \frac{x}{n} \rfloor = \lfloor q + r/n + \epsilon/n \rfloor$. Since $r + \epsilon < n$ and q is an integer, we must have $\lfloor q + (r + \epsilon)/n \rfloor = q$. ∎

———————————— **Exercises for Section 2.1** ————————————

1. What are the divisors of 1?

2. Prove the following:

   (a) If $a \mid b$ and $b \mid c$, then $a \mid c$.

   (b) If $a \mid b$ and $a \mid c$, then $a \mid mb + nc$ for all integers m and n.

   (c) If $a \mid b$, then $a \mid kb$ for every integer k.

3. What condition must a and b satisfy so that $a \mid b$ and $b \mid a$?

4. Determine all positive numbers $n \leq 50$ that are prime.

5. Apply the division theorem in the following cases, and find q and r in $a = bq + r$, $0 \leq r < |b|$.

   (a) $a = 300, b = 17$ (b) $a = 729, b = 31$

   (c) $a = 300, b = -17$ (d) $a = -729, b = 31$

6. Prove that the product of two consecutive integers is always divisible by 2, and the product of three consecutive integers is divisible by 3.

7. Show that the sum of two odd numbers is even, the sum of an odd and an even number is odd, and the sum of even numbers is even.

8. Show that when n is odd, $n^2 - 1$ is a multiple of 8.

9. Show that a perfect square is never of the form $3k + 2$ for any k. Conclude that if p is prime and $p \geq 5$, then $p^2 + 2$ is always composite.

10. Show that property (d) of Lemma 2.1.13 does not hold when n is negative.

11. Prove properties (a) and (c) of $\lfloor x \rfloor$ stated in Lemma 2.1.13.

12. Show that $\lfloor -x \rfloor = -\lceil x \rceil$.

13. Suppose a and b are two real numbers. Give a formula for the number of integers in the open interval (a, b) using the floor and ceiling functions. Similarly, derive formulas for the number of integers in the half-open intervals $(a, b]$ and $[a, b)$.

14. Determine the number of multiples of d in the interval $[n, m]$.

15. Find the number of integers less than or equal to 400 that are divisible by 3, by 5, and by 15.

16. Determine the number of integers between 200 and 400 divisible by 3, by 7, and by 3 or 7.

17. Give another proof of the division theorem as follows: If $b > 0$, let $q = \lfloor a/b \rfloor$, and if $b < 0$, take $q = \lceil a/b \rceil$. Show that $r = a - bq$ satisfies $0 \leq r < |b|$.

18. Show that for any real number x and integers m and n,

$$\left\lfloor \frac{x+n}{m} \right\rfloor = \left\lfloor \frac{\lfloor x \rfloor + n}{m} \right\rfloor.$$

19. Derive a formula for the number of digits in the decimal expansion of an integer using the floor function.

20. State properties of the ceiling function that are analogous to those of the floor function in Lemma 2.1.13 and prove them.

21. For any prime p, show that p divides the binomial coefficients $\binom{p}{k}$, $k = 1, \ldots, p - 1$. (The binomial coefficients are described in Appendix B.)

22. Show that $6 \mid n^3 - n$ for all integers n. Does $4 \mid n^4 - n$ for all integers n?

23. Show that $5 \mid n^5 - n$ for all integers n. Actually, $n^5 - n$ is a multiple of 30 for every n. Prove this assertion.

24. We want to generalize the results of the previous two exercises. Fix an integer m between 5 and 20 and make a table of values $n^m - n$ for different n, say n in the range 2 through 25. Does m divide all of these? Do this for all m satisfying $6 \leq m \leq 20$. Make a reasonable conjecture about the numbers m for which $n^m - n$ is always divisible by m. Can you prove your conjecture?

25. The Fibonacci numbers are defined by:

$$f_1 = 1$$
$$f_2 = 1$$
$$f_n = f_{n-1} + f_{n-2}.$$

Hence $f_3 = f_2 + f_1 = 1 + 1 = 2$, $f_4 = f_3 + f_2 = 2 + 1 = 3$, and so on. Make a table of the first 50 Fibonacci numbers. Which ones are even?

Similarly, identify the multiples of 3, of 5, and of other odd numbers. Prove your conjecture in a few cases. If m is an odd number, there may not seem a way to identify all Fibonacci numbers divisible by m. But in the special case when m is a Fibonacci number f_d, it is possible to find the Fibonacci numbers that are multiples of f_d. Determine the multiples of f_d and prove your conjecture.

26. Make a list of a few odd primes p that can be written as a sum of two squares (i.e., as $p = x^2 + y^2$); for example, $5 = 2^2 + 1^2$, $13 = 3^2 + 2^2$. What is the remainder when these primes are divided by 4? Prove your guess about the form of the remainder by first showing that one of x or y must be odd and the other even.

    (This exercise excludes certain types of primes from being of the form $x^2 + y^2$; it is more difficult to show that the remaining ones can be expressed in this form.)

27. Show that p, $p + 2$, and $p + 4$ cannot all be prime except when $p = 3$.

28. Determine all primes of the form $n^3 - 1$.

29. The goal of this exercise is to look at numbers of the form $a^n - 1$ and determine when they can be prime. It will be convenient to use the formula for the sum of a geometric progression

$$(r - 1)(1 + r + \cdots + r^{n-1}) = r^n - 1.$$

    (a) Show that for $n > 1$, $x^n - 1$ cannot be prime unless $x = 2$.

    (b) Recall the technique of long division of polynomials. Determine the remainder and quotient when $x^{16} - 1$ is divided by $x^3 - 1$. Explain why $x^2 - 1 \nmid x^{15} - 1$, but $x^3 - 1 \mid x^{15} - 1$. Do you see that $x^4 - 1 \nmid x^{15} - 1$, but $x^4 - 1 \mid x^{16} - 1$? Try a few other examples of division of polynomials of the form $x^n - 1$ by polynomials of the form $x^m - 1$ for $m < n$. What can you say about the remainder and the quotient? Do you see a general pattern? Use this technique with $x = 2$ to show that $2^n - 1$ is composite when n is composite.

    (c) Parts (a) and (b) show that the only primes of the form $a^n - 1$ are of the form $2^p - 1$ for p prime. Are all numbers of the form $2^p - 1$ prime for p prime? Prove or disprove.

2.2 Primes

The definition of a prime number is a negative assertion; that is, if we fail to find any nontrivial divisors of a number, then it must be prime. This makes it hard to show that a number is prime; but to show that a number is composite we just need to exhibit a divisor. We can easily check that 5, 7, and 31 are prime. It may take a bit longer to determine that 911 is prime, but the use of the definition is impractical for much larger numbers. One of our primary goals is to develop better methods for testing primality. But before we do that, we must demonstrate the importance of prime numbers. Primes are the building blocks of the integers. We will show that most questions about divisibility reduce to questions about prime numbers.

To demonstrate that the primes are the building blocks of the integers, we observe that any composite number n has a nontrivial divisor a (different from 1 or n) and can be written as $n = ab$.

Proposition 2.2.1. *Every positive integer can be decomposed as a product of prime numbers.*

PROOF. The proof will have to be by contradiction as there is no known efficient method to find a prime factor of an arbitrary integer. Assume, for the sake of contradiction, that there are positive integers that are not products of primes. Suppose n is the smallest positive integer that cannot be written as a product of primes; then n is not prime (otherwise, it is a trivial product of primes consisting of one prime alone); hence $n = ab$, $1 < a, b < n$. Both a and b are smaller than n, and by hypothesis a and b are products of primes; hence n is a product of prime numbers.   ∎

Note that the proof works with positive numbers because we can choose the smallest number without a factorization and derive a contradiction. Once proved for positive integers, negative integers can also be written as products of prime numbers and -1.

The fundamental role played by primes in arithmetic is due to the factorization into primes and its uniqueness; that is, there is only one way to write a number as a product of primes (proved in Section 2.3).

To learn more about prime numbers, we begin with the classic result of Euclid on the number of primes.

Theorem 2.2.2 (Euclid). *There are infinitely many prime numbers.*

PROOF. Assume for the sake of contradiction that there are only a finite number of primes. Let $p_1, p_2, \ldots, p_k$ be the complete list of primes. We will show that this list cannot be complete by demonstrating the existence of a prime not in the list. Consider $N = p_1 p_2 \cdots p_k + 1$. Now N is a product of primes and hence some p_i divides it. Since $p_i \mid (p_1 \cdots p_i \cdots p_k)$, Lemma 2.1.3 implies that $p_i \mid (N - p_1 \cdots p_k)$, that is, $p_i \mid 1$. This is not possible as $p_i > 1$. We have derived a contradiction because the only divisors of 1 are ± 1. Therefore, any prime factor of N is not one of $p_1, \ldots, p_k$, so there must be an infinite number of primes.   ∎

Exercise. As a variation on the proof, we can replace the number N constructed in the proof by any other number for which we can show that none of the p_i divide it. For example, show that taking $N = n!$ for some integer $n > p_k$ (where p_k is the largest prime) proves the infinitude of primes.

There are many proofs of the infinitude of primes; we will give a few more in later sections.

An immediate problem raised by the factorization of integers into primes is of a practical nature; how do we find the prime factorization of an integer? First, we need a test to identify prime numbers. To check if n is prime, we could check if any number between 1 and n is a divisor of n. This would require about n divisions. If n is composite and $n = ab$, then both a and b cannot be larger than $\sqrt{n}$; otherwise ab would be larger than n. Therefore, a composite number n has a divisor less than or equal to $\sqrt{n}$, so it suffices to test for divisibility by integers up to $\sqrt{n}$, requiring about $\sqrt{n}$ divisions.

Proposition 2.2.3 (Primality Test). *A number p is prime if and only if it is not divisible by any prime q, $1 < q \le \sqrt{p}$.*

PROOF. This follows from the above observation that if p is composite, it has a divisor less than or equal to $\sqrt{p}$, hence a prime divisor less than or equal to $\sqrt{p}$. ■

Example 2.2.4. Let's prove that 101 is prime. Because $\lfloor \sqrt{101} \rfloor = 10$ we check if 101 is divisible by any of the primes $2, 3, 5, 7$. Now,

$$101 = 2 \cdot 50 + 1$$
$$101 = 3 \cdot 33 + 2$$
$$101 = 5 \cdot 20 + 1$$
$$101 = 7 \cdot 14 + 3,$$

so 101 is not divisible by any of the primes smaller then $\sqrt{101}$, proving the primality.

Exercise. Is 127 prime?

To prove the primality of a number by this method, we need a table of small primes. Such a table can be constructed using the Sieve of Eratosthenes. The procedure is as follows. To find all the primes less than or equal to n, we write the integers from 2 through n. Remove the multiples of 2 larger than 2. The next number in the list is 3; we keep 3 and remove all its multiples, $3k$, $k > 1$. The next number is 5 and we remove all multiples of 5 larger than 5. (The first remaining multiple of 5 is 25 because 10, 15, and 20 have already been removed.) We carry out this procedure until we reach $\sqrt{n}$; then all the remaining numbers up to n must be prime. (Why?)

Let us illustrate this by finding all the primes less than 50. The integers from 1 through 50 are written in an array. In the first pass, we remove all even numbers except 2; in the second pass, the remaining multiples of 3: $9, 15, 21, 27, 33, 39$, and 45 are removed. Then the remaining multiples of 5

Eratosthenes: (c. 230 B.C.)

Eratosthenes was a native of Cyrene. Exact details about his early life do not seem to be known. He excelled in many fields—poetry, astronomy, history, and mathematics. Eratosthenes was appointed the Librarian of the University in Alexandria by Ptolemy III. His most important accomplishment was a measurement of the circumference of the earth.

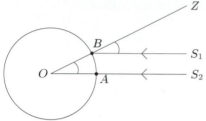

Eratosthenes observed that at noon on the day of the summer solstice, the sun shone directly down a deep well at Syene, now known as Aswan, (point A in the figure). At the same time at Alexandria (point B in figure), the sun was found to cast a shadow indicating that at Alexandria, the sun's angular distance from the Zenith, Z, was one fiftieth of a circle. This implies that the circumference of the Earth is fifty times the distance between Syene and Alexandria, about 500 miles. This gives a value of $25,000$ miles for the circumference of the Earth, while the actual value is $24,818$ miles.

are 25 and 35, and 49 is the only remaining multiple of 7. We can stop now because $\lfloor\sqrt{50}\rfloor = 7$. Below is the array with the integers that are removed by the sieve crossed out.

$$
\begin{array}{cccccccccc}
1 & 2 & 3 & \cancel{4} & 5 & \cancel{6} & 7 & \cancel{8} & \cancel{9} & \cancel{10} \\
11 & \cancel{12} & 13 & \cancel{14} & \cancel{15} & \cancel{16} & 17 & \cancel{18} & 19 & \cancel{20} \\
\cancel{21} & \cancel{22} & 23 & \cancel{24} & \cancel{25} & \cancel{26} & \cancel{27} & \cancel{28} & 29 & \cancel{30} \\
31 & \cancel{32} & \cancel{33} & \cancel{34} & \cancel{35} & \cancel{36} & 37 & \cancel{38} & \cancel{39} & \cancel{40} \\
41 & \cancel{42} & 43 & \cancel{44} & \cancel{45} & \cancel{46} & 47 & \cancel{48} & \cancel{49} & \cancel{50}.
\end{array}
$$

Hence the primes less than 50 are

$$2,\ 3,\ 5,\ 7,\ 11,\ 13,\ 17,\ 19,\ 23,\ 29,\ 31,\ 37,\ 41,\ 43,\ 47.$$

Algorithm 2.2.5 (Sieve of Eratosthenes). This algorithm constructs a table of primes up to n.

1. [Initialize] Let $P = \{2, 3, 4 \ldots, n\}$, $i = 1$. (We will use p_i to denote the ith element of P, where P is always arranged in increasing order.)

2. [Are we done?] If $p_i > \sqrt{n}$, go to step 4.

3. [Remove multiples of p_i] Remove kp_i from P for $2 \le k \le \lfloor n/p_i \rfloor$. Increment $i = i + 1$ and return to step 2.

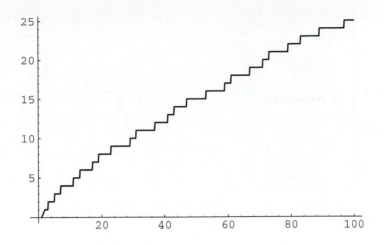

Figure 2.2.1 Graph of $\pi(x), 1 \leq x \leq 100$

4. [Output] Store P and terminate algorithm.

The algorithm can be made more efficient by noticing that some multiples of p_i have already been removed, so we have fewer and fewer numbers to delete. In the example above, when we are removing multiples of 5, the first multiple is 25, and for 7, the first multiple remaining is 49. It is also possible to modify the sieve of Eratosthenes to produce a list of prime numbers in the interval $[n, m]$.

On a computer system, this sieve needs to be performed only once, and a reasonably large table of primes can be stored in memory. For practical uses, say factoring algorithms, one needs a table of primes up to 10^6. Efficient methods for storing and retrieving a table of primes are discussed by Riesel.[1]

Using a table of primes, our algorithm for primality testing is as follows:

Algorithm 2.2.6. Given $n \leq 10^{12}$, this method determines if n is prime.

1. [Initialize] Input a table of primes up to 10^6 into P. Let p_i be the ith element of P. Let $i = 1$.

2. [Are we done?] If $p_i > \sqrt{n}$, return n is prime.

3. [Test if composite] If $p_i \mid n$, return n is composite.

4. [Get next prime] $i = i + 1$ and go to step 2.

---

[1]H. Riesel. *Prime Numbers and Computer Methods of Factorization.* Boston-Basel-Stuttgart: Birkhauser, 1985.

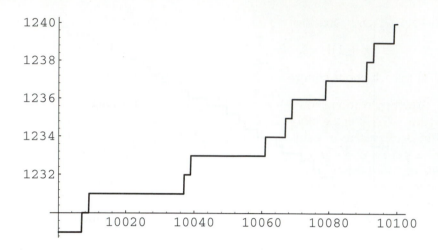

Figure 2.2.2 Graph of $\pi(x) : 10000 \leq x \leq 10100$

For an algorithm to be practical, it should not take too long to produce a result. How many steps are performed in the previous algorithm? A number with small prime factors will be found composite in a few steps, but if n is prime, we have to divide by all the primes up to $\sqrt{n}$ before we are certain of its primality. This is the maximum number of operations performed by the algorithm. To gain a better understanding of the number of steps in the algorithm, we need to know the number of primes in the interval $[1, \sqrt{n}]$.

Definition 2.2.7. The number of prime numbers less than or equal to x is denoted by $\pi(x)$ and is called the **prime number function.**

Example 2.2.8. $\pi(10) = 4$, $\pi(50) = 15$, (from the example above). Using the Sieve of Eratosthenes, we can compute $\pi(100) = 25$, $\pi(200) = 46$. A graph of $\pi(x)$ for $2 \leq x \leq 100$ is given in Figure 2.2.1.

Euclid's Theorem on the infinitude of primes is equivalent to the assertion that $\lim_{x \to \infty} \pi(x) = \infty$.

A complete understanding of the prime number function is one of the great mysteries of mathematics. The function is highly irregular, with large flat regions and sudden jumps. (See Figure 2.2.2.) It is not possible to give an elementary formula to describe the prime number function. To illustrate the irregularity of $\pi(x)$, we prove:

Proposition 2.2.9. *There are arbitrarily large gaps in the sequence of prime numbers; that is, given any positive integer k, there exists a sequence of k consecutive composite numbers.*

PROOF. Given k, consider the k consecutive numbers

$$(k+1)! + 2, \ (k+1)! + 3, \ldots, (k+1)! + (k+1).$$

Clearly, each one is composite as $2, 3, \ldots, k+1$ occur in $(k+1)!$. ∎

The proposition shows that $\pi(x)$ can remain flat for long stretches. (See Figure 2.2.2.) On the other hand, it is expected that there are infinitely many **twin primes**, that is, pairs of primes differing by 2. Some examples of twin prime pairs are: 3 and 5, 5 and 7, 11 and 13, 17 and 19, 41 and 43. The infinitude of twin prime pairs will mean that $\pi(x)$ has infinitely many jumps over intervals of length 2, implying along with Proposition 2.2.9, a very irregular behavior. The conjecture regarding twin primes is among the many famous problems regarding prime numbers. The exercises list a few more conjectures involving primes numbers.

We give a table of some values of $\pi(x)$ in Table 2.2.1. While it is possible to use the Sieve of Eratosthenes to compute the value of $\pi(x)$ for x with 9 or 10 digits, for larger numbers this task becomes impossible. Legendre was the first mathematician to devise a method to compute $\pi(x)$ without enumerating the primes (see Section 16.1). Using a formula based on the idea of the sieve (but which does not require one to do the sieve), Legendre computed $\pi(10^6)$ to be $78,526$; the correct value is actually $78,498$. Meissel improved Legendre's formula and computed several values including $\pi(x)$ at 10^7, 10^8, and 10^9. Meissel's computations were not free of errors either, pointing to the difficulties in successfully carrying out the method. These methods have been improved for implementation on computers, and $\pi(x)$ has been computed for larger values.

The largest known value of $\pi(x)$ is $\pi(10^{18}) = 24,739,954,287,740,860$ due to M. Deleglise and J. Rivat in 1994.

Based on the tables for $\pi(x)$, Legendre conjectured that

$$\pi(x) \sim \frac{x}{\log x - 0.83},$$

and Gauss conjectured that

$$\pi(x) \sim \frac{x}{\log x}.$$

Gauss's conjecture was in terms of the function Li (x); the logarithmic integral defined as Li $(x) = \int_2^x dt/\log t$. Gauss conjectured that $\pi(x) \sim$ Li (x). (The notation $f \sim g$ means that $\frac{f}{g} \to 1$ as $x \to \infty$.)

Chebyshev was the first mathematician (after Euclid) to prove something about $\pi(x)$. In 1852 he showed that

$$0.929 \frac{x}{\log x} < \pi(x) < 1.1 \frac{x}{\log x},$$

Table 2.2.1 Some val-
ues of $\pi(x)$

n	$\pi(10^n)$
3	168
4	1229
5	9592
6	78498
7	664579
8	5761455
9	50847534
10	455051511
11	4118054813
12	37607012018

and also that if $\dfrac{\pi(x)}{x/\log x}$ has a limit, then it must be 1. We give a proof of
Chebyshev's Theorem in Chapter 16.

In 1898, Hadamard and De La Valleé Poussin independently succeeded
in proving the Prime Number Theorem, which asserts that

$$\pi(x) \sim \frac{x}{\log x} \quad \text{or} \quad \lim_{x\to\infty} \frac{\pi(x)}{x/\log x} = 1.$$

The difficult part of their proof is to show that the limit actually exists.
The method used by Hadamard and De La Valleé Poussin involves analysis
of functions of one complex variable. An elementary proof, without using
complex analysis, was discovered by Erdös and Selberg in 1952.

Using these results on $\pi(x)$, we can estimate the number of steps performed in the primality test. The maximum number of divisions in the test
for primality of n is $\pi(\sqrt{n})$; for n large we have

$$\pi(\sqrt{n}) \approx \frac{\sqrt{n}}{\log \sqrt{n}} = \frac{2\sqrt{n}}{\log n}.$$

If $n = 10^6$, we have to perform approximately $10^3/(3\log 10) \approx 10^2$ steps,
whereas if $n = 10^{20}$, there are $10^{10}/(5\log 10)$ steps. Assuming a fast computer can perform 10^5 divisions per second, this process would still take over
10^5 seconds (about a day). It is clear that the method is not practical for larger
integers. Another disadvantage of the method is that the only way we can verify our result is to repeat the same computation. We will describe improved
primality tests in later chapters, and implementations of these can easily test

Table 2.2.2 Some Mersenne primes and their discoverers

Year	Prime	Discoverer
1588	$2^{17} - 1$	P. A. Cataldi
1750	$2^{31} - 1$	L. Euler
1876	$2^{127} - 1$	E. Lucas
1952	$2^{2281} - 1$	R. M. Robinson
1963	$2^{11213} - 1$	D. B. Gillies
1978	$2^{21701} - 1$	L. C. Noll and L. Nickel
1985	$2^{216091} - 1$	D. Slowinski
1994	$2^{859433} - 1$	D. Slowinski and P. Gage
1996	$2^{1257787} - 1$	D. Slowinski and P. Gage

primality of numbers up to 10^{200} in a matter of seconds. For much larger numbers, there are primality tests for special types of numbers.

Even though we know that there are infinitely many primes, finding large primes is not a trivial task. A class of numbers that usually lead to some extremely large primes are the **Mersenne numbers**. From Exercise 2.1.29, we know that for $a^n - 1$ to be prime, a must be 2 and n a prime. Numbers of the form $2^p - 1$ for a prime p are called Mersenne numbers. If $2^p - 1$ is prime, then it is called a Mersenne prime. In his *Cogitata Physico Mathematica* of 1644, Mersenne conjectured that $2^p - 1$ is prime when p is one of $2, 3, 5, 7, 13, 17, 19, 31, 67, 127, 257$ and composite for all others. Euler showed that $2^{31} - 1$ was prime in 1772. Lucas proved that $2^{67} - 1$ is composite, and it was found that $2^{257} - 1$ is composite. Mersenne did not include 61, 89, and 107 in his list of numbers that yield Mersenne primes. His conjecture is quite astonishing given the size of the numbers involved and the methods available at the time. Mersenne numbers occur as examples of the largest known primes, primarily because there is a special test for their primality, the Lucas–Lehmer test. The test is described in a project at the end of Chapter 17. Mersenne numbers have a surprising connection with perfect numbers. This connection is explored in one of the projects at the end of this chapter.

At the time of this writing, there are 34 known Mersenne primes, and it is conjectured that there are infinitely many. We give a brief table of Mersenne primes and their discoverers. In most cases they were also the largest prime known at the time.

The study of primes in special sequences is an important part of Number Theory. We can ask if there are infinitely many primes in sequences of subsets of integers, for example, arithmetic progressions or numbers of the form $n^2 + 1$ or other quadratic polynomials. These questions turn out to be quite difficult, and very little is known about primes of the form $n^2 + 1$ or $2^n - 1$. The only known results about subsequences of primes are those in arithmetic progressions of integers. Dirichlet proved that in every suitable arithmetic progression there are infinitely many primes. An outline of the proof of Dirichlet's Theorem is given in Chapter 15. While the general proof of Dirichlet requires analytic methods of infinite series and products, special cases are accessible by elementary methods. Exercise 6 asks you to prove the infinitude of primes in the sequence $1, 5, 9, 13, \ldots$ by imitating Euclid's argument.

Exercises for Section 2.2

1. In the Sieve of Eratosthenes, when we reach the stage of deleting the multiples of a prime p, what is the first multiple of p that is still in the table?

2. Modify Algorithm 2.2.5 to build a table of primes between two integers n and m. Assume that you have a table of small primes to carry out the sieve.

3. Implement the sieve of Eratosthenes to produce a table of primes of up to 10^6. A table of primes is necessary for many of the later exercises and factoring methods.

4. Implement the primality test algorithm described in the text. Use it to verify if the following numbers are prime.

   (a) $10^8 - 10^4 + 1 = 99990001$.

   (b) $10^{10} - 10^5 + 1 = 9999900001$.

   (c) 9091, 909091, 90909091, and 9090909091.

5. What are the possible remainders when an odd prime is divided by 4? Show that the product of numbers of the form $4k + 1$ is again of the form $4k + 1$.

6. This exercise proves the infinitude of primes of the form $4n + 3$. Imitating Euclid's proof, let $p_1, \ldots, p_k$ be the finite number of such primes (excluding 3). Consider

$$N = 4p_1 \cdots p_k + 3.$$

   Use the previous exercise to show that N has a prime divisor of the form $4n + 3$. Deduce the infinitude of primes in the progression $3, 7, 11, 15, \ldots$.

   (It is more difficult to show that there are infinitely many primes in the sequence $4n + 1$. See Exercise 9.1.7.)

7. Show that there are infinitely many primes in the sequence $3n + 2$. To what other sequences can this argument be applied?

8. Compute the values of the polynomials $f(x) = x^2 + x + 11$ and $g(x) = x^2 + x + 17$ for integral values of x satisfying $-20 \le x \le 20$. For which values do they give prime numbers?

9. Repeat Exercise 8 for the polynomial $f(x) = x^2 + x + 41$. Does it always give prime numbers for different values of x? This is one of the polynomials such that $f(x)$ is prime for many values of x. What is the relative occurrence of primes in the values of $f(x)$? Can you find other polynomials that give many prime numbers?

10. Suppose $f(x) = a_n x^n + \cdots + a_1 x + a_0$. Show that if $a_0 \ne 1$, then there exists an integer x such that $f(x)$ is composite. If $a_0 = 1$, can you prove that $f(x)$ cannot be prime for all x?

11. Let $f(x) = \pi(x) - \pi(x-1)$. When is $f(x)$ nonzero?

12. How many digits are in the base 10 representation of the Mersenne number $2^p - 1$? Determine the number of digits in the base 10 representation of $2^{127} - 1$, $2^{8191} - 1$, and $2^{859433} - 1$.

13. Use a table of primes to compute $\pi(x)$ in intervals of $10,000$. Graph $\pi(x)$, Legendre's approximation $x/(\ln x - 1.08366)$, and Gauss's approximations $x/\ln x$ and $\mathrm{Li}\,(x) = \int_2^x (dt/\log t)$. Which of the three functions is a good approximation of $\pi(x)$? It is helpful to graph the differences: $\pi(x) - \mathrm{Li}\,(x)$, $\pi(x) - x/\ln(x)$, and $\pi(x) - x/(\ln x - 0.93)$. What do you think is the relationship between $\pi(x)$ and $\mathrm{Li}\,(x)$?

14. A **twin prime pair** is a pair p, $p+2$ such that both p and $p+2$ are prime, for example, 3 and 5, or 17 and 19. A famous unsolved problem is to determine if there are infinitely many twin prime pairs. We will use a table of primes to study the distribution of twin primes. Count the number of twin primes less than or equal to x, $T(x)$, in intervals of $10,000$, and draw a graph of $T(x)$ as a function x. It may seem that $T(x)$ gets larger steadily, and we would like to find a function $f(x)$ to describe $T(x)$ for large x such that $T(x) \sim cf(x))$; that is, $T(x)/f(x)$ has nonzero constant limit c as x approaches ∞. Which of the following functions seems to satisfy this condition?

$$f_1(x) = \frac{x}{\ln x}, \quad f_2(x) = \frac{x}{\ln \ln x}, \quad f_3(x) = \frac{x}{(\ln x)^2}.$$

15. Bertrand conjectured that there is always a prime between n and $2n$ for sufficiently large n. Investigate the truth of this conjecture. Are there always at least two primes between n and $2n$? Is there always a prime in shorter intervals like $[n, 3n/2]$ and $[n, 6n/5]$? Use a table of primes to investigate.

16. A famous unsolved conjecture of Goldbach states that every even number is the sum of two odd primes. Write a program to verify this conjecture for an integer n. What is the largest n for which your algorithm is guaranteed

to give a result? Verify that the conjecture is true for all even numbers up to 1000. Is a table of primes up to 800 sufficient to verify Goldbach's conjecture for even numbers less than 1000?

17. Exercise 6 showed that there are infinitely many primes of the form $4n + 3$. One can also show that there are infinitely many primes of the form $4n + 1$. (See Example 4.1.6.) Let $\pi_1(x)$ denote the number of primes of the form $4n + 1$ that are less than or equal to x; similarly, define $\pi_3(x)$. Use the table of primes to compute $\pi_1(x)$ and $\pi_3(x)$ for various values of x. What do you conjecture about the relationship between the two functions?

18. In 1848, de Polignac conjectured that every odd number is the sum of a prime and a power of 2; for example, $107 = 2^6 + 43$, $33 = 2^4 + 17$. Test the conjecture to see if it is true. You can carry out the search by using a table of primes and another table of powers of 2.

19. Proposition 2.2.9 states that a gap of k in the table of primes occurs starting at $(k + 1)!$. This number is much too large. For example, a gap of 5 occurs at $24 - 28$, whereas $5! = 120$. Use the table of primes to investigate the first occurrence of k composite numbers (or a gap of $k + 1$ in the sequence of primes).

2.3   Unique Factorization

The decomposition of an integer into a product of primes was established in the previous section. Consider a few factorizations:

$$21 = 3 \cdot 7$$
$$60 = 2^2 \cdot 3 \cdot 5$$
$$108 = 2^2 \cdot 3^3.$$

It is clear that these factorizations are unique; that is, there is no other way to write these numbers as the product of primes, except to write the primes in a different order. We might be tempted to expect that the factorization into primes is unique for all integers. Our expectation is indeed true, but it is by no means obvious.

The uniqueness of the factorization seems to have been noticed first by Euclid. Its importance was missed by generations of mathematicians who took the uniqueness as obvious. Even Euler and Legendre seem to have taken it for granted. Gauss, who was always meticulous about the rigorousness of his demonstrations, was the first to identify the importance of the uniqueness. Mathematicians soon discovered number systems where factorization into primes holds but not unique factorization; in fact, the unique factorization property seems quite rare among number systems. (See the projects at the end of this chapter for an example of a number system in which unique factorization fails.) Unique factorization in the integers is a fundamental property of the integers and essential to the development of number theory.

Fundamental Theorem of Arithmetic. *Every positive integer greater than* 1 *can be factored* **uniquely** *into a product of primes.*

PROOF. The proof will be by contradiction, so we assume that there are positive integers that do not have unique factorization. Suppose n is the smallest positive number that can be written in two different ways as a product of primes:

$$n = p_1 p_2 \cdots p_k = q_1 q_2 \cdots q_r.$$

Observe that no p_i occurs among the q_i's, and vice versa; otherwise, by cancellation, we get the factorization of a smaller number in two different ways. Also, n is not prime, so there is more than one prime in each factorization.

Let p_1 and q_1 be the smallest primes in the sets $\{p_1, p_2, \ldots, p_k\}$ and $\{q_1, q_2, \ldots, q_r\}$, respectively. We claim that $n - p_1 q_1$ is positive; this is because p_1 and q_1 are the smallest in each prime factorization; hence $p_1 \leq \sqrt{n}$ and $q_1 \leq \sqrt{n}$, and one of these is strictly less than $\sqrt{n}$. So $p_1 q_1 < n$.

Now, p_1 and q_1 both divide $n - p_1 q_1$. Suppose $n - p_1 q_1 = p_1 m$ for some m. We claim that $q_1 \mid m$ because the number $n - p_1 q_1$ is smaller than n and can be uniquely factored into primes by the hypothesis on n. We have shown that $p_1 q_1$ divides n; hence we can write

$$p_1 q_1 m = n - p_1 q_1$$
$$p_1 q_1 m = p_1 (p_2 p_3 \cdots p_k - q_1).$$

Canceling p_1, we obtain $q_1 m - p_2 p_3 \cdots p_k = q_1$; hence q_1 divides $p_2 \cdots p_k$ (but does not occur in it). The product $p_2 \cdots p_k$ is a smaller number than n; hence it factors uniquely into a product of primes, which implies that q_1 is one of $p_2, \ldots, p_k$, a contradiction. Hence n factors into a product of primes in only one way. ∎

Exercise. The reader will benefit from a careful study of the proof. Identify all the places where the minimality of n is used. Why did we have to show that $n - p_1 q_1 > 0$?

For negative integers we obtain the unique factorization by multiplying by -1 and using the theorem. (Why did we restrict the proof of the theorem to positive integers?)

By the Fundamental Theorem, every positive integer can be decomposed in only one way as $n = p_1^{e_1} \cdots p_k^{e_k}$, with $e_i > 0$, $i = 1, \ldots, k$ where the p_i are distinct primes. We take this to be the canonical form of the factorization. We assume that whenever an integer is written this way, the primes occurring in the factorization are distinct. We stress again that it is the uniqueness of the factorization that is fundamental, allowing us to derive many important consequences.

Lemma 2.3.1. *Let a have the prime factorization $a = p_1^{a_1} \ldots p_k^{a_k}$. A positive integer b divides a if and only if it has a prime factorization of the form $b = p_1^{b_1} \ldots p_k^{b_k}$, with $0 \le b_i \le a_i$ for $1 \le i \le k$.*

PROOF. Suppose $b \mid a$ and has the factorization $b = q_1^{b_1} \cdots q_r^{b_r}$. If $b \mid a$, then $a = bc$ for some c; hence $p_1^{a_1} \cdots p_k^{a_k} = q_1^{b_1} \cdots q_r^{b_r} c$. By the uniqueness of the factorization, the primes q_i must occur on the left-hand side, so q_1 must be one of the p_i's, say $p_1 = q_1$. This implies that $p_1^{b_1} \mid p_1^{a_1}$, so we must have $b_1 \le a_1$. (Why?). Repeating this process for each prime, we see that $p_i = q_i$ and $b_i \le a_i$ for $1 \le i \le k$. Hence, any divisor b of a has the form

$$b = p_1^{b_1} \cdots p_r^{b_r}, \quad 0 \le b_i \le a_i. \qquad \blacksquare$$

Proposition 2.3.2. *Let n be a positive integer with prime factorization $n = p_1^{e_1} \cdots p_k^{e_k}$. The number of positive divisors $\nu(n)$ of n (including 1 and n) is*

$$\nu(n) = (e_1 + 1) \cdots (e_k + 1).$$

PROOF. By the previous lemma, any divisor d of n has the form

$$d = p_1^{d_1} \cdots p_k^{d_k} \quad \text{with } d_i \le e_i.$$

There are $e_i + 1$ possibilities for each d_i, and each combination of these d_i's gives a distinct divisor; hence the total number of divisors is the product of the numbers $(e_1 + 1) \ldots (e_k + 1)$. $\qquad \blacksquare$

Example 2.3.3. Suppose $n = p_1^{e_1} \ldots p_k^{e_k}$ and $\nu(n) = 12$. We consider the different possible values of n based on the fact that it has only 12 divisors. If $k = 1$, then $12 = (e_1 + 1)$; hence $e_1 = 11$, and $n = p_1^{11}$. If $k = 2$, then $12 = 6 \cdot 2$ or $12 = 4 \cdot 3$; that is, $n = p_1^5 p_2$ or $n = p_1^3 p_2^2$. If $k = 3$, then $12 = 3 \cdot 2 \cdot 2$ implies that $n = p_1^2 p_2 p_3$. The number of distinct prime factors of n cannot be more than 3, as $v(n) = 12$ can be written as a product of at most 3 integers.

Proposition 2.3.4. *Let a and b be integers. If p is a prime number such that $p \mid ab$, then $p \mid a$ or $p \mid b$.*

PROOF. Let $a = p_1^{a_1} \cdots p_k^{a_k}$ and $b = q_1^{b_1} \cdots q_r^{b_r}$. Then there exists c such that $ab = pc$. In terms of the prime factorization we obtain

$$p_1^{a_1} \cdots p_k^{a_k} q_1^{b_1} \cdots q_r^{b_r} = pc.$$

The uniqueness of the factorization of pc implies that p is one of the p_i's or one of the q_i's (or both); hence $p \mid a$ or $p \mid b$. $\qquad \blacksquare$

The proposition above states a fundamental property of prime numbers. This property can be used as a definition of primes. It is easy to construct examples of a composite number n dividing a product ab, but n divides neither a nor b.

We give a few other applications of unique factorization.

Proposition 2.3.5. *The number $\sqrt{2}$ is irrational.*

PROOF. Assume to the contrary that $\sqrt{2} = a/b$ is a rational number, with a and b having the prime factorizations $p_1^{a_1} \ldots p_k^{a_k}$ and $q_1^{b_1} \ldots q_r^{b_r}$, respectively. Now, $2 = \dfrac{a^2}{b^2}$, or $a^2 = 2b^2$. Using the factorization of a and b, we have

$$p_1^{2a_1} \cdots p_k^{2a_k} = 2q_1^{2b_1} \cdots q_r^{2b_r},$$

in which 2 occurs on the left to an even exponent and on the right to an odd exponent, contradicting the Fundamental Theorem of Arithmetic; hence $\sqrt{2}$ is not a rational number. ■

Example 2.3.6. If $a^2 \mid b^2$, then $a \mid b$.

PROOF. Let $a = p_1^{a_1} \cdots p_k^{a_k}$ and $b = p_1^{b_1} \cdots p_k^{b_k}$. We can use the same primes for both by allowing the exponent to be 0 if the prime does not occur in the factorization. Then $a^2 \mid b^2$ implies that there exists a c satisfying $b^2 = a^2 c$; hence

$$p_1^{2b_1} \cdots p_k^{2b_k} = p_1^{2a_1} \cdots p_k^{2a_k} c.$$

This implies that $p_i^{2a_i} \mid p_i^{2b_i}$ for $i = 1, \ldots, k$, that is, $2b_i \geq 2a_i$; hence $b_i \geq a_i$, which implies that $a \mid b$. ■

We determine the prime factorization of $n!$, a formula which has many applications. First, consider an example:

Example 2.3.7. Let us compute the prime factorization of 25!. Only the primes less than 25 can occur in it. Let's compute the exponent of 2 in the factorization of 25!. Figure 2.3.1 shows the even integers less than 25, and the number of factors of 2 in each are represented by the dots. We count the total number of times 2 occurs by counting the dots in each row. There is a dot in the first row for every number as 2 occurs in all the even numbers, of which there are 12. But 2 also occurs twice in the multiples of 4 $(4, 8, 12, \ldots)$, of which there are 6. We have to include these once more. The exponent of 2 is 3 in the multiples of 8 $(8, 16, 24)$, and since these are also multiples of 4, we have included them twice so far. So we have to include these once more for the exponent of 2. The exponent of 2 is 4 in multiples of 16 of which there is only one less than 25; therefore, the exponent of 2 is $12 + 6 + 3 + 1 = 22$.

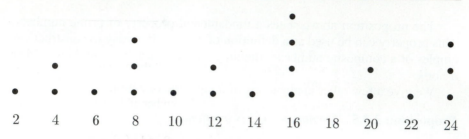

Figure 2.3.1 Exponent of 2 in 25!

The next prime, 3, occurs at least once in each of the eight numbers $3, 6, 9, 12, 15, 18, 21, 24$ and it occurs twice in $9, 18$; hence the exponent is $8 + 2 = 10$. Similarly, the exponent of 5 is 6, the exponent of 7 is 3, the exponent of 11 is 2, and the exponent of all the other primes less than 25 is 1 (as $13 \cdot 2 = 26 > 25$). Then

$$25! = 2^{22} \cdot 3^{10} \cdot 5^6 \cdot 7^3 \cdot 11^2 \cdot 13^1 \cdot 17^1 \cdot 19^1 \cdot 23^1.$$

Exercise. Determine the number of zeros at the end of the decimal representation of 25!.

Exercise. Write down the prime factorization of 12!.

In the example, we had to count the number of multiples of p, p^2, and so on. The number of multiples of d that are less than or equal to n is given by $\lfloor n/d \rfloor$. This result allows us to prove the following formula.

Proposition 2.3.8. *If $p \leq n$, the exponent of p in the factorization of $n!$ is*

$$\left\lfloor \frac{n}{p} \right\rfloor + \left\lfloor \frac{n}{p^2} \right\rfloor + \left\lfloor \frac{n}{p^3} \right\rfloor + \cdots .$$

PROOF. To find the exponent, we determine how many times p occurs in the factorization of integers less than or equal to n.

There are $\lfloor n/p \rfloor$ multiples of p that are less than or equal to n; hence each contributes 1 to the exponent of p. But p also occurs to a degree of 2 or more in the multiples of p^2. One occurrence has already been accounted for, so we add $\lfloor n/p^2 \rfloor$, the number of multiples of p^2, to the sum. Next, we look at the number of integers which are divisible by p^3. Each of these contributes a factor of p^3 to the product. Two factors of p have already been included, so we add one more factor for each multiple of p^3 by adding $\lfloor n/p^3 \rfloor$. Continuing this argument, we obtain the desired formula. Note that the sum is finite as $p^k \geq n$ for some k; hence $\lfloor n/p^r \rfloor = 0$ for $r > k$. ■

Exercises for Section 2.3

1. Show that a number is a square if and only if all the exponents in its prime factorization are even.

2. Determine the set of integers for which the number of divisors is odd. Make a general conjecture and prove your claim.

3. Characterize the integers n such that $\nu(n) = 2$ and $\nu(n) = 4$.

4. Determine the smallest positive integer satisfying $\nu(n) = 6$, $\nu(n) = 20$, and $\nu(n) = 100$.

5. Determine all primes p such that $11p + 1$ is a perfect square.

6. Show that if $a^3 \mid b^2$, then $a \mid b$.

7. If a and b are integers, does $a^2 \mid b^3$ imply that $a \mid b$? Prove or disprove.

8. Show that if p is prime and $p \mid a^n$, then $p \mid a$.

9. If p is a prime number, determine the sum of all the divisors of p^n.

10. Prove that $\sqrt{n}$ is irrational if n is not a perfect square.

11. Show that any positive integer n can be written uniquely as $n = 2^r s$, where s is odd.

12. Determine the prime factorization of 61! and 100!.

13. Determine the number of zeros at the end of the decimal representations of 200! and 153! .

14. Use $\lfloor x + y \rfloor \geq \lfloor x \rfloor + \lfloor y \rfloor$ and the decomposition of $n!$ to prove that the binomial coefficient

$$\binom{n}{r} = \frac{n!}{r!(n-r)!}$$

is an integer. Extend the result and show that the multinomial coefficient

$$\frac{n!}{r_1! r_2! \ldots r_k!} \quad \text{with} \quad r_1 + r_2 + \cdots + r_k = n$$

is an integer.

15. Use the prime factorization of $n!$ to show that if p is prime, then $p \mid \binom{p}{k}$, $1 \leq k \leq p - 1$, and $p \nmid \binom{p^k x}{p^k}$ for $p \nmid x$.

16. Show that if $ab = m^2$ and a and b have no prime factor in common, then a and b must be perfect squares.

17. Show that if n has t divisors, then the product of all the positive divisors of n is $n^{t/2}$.

18. Let p be a prime. The expression $p^\alpha \,\|\, a$ (p^α **exactly divides** a) means that $p^\alpha \,|\, a$, but $p^{\alpha+1} \nmid a$. This means that α is the exponent of p in the prime factorization of a.

    (a) Show that $p^\alpha \,\|\, a$ and $p^\beta \,\|\, b$ implies $p^{\alpha+\beta} \,\|\, ab$.

    (b) Show that if $\alpha < \beta$ and $p^\alpha \,\|\, a$ and $p^\beta \,\|\, b$, then $p^\alpha \,\|\, a+b$. Naturally, a similar statement is true for $\beta < \alpha$. Investigate what happens for $\alpha = \beta$ by considering several examples. What can be said about the largest power of p dividing $a + b$ in this case?

19. Let $N_j(x)$ denote the number of positive integers that are less than or equal to x and that have no prime factor greater than p_j, the jth prime.

    (a) Show that any $n \le x$ with no prime factors greater than p_j can be written as

    $$n = m^2 p_1^{e_1} p_2^{e_2} \cdots p_j^{e_j} \quad \text{with } e_i = 1 \text{ or } 0,\ 1 \le i \le j.$$

    (b) Show that $N_j(x) \le \sqrt{x}2^j$.

    (c) If $\pi(x) = k$, then $N_k(x) = x$: conclude that $\pi(x) \ge \ln x/(2\ln 2)$, and, hence, the number of primes is infinite.

    (d) Show that the jth prime, p_j, satisfies $p_j < 4^j$.

    (e) Show that the series $\sum \dfrac{1}{p}$ diverges, where the sum is taken over the prime numbers.

2.4 Elementary Factoring Methods

The problem of distinguishing prime numbers from composite numbers and of resolving the latter into their prime factors is known to be one of the most important and useful in arithmetic .... The dignity of the science itself seems to require that every possible means be explored for a solution of a problem so elegant and so celebrated.

— C. F. Gauss

We consider the problem of determining the prime factorization of an integer n. The simplest method is to divide n by primes until a divisor p is found. Then we apply this method to n/p, a smaller number, continuing until we have obtained all the factors of n. A table of primes is necessary to implement this method. Suppose we construct a table of primes up to n. Then we can completely factor all the numbers up to n^2. Trial division also applies to larger integers, where instead of obtaining a complete factorization, we isolate the small prime factors. In practice, it seems reasonable to have a table of primes up to 10^6, allowing factorization of integers with at most 11 digits. The method is simple to implement and reasonably fast. It is always the first method to use; more complex methods become necessary for larger numbers.

Algorithm 2.4.1. Factoring by Trial Division
Given an integer n, the procedure outputs its factorization. A table of primes is stored in P. We use a list called *factors* to store the factorization of n in the form of a list consisting of elements $\{p, e\}$, where p^e is the exact power of p appearing in the factorization of n.

1. [Initialize] Let $i = 1$, *factors*$= \{\}$, $p = P[i]$, $e = 0$.

2. [Check if prime] If $p > \sqrt{n}$, then add $\{n, 1\}$ to *factors* and return *factors*.

3. [Does p divide n?] If p divides n, then $e = e + 1$, $n = n/p$; repeat step 3.

4. [Add a factor] If $e \neq 0$, add $\{p, e\}$ to *factors*.

5. [Is n factored?] If $n = 1$, return *factors*.

6. [Get next prime] Let $i = i + 1$. If $i <$ number of primes, let $p = P[i]$, $e = 0$, and go to step 2. Otherwise, report that n may not be completely factored and return *factors* along with n.

Note that in step 2, if $p > \sqrt{n}$, then we stop, as the remaining number must be prime (why?). For large numbers, the algorithm finds all the small prime factors.

Example 2.4.2.   1. Suppose $n = 217$. Division by 2, 3, and 5 yield nonzero remainders, but 7 divides 217 with a quotient of 31. Since $7 > \sqrt{31}$, 31 must be prime! Therefore, $217 = 7 \cdot 31$.

2. Consider $n = 9361$. It is not divisible by 2, 3, 5, or 7. We see that $11 \mid 9361$ with the quotient $9361/11 = 851$, which is not divisible by 11. Because $\sqrt{851} = 29.17 > 11$, we cannot conclude that 851 is prime. We continue and find that 23 divides 851: $851/23 = 37$, $\sqrt{37} < 23$, so 37 must be prime. Hence, $9361 = 11 \cdot 23 \cdot 37$.

3. Let $n = 732697$. Division by 2, 3, and 5 fails. We see that 7 divides twice giving a quotient of 14953. As $\sqrt{14953} = 122.2$, we continue the trial division algorithm. The next prime that divides $n = 14953$ is 19, and the quotient is 787. Now, 23 does not divide 787, and the next prime is 29, but $29 > \sqrt{787}$, so 787 is prime, and we have obtained $732697 = 7^2 \cdot 19 \cdot 787$.

Exercise. Factor $n = 1771$ using the algorithm.

The above method requires a storage of $\pi(\sqrt{n})$ primes, approximately $\sqrt{n}/\log\sqrt{n}$ words (slightly more than this, as large primes require more storage space). This can be avoided by generating the primes as necessary (by a sieve method) or by not dividing by primes at all, but by a larger set of odd numbers. For example, one can divide by 2, 3, and by all the numbers of the form $6k \pm 1$ $(5, 7, 11, 13, 17, 19, 23, 25, \ldots)$. Every prime other than 2 and 3 is in one of the sequences $6k + 1$ or $6k - 1$ ($\sim 6k + 5$). In this method, we will be dividing by a third of the integers up to $\sqrt{n}$. For example, if $\sqrt{n} = 10^6$, then we do $10^6/3 \simeq 333,000$ divisions, whereas there are at most 78498 divisions if we divide by primes.

Example 2.4.3.　　1. Consider $n = 6264947$. Division by 2, 3, 5, 7, 11 gives nonzero remainders. The first factor obtained is 13, which divides n with quotient 481919. The next prime that divides n is 23, dividing it twice. The quotient is 911 and $\sqrt{911} = 30.18$. Next, we divide by 25 and 29 unsuccessfully. The next number to consider is 31, which is larger than $\sqrt{911}$, so 911 is prime and $6264947 = 13 \cdot 23^2 \cdot 911$.

2. Suppose $n = 1723411$. After 288 unsuccessful divisions, we find that 863 divides n with quotient 1997; $\sqrt{1997} = 44.6$, so 1997 is prime, and therefore, $1723411 = 863 \cdot 1997$.

It is clear that the first method is faster, but its analysis is more complicated because we have to use the prime number function. It is easier to analyze the second method. Let $n = p_1 p_2 \cdots p_k$ be the prime factorization of n, and let p_k be the largest prime occurring in the factorization and p_{k-1} the second largest (p_{k-1} can equal p_k if it occurs more than once). Suppose $p_{k-1} = p_k$. Then $p_k^2 \mid n$ implies that $p_k < \sqrt{n}$; hence we need at most p_{k-1} (or p_k) steps to find the factorization. If $p_{k-1} < p_k$, then after dividing by $d = p_{k-1}$ in the algorithm, p_k is the remaining part to be factored. If $d > \sqrt{p_k}$, we stop, as p_k must be prime. If $d < \sqrt{p_k}$, then we must continue until we reach $\sqrt{p_k}$ to see that p_k is prime. Hence the algorithm seems to require $\max(p_{k-1}, \sqrt{p_k})$ steps. In order to understand the average behavior of the algorithm, we need more information regarding the distribution of primes. This analysis is beyond the level of our study. For example, it has been established that the probability that n has a prime factor greater than $\sqrt{n}$ is $\ln 2$, approximately 69%.[2] This implies that for a significant proportion of numbers, the algorithm must work hard (using the maximum number of operations). The worst case encountered by the method occurs when n is prime. It is possible to avoid applying the algorithm for primes by checking

---

[2] See D. E. Knuth. *The Art of Computer Programming.* Vol. 2. Reading, MA: Addison Wesley, 1980.

for primality using one of the better methods discussed in Sections 6.1 and 6.2.

Fermat's factoring method: Fermat's method is the first method not to make use of trial division to factor integers. While the method may seem impractical, extensions of it have been used to construct special purpose factoring machines. (See Section 12.2 .)

Suppose the integer to be factored, n, can be written as the difference of two squares $n = a^2 - b^2$. Then the identity

$$n = a^2 - b^2 = (a - b)(a + b)$$

gives a factorization of n if $a - b \neq 1, n$.

To find an expression of an integer as a difference of squares, we proceed as follows. Find the first perfect square a^2 larger than n; compute $a^2 - n$. If $a^2 - n = b^2$ for some b, then we have found the desired expression. Otherwise, we find the next larger perfect square and continue the process. Consider a couple of examples, which will also give you a clue as to when to stop the algorithm.

Example 2.4.4.   1. Take $n = 221$. The first perfect square larger than n is $225 = 15^2$.
Now $15^2 - 221 = 4 = 2^2$; hence $221 = 15^2 - 2^2 = (15 - 2)(15 + 2) = 13 \cdot 17$.

2. Consider $n = 145$. The first perfect square larger than n is $169 = 13^2$. We compute the sequence of squares a^2 and the differences $a^2 - n$.

$$169 - 145 = 24$$
$$196 - 145 = 51$$
$$225 - 145 = 80$$
$$256 - 145 = 111$$
$$289 - 145 = 144 = 12^2;$$

therefore, $145 = 17^2 - 12^2 = 5 \cdot 29$.

Exercise. Factor 517 using Fermat's method.

In the first example we were lucky; in the second, we succeeded after five steps. Trial division would have required divisions by $2, 3$, and 5 and a square root computation. Consider another example.

Example 2.4.5. Let's try to factor $n = 101$ using Fermat's method. We list the differences $a^2 - n$ between perfect squares and n.

$$121 - 101 = 21,$$
$$144 - 101 = 43,$$
$$169 - 101 = 68,$$

and so on. When do we stop? We continue until we get to $51^2 - 101 = 50^2$. This implies that $101 = 1 \cdot 101$; hence 101 is a prime number!

This is the worst case for the method, requiring $[(n+1)/2] - \sqrt{n}$ steps, much larger than trial division. The best case occurs when $n = ab$, with a and b of about the same size. In this case, the number of steps required can be considerably smaller than trial division.

In the algorithm, we have to test if $a^2 - n$ is a perfect square. One way to test if an integer k is a perfect square is to compute $\sqrt{k}$, find the integer part, $\lfloor \sqrt{k} \rfloor$, and square it back to see if it equals k. This is a time-consuming operation and can be avoided in many cases by removing obvious nonsquares. For example, the last digit of a perfect square must be $0, 1, 4, 5, 6,$ or 9. (See Exercise 7.) Using the last two digits, we can eliminate many more numbers.

In the algorithm, if $a^2 - n$ is not a perfect square, we have to try again with $(a+1)^2$. This can be easily computed by adding $2a + 1$ to a^2, using $(a+1)^2 = a^2 + 2a + 1$.

Fermat's factorization method does not apply to integers n such that $n \bmod 4 = 2$. An integer of the form $4k + 2$ cannot be expressed as a difference of squares. In this case, n is even, so we can divide by two and the result will be an odd number. Fermat's method can be applied to this odd number to find a factor.

Algorithm 2.4.6 (Fermat's Factoring Method). This method finds a factor of an integer n. The algorithm will stop after a prespecified number of iterations *maxsteps* if no factor has been found.

1. [Initialize] Let $u = \lfloor \sqrt{n} \rfloor + 1, i = 1, a = u^2$.

2. [Check if difference is a perfect square] If $a - n = b^2$, a perfect square, then n has been factored. Return $u - b$ and $u + b$ as the factors of n.

3. [Get next square] Let $a = a + 2u + 1, u = u + 1, i = i + 1$.

4. If $i > $ *maxsteps*, report that the algorithm is unable to factor n.

Many improvements are possible in Fermat's method. The simplest is to use divisibility conditions to check when $a^2 - n$ is likely to be a square. This reduces the number of a's which have to be considered. These methods and generalizations are discussed in Chapter 12.

——————————— **Exercises for Section 2.4** ———————————

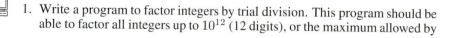 1. Write a program to factor integers by trial division. This program should be able to factor all integers up to 10^{12} (12 digits), or the maximum allowed by

your computer. The procedure should also identify the small prime factors of larger numbers.

2. Describe the algorithm to do trial division by 2, 3, and the numbers in the sequences $6k - 1$ and $6k + 1$, $k = 1, 2, \ldots$.

3. Use the program to factor the Mersenne numbers $M_n = 2^n - 1$ for n an odd prime. Your program should be able to factor all but one of the Mersenne numbers $M_3, M_5, \ldots, M_{57}$.

4. Determine the factorizations of the numbers $2^n + 1$ for $n = 1, \ldots, 50$. Can you factor $2^{58} + 1$ by trial division?

5. Determine the factorizations of the "repunits" $R_n = (10^n - 1)/9$, for $n = 3, 4, 7, \ldots, 25$. You should be able to factor all but 3 of the numbers.

6. Verify the Aurifeuillian factorization

$$2^{4n+2} + 1 = (2^{2n+1} - 2^{n+1} + 1)(2^{2n+1} + 2^{n+1} + 1).$$

Use this identity to obtain a factorization of $2^{58} + 1$. (Without the aid of this identity, it would be very difficult to factor this number. The mathematician Landry devoted his life to the study of the numbers of the form $2^n + 1$. He factored $2^{58} + 1$ in 1869 after great labor. Ten years later, Aurifeuille discovered this identity which makes the factorization of $2^{58} + 1$ quite trivial.)

7. Prove that the last digit of a perfect square can only be 0, 1, 4, 5, 6, or 9. What are the possibilities for the last two digits of perfect squares?

8. Determine the possible remainders of a perfect square when it is divided by 16.

9. Show that if $n \bmod 4 = 2$, then n cannot be expressed as a difference of two squares.

10. Implement Fermat's factoring method. You should use a simple test to eliminate some numbers that are not perfect squares, for example, by looking at the remainders upon division by 16 or by 100. This will speed up the algorithm considerably.

11. In many cases, we may need to use a combination of methods to obtain a complete factorization. Consider $n = 10^{22} + 1$. First use trial division to find its small prime factors and a large cofactor m with no prime factors less than 10^6. Then apply Fermat's factoring method to this number to write it as a product of two numbers a and b. Now use the trial division method to each of these numbers to factor $10^{22} + 1$ completely.

12. Consider the numbers $10^2 - 10 + 1$, $10^4 - 10^2 + 1$, $10^6 - 10^3 + 1, \ldots$. Determine which are prime or composite up to whatever range your computer will allow. Is there a pattern?

13. A number is **squarefree** if no square divides it, that is, the prime factorization of the number has all the exponents equal to 1. We want to estimate the probability that a randomly chosen integer is squarefree. First, write a procedure using the factoring algorithm to test if an integer is squarefree. Pick 50 random numbers in each interval $[1000, 2000]$, $[2000, 3000]$, $[3000, 4000], \ldots, [10000, 11000]$. Determine how many are squarefree, and use these data to estimate the probability that a randomly chosen integer is squarefree. Increase the sample size or pick different intervals and repeat the experiment! Does the probability change? Is the result surprising?

2.5   GCD and LCM

A quantity crucial for further study is the greatest common divisor.

Definition 2.5.1. The **greatest common divisor** (GCD) of two numbers a and b, not both zero, is the largest integer dividing both a and b. It will be denoted by $\gcd(a, b)$ or (a, b).

Remark. Every positive integer divides 0; hence $(0, 0)$ is undefined.

Example 2.5.2.   1. For any integer $a \neq 0$, $(a, 0) = |a|$, as $|a|$ is the largest divisor of a.

2. It can be seen by enumerating the divisors of 24 and 36 that $(36, 24) = 12$.

3. The numbers 6 and 35 have no nontrivial divisor in common; hence $(6, 35) = 1$.

Exercise. Determine $(24, 90)$.

Definition 2.5.3. Two integers a and b are said to be **relatively prime** (or **coprime**) if $(a, b) = 1$.

Lemma 2.5.4. *The greatest common divisor of two numbers satisfies the following properties:*

(a) $(a, b) = (-a, b)$.

(b) $(a, b) = (a - b, b)$.

(c) *If $(a, b) = d$, then $(a/d, b/d) = 1$.*

PROOF.   (a) The assertion is clear because the set of divisors of a is the same as that of $-a$.

(b) If $d \mid (a - b)$ and $d \mid b$, then $d \mid (a - b + b)$, that is, $d \mid a$. If $d \mid a$ and $d \mid b$, then $d \mid a - b$; hence the common divisors on both sides are the same.

(c) If $(a/d, b/d) = d' \geq 1$, then $d' \mid (a/d)$, and $d' \mid (b/d)$. This implies that $a/d = d'k$ for some k, or, equivalently, $a = dd'k$, that is, $dd' \mid a$. Similarly, $dd' \mid b$; hence $dd' \leq d$, as d is the largest common divisor of a and b. The two inequalities on d' are satisfied only when $d' = 1$. ∎

We use property (c) to reduce fractions to their lowest forms. We observe that $a - kb$ and b have the same common divisors as a and b. Hence property (b) can be extended to include any linear combination of a and b. If $d \mid a$ and $d \mid b$, then $d \mid ma + nb$ for any m and n. Hence the greatest common divisor of a and b divides every linear combination $ma + nb$. It is a fundamental property that the converse is true; that is, every multiple of d is a linear combination of a and b.

One way to show that two numbers are relatively prime is to show that 1 can be expressed as a linear combination of the two numbers.

Example 2.5.5. The numbers $6k + 5$ and $5k + 4$ are relatively prime for every $k \geq 1$. This follows from the linear combination $5(6k + 5) - 6(5k + 4) = 1$; hence any common divisor of $6k + 5$ and $5k + 4$ must divide 1, so the greatest common divisor is 1.

Next we prove the important fact that the greatest common divisor of two numbers is always a linear combination of them. The property is used frequently in later chapters.

Theorem 2.5.6. *For any two integers a and b there exist m and n such that $ma + nb = (a, b)$.*

PROOF. We may assume that a and b are both positive. Since every linear combination of a and b is a multiple of $d = (a, b)$, if d can be expressed in this form, then it must be the smallest positive integer with this property. Consider the set of all linear combinations

$$S = \{ma + nb \mid m, n \in \mathbb{Z}\}.$$

Let c be the smallest positive integer in S, and $c = m_1 a + n_1 b$ for some m_1 and n_1. We must have $c \leq a$ and $c \leq b$ (why?). We will show that c is a divisor of both a and b; hence it must be the greatest common divisor. Use the division algorithm to write

$$a = qc + r, \quad 0 \leq r < c.$$

Therefore,

$$r = a - qc = a - q(m_1 a + n_1 b)$$
$$= (1 - qm_1)a - qn_1 b.$$

If $r \neq 0$, we have expressed a positive integer smaller than c as a linear combination of a and b, contradicting the choice of c. Hence $r = 0$; that is, $a = qc$, or $c \mid a$. Similarly, $c \mid b$. Therefore, c must equal d as $d \mid c$, and c is a divisor of a and b.   ∎

Example 2.5.7.   1. We can express $(36, 24)(= 12)$ as the linear combination $1 \cdot 36 - 1 \cdot 24 = 12$.

2. Clearly, $(6, 15) = 3$, and $15 - 2 \cdot 6 = 3$ is a linear combination of 6 and 15 which results in their GCD.

Exercise. The linear combinations given in the above examples are not the only ones possible. Can you give a few more examples of $6m + 15n = 3$?

A quantity closely related to the GCD is the Least Common Multiple (LCM).

Definition 2.5.8. The **least common multiple** (denoted $[a, b]$) of two integers a and b is the smallest positive integer divisible by both a and b.

Example 2.5.9. The following properties of the least common multiple follow easily from the definition.

1. $[0, 0] = 0$.

2. $[a, a] = |a|$.

3. If $a \mid b$, then $[a, b] = |b|$.

4. $[a, b] \leq |ab|$, as $|ab|$ is a common multiple.

From the definition, it would seem that we have to compute the GCD by enumerating the divisors of the two numbers. Enumerating the divisors of a large number is by no means trivial, because it requires the prime factorization of the integer. The following proposition describes the GCD and LCM in terms of the prime factorization. While it doesn't make the computation of the GCD any simpler, it is very useful to prove properties of the GCD and LCM.

Proposition 2.5.10. *Suppose $a = p_1^{a_1} \cdots p_k^{a_k}$ and $b = p_1^{b_1} \cdots p_k^{b_k}$ with $a_i, b_i \geq 0$. Then*

$$(a, b) = p_1^{\min(a_1, b_1)} p_2^{\min(a_2, b_2)} \cdots p_k^{\min(a_k, b_k)} \tag{1}$$
$$[a, b] = p_1^{\max(a_1, b_1)} p_2^{\max(a_2, b_2)} \cdots p_k^{\max(a_k, b_k)}. \tag{2}$$

PROOF. Any divisor of a and b is of the form $p_1^{c_1} \cdots p_k^{c_k}$ with $c_i \leq a_i, b_i$. Hence $c_i \leq \min(a_i, b_i)$, and the GCD must be the expression given above. The proof of (2) follows by reversing the inequalities. ∎

Remark. In the above proposition we take the same prime factors for a and b by allowing the exponent to be 0 if it doesn't appear in the factorization.

Example 2.5.11. Let $a = 3593700$ and $b = 15246$. Their prime factorizations are:

$$a = 2^2 \cdot 3^3 \cdot 5^2 \cdot 7^0 \cdot 11^3$$
$$b = 2 \cdot 3^2 \cdot 5^0 \cdot 7^1 \cdot 11^2.$$

Euclid (323–285 B.C.)

Euclid

Euclid is the author of the *Elements of Geometry*, the most celebrated textbook on mathematics ever written. Nothing is known of his life except that he lived and taught in Alexandria under the reign of Ptolemy (306–283 B.C.). His work was the first great rigorous textbook on mathematics. The book was in use throughout Europe and America into the twentieth century. Books VII, VIII, and IX of the *Elements* were devoted to arithmetic. Book VII contains what we know today as the Euclidean algorithm for the greatest common divisor. Book IX includes both Euclid's proof of the infinitude of primes and his formula for perfect numbers. Euclid stated his result on perfect numbers as follows:

If as many numbers as we please, beginning from unity, be set out continuously in double proportion until the sum of all becomes prime, and if the sum is multiplied by the last, the product will be perfect.

On the life of Euclid, consult *The Dictionary of Scientific Biography*. New York: Scribners, 1970.

Using the proposition, we obtain

$$(a, b) = 2 \cdot 3^2 \cdot 5^0 \cdot 7^0 \cdot 11^2$$
$$[a, b] = 2^2 \cdot 3^3 \cdot 5^2 \cdot 7^1 \cdot 11^3.$$

Corollary 2.5.12. $(a, b)[a, b] = |ab|$.

PROOF. It follows from the above proposition and the identity

$$\min(a, b) + \max(a, b) = a + b.$$

This identity is established by observing that if $a \leq b$, then $\min(a, b) = a$ and $\max(a, b) = b$.   ∎

Corollary 2.5.13. *If $a \mid bc$ and $(a, c) = 1$, then $a \mid b$.*

PROOF. Since $(a, c) = 1$, a and c have no prime factors in common. Since $bc = ak$ for some k, unique factorization implies that every prime factor of a appears in that of b, that is, $a \mid b$.

Another way to prove the assertion is to observe that $(a, c) = 1$ implies that $am + cn = 1$ for some m and n. But then $abm + bcn = b$. As a divides each term on the left side, we must have $a \mid b$.   ∎

Determining the prime factorization of large numbers is an extremely time-consuming process, and this would make the evaluation of the greatest common divisor all but impossible. Fortunately, there is an algorithm that allows us to compute the GCD rapidly, even for numbers with thousands of digits. This method was discovered before 300 B.C. in Greece and is known as the Euclidean algorithm. The same process was discovered in India in the fifth century by Aryabhata and was known as the "kuttaka" (or the pulverizer). The algorithm is based on the fact that $(a, b) = (a - b, b) = (a - 2b, b) = \cdots$. If a is larger than b, we subtract b from a, and viceversa, thus reducing the size of the numbers involved.

Example 2.5.14. We compute the GCD of 54 and 21 by this method. Then $(54, 21) = (33, 21) = (12, 21) = (12, 9) = (3, 9) = (3, 6) = (3, 3) = 3$.

We can assume that a and b are positive. If we subtract the smaller integer, say b, from the larger integer, say a, eventually we will reach a negative number. This will happen after $q = \lfloor a/b \rfloor$ steps, and the number remaining is r. Hence we can reduce the number of subtractions we have to do by performing one division. This is the content of the following proposition.

Proposition 2.5.15. *Given two integers a and b, if $a = bq + r$, and $0 \leq r < b$, then $(a, b) = (b, r)$.*

The proof is left to the reader.

Algorithm 2.5.16 (Euclid, Aryabhata). Given positive integers a and b, this procedure finds their greatest common divisor. Assume a and b are not both 0.

1. [Initialize] Set $x = a$, and $y = b$.

2. [Are we done?] If $y = 0$, return x as the GCD.

3. [Take the remainder] Set $r = x \bmod y$, $x = y$, $y = r$, and go to step 2.

Example 2.5.17.   1. We compute $(54, 21)$ using the Euclidean algorithm. Compare with Example 2.5.14.

$$54 = 2 \cdot 21 + 12$$
$$21 = 1 \cdot 12 + 9$$
$$12 = 1 \cdot 9 + 3$$
$$9 = 3 \cdot \boxed{3} + 0.$$

The last nonzero remainder is the GCD.

2. We can express $3 = (54, 21)$ as a linear combination $54m + 21n$ of 54 and 21 using the Euclidean algorithm. This is done by working backwards from the last equation with a nonzero remainder in the above computation. We write $3 = 12 - 1 \cdot 9$, and then substitute $9 = 21 - 1 \cdot 12$ (obtained from the equation with remainder 9), to get $3 = 2 \cdot 12 - 1 \cdot 21$. Next substitute the expression for 12 from the first equation to get $3 = 2 \cdot 54 - 5 \cdot 21$.

3. Consider $a = 693$ and $b = 147$. Applying the Euclidean algorithm gives the following equations.

$$693 = 4 \cdot 147 + 105$$
$$147 = 1 \cdot 105 + 42$$
$$105 = 2 \cdot 42 + 21$$
$$42 = 2 \cdot \boxed{21} + 0.$$

Hence the GCD is 21.

4. Let $a = 82$ and $b = 97$.

$$82 = 0 \cdot 97 + 82$$
$$97 = 1 \cdot 82 + 15$$
$$82 = 5 \cdot 15 + 7$$
$$15 = 2 \cdot 7 + 1$$
$$7 = 7 \cdot \boxed{1} + 0.$$

The GCD is 1.

The Euclidean algorithm is extremely efficient. The worst case of the algorithm, that is, requiring the most steps, occurs when the inputs are two consecutive Fibonacci numbers (see Exercise 24). Here the number of divisions does not exceed $4 \cdot 8 \log_{10} n$. If n is the smallest of the two numbers, the average number of division steps is approximately $(12 \ln 2/\pi^2) \ln n$. The worst case estimate is not difficult, but the average case requires more advanced techniques.

The greatest common divisor of a set of numbers $\{a_1, \ldots, a_n\}$ is the largest integer dividing all the numbers; it is denoted by $(a_1, a_2, \ldots, a_n)$. This greatest common divisor can be computed recursively by using the Euclidean algorithm and the identity

$$(a_1, a_2, \ldots, a_n) = (a_1, a_2, \ldots, a_{n-2}, (a_{n-1}, a_n)).$$

The properties of the greatest common divisor generalize easily to that of more than two numbers. For example, if $d = (a_1, a_2, \ldots, a_n)$, every common divisor of all the a_i's divides d, and d is a linear combination of the numbers, $d = a_1 x_1 + \cdots + a_n x_n$ for some $x_1, \ldots, x_n$.

If $(a_i, a_j) = 1$ for $i \neq j$, then these numbers are said to be **pairwise relatively prime**.

Example 2.5.18.   1. The numbers $15, 26, 77$ are pairwise relatively prime; hence $(15, 26, 77) = 1$. In this case, the GCD can be expressed as a linear combination of the three numbers because any two can be used to express 1 as their linear combination.

2. Consider the numbers $15, 39$, and 65. No two are relatively prime, but the greatest common divisor of the three numbers is 1. In order to express 1 as a combination of these three, we first write $(15, 39)$ in the form $15x + 39y$. Using the Euclidean algorithm, we compute $2 \cdot 39 - 15 \cdot 5 = 3$. Next we can write $(3, 65)$ in terms of 65 and 3 as $3 \cdot 22 - 65 = (3, 65) = 1$. Combining these expressions, we obtain $39 \cdot 44 - 15 \cdot 110 - 65 = 1$.

———————————— **Exercises for Section 2.5** ————————————

1. Compute the greatest common divisor of the following pairs of numbers by (i) computing their prime factorization, and (ii) the Euclidean algorithm.

     (a) $(781, 994)$ (b) $(5950, 13300)$

     (c) $(6963, 7385)$ (d) $(408, 1071)$

     (e) $(781, 991)$ (f) $(67457, 43521)$

2. Express the greatest common divisor as a linear combination of the numbers in parts (a) and (b) of Exercise 1.

3. Prove that $(n, n + 1) = 1$ for all n. For which n does $(n, n + 2) = 1$ hold?

4. Show that if $(a, b) = 1$, then $(a + b, a - b) = 1$, or 2.

5. Suppose $(a, b) = 1$ and $(b, c) = 1$. Does this imply that $(a, c) = 1$?

6. Verify that 30 satisfies the following property:

$$1 < k < 30, \ (k, 30) = 1 \ \Rightarrow \ k \text{ is prime}.$$

   (Incidentally, 30 is the largest integer with this property.)

7. Show that the numbers $6k + 5$ and $7k + 6$ are relatively prime for every $k \geq 1$.

8. Let p be a prime such that $p^n \mid ab$ and $p \nmid (a, b)$. Prove that $p^n \mid a$ or $p^n \mid b$. Give an example to show that p has to be prime for the conclusion to be valid.

9. Generalize Proposition 2.5.10 to determine the greatest common divisor and least common multiple of n numbers $a_1, \ldots, a_n$ in terms of their prime factorizations.

10. Prove the following identities:

    (a) $(a, [b, c]) = [(a, b), (a, c)]$.

    (b) $[a, (b, c)] = ([a, b], [a, c])$.

    (c) $([a, b], [a, c], [b, c]) = [(a, b), (a, c), (b, c)]$.

11. Give an example to show that $[a, b, c](a, b, c) = |abc|$ is false in general. Show that
$$[a, b, c] = \frac{|abc|(a, b, c)}{(a, b)(a, c)(b, c)}.$$

    Guess the formula for the LCM of four or more numbers. Verify that your guess is right by using suitable examples. Can you prove that your formula is correct?

12. Prove that $6n - 1$, $6n + 1$, $6n + 2$, $6n + 3$, and $6n + 5$ are pairwise relatively prime.

13. Show that if m and n are relatively prime, then
$$\sum_{k=1}^{n-1} \left\lfloor \frac{mk}{n} \right\rfloor = \frac{(n-1)(m-1)}{2}.$$

14. Give an example of numbers $a_1, \ldots, a_k$ for any k such that $(a_1, \ldots, a_k) = 1$, but no two are pairwise relatively prime.

15. Let $F_n = 2^{2^n} + 1$ be the nth Fermat number. Use the identity $a^2 - b^2 = (a - b)(a + b)$ to show that

$$F_n - 2 = F_0 F_1 F_2 \cdots F_{n-1}.$$

Conclude that $(F_n, F_m) = 1$ for all $n \neq m$. Show that this implies the infinitude of primes.

16. Write a computer program that finds the GCD using the Euclidean algorithm. Test it on the pairs given in Exercise 1.

17. A point in the plane with integer coordinates x and y is said to be **visible** if $(x, y) = 1$. These are the points that can be seen if one stands at the origin; for example, in Figure 2.5.1, $(3, 2)$ and $(5, 3)$ are visible, but $(6, 4)$ is not. This exercise investigates the probability that a randomly chosen point is visible. Find the number of visible points, $V(n)$, in the square $\{(x, y) \mid 1 \leq x \leq n, 1 \leq y \leq n\}$ for $n = 10, 20, 30, 40, 50$. What is the ratio $V(n)/n^2$? Evaluate for some larger values of n. Does the ratio $V(n)/n^2$ have a limit as $n \to \infty$? If so, is the result surprising?

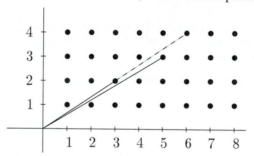

Figure 2.5.1 Visible points in the plane

18. There is another method for computing the greatest common divisor that does not involve any divisions. This is convenient for large integers, for which division is a more time-consuming operation. The method involves only division by 2 (which is easy on a binary computer) and subtractions. The idea of the method, called binary GCD, is that if both a and b are even, then $(a, b) = 2(a/2, b/2)$. If only one is even, say a, then $(a, b) = (a/2, b)$. If both are odd, we can reduce the numbers using $(a, b) = (a - b, b)$ if $a > b$. Write a program to find the GCD using the binary GCD algorithm.

19. Show that $(f_n, f_{n+1}) = 1$ for consecutive Fibonacci numbers f_n and f_{n+1}.

20. Extend the GCD program so that it can find the greatest common divisor of any set of numbers.

21. Investigate the greatest common divisor of f_m and f_n, the mth and nth Fibonacci numbers, respectively. Can you guess the value of (f_m, f_n)? Exercise 2.1.25 on the divisibility properties of Fibonacci numbers may be helpful to get started.

22. Given two integers a and b, show that we can find integers x, y, u, and v such that $a = xu$, $b = yv$, with $(x, y) = 1$, and $xy = \text{LCM}(a, b)$.

23. Use the Euclidean algorithm to show that $(a^m - 1, a^n - 1) = a^{(m,n)} - 1$.

24. This exercise deals with the worst case scenario in the Euclidean algorithm.

    (a) We want to find the smallest pair of integers a and b for which the Euclidean algorithm takes n steps. Explain why the quotients and the GCD have to be 1. Write

    $$a = 1 \cdot b + r_{n-1}$$
    $$b = 1 \cdot r_{n-1} + r_{n-2}$$

    $$\vdots$$

    $$r_3 = 1 \cdot r_2 + r_1$$
    $$r_2 = 1 \cdot r_1 + r_0$$

    with $r_1 = 1$ and $r_0 = 0$. Write $b = r_n$ and $a = r_{n+1}$. Show that $r_n = f_n$, the nth Fibonacci number. Conclude that if $0 \leq a, b < N$, the maximum number of steps, n, occurs when $a = f_{n+1}$ and $b = f_n$, with n as large as possible so that $f_{n+1} < N$.

    (b) Let $\phi = \dfrac{1 + \sqrt{5}}{2}$ be the Golden Ratio. Show that

    $$f_n = \frac{1}{\sqrt{5}} \left\{ \left(\frac{1 + \sqrt{5}}{2} \right)^n - \left(\frac{1 - \sqrt{5}}{2} \right)^n \right\}.$$

    (c) Use the formula for f_n above to show that if $f_n < N$ then $n + 1 < \log_\phi(\sqrt{5}N)$. Show that $\log_\phi(\sqrt{5}N) \approx 2.078 \log(N) + 1.672$; hence this is the maximum number of steps required in the Euclidean algorithm for the GCD of a and b, $0 \leq a, b < N$.

25. The rational numbers h/k, $0 \leq h \leq k \leq n$, $(h, k) = 1$ listed in increasing order are called the **Farey sequence of order n**. The Farey sequence of order 4 is

    $$0 = \frac{0}{1}, \frac{1}{4}, \frac{1}{3}, \frac{1}{2}, \frac{2}{3}, \frac{3}{4}, \frac{1}{1} = 1$$

    (a) Write down the Farey sequences of order 5 and 6. (Remember to place them in increasing order.)

    (b) Show that if k and k' are positive and $h/k < h'/k'$, then

    $$\frac{h}{k} < \frac{h + h'}{k + k'} < \frac{h'}{k'}.$$

    Use this inequality, and explain how to construct the Farey sequence of order n from the Farey sequence of order $n - 1$. (First, try to construct a few examples, say, sequences of order 5 and 6 from previous ones.)

(c) Show that if h/k and h'/k' are adjacent fractions in some Farey sequence, with $h/k < h'/k'$, then $h'k - hk' = 1$. (Use induction on the length of the sequences.)

2.6   Linear Diophantine Equations

A problem that arises frequently is to find integers x and y satisfying

$$ax + by = m,$$

where a, b and m are integers. This equation is known as a linear Diophantine equation. By a Diophantine equation, we mean an equation for which the solutions are sought in the integers. These are named after Diophantus of Alexandria (c. 250 A.D.) though Diophantus only considered rational solutions of equations, not integral solutions. The first mathematician to systematically find integer solutions seems to have been Aryabhata (c. 495 A.D.). The method used by Aryabhata for linear equations is essentially what we know today as the Euclidean algorithm. The method was rediscovered in Europe in the seventeenth century by Bachet.

The equation $ax + by = m$ has infinitely many solutions in the rational, or real number systems. This follows from the formula $by = m - ax$. Unless $b = 0$, we can solve for y for any given value of x. But solutions in the integers are a completely different matter; there may be infinitely many solutions or none at all. For example, $2x + 2y = m$ has infinitely many solutions for m even, but none for m odd.

For integers a and b, if $ax + by = m$, it is clear that (a, b) must divide m. We showed in Section 2.5 that (a, b) [and hence every multiple of (a, b)] is expressible as a linear combination of a and b. This allows the description of all the solutions.

Theorem 2.6.1 (Brahmagupta). *The equation $ax + by = m$ is solvable if and only if $(a, b) \mid m$. In this case, if (x_0, y_0) is any solution, then all the solutions are obtained by*

$$x = x_0 + \frac{b}{(a, b)}k, \qquad y = y_0 - \frac{a}{(a, b)}k \qquad \text{for any } k \in \mathbb{Z}.$$

PROOF. Let $d = (a, b)$. It is clear that if $ax + by = m$ has a solution, then d divides m, because d divides a and b. We now show that if $d \mid m$, then $ax + by = m$ has a solution. From Theorem 2.5.6, there exist x' and y' such that $ax' + by' = d$. If $m = dc$, then $x_0 = x'c$ and $y_0 = y'c$ satisfy

$$ax_0 + by_0 = m. \tag{1}$$

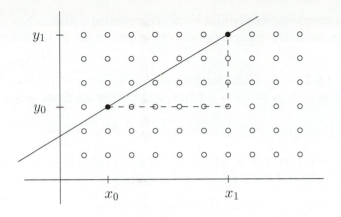

Figure 2.6.1 Solutions to $ax + by = c$

We have shown that the desired equation has a solution. We need to find all the solutions. If x_1, y_1 is another solution, then

$$ax_1 + by_1 = m. \tag{2}$$

Subtracting (2) from (1), we obtain

$$a(x_0 - x_1) + b(y_0 - y_1) = 0$$

or, by rearranging,

$$a(x_0 - x_1) = -b(y_0 - y_1).$$

Equivalently, the slope of the line $ax + by = c$ is $-a/b$ (see Figure 2.6.1). We cancel the common divisor d on both sides and write $a(x_0 - x_1)/d = -b(y_0 - y_1)/d$. As a/d and b/d have no factors in common, we deduce that

$$\frac{a}{d} \mid (y_0 - y_1), \quad \frac{b}{d} \mid (x_0 - x_1).$$

Write

$$y_0 - y_1 = k\frac{a}{d} \quad \text{and} \quad x_0 - x_1 = -k\frac{b}{d};$$

therefore,

$$y_1 = y_0 - k\frac{a}{d}, \quad x_1 = x_0 + k\frac{b}{d}.$$

We have shown that any solution can be constructed from a single solution using the formula in the theorem. Also, if x_1 and y_1 are of this form, then it is a solution because

$$ax_1 + by_1 = ax_0 + \frac{abk}{d} + ay_0 - \frac{abk}{d} = m.$$

Hence we have obtained all the solutions to the equation. ∎

The theorem shows that if a linear Diophantine equation has a solution, then it has infinitely many, and we can construct all of them for any one solution.

Example 2.6.2.   1. Consider the equation $3x + 2y = 5$. Since $(3, 2) = 1$ and $x_0 = 1, y_0 = 1$ is a solution, all possible solutions are of the form

$$x = 1 + 2k, \quad y = 1 - 3k, \quad k \in \mathbb{Z}.$$

2. The equation $21x + 111y = 7$ has no solution in the integers, as $(21, 111) = 3$ and $3 \nmid 7$.

Exercise. Does the equation $21x + 91y = 18$ have integer solutions?

To use the theorem, one must first express the greatest common divisor as a linear combination of a and b. This linear combination can be obtained via the Euclidean algorithm. The idea is that the remainder at each step is a linear combination of a and b, and we eventually obtain the GCD as a remainder. We illustrate with an example and then describe the algorithm.

Example 2.6.3. Express $(119, 42)$ as a linear combination of 119 and 42.
We apply the Euclidean algorithm,

$$119 = 2 \cdot 42 + 35$$
$$42 = 1 \cdot 35 + 7$$
$$35 = 5 \cdot 7.$$

From the first equation, the first remainder, 35, is clearly a linear combination; $35 = 119 - 2 \cdot 42$. We use this in the second equation to express 7 as a linear combination, $42 = 1(119 - 2 \cdot 42) + 7$; therefore,

$$7 = 42 + 2 \cdot 42 - 1 \cdot 119 = 3 \cdot 42 - 1 \cdot 119.$$

Exercise. Express $(115, 20)$ as a linear combination of 115 and 20.

Let us write down the general procedure, which was illustrated in the example above. We express each remainder $r_0, r_1, \ldots$ in the Euclidean algorithm as a linear combination and eventually obtain the desired representation of the GCD. We have $r_0 = a = a \cdot 1 + b \cdot 0$, $r_1 = b = a \cdot 0 + b \cdot 1$. Suppose $m = as + bt$ and $n = au + bv$ are successive remainders that have been expressed as linear combinations. Then $m = qn + r$, and r is the next remainder. If $r = 0$, then n is the GCD and has already been represented. Otherwise,

$$r = m - qn = as + bt - q(au + bv)$$
$$= a(s - qu) + b(t - qv)$$

Brahmagupta (c. 600 A.D.)

Brahmagupta, the most prominent Indian mathematician of the seventh century, lived and worked in Ujjain, in central India. In 628 A.D., he wrote the *Brahma-Sphuta-Siddhanta* ("the revised system of Brahma"), a work on astronomy of which two chapters dealt with mathematics.

Among his accomplishments are the formula for the area of a quadrilateral of sides a, b, c, d that can be inscribed in a circle. His formula is

$$\text{Area} = \sqrt{(s-a)(s-b)(s-c)(s-d)}$$

where $2s = a + b + c + d$. Brahmagupta was the first mathematician to give a systematic treatment of negative numbers and zero, though he erred in stating that $0 \div 0 = 0$. In number theory, he was the first to give the general solution to linear Diophantine equations. He also solved the equation $x^2 - dy^2 = 1$ in many cases. The latter equation was completely solved by Bhaskara in the eleventh century.

is a linear combination of a and b. We can continue in this manner until the GCD is reached. This method is known as the Extended Euclidean Algorithm.

Algorithm 2.6.4 (Extended GCD). The procedure finds the greatest common divisor of a and b, and integers x, y, such that $ax + by = (a, b)$.

1. [Initialize first two terms] $\{m, s, t\} = \{a, 1, 0\}$, $\{n, u, v\} = \{b, 0, 1\}$.

2. Let $q = \lfloor \frac{m}{n} \rfloor$ and $r = m - nq$.

3. [Are we done?] If $r = 0$, return $\{n, u, v\}$.

4. [Find the representation of r and use n and r as the new m and n.] Let

$$\{m', s', t'\} = \{n, u, v\}$$
$$\{n, u, v\} = \{r, s - qu, t - qv\}$$
$$\{m, s, t\} = \{m', s', t'\}.$$

Go to step 2.

Note that the algorithm makes use of an auxiliary variable in step 4 to switch from the previous two remainders, m and n, to the next two, n and r.

In many cases, we are only interested in positive solutions to the linear Diophantine equation $ax + by = d$. A positive solution is a pair (x, y) with both x and y positive and $ax + by = d$. If a and b are of the same sign then the slope of the line $ax + by = c$ is negative, so there are at most a finite number of positive solutions. If a and b have opposite signs, then the equation always has an infinite number of positive solutions.

Example 2.6.5. Let x_0, y_0 be any solution to the equation $ax + by = m$ with a and b positive. Then the number of positive solutions to the equation is the number of integers k satisfying

$$(a, b)\frac{-x_0}{b} < k < \frac{y_0}{a}(a, b).$$

If x_0, y_0 is a solution to $ax + by = d$, then any other solution x_1, y_1 is

$$x_1 = x_0 + \frac{b}{(a, b)}k, \quad y_1 = y_0 - \frac{a}{(a, b)}k.$$

We find k for which this solution is positive, that is,

$$x_0 + \frac{b}{(a, b)}k > 0, \quad y_0 - \frac{a}{(a, b)}k > 0;$$

then $k > -x_0(a, b)/b$ and $k < y_0(a, b)/a$. On the other hand, every k satisfying these conditions clearly yields a positive solution.

Example 2.6.6 (Euler). A farmer lays out the sum of 1770 crowns in purchasing horses and oxen. He pays 31 crowns for each horse and 21 crowns for each ox. How many horses and oxen did the farmer buy?

Solution: If x and y represent the number of horses and oxen bought respectively, then we are interested in positive solutions to the Diophantine equation

$$31x + 21y = 1770.$$

Now, $(31, 21) = 1$ and the Euclidean Algorithm yields

$$1 = -2 \cdot 31 + 3 \cdot 21.$$

Therefore,

$$1770 = (-2 \cdot 1770)31 + (3 \cdot 1770)21.$$

We take $x_0 = -2 \cdot 1770$ and $y_0 = 3 \cdot 1770$, and then the number of positive solutions is the number of integers k satisfying

$$\frac{2 \cdot 1770}{21} < k < \frac{3 \cdot 1770}{31},$$

or, equivalently,

$$168.4 < k < 171.29$$

So k can be 169, 170, or 171. These values of k give the three solutions:

$$x = -2 \cdot 1770 + 21 \cdot 169 = 9,$$
$$y = 3 \cdot 1770 - 31 \cdot 169 = 71,$$

$$x = -2 \cdot 1770 + 21 \cdot 170 = 30,$$
$$y = 3 \cdot 1770 - 31 \cdot 170 = 40,$$

$$x = -2 \cdot 1770 + 21 \cdot 171 = 51,$$
$$y = 3 \cdot 1770 - 31 \cdot 171 = 9.$$

─────────────── **Exercises for Section 2.6** ───────────────

1. Solve the following linear Diophantine equations. Determine in each case whether positive solutions exist, and if so, how many.

    (a) $144x + 34y = 20$

    (b) $39x + 51y = 7$

    (c) $63x - 37y = 3$

    (d) $119x - 29y = 8$

2. Write a computer program to implement the extended Euclidean Algorithm and use it to verify your solutions in Exercise 1.

3. Let $a, b > 0$ and $a \nmid b$, $b \nmid a$. Show that there exist integers x and y that are solutions to

    $$ax - by = (a, b) \quad 1 \le x \le b - 1, \ 1 \le y \le a - 1.$$

4. Find the smallest integer m such that $11x + 29y = m$ has exactly one solution in the nonnegative integers.

5. Find all m such that $7x + 31y = m$ has exactly two positive solutions.

6. A square of side 1 is divided into a strips by $a - 1$ equally spaced red lines parallel to a side and into b strips by $b - 1$ equally spaced blue lines parallel to the red lines. Suppose $a \nmid b$ and $b \nmid a$. What is the smallest distance between a red line and a blue line?

7. Diophantus passed one sixth of his life in childhood, one twelfth in youth, and one seventh more as a bachelor. Five years after his marriage, a son was born, who died four years before his father, at half his age.

    Use this epigram to determine Diophantus's age when he died.

8. A child has \$6.50 in dimes and quarters. Determine the number of possible combinations of coins he/she can have.

9. Find a number n, which, when divided by 29, leaves a remainder of 17; and when divided by 78, leaves a remainder of 37.

10. A necessary condition that the equation $ax + by + cz = d$ have solutions is that $(a, b, c) \mid d$. Assume that this is the case. First find solutions to the equation $by + cz = (b, c)r$ for some integer r, and then combine each with the solution of $ax + (b, c)r = d$ to find all integral solutions to the equation $ax + by + cz = d$.

11. Find all solutions to the linear equation $10x + 11y + 12z = 20$ in the integers.

12. (Bhaskara) The quantity of rubies without flaw, sapphires, and pearls belonging to one person is five, eight and seven respectively; the number of like gems appertaining to another is seven, nine and six; in addition, one has ninety-two gold coins and the other sixty-two and they are equally rich. Tell me quickly then, intelligent friend, who are conversant with algebra, the prices of each sort of gem.

13. (Bachet) A party of 41 persons, men, women, and children, take part in a meal at the inn. The bill is 40 sous and each man pays 4 sous, each woman 3, and each child 1/3 sou. How many men, women, and children were there?

14. (Euler) Find integers x, y, and z satisfying the following equations:

$$3x + 5y + 7z = 560$$
$$9x + 25y + 49z = 2920$$

Projects for Chapter 2

1. **Pythagorean Triples** The theorem of Pythagoras asserts that for a right-angled triangle, we have the relation $x^2 + y^2 = z^2$, where x and y are the legs, and z, the hypotenuse of the triangle. We call $\{x, y, z\}$ a **Pythagorean triple** if x, y, and z are integers satisfying $x^2 + y^2 = z^2$.

    (a) Let $d = (x, y, z)$ and suppose that $\{x, y, z\}$ is a Pythagorean triple. Show that $\{x/d, y/d, z/d\}$ is also a Pythagorean triple with $(x/d, y/d, z/d) = 1$.

    A Pythagorean triple $\{x, y, z\}$ satisfying $(x, y, z) = 1$ is called **primitive**. To find all Pythagorean triples, it is enough to determine the primitive triples. (Why?)

    (b) Show that if $\{x, y, z\}$ is a primitive triple, then x, y, z are relatively prime in pairs. Hence x and y cannot both be even.

(c) Suppose both x and y are odd; then z^2 is even. Show that z must be even; hence $4 \mid z^2$. If x is odd, what is the remainder when x^2 is divided by 4? Similarly, for y^2. What is the remainder when $x^2 + y^2$ is divided by 4? Conclude that x and y cannot be both odd.

(d) Suppose x is even, y and z odd, and $\{x, y, z\}$ is a primitive Pythagorean triple. Write
$$x^2 = z^2 - y^2 = (z - y)(z + y).$$
Show that both $z - y$ and $z + y$ are even; hence we can write
$$\left(\frac{x}{2}\right)^2 = \frac{x^2}{4} = \left(\frac{z - y}{2}\right)\left(\frac{z + y}{2}\right).$$
Let $r = \dfrac{z - y}{2}$ and $s = \dfrac{z + y}{2}$. Show that $(r, s) = 1$. (Use $(z, y) = 1$.)

Since $\left(\dfrac{x}{2}\right)^2 = rs$ with $(r, s) = 1$, show that both r and s must be perfect squares; that is, there exist integers m and n, $(m, n) = 1$ such that $r = m^2$, $s = n^2$. Show that
$$z = m^2 + n^2$$
$$y = n^2 - m^2$$
$$x = 2mn.$$

(e) Conversely, show that if x, y, z are given by the formula above, for relatively prime integers m and n of different parity, then $\{x, y, z\}$ forms a primitive Pythagorean triple.

(f) Determine all Pythagorean triples with the hypotenuse ≤ 25.

(g) Compute the product xyz of the elements of the triples in part (f). What can you conjecture about the product? Prove your conjecture.

2. **The sum of divisors function $\sigma(n)$.** For an integer n, define the sums of divisors function $\sigma(n)$ as
$$\sigma(n) = \sum_{d \mid n} d,$$
where the sum is over all positive divisors of n. For example,
$$\sigma(6) = 1 + 2 + 3 + 6 = 12$$
$$\sigma(8) = 1 + 2 + 4 + 8 = 15.$$

(a) Determine $\sigma(p^n)$ for p prime.

(b) Show that if $d \mid mn$, $(m, n) = 1$, then d can be written as $d = d_1 d_2$, with $(d_1, d_2) = 1$ and $d_1 \mid m$, $d_2 \mid n$. Conversely, if $d = d_1 d_2$, with $(d_1, d_2) = 1$ and $d_1 \mid m$ and $d_2 \mid n$, then $d \mid mn$.

(c) Use part (b) to show that $\sigma(mn) = \sigma(m)\sigma(n)$ if $(m, n) = 1$. [Write
$$\sigma(mn) = \sum_{d \mid mn} d = \sum_{d_1 d_2 \mid mn} d_1 d_2, \text{ and break the second sum into a}$$
product of two sums.]

    (d) Use parts (a) and (c) to give a formula for $\sigma(n)$ using the prime factorization of n.

3. **Perfect numbers** A number n is called **perfect** if $\sigma(n) = 2n$. For example, 6 is perfect. [See the previous project for the definition of $\sigma(n)$.]

    (a) Suppose $n = 2^{p-1}(2^p - 1)$ where both p and $2^p - 1$ are prime. Show that n is perfect.

    (b) Conversely, suppose n is an even perfect number. Write $n = 2^r s$, s odd. Show that $2^{r+1} \mid \sigma(s)$ and $2^{r+1} - 1 \mid s$.

    (c) Let $s = (2^{r+1} - 1)t$. Show that if $t > 1$, then

$$\sigma(s) \geq 1 + 2^{r+1} - 1 + t + (2^{r+1} - 1)t;$$

    that is, $\sigma(s) \geq 2^{r+1}(1 + t)$.

    (d) Derive a contradiction to the fact that $t > 1$; hence $t = 1$ and $s = 2^{r+1} - 1$. Show that $\sigma(s) = 2^{r+1}$ implies that s is prime; hence $r + 1$ must be prime; that is, $n = 2^{p-1}(2^p - 1)$ for a prime p such that $2^p - 1$ is prime.

    (e) Show that this characterization is the same as Euclid's statement.

> If as many numbers as we please, beginning from unity, be set out continuously in double proportion until the sum of all becomes prime, and if the sum is multiplied by the last, the product will be perfect.

Part (a) was known to Euclid. The characterization of even perfect numbers is due to Euler. At the time of this writing, there are 34 known perfect numbers corresponding to the 34 known Mersenne primes. No one has found an odd perfect number or shown that they don't exist. It has been established that an odd perfect number if it exists must be greater than 10^{300}.[3] See the exercises of Section 15.1 for some results on odd perfect numbers.

4. This project gives an example of a number system where unique factorization into primes fails.

We consider numbers of the form $a + b\sqrt{-5}$, where a and b are integers. Addition and multiplication are defined as follows:

$$(a + b\sqrt{-5}) + (c + d\sqrt{-5}) = a + c + (b + d)\sqrt{-5}$$
$$(a + b\sqrt{-5})(c + d\sqrt{-5}) = ac - 5bd + (ad + bc)\sqrt{-5}.$$

We denote by $\mathbb{Z}[\sqrt{-5}]$ the set of these numbers with addition and multiplication defined as above. This is analogous to the addition and multiplication of complex numbers with $\sqrt{-5}$ playing the role of i.

Once addition and multiplication are defined, divisibility and primes can be defined just as for the integers. Again, ± 1 are not primes.

We say that $a - b\sqrt{-5}$ is the **conjugate** of $a + b\sqrt{-5}$.

---

[3] See R. K. Guy. *Unsolved Problems in Number Theory.* 2d ed. New York: Springer–Verlag, 1994.

(a) Show that $(x + y\sqrt{-5})(x - y\sqrt{-5}) = x^2 + 5y^2$.

(b) Show that if $x^2 + 5y^2 = 1$, then $x = \pm 1$.

(c) Suppose $x^2 + 5y^2 = p$, a prime number in the integers. Show that $(x + y\sqrt{-5})$ must be prime in the number system $\mathbb{Z}[\sqrt{-5}]$ (similarly $x - y\sqrt{-5}$ must be prime). Find all primes less than 50 in $\mathbb{Z}$ that can be written in the form $x^2 + 5y^2$.

(d) Suppose p is prime in the integers and it cannot be written as $x^2 + 5y^2$. Show that p is also prime in the system $\mathbb{Z}[\sqrt{-5}]$. Find all primes $p \le 50$ that cannot be written as $x^2 + 5y^2$.

(e) Find all representations of $21 = 7 \cdot 3$ in the form $x^2 + 5y^2$. Now factor each representation as in (1) above and show using (c) and (d) that the components are prime. This shows that a number can be written in more than one way as a product of primes.

(f) Find similar examples of number systems where unique factorization fails.

[In (c), (d), and (e) you will need to use the unique factorization theorem for the integers. Suppose $x + y\sqrt{-5} = (a + b\sqrt{-5})(c + d\sqrt{-5})$. Show that $x^2 + 5y^2 = (a^2 + 5b^2)(c^2 + 5d^2)$; hence the factors on the left and right should match. Then use (c) and (d) to show the primality of the components in (e).]

5. **Goldbach's Conjecture** In a letter to Euler in 1742, Goldbach stated that

> Every integer greater than 5 is the sum of three primes.

Euler replied that this was equivalent to

> Every even integer greater than or equal to 4 is the sum of two primes.

(a) Show that the two statements are equivalent.

(b) Write a computer program to verify Goldbach's conjecture for an even integer.

(c) Let $R_2(n)$ be the number of representations of n as a sum of two primes (counting $p + q$ and $q + p$ as distinct). Modify your program to compute $R_2(n)$ and plot the graph of $R_2(n)$ for $4 \le n \le 5000$. The graph shows that while $R_2(n)$ is very random, there is a distinct lower bound. Verify that $C\dfrac{n}{(\log n)^2} \log \log n$ provides an upper bound for $C = 1.32$ and a lower bound $C = 0.28$.

(d) An even better estimate on $R_2(n)$ is obtained by the following formula. Let

$$P_2(n) = \prod_{p < 2,\, p \mid n} \frac{p-1}{p-2} \quad \text{and} \quad \mathrm{Li}_2(x) = \int_2^n \frac{dx}{(\log x)^2}.$$

The extended Goldbach conjecture states that

$$\frac{R_2(x)}{P_2(x)} \sim c_2 \operatorname{Li}_2(x),$$

where $c_2 = 1.32\ldots$. Verify this estimate for x less that 5000.[4]

6. **Quotients in the Euclidean Algorithm** This project studies the relative occurrence of various integers as quotients in the Euclidean Algorithm.

   (a) Write a computer program to output a list of the quotients that occur when the Euclidean Algorithm is applied to a pair of integers, a and b.

   (b) Take a large random sample of pairs of integers a and b, $0 < a < b$, and compute all the quotients that occur in the Euclidean Algorithm applied to these pairs. The probability that a quotient is 1 is defined as the the number of times 1 occurs divided by the total number of quotients. What is the probability that a quotient is 1? Repeat this for a different sample of pairs of integers. Is the probability the same?

   (c) Repeat the exercise to find the probability that 2, 3, 4, etc., occur as quotients in the Euclidean Algorithm.

   (d) Estimate (i) the probability that the second quotient is 1, and (ii) the probability that the third quotient is 1.

These probabilities were estimated by Gauss, who conjectured a formula for the probability that a quotient is m. R. O. Kuz'min proved Gauss's conjecture in 1928. See the projects at the end of Chapter 11 for further discussion of this problem.

---

[4]For further information, consult P. Ribenboim. *The Little Book of Big Primes*. New York: Springer–Verlag, 1991.

Chapter 3

Modular Arithmetic

*The invention of the symbol ≡ by Gauss affords a
striking example of the advantage which may be
derived from an appropriate notation, and marks an
epoch in the development of the science of arithmetic.*

— G. B. MATHEWS

3.1 Congruences

The basic idea of congruences is to do arithmetic with the remainders obtained upon division by different integers. The reason being that all the information regarding divisibility questions is contained in the remainders obtained from divisions. The notation of congruences expresses the properties of remainders in a compact way, permitting their manipulation to deduce interesting facts about the integers. The notation and basic facts were given by Gauss in his great work, the *Disquisitiones Arithmeticae* published in 1801.[1]

The notion of congruences formalizes a natural idea. If we wish to find the remainder upon division by 10 when we add 193 and 438, we simply take the sum of the last digits, $3 + 8 = 11$, and its last digit is the last digit of the sum of 193 and 438. It is not necessary to do the entire addition to know the last digit. To know the last digit when we multiply two numbers, we simply multiply their last digits and take the last digit of the result. If you are familiar with arithmetic in different bases, you know that the same is

---

[1]C. F. Gauss. *Disquisitiones Arithmeticae*. Translated by A. A. Clarker. Revised by W. C. Waterhourse. New York: Springer–Verlag, 1986.

true in any base. The notion of congruences makes this idea into a powerful tool. Recall that we introduced the notation $a \bmod m$ to denote the remainder when a is divided by m. We can write the rule for the remainder when $a + b$ is divided by m as

$$(a + b) \bmod m = (a \bmod m + b \bmod m) \bmod m.$$

It is not correct to just write $a \bmod m + b \bmod m$ on the right as the sum can be larger than m. To simplify the computation with the remainders, which are also called residues, we introduce the notion of congruences. The principle of congruences is that the result of the computation does not depend on a and b, but only their residues modulo m. Replacing a and b by other numbers giving the same remainder will not affect the computation.

Definition 3.1.1. If a, b, and m are integers, we say that a **is congruent to** b **modulo** m [denoted by $a \equiv b \pmod{m}$] if $m \mid a - b$. If $m \nmid a - b$, we write $a \not\equiv b \pmod{m}$ and say that a is not congruent, or **incongruent to** b **modulo** m.

Example 3.1.2.   1. As $9 = 23 - 14$, the definition implies that $23 \equiv 14 \pmod{9}$. In fact, any two numbers in the list $\{\ldots, -4, 5, 14, 23, \ldots\}$ are congruent modulo 9.

2. The congruence $a \equiv b \pmod{1}$ is true for all integers a and b.

3. Clearly, $a \equiv b \pmod{m}$ is the same as $a \equiv b \pmod{(-m)}$. Hence we consider only positive moduli.

Remark. The congruence $a \equiv b \pmod{m}$ means that both a and b yield the same remainder when divided by m. This follows by writing $a = mq + r$ and $b = mq' + r'$ where $r = a \bmod m$ and $r' = b \bmod m$. Therefore, $a - b = m(q - q') + (r - r')$. Since r and r' are less than m, so is $(r - r')$. But $a - b$ is a multiple of m, so $r - r' = 0$.

Remark. In the previous chapter we defined the operation $a \bmod b$ to mean the remainder obtained when a is divided by b in accordance with the division algorithm. While the notation for congruences and the remainder are similar, they are different concepts. If we write $x = a \bmod b$, then x is the remainder when a is divided by b, but $x \equiv a \pmod{b}$ means that both x and a give the same remainder when divided by b. Naturally, $x = a \bmod b$ implies that $x \equiv a \pmod{b}$, but the converse is not necessarily true. If $x \equiv a \pmod{b}$, and $x < 0$ or $x \geq b$, then $x \neq a \bmod b$.

Exercise.   (a) What are all the integers n such that $n \equiv 3 \pmod{8}$?

(b) Can an odd number p be congruent to 2 mod 4?

(c) Is $23 = 41$ mod 9? Is $23 \equiv 41 \pmod 9$?

Proposition 3.1.3 (Basic Properties of Congruences). *If a, b, and c are integers, then*

(a) $a \equiv a \pmod m$ *for every a.*

(b) $a \equiv b \pmod m$ *if and only if $b \equiv a \pmod m$.*

(c) $a \equiv b \pmod m$ *and $b \equiv c \pmod m$ imply that $a \equiv c \pmod m$.*

PROOF. The first two statements are clear from the definition. To prove the third assertion, note that $m \mid a - b$ and $m \mid b - c$ imply that m divides their sum, $m \mid (a - b) + (b - c) = a - c$. Hence $a \equiv c \pmod m$. ∎

The virtue of congruences lies in their arithmetic properties of addition and multiplication. First, let us illustrate the statements made before the definition of congruences.

Example 3.1.4. Suppose $n \equiv 3 \pmod{17}$ and $m \equiv 8 \pmod{17}$. Does it follow that $n + m \equiv 11 \pmod{17}$? The answer is "yes," which we can see as follows.

We can write $n = 17k + 3$ and $m = 17l + 8$ for some k, l. Then $n + m = 17k + 3 + 17l + 8 = 17(k + l) + 11$; hence $(n + m \bmod 17) = 11$. In congruence notation, $n \equiv 3 \pmod{17}$ and $m \equiv 8 \pmod{17}$; adding these two, we get $n + m \equiv 3 + 8 \equiv 11 \pmod{17}$.

Suppose we want to find the remainder when nm is divided by 17. Multiplying the two expressions, we obtain, $nm = (17k + 3)(17l + 8) = 289kl + 51l + 136k + 24 = 17(17kl + 3l + 8k + 1) + 7$. By the uniqueness of the remainder in the division algorithm, the remainder obtained is 7. In order to do the same computation in terms of congruences, we write $n \equiv 3 \pmod{17}$ and $m \equiv 5 \pmod{17}$, which imply $nm \equiv 3 \cdot 8 \equiv 7 \pmod{17}$. Hence we have $7 = nm \bmod 17$.

Exercise. With n and m as in the previous example, what is the remainder obtained after dividing $n - m$ by 17? What about $m - n$?

The following proposition gives the basic arithmetic properties of congruences. These allow us to manipulate congruences without having to express the integers in terms of the quotient and remainders as in Example 3.1.4. The proposition shows that congruences with a fixed modulus can be manipulated using the "usual" rules of algebra.

Proposition 3.1.5. *If a, b, c, and d are integers, then*

(a) $a \equiv b \pmod m$ *implies that $ac \equiv bc \pmod m$ for all c.*

(b) $a \equiv b \pmod{m}$ *and* $c \equiv d \pmod{m}$ *imply that* $a + c \equiv b + d$ $\pmod{m}$.

(c) $a \equiv b \pmod{m}$ *and* $c \equiv d \pmod{m}$ *imply that* $ac \equiv bd \pmod{m}$.

(d) $a \equiv b \pmod{m}$ *implies that* $a^k \equiv b^k \pmod{m}$ *for all positive integers* k.

PROOF.   (a) This follows from the fact that if m divides $a - b$, then it divides any multiple of $a - b$.

(b) If $m \mid a - b$ and $m \mid c - d$, then $m \mid (a - b + c - d)$, and rearranging, we obtain $m \mid (a + c - (b + d))$. This proves the second assertion.

(c) If $m \mid a - b$ and $m \mid c - d$, then $m \mid (a - b)c + b(c - d)$; that is, $m \mid ac - bd$ or $ac \equiv bd \pmod{m}$.

(d) Multiplying the congruence $a \equiv b \pmod{m}$ with itself gives $a^2 \equiv b^2$ $\pmod{m}$. Repeating this process, we get $a^3 \equiv b^3 \pmod{m}$, ..., $a^k \equiv b^k \pmod{m}$ for any positive integer k.   ∎

Example 3.1.6.   1. Since $16 \equiv -1 \pmod{17}$, then it follows that $16^2 \equiv 256 \equiv 1 \pmod{17}$.

2. Let's compute $2^{4k} \bmod 5$. As $2^4 \equiv 16 \equiv 1 \pmod{5}$, we have $(2^4)^2 \equiv 2^8 \equiv 1 \pmod{5}$, $2^{12} \equiv 2^8 2^4 \equiv 1 \pmod{5}$, and so on. For any $k \geq 1$, we have $(2^{4k} \bmod 5) = 1$.

3. We can determine $2^{32} \bmod 17$ without evaluating 2^{32}. The computation is done using the arithmetic properties of congruences as follows.

$$2^3 \equiv 8 \pmod{17}, \quad 2^4 \equiv 16 \pmod{17}, \quad 2^5 \equiv 32 \equiv 15 \pmod{17},$$
$$2^{10} \equiv 15^2 \equiv 4 \pmod{17}, \quad 2^{30} \equiv 64 \equiv 13 \pmod{17},$$
$$2^{32} \equiv 2^{32} 2^2 \equiv 13 \cdot 4 \equiv 52 \equiv 1 \pmod{17}.$$

This calculation can be done in many different ways. Notice that $16 \equiv -1 \pmod{17}$; that is, $2^4 \equiv -1 \pmod{17}$; therefore, by squaring, $2^8 \equiv 1 \pmod{17}$, so $2^{16} \equiv 1 \pmod{17}$, and finally, $2^{32} \equiv 2^{16} 2^{16} \equiv 1 \pmod{17}$.

This example shows the advantage of congruences in doing computations. One virtue of the method is that the integers occurring in the computation are about the same size as the modulus. While 2^{32} is a large integer, we never evaluate any large values in the example.

Exercise. Determine the remainder when 2^{36} is divided by 11.

The proposition above only gives addition and multiplication properties. Naturally, we would like to see if any division properties hold for congruences. It is easy to construct examples where division properties fail. For example, $ac \equiv bc \pmod{m}$ does not imply that $a \equiv b \pmod{m}$; that is, we cannot "cancel c." For example, $6 \equiv 14 \pmod{8}$ or $2 \cdot 3 \equiv 2 \cdot 7 \pmod{8}$, but $3 \not\equiv 7 \pmod{8}$. This happens because 2 (the factor common to both sides) and 8 (the modulus) have a nontrivial common factor.

Proposition 3.1.7. (a) *If $a \equiv b \pmod{m}$ and $d \mid m$, then $a \equiv b \pmod{d}$.*

(b) $ac \equiv bc \pmod{m}$ *implies that $a \equiv b \pmod{m/(c,m)}$.*

PROOF. (a) If $d \mid m$ and $m \mid a - b$, then $d \mid a - b$. This proves the first assertion.

(b) Let $d = (c, m)$. We can write $c = dc'$ and $m = dm'$ with $(c', m') = 1$; then $dm' \mid c(a - b)$ and $c(a - b) = dc'(a - b)$ imply that $m' \mid c'(a - b)$; since m' and c' have no factors in common, we must have $m' \mid a - b$; that is, $a \equiv b \pmod{m/(c, m)}$. ∎

A special case of the previous proposition occurs when $(c, m) = 1$. If this is the case, then $ac \equiv bc \pmod{m}$ implies that $a \equiv b \pmod{m}$.

A virtue of congruences is that we can deduce results about the arithmetic of very large numbers while performing the actual computations with much smaller numbers. In the example above, we computed $2^{32} \bmod 17$ without evaluating 2^{32}. Any question about divisibility modulo m can be reduced to a calculation involving the numbers $0, 1, \ldots, m - 1$. We illustrate with more examples.

Example 3.1.8. Show that $3 \mid n^3 - n$ for all n.

We are asked to show that $n^3 - n \equiv 0 \pmod{3}$ for all n; that is, $n^3 \equiv n \pmod{3}$. Now, any number n is congruent to 0, 1, or 2 modulo 3; hence we only need to verify the statement for the three numbers 0, 1, and 2.

$$0^3 \equiv 0 \pmod{3}, \quad 1^3 \equiv 1 \pmod{3}, \quad \text{and} \quad 2^3 \equiv 8 \equiv 2 \pmod{3}.$$

Hence $3 \mid n^3 - n$ for all n.

Exercise. Show that $6 \mid n^3 - n$ for all n; that is, 6 divides the product of any three consecutive integers.

Example 3.1.9. Congruences have applications to many systematic ways of shuffling a deck of cards. The analysis of these shuffles using congruences shows that many standard shuffling techniques do not really shuffle the deck at all and after a few repeated shuffles the deck returns to its original arrangement. As an illustration, consider the riffle shuffle. A deck of 52 cards (numbered sequentially from the bottom) is divided into two hands with 26 cards

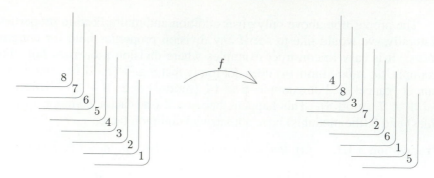

Figure 3.1.1 Riffle shuffle

each, one containing cards $1 - 26$, and another cards $27 - 52$. The two hands are then interlaced by alternately placing a card from each hand. (See Figure 3.1.1 for the result of the shuffle on a deck of 8 cards.) The arrangement, from the bottom, after the shuffle is the following:

$$27, 1, 28, 2, 29, 3, 30, 4, \dots .$$

The first card goes to the second place, the second to the fourth, and so on. Let $f(i)$ denote the position of the ith card after the shuffle. We expect that $f(i) = 2i$. This is correct for $1 \leq i \leq 26$. The 27th card goes to first place, 28th to third place, and so on. Observe that $(2 \cdot 27) \bmod 53 = 1$, and $2 \cdot 28 \bmod 53 = 3$ and so on, where $a \bmod b$ denotes the remainder when a is divided by b. So f is given by $f(i) = 2i \bmod 53$.

If we shuffle the deck twice, the position of the ith card after 2 shuffles satisfies $f(f(i)) \equiv 2(2i) \equiv 2^2 i \pmod{53}$. Repeating this process, we see that the position of the ith card after k shuffles is $2^k i \bmod 53$.

Exercise. Show that $2^{52} \equiv 1 \pmod{53}$; hence the deck returns to its original order after 52 riffle shuffles.

We will need to combine congruences modulo different numbers. If $a \equiv b \pmod{m}$ and $a \equiv b \pmod{n}$, it is not true in general that $a \equiv b \pmod{mn}$; for example,

$$2 \equiv 10 \pmod{8}, \quad 2 \equiv 10 \pmod{4}, \quad \text{but } 2 \not\equiv 10 \pmod{32}.$$

But there is one case in which it is true.

Proposition 3.1.10. *If* $(m, n) = 1$, *then* $a \equiv b \pmod{m}$ *and* $a \equiv b \pmod{n}$ *if and only if* $a \equiv b \pmod{mn}$.

PROOF. If $m \mid a - b$, then $a - b = km$ for some integer k. Now $n \mid a - b$, but $(n, m) = 1$; hence $n \mid k$; therefore, $nm \mid a - b$, or $a \equiv b \pmod{mn}$. Conversely, it is clear that if $a \equiv b \pmod{mn}$, then $a \equiv b \pmod{d}$ for any divisor d of mn, in particular, for m and n. ■

This proposition allows us to analyze a congruence modulo a composite number m by studying the congruences modulo the prime powers occurring in the factorization of m. This is needed because it is easier to answer many questions for prime numbers, and then we can combine the congruences to answer the corresponding question for composite numbers.

Example 3.1.11. 1. $a \equiv b \pmod{12}$ is equivalent to $a \equiv b \pmod 4$ and $a \equiv b \pmod 3$.

2. If p and q are distinct primes and a is an integer such that $a^2 \equiv 1 \pmod{pq}$. This congruence is satisfied if and only if $a^2 \equiv 1 \pmod p$ and $a^2 \equiv 1 \pmod q$.

Certainly, $a \equiv b \pmod 8$ implies $a \equiv b \pmod 4$ and $a \equiv b \pmod 2$, but the converse is not true. (Why? Can you give an example?) In general, we cannot reduce a congruence modulo a prime power into smaller moduli.

Here are some more examples of the use of congruences to investigate divisibility properties.

Example 3.1.12. Suppose we attempt to solve the equation $x^2 + y^2 + z^2 = xyz$ in the integers. A few solutions are $(x, y, z) = (3, 3, 3)$ or $(3, 3, 6)$ or $(3, 6, 15)$. We notice that in every solution (x, y, z), each of x, y, and z is divisible by 3. To prove this, we can use congruences. Any integer is congruent to 0, 1, or 2 modulo 3. If $x \equiv 0 \pmod 3$, then $x^2 \equiv 0 \pmod 3$, and if $x \equiv 1 \pmod 3$, then $x^2 \equiv 1 \pmod 3$. If $x \equiv 2 \pmod 3$, then $x^2 \equiv 1 \pmod 4$. If (x, y, z) forms a solution to $x^2 + y^2 + z^2 = xyz$, then we have the congruence $x^2 + y^2 + z^2 \equiv xyz \pmod 3$. Let us make a table of values of both sides of the congruence. The values given are modulo 3.

x^2	y^2	z^2	$(x^2 + y^2 + z^2)$	xyz
1	1	1	0	1 or 2
1	1	0	2	0
1	0	0	1	0
0	0	0	0	0

We see that the only case in which $x^2 + y^2 + z^2 \equiv xyz \pmod 3$ occurs when $x^2 \equiv y^2 \equiv z^2 \equiv 0 \pmod 3$, which is possible only when $x \equiv y \equiv z \equiv 0 \pmod 3$. This shows that all the entries in any integer solution are divisible by 3.

The basic properties of congruences (Proposition 3.1.3) say that the relation $a \equiv b \pmod m$ on the set of integers is an equivalence relation. The three properties, reflexivity, symmetry, and transitivity of an equivalence relation are the three parts of Proposition 3.1.3. This relation partitions the set of integers into equivalence classes. The set of equivalence classes is denoted by $\mathbb{Z}/m\mathbb{Z}$.

Example 3.1.13. The five equivalence classes of integers modulo 5 are:

$$\{\ldots, -5, 0, 5, 10, \ldots\}$$
$$\{\ldots, -4, 1, 6, 11, \ldots\}$$
$$\{\ldots, -3, 2, 7, 12, \ldots\}$$
$$\{\ldots, -2, 3, 8, 13, \ldots\}$$
$$\{\ldots, -1, 4, 9, 14, \ldots\}.$$

These classes are sequences given by the formula $5k + r$ and are determined by the remainder r. We can write $[r]$ to denote the class $\{5k + r\}$; that is, all integers a such that $a \equiv r \pmod{5}$. Then $\mathbb{Z}/5\mathbb{Z}$ consists of the classes $[0]$, $[1]$, $[2]$, $[3]$, and $[4]$. When there is no confusion, these classes can be written as 0, 1, 2, 3, and 4.

Proposition 3.1.5 defines addition, subtraction, and multiplication on the set of equivalence classes. The arithmetic does not depend on the choice of an element in a class. Instead of working with whole classes, most of the time it is more convenient to choose a number in each class.

Definition 3.1.14. A **complete residue system** modulo m is a set S of integers such that every integer is congruent to exactly one element of S.

Example 3.1.15. 　　1. The **standard residue system** modulo m is

$$S = \{0, 1, 2, \ldots, m - 1\}.$$

2. The standard residue system modulo 7 is $\{0, 1, 2, 3, 4, 5, 6\}$. Another complete residue system is $\{0, 2, 4, 6, 8, 10, 12\}$. This is because no two of these seven integers are congruent modulo 7, and hence every integer is congruent to exactly one of these seven integers.

3. In some cases, it will be convenient to use both positive and negative numbers.

   For m even, $\left\{ \dfrac{-m}{2} + 1, \dfrac{-m}{2} + 2, \ldots, -1, 0, 1, \ldots, \dfrac{m}{2} \right\}$ is a complete residue system.

   For m odd, $\left\{ \dfrac{-(m - 1)}{2}, \ldots, -1, 0, 1, \ldots, \dfrac{m - 1}{2} \right\}$ is a complete residue system.

   We will refer to these two as the **absolutely least residue systems**.

For computational convenience, we usually choose the standard residue system, but any other system will do just as well.

──────────────── **Exercises for Section 3.1** ────────────────

1. Determine if the following assertions are true.

   (a) $-2 \equiv 31 \pmod{11}$.
   (b) $77 \equiv 5 \pmod{12}$.
   (c) $1111 \equiv 11 \pmod{111}$.

2. Compute the following quantities.

   (a) $2^{83} \bmod 17$.   (b) $3^{29} \bmod 31$.

   (c) $9^{99} \bmod 100$.   (d) $19^{19} \bmod 7$.

3. For which positive integers m does the congruence $75 \equiv 19 \pmod{m}$ hold?

4. Find complete residue systems modulo 11 using only even numbers or only odd numbers.

5. Prove or disprove:

   $$a \equiv b \pmod{m} \text{ implies that } a^2 \equiv b^2 \pmod{m^2}.$$

6. Determine the possible remainders when a perfect square is divided by 4. What if it is divided by 8?

7. Show that if n is odd, then $n^2 \equiv 1 \pmod{8}$.

8. If n and m are odd and not divisible by 3, show that $24 \mid n^2 - m^2$.

9. Show that $3^{2n+5} + 2^{4n+1}$ is divisible by 7 for every $n \geq 1$.

10. Example 3.1.8 shows that the product of three consecutive integers is divisible by 3. It is easy to extend this to show that the product of three consecutive integers is divisible by 6. Show that the product of four consecutive integers is divisible by 24. What about the product of five consecutive integers? Is there a natural generalization to the product of k consecutive integers? Can you prove your guess?

11. Is the sum of three consecutive cubes always divisible by 9?

12. Show that if p is prime and $x^2 \equiv y^2 \pmod{p}$, then $x \equiv \pm y \pmod{p}$.

13. One method to compute $a^k \bmod m$ is to evaluate $a^i \bmod m$ for $i = 2, 3, \ldots$ until we reach $a^k \bmod m$. Is there a better method?

14. Prove that if $a \equiv b \pmod{m}$, then $(a, m) = (b, m)$.

15. Prove or disprove: if $a \equiv b \pmod{m}$ and $c \equiv d \pmod{m}$, and $c \mid a$ and $d \mid b$, then $\dfrac{a}{c} \equiv \dfrac{b}{d} \pmod{m}$.

16. The condition $(m, n) = 1$ in Proposition 3.1.10 can be removed with a suitable change in the assertion. Investigate how to combine congruences when the moduli are not relatively prime, make a conjecture, and prove it.

17. The following exercises deal with divisibility tests for integers expressed in the decimal notation. These tests have been known for a long time but can only be rigorously established using congruences.

    An integer $n = a_m 10^m + a_{m-1} 10^{m-1} + \cdots + a_1 10 + a_0$, with $0 \le a_i \le 9$, $1 \le i \le m$, has the decimal representation $n = a_m a_{m-1} \ldots a_1 a_0$. Use this in the following exercises.

    (a) Prove that n is divisible by 2 (respectively 5) if and only if a_0 is divisible by 2 (respectively 5).

    (b) Determine the divisibility criterion for division by 3 and 9 in terms of the digits $a_0, a_1, \ldots, a_m$.

    (c) Show that n is divisible by 11 if and only if

    $$a_0 + a_2 + a_4 + \cdots \equiv a_1 + a_3 + a_5 + \cdots \pmod{11}.$$

    (d) Formulate a divisibility criterion for division by 7 and for division by 13.

18. In Example 3.1.9 we described the riffle shuffle. Magicians call this an in-shuffle because the top and the bottom cards fall inside the shuffled deck, second from the top and bottom, respectively. There is an analog of this, called the out-shuffle, in which the bottom card falls first. Give a formula for the position of the ith card after an out-shuffle on a deck of $2n$ cards. How many shuffles are needed to restore a deck of 52 cards to its original order? Compare with the number of in-shuffles needed to restore the same deck to its original order.

19. Show that $1 + 2 + 3 + \cdots + (n-1) \equiv 0 \pmod{n}$ if and only if n is odd. Determine the corresponding condition on n so that $1^2 + 2^2 + 3^2 + \cdots + (n-1)^2 \equiv 0 \pmod{n}$.

20. Use Exercise 6 to show that if $m \equiv 3 \pmod{4}$, then m cannot be a sum of two squares.

21. Use Exercise 6 to investigate if there is a congruence condition modulo 4 or modulo 8 on m that determines when m cannot be written as the sum of three squares, $m = x^2 + y^2 + z^2$.
    (It is much more difficult to determine which numbers can be written as the sum of three squares.)

22. (a) Let p be a prime number. What are the possible values of p in the standard residue system modulo 6?

(b) What are the possible values of $n^2 \bmod 6$ for an odd integer n?

(c) Show that if $2^n + n^2$ is prime, then $n \equiv 3 \pmod 6$.

23. Use a computer to investigate the least nonnegative residue of $(m-1)! \bmod m$. What can you conjecture about the value of $(m-1)! \bmod m$ for $m > 2$? When is it equal to 0? To $m-1$?

24. Suppose x and y are integers such that $x^2 \equiv y^2 \pmod n$. Suppose $x \not\equiv y \pmod n$ and $x \not\equiv -y \pmod n$. Then show that $(x - y, n)$ or $(x + y, n)$ is a proper factor of n (i.e., not equal to 1 or n). (This property will be very useful later in sections on factoring.)

3.2 Inverses Modulo m and Linear Congruences

Application of congruences to divisibility problems lead to congruences involving one or more unknown variables. For example, if we want to find all numbers of the form $7k + 5$ that are multiples of 13, then we have to solve the congruence $7k + 5 \equiv 0 \pmod{13}$. If we want to find all numbers n such that the last two digits of n and n^2 are the same, then we have to solve $n^2 \equiv n \pmod{100}$. In this section, we will solve the linear congruence $ax \equiv b \pmod m$ for the unknown variable x. Higher degree congruences are studied in Section 3.4.

In the rational number system, we solve $ax = b$ by dividing by a, or equivalently, by multiplying both sides by the reciprocal (multiplicative inverse) of a. We do the same to solve a linear congruence modulo m, but first we need the notion of the reciprocal or multiplicative inverse modulo m.

Definition 3.2.1. A number a' is called the **inverse of a modulo** m if $aa' \equiv 1 \pmod m$. We say a is **invertible modulo** m if it has an inverse.

We will denote the inverse by $a^{-1} \bmod m$ and omit the modulus when it is clear and there is no confusion with $1/a$, the inverse of a in the rational number system.

Example 3.2.2.   1. As $2 \cdot 6 \equiv 1 \pmod{11}$, the inverse of 6 modulo 11 is 2 and the inverse of 2 modulo 11 is 6.

2. The inverse of 3 modulo 8 is 3 because $3 \cdot 3 \equiv 1 \pmod 8$.

3. There is no inverse for 2 modulo 8, as $2x \equiv 1 \pmod 8$ for some x implies that $8 \mid 2x - 1$, an impossibility, as $2x - 1$ is an odd number.

The last example shows that an integer a need not have an inverse modulo m. The same argument shows that if a and m have a common factor, then a cannot have an inverse modulo m.

Exercise. What are the inverses of 5 and 4 modulo 7? Which numbers do not have inverses modulo 7? What about modulo 14?

Proposition 3.2.3. *An integer a is invertible modulo m if and only if $(a, m) = 1$. If a has an inverse, then it is unique modulo m.*

PROOF. If a has an inverse a', then $aa' \equiv 1 \pmod{m}$ means that $aa' - 1 = km$ for some k. This implies that the greatest common divisor $(a, m) \mid 1$ because it divides the other two terms. Hence $(a, m) = 1$.

Conversely, if $(a, m) = 1$, then there exist x and y such that $ax + my = 1$, that is, $ax \equiv 1 \pmod{m}$. The inverse of a is x modulo m.

From the classification of solutions to linear Diophantine equations, any other solution (x', y') satisfying $ax' + my' = 1$ is such that $x' = x + cm$ for some c. Hence $x' \equiv x \pmod{m}$. This proves the uniqueness of the inverse modulo m. ∎

The proof shows that the inverse can be found by applying the extended Euclidean Algorithm. We express (a, m) as a linear combination of a and m; if $(a, m) \neq 1$, then a is not invertible mod m. Otherwise, any x such that $ax + my = 1$ is an inverse of a modulo m.

Example 3.2.4. Let's compute the inverse of 11 modulo 31. The extended Euclidean algorithm applied to 11 and 31 gives

$$5 \cdot 31 - 14 \cdot 11 = 1.$$

Hence $-14 \cdot 11 \equiv 1 \pmod{31}$, so -14 is an inverse of 11 modulo 31. Since $17 = -14 \bmod 31$, 17 is also an inverse of 11 modulo 31.

Example 3.2.5. An important observation is that modulo a prime p, every number that is not a multiple of p has an inverse. In a complete residue system modulo p, every number not congruent to 0 has an inverse. This shows that the integers modulo p form a **finite field**. A complete residue system modulo p consists of p elements, so it is also denoted by $\mathbb{F}_p$, the finite field of p elements. [2]

Remark. We can define a^n for a negative integer n by using the inverse. Let $a^{-n} \equiv (a^{-1})^n \pmod{m}$ for $n > 0$. It can be verified that the property $a^{x+y} \equiv a^x a^y \pmod{m}$ holds for all integers x and y.

Next, we discuss the solution of linear congruences. To solve $ax \equiv b \pmod{m}$, we multiply by $a^{-1} \bmod m$ if a has an inverse. Then $a^{-1}ax \equiv a^{-1}b \pmod{m}$; that is, $x \equiv a^{-1}b \pmod{m}$ is the solution.

---

[2]For a discussion of algebraic structures, see a book on algebra, such as *Algebra* by M. Artin. Upper Saddle River, NJ: Prentice-Hall, 1991.

Example 3.2.6. Solve $3x \equiv 5 \pmod{13}$.

Applying the extended Euclidean Algorithm, we get $3 \cdot 9 - 13 \cdot 2 = 1$. So 9 is an inverse of 3 modulo 13. Multiplying both sides of the congruence by 9, we obtain

$$9 \cdot 3x \equiv 9 \cdot 5 \equiv 6 \pmod{13},$$
$$x \equiv 6 \pmod{13}.$$

If $(a, m) \neq 1$, we can still solve the congruence in some cases. In this case, there may be more than one solution or none at all.

Proposition 3.2.7. *The linear congruence* $ax \equiv b \pmod{m}$ *has exactly* $d = (a, m)$ *solutions if* $d \mid b$, *and no solutions if* $d \nmid b$.

If $d \mid b$ *and* x_0 *is a solution, then the* d *distinct solutions modulo* m *are* $x_0 + (m/d)i \bmod m$ *for* $i = 0, 1, \ldots, d - 1$.

PROOF. If $d = 1$, the equation has a solution, as a has an inverse modulo m. The solution is unique modulo m because $ax_1 \equiv b \pmod{m}$ and $ax_2 \equiv b \pmod{m}$ imply that $ax_1 \equiv ax_2 \pmod{m}$. We can cancel a, because $(a, m) = 1$, to obtain $x_1 \equiv x_2 \pmod{m}$.

Next, if $d > 1$, then it is clear that if $d \nmid b$, there is no solution because $m \nmid ax - b$ for any x. If $d \mid b$, then we solve

$$\frac{a}{d}x \equiv \frac{b}{d} \pmod{\frac{m}{d}}.$$

This has a unique solution x_0 modulo m/d because a/d has an inverse modulo m/d (why?). Write $(a/d)x_0 - (b/d) = k(m/d)$ for some k and eliminate d to get $ax_0 - b = km$. Then x_0 is a solution to $ax \equiv b \bmod m$. Any other solution x must satisfy $x \equiv x_0 \pmod{m/d}$ (why?); that is, $x - x_0 = i(m/d)$. Write $x_i = x_0 + i(m/d)$. The solutions x_i and x_j are distinct modulo m if and only if $i \not\equiv j \pmod{d}$. To see this, suppose $x_i \not\equiv x_j \pmod{m}$. This implies that $i(m/d) \not\equiv j(m/d) \pmod{m}$. Since m/d is a divisor of m, we can divide every term by m/d to obtain $i \not\equiv j \pmod{d}$. Conversely, it is clear that if $i \not\equiv j \pmod{d}$, then $x_i \not\equiv x_j \pmod{m}$. This shows that there are exactly d distinct solutions. ■

The proof shows that the solutions to a linear congruence can be found by applying the extended Euclidean Algorithm.

Example 3.2.8.   1. In Example 3.2.6, the solution is unique modulo 13 because $(3, 13) = 1$.

2. Solve $4x \equiv 10 \pmod{30}$.
   Since $(4, 30) = 2$ and $2 \mid 10$, there are two solutions. We have $8 \cdot 4 - 1 \cdot 30 = 2$, so $40 \cdot 4 - 5 \cdot 30 = 10$. Hence $40 \cdot 4 \equiv 10 \pmod{30}$, so 10 is a solution modulo 30. The other solution is $10 + (30/2) \equiv 25 \pmod{30}$.

3. The equation $21x \equiv 49 \pmod{119}$ has 7 solutions modulo 119 because $(21, 119) = 7$ and $7 \mid 49$. All the solutions are congruent modulo $119/7 = 17$. Applying the extended Euclidean Algorithm, we see that $6 \cdot 3 - 1 \cdot 17 = 1$, so the inverse of 3 modulo 17 is 6. Hence the solution to the congruence $3x \equiv 7 \pmod{17}$ is $x \equiv 8 \pmod{17}$. Then the solutions to $21x \equiv 49 \pmod{119}$ are $8, 25, 42, 59, 76, 93$, and 110.

———————————— **Exercises for Section 3.2** ————————————

1. Determine the invertible elements modulo 15, 17, and 32.

2. Determine the inverse of 67 modulo 119.

3. Determine all solutions to the congruences $7x \equiv 5 \pmod{13}$ and $4x \equiv 6 \pmod{18}$.

4. Solve the following linear congruences.

    (a) $11x \equiv 28 \pmod{37}$.

    (b) $42x \equiv 90 \pmod{156}$.

5. If a^{-1} is the inverse of a modulo m and b^{-1} the inverse of b modulo m, show that $a^{-1}b^{-1}$ is the inverse of ab modulo m.

6. Let p be an odd prime.

    (a) What are the possible values of $x \bmod p$ so that x is its own inverse modulo p?

    (b) Consider the product $1 \cdot 2 \cdots (p-1) \equiv (p-1)! \pmod{p}$. Show by pairing the elements with their inverses that $(p-1)! \equiv -1 \pmod{p}$. This result is known as Wilson's Theorem. (Does this match with your conjecture in Exercise 3.1.23?)

7. If m is composite, what is the value $(m-1)! \bmod m$ in the standard residue system? Conclude that $(m-1)! \equiv -1 \pmod{m}$ implies that m is prime.

8. Use Exercise 6 to answer the following questions:

    (a) Determine $65! \bmod 67$.

    (b) If p is a prime number, what is $(p-2)! \bmod p$?

9. Show that if $p = 4k+1$ is prime, then $[(2k)!]^2 \equiv -1 \pmod{p}$.

10. Write a computer program that determines if an element a is invertible modulo m, and if it is, the program computes the inverse.

11. Write a computer program to solve linear congruences modulo m.

12. Let $R_m = \{0, 1, 2, \ldots, m - 1\}$ be a complete residue system modulo m. Let a be an integer. When is $a \cdot R_m = \{0, 1 \cdot a, 2 \cdot a, \ldots, a \cdot (m - 1)\}$ a complete residue system modulo m? Prove your conjecture. (Try $m = 15$ and various values of a.)

13. Show that if p is prime, then the simultaneous linear congruence

$$ax + by \equiv u \pmod{p}$$
$$cx + dy \equiv v \pmod{p}$$

has a unique solution x, y modulo p when $ad - bc \not\equiv 0 \pmod{p}$.

14. Let a, b, and m be integers such that $(a, m) = (b, m) = 1$. Determine the number of pairs (x, y) with x and y in a complete residue system, satisfying the congruence $ax + by \equiv 0 \pmod{m}$. What can you say about the number of solutions if the conditions $(a, m) = 1$ or $(b, m) = 1$ are dropped?

15. (Clement, 1949) Let $n \geq 2$.

    (a) Show that $4[(n - 1)! + 1] + n \equiv 0 \pmod{n}$ if and only if n is prime.
    (b) Show that $2(n - 1)! \equiv -1 \pmod{n + 2}$ if and only if $n + 2$ is prime. Hence show that $4[(n - 1)! + 1] + n \equiv 0 \pmod{n + 2}$ if and only if $n + 2$ is prime.
    (c) Show that the integers n and $n + 2$ form a pair of twin primes if and only if

$$4[(n - 1)! + 1] + n \equiv 0 \pmod{[n(n + 2)]}.$$

3.3 Chinese Remainder Theorem

An important problem is to find integers satisfying many different divisibility conditions. Also, in many applications, we reduce a computation modulo composite numbers to a computation over its prime factors. The Chinese Remainder Theorem is the fundamental tool that allows us to combine congruences to reach conclusions about the original problem. The idea first appears in the writings of the Chinese mathematician Sun–Tzu, who lived in the third century. The method was further developed by Chin Chiu–Shao in the thirteenth century. In the West, Euler seems to have been the first to study it extensively. The method is fundamental and has numerous applications. We begin with a motivating example.

Example 3.3.1. Determine the smallest positive integer that gives a remainder of 2 upon division by 3, a remainder of 1 upon division by 5, and a remainder of 6 upon division by 7.

Let x be a solution; then the conditions require that

$$x \equiv 2 \pmod{3} \tag{1}$$
$$x \equiv 1 \pmod{5} \tag{2}$$
$$x \equiv 6 \pmod{7}. \tag{3}$$

This is a system of three congruences on x. We can solve it as follows. The first congruence implies that $x = 3k + 2$ for some k. Using this in equation (2) gives a condition on k.

$$3k + 2 \equiv 1 \pmod 5,$$
$$3k \equiv 1 - 2 \equiv 4 \pmod 5. \tag{4}$$

We can solve for k by multiplying by 3^{-1} mod 5, which is 2. Hence $k \equiv 2 \cdot 4 \equiv 8 \equiv 3 \pmod 5$. Hence there exists an integer r such that $k = 5r + 3$, so $x = 15r + 9 + 2$. Use this value of x in (3) to obtain

$$15r + 11 \equiv 6 \pmod 7$$
$$15r \equiv 6 - 11 \equiv 2 \pmod 7.$$

Since $15 \equiv 1 \pmod 7$, we have $r \equiv 2 \pmod 7$. Therefore, there exists an integer s so that $r = 7s + 2$; hence $x = 15(7s + 2) + 11 = 105s + 41$. It can be verified that every x of this form is a solution. The smallest positive solution is $x = 41$, and it is the only solution in the interval $0 \leq x < 105$. Observe that 105 is the product of 3, 5, and 7, the three moduli.

Exercise. Find the smallest positive solution to the following simultaneous congruence.

$$x \equiv 3 \pmod 7$$
$$x \equiv 8 \pmod{11}.$$

Let us apply the technique of the above example to find the general solution in the case of two simultaneous congruences. Suppose m_1 and m_2 are two relatively prime integers. Given a_1 and a_2, we want to find x so that $x \equiv a_i \pmod{m_i}$, $i = 1, 2$. The first congruence implies $x = m_1 r + a_1$ for some r. Using this in the second congruence gives

$$m_1 r + a_1 \equiv a_2 \pmod{m_2},$$
$$\text{or} \quad m_1 r \equiv a_2 - a_1 \pmod{m_2}$$

Let u be an inverse of m_1 modulo m_2. Then $r \equiv u(a_2 - a_1) \pmod{m_2}$; that is, there exists an integer s so that $r = m_2 s + u(a_2 - a_1)$; hence $x = m_1 m_2 s + m_1 u a_2 - m_1 u a_1 + a_1$.

Instead of starting with the first equation (with respect to m_1), we could have started with the second and obtained a different form for the solution. Since the solution is the same, we can write it in a more symmetric form. Because, $m_1 u \equiv 1 \pmod{m_2}$, we can write $1 - m_1 u = v k m_2$ for some v. Therefore $x = m_1 m_2 s + m_1 u a_2 + m_2 v a_1$. Now, $x \equiv a_1 \pmod{m_1}$ implies that $m_2 v \equiv 1 \pmod{m_1}$; that is, v is the inverse of $m_2 \pmod{m_1}$.

The numbers u and v can be found by the extended Euclidean Algorithm. If $m_1 u + m_2 v = 1$, then we have shown that every solution to the two congruences satisfies $x \equiv u m_1 a_2 + v m_2 a_1 \pmod{m_1 m_2}$. Conversely, it is easy to show that any such x is a solution to the congruence. The following exercise extends this technique to three equations. The reader is urged to try the following exercise before reading the proof of the Chinese Remainder Theorem.

Exercise. Let m_1, m_2, and m_3 be integers that are relatively prime in pairs. Suppose u and v are integers such that $u m_1 + v m_2 = 1$.

(a) Show that three congruences $x \equiv a_i \pmod{m_i}$, $i = 1, 2, 3$ can be reduced to the two congruences

$$x \equiv u m_1 a_2 + v m_2 a_1 \pmod{m_1 m_2}$$
$$x \equiv a_3 \pmod{m_3}.$$

Now, apply the technique for the solution of two congruences to this pair. Let y and z be integers such that $y m_1 m_2 + z m_3 = 1$. Show that every solution to the simultaneous congruence is of the form

$$x \equiv y m_1 m_2 a_3 + u z m_1 m_3 a_2 + v z m_2 m_3 a_1 \pmod{m_1 m_2 m_3}.$$

(b) Show that y is the inverse of $m_1 m_2 \pmod{m_3}$, uz is the inverse of $m_1 m_3 \pmod{m_2}$, and vz is the inverse of $m_2 m_3 \pmod{m_1}$.

(c) Use this to find the general solution in Example 3.3.1.

(d) Do you see a pattern in the general solution to two and three congruences? Generalize to an arbitrary number of congruences.

Chinese Remainder Theorem. *Let $m_1, m_2, \ldots, m_r$ be pairwise relatively prime integers. Then the simultaneous congruence*

$$x \equiv a_1 \pmod{m_1}$$
$$x \equiv a_2 \pmod{m_2}$$
$$\vdots$$
$$x \equiv a_r \pmod{m_r}$$

has a unique solution modulo the product $m_1 m_2 \cdots m_r$.

PROOF. The computations performed above for two and three simultaneous congruences show the general pattern. We give an explicit formula for the solution and then show its uniqueness.

Let $M = m_1 m_2 \cdots m_r$, and $M_i = M/m_i$. Since the m_i's are pairwise relatively prime, M_i is relatively prime to m_i, so it has an inverse x_i modulo m_i; that is, $M_i x_i \equiv 1 \pmod{m_1}$. If $i \neq j$, then $m_i \mid M_j$; that is, $M_j \equiv 0 \pmod{m_i}$. Then

$$x \equiv a_1 M_1 x_1 + a_2 M_2 x_2 + \cdots + a_r M_r x_r \pmod{M}$$

is a solution to the system because

$$a_1 M_1 x_1 + a_2 M_2 x_2 + \cdots + a_r M_r x_r \equiv a_i M_i x_i \pmod{m_i}$$
$$\equiv a_i \cdot 1 \pmod{m_i}$$
$$\equiv a_i \pmod{m_i}.$$

Suppose x and y are two distinct solutions. Now $x \equiv y \pmod{m_i}$ for each i; that is, $m_i \mid (x - y)$. The m_i are pairwise relatively prime, hence, by unique factorization, the product $m_1 m_2 \cdots m_r \mid (x - y)$; that is, $x \equiv y \pmod{m_1 \cdots m_r}$. This proves the uniqueness. ∎

The virtue of the proof is that it gives an explicit formula for the solution.

Example 3.3.2. Let's solve the congruence in Example 3.3.1 using the formula given in the proof of the theorem.

First, we have to compute the quantities M_i. From their definition, $M = 3 \cdot 5 \cdot 7 = 105$, $M_1 = 35$, $M_2 = 21$, and $M_3 = 15$. We can now determine the inverses, either by the Euclidean Algorithm or by inspection, as the numbers involved are small. The inverses x_i of M_i are:

$$x_1 \equiv 2 \pmod{3}, \quad x_2 \equiv 1 \pmod{5}, \quad x_3 \equiv 1 \pmod{7}.$$

Therefore,

$$x \equiv 2 \cdot 35 \cdot 2 + 1 \cdot 21 \cdot 1 + 6 \cdot 15 \cdot 1 \pmod{105}$$
$$\equiv 140 + 21 + 90 \pmod{105}$$
$$\equiv 251 \pmod{105}$$
$$\equiv 41 \pmod{105}.$$

We verify: $41 \equiv 2 \pmod{3}$, $41 \equiv 1 \pmod{5}$ and $41 \equiv 6 \pmod{7}$.

Since the solution is unique modulo the product, all other solutions are of the form $x = 41 + 105k$, where k is an integer.

Exercise. Find the smallest positive number satisfying the following congruences.

$$x \equiv 3 \pmod{8}$$
$$x \equiv 2 \pmod{5}$$
$$x \equiv 6 \pmod{9}.$$

Suppose $m = p_1^{e_1} \ldots p_k^{e_k}$ is the prime factorization of m. The Chinese Remainder Theorem states that any a, $1 \le a \le m$, is completely determined by knowing the remainders r_i upon division by $p_i^{e_i}$; that is, a is the unique solution, satisfying $1 \le a \le m$, to the set of congruences $x \equiv r_i \pmod{p_i^{e_i}}$ for $i = 1, \ldots, k$. This is the most useful feature of the Chinese Remainder Theorem.

A natural question is to try to extend the Chinese Remainder Theorem to moduli that are not necessarily relatively prime.

Example 3.3.3. Consider the congruences

$$x \equiv 3 \pmod 8$$
$$x \equiv 7 \pmod{12}.$$

As 12 is not a prime power, we can break the second congruence into $x \equiv 7 \pmod 3$ and $x \equiv 7 \pmod 4$; that is, $x \equiv 1 \pmod 3$ and $x \equiv 3 \pmod 4$. We now have three congruences

$$x \equiv 3 \pmod 8$$
$$x \equiv 7 \pmod 4$$
$$x \equiv 7 \pmod 3.$$

Now, $x \equiv 3 \pmod 8$ implies $x \equiv 3 \pmod 4$; hence $x \equiv 3 \pmod 4$ is not necessary. We have two remaining congruences

$$x \equiv 3 \pmod 8$$
$$x \equiv 1 \pmod 3.$$

Now the moduli are relatively prime and we can solve this pair of congruences using the previous technique to obtain $x \equiv 19 \pmod{24}$ as the unique solution.

We can also solve the congruences by successive substitution; the congruence $x \equiv 3 \pmod 8$ means that $x = 3 + 8k$ for some integer k. Substitute this into the second equation to obtain $3 + 8k \equiv 7 \pmod{12}$, or $8k \equiv 4 \pmod{12}$. As $(8, 12) = 4$ and $4 \mid 4$, we obtain $2k \equiv 1 \pmod 3$. As 2 is invertible modulo 3, $k \equiv 2 \pmod 3$; that is, $k = 2 + 3r$ and $x = 3 + 8(2 + 3r) = 19 + 24r$ for some integer r.

Exercise. Does the congruence

$$x \equiv 5 \pmod 8$$
$$x \equiv 7 \pmod{12}$$

have a solution?

Ch'in Chiu-Shao (c. 1202-1261)

Ch'in Chiu-Shao was born in 1202 in the province of Szechwan. He seems to have been extremely bright and excelled in many fields. He served in the military and later as a Governor, a position in which he is supposed to have accumulated immense wealth.

   In 1247, he published the *Shu-Shu Chiu-Chang* ("nine sections of mathematics"). The work includes his solution to simultaneous congruences, a method essentially the same as what we use today. In addition, Ch'in Chiu-Shao was the first to find a method to approximate roots of polynomial equations. This is the same method published by Horner in 1819 and is now known as Horner's method.

The following theorem generalizes the Chinese Remainder Theorem to arbitrary moduli.

Theorem 3.3.4. *Let $m_1, \ldots, m_r$ be integers; then the system of congruences $x \equiv a_i \pmod{m_i}$, $i = 1, \ldots, r$ has a solution if and only if for all $i \neq j$, $(m_i, m_j) \mid a_i - a_j$. The solution is unique modulo $[m_1, \ldots, m_r]$.*

PROOF. It suffices to prove the theorem for two integers m_1 and m_2. First, consider the case where $m_i = p^{e_i}$ for some prime p. For $i \neq j$, if $e_j \geq e_i$, then $x \equiv a_j \pmod{p^{e_j}}$ implies that $x \equiv a_j \pmod{p^{e_i}}$; on the other hand, $x \equiv a_i \pmod{p^{e_i}}$; therefore, $a_i \equiv a_j \pmod{p^{e_i}}$, or $p^{e_i} \mid a_i - a_j$. [Here $e_i = \min(e_i, e_j)$, so $(p^{e_i}, p^{e_j}) = p^{e_i}$.]

   In general, let $m_1 = p_1^{e_1} \cdots p_k^{e_k}$ and $m_2 = p_1^{d_1} \cdots p_k^{d_k}$. Then $x \equiv a_1 \pmod{m_1}$ is equivalent to $x \equiv a_1 \pmod{p_i^{e_i}}$ for $i = 1, \ldots, k$, and $x \equiv a_2 \pmod{m_2}$ is equivalent to $x \equiv a_2 \pmod{p_i^{d_i}}$ for $i = 1, \ldots, k$. Looking at the congruence modulo powers of p_i, we see that $a_1 \equiv a_2 \pmod{p_i^{\min(e_i, d_i)}}$, or equivalently, $a_1 \equiv a_2 \pmod{(m_1, m_2)}$. This is the necessary condition for the set of simultaneous congruences to have a solution.

   Conversely, if $(m_i, m_j) \mid a_i - a_j$ for all $i \neq j$, then we can solve the congruences by reducing to the case of relatively prime moduli. If $e_i \geq d_i$, then $a_1 \equiv a_2 \pmod{(m_1, m_2)}$ and $x \equiv a_1 \pmod{p_i^{e_i}}$ imply that $x \equiv a_2 \pmod{p_i^{d_i}}$, and we can drop the second congruence. (Similarly, if $e_i \leq d_i$.) Hence we have a set of congruences

$$x \equiv a_i \pmod{p_i^{\max(e_i, d_i)}}, \quad i = 1, \ldots, k,$$

where a_i is a_1 or a_2 depending on where the maximum occurs. Now we can apply the Chinese Remainder Theorem to obtain a unique solution modulo the LCM. ∎

Example 3.3.5. Consider the system

$$x \equiv 5 \quad (\mathrm{mod}\ 8)$$
$$x \equiv 7 \quad (\mathrm{mod}\ 14)$$
$$x \equiv 21 \quad (\mathrm{mod}\ 35).$$

Now, $(8, 14) \mid (7 - 5)$ and $(35, 14) \mid (21 - 7)$, so the system has a unique solution modulo $[8, 14, 35] = 280$.

To solve the system, we reduce to looking at prime powers:

$$x \equiv 5 \quad (\mathrm{mod}\ 8) \tag{1}$$
$$x \equiv 7 \quad (\mathrm{mod}\ 2) \tag{2}$$
$$x \equiv 7 \quad (\mathrm{mod}\ 7) \tag{3}$$
$$x \equiv 21 \quad (\mathrm{mod}\ 5) \tag{4}$$
$$x \equiv 21 \quad (\mathrm{mod}\ 7). \tag{5}$$

Now, (1) implies (2), so we can drop (2). Equations (3) and (5) are the same; therefore, we have a system of three congruences,

$$x \equiv 5 \quad (\mathrm{mod}\ 8),$$
$$x \equiv 0 \quad (\mathrm{mod}\ 7),$$
$$x \equiv 1 \quad (\mathrm{mod}\ 5),$$

with relatively prime moduli.

Remark. The Chinese Remainder Theorem can also be proved by the following algebraic interpretation. This proof shows the existence of a solution without giving any method for finding one. Recall that $\mathbb{Z}/m\mathbb{Z}$ is the set of equivalence classes of integers that are congruent modulo m.

Let $(m, n) = 1$, consider the product $\mathbb{Z}/m\mathbb{Z} \times \mathbb{Z}/n\mathbb{Z}$ consisting of pairs (x, y) with $x \in \mathbb{Z}/m\mathbb{Z}$ and $y \in \mathbb{Z}/n\mathbb{Z}$. We can define a function from $\mathbb{Z}/mn\mathbb{Z}$ to $\mathbb{Z}/m\mathbb{Z} \times \mathbb{Z}/n\mathbb{Z}$ as follows: if $a \in \mathbb{Z}/mn\mathbb{Z}$: define

$$f(a) = (a \bmod m, a \bmod n).$$

It can be easily checked that f preserves addition and multiplication; that is,

$$f(a + b) = f(a) + f(b)$$
$$f(ab) = f(a)f(b).$$

If $f(a) = 0$, that is, $a \equiv 0 \pmod{m}$ and $a \equiv 0 \pmod{n}$, then a is divisible by mn, that is, $a \equiv 0 \pmod{mn}$. This is equivalent to the fact that f is one-to-one. Now, $\mathbb{Z}/mn\mathbb{Z}$ has mn elements, and $\mathbb{Z}/m\mathbb{Z} \times \mathbb{Z}/n\mathbb{Z}$ also has

mn elements. Since the number of elements is the same, the function f must also be onto. Equivalently, f is an isomorphism (of rings)

$$\mathbb{Z}/mn\mathbb{Z} \approx \mathbb{Z}/m\mathbb{Z} \times \mathbb{Z}/n\mathbb{Z}.$$

The fact that f is onto is equivalent to the Chinese Remainder Theorem because if we are given $a \in \mathbb{Z}/m\mathbb{Z}$ and $b \in \mathbb{Z}/n\mathbb{Z}$, then there is a unique $x \in \mathbb{Z}/mn\mathbb{Z}$ such that $a \equiv x \pmod{m}$ and $b \equiv x \pmod{n}$.

──────────────── **Exercises for Section 3.3** ────────────────

1. Solve the following simultaneous congruences.

   (a) $x \equiv 1 \pmod{2}$ (b) $x \equiv 7 \pmod{9}$
   $x \equiv 2 \pmod{3}$ $x \equiv 0 \pmod{10}$
   $x \equiv 4 \pmod{5}$ $x \equiv 3 \pmod{7}.$
   $x \equiv 2 \pmod{7}.$

2. Determine if the following simultaneous congruences have a solution, and find the smallest positive solution if it exists.

   (a) $x \equiv 3 \pmod{8}$ (b) $x \equiv 4 \pmod{6}$
   $x \equiv 7 \pmod{12}$ $x \equiv 8 \pmod{12}$
   $x \equiv 4 \pmod{15}.$ $x \equiv 12 \pmod{18}.$

3. Write a computer program to solve a set of simultaneous congruences when the moduli are relatively prime. You can use the Extended Euclidean Algorithm to find the solution for two moduli and then use this inductively for an arbitrary set of moduli. This program will be used frequently in later sections, so test it well.

4. [Bhaskara] There are n eggs in a basket. If eggs are removed from the basket 2, 3, 4, 5, and 6 at a time, there remain 1, 2, 3, 4, and 5 eggs in the basket respectively. If eggs are removed from the basket 7 at a time, no eggs remain in the basket. What is the smallest possible number of eggs the basket could have contained?

5. [Ch'in Chiu–Shao] Three farmers divide equally the rice that they have grown. One goes to a market where an 83-pound weight is used, another to a market that uses a 110-pound weight, and the third to a market using a 135-pound weight. Each farmer sells as many full measures as possible, and when the three return home, the first has 32 pounds of rice left, the second 70 pounds, and the third 30 pounds. Find the total amount of rice they took to the market.

6. You are asked to design a system for numbering TV programs to facilitate the programming of VCRs. Each program should be assigned a number so

that a VCR can determine the day of the week, starting time, ending time and the channel of the program. The system should be efficient, that is, use as few numbers as possible, and also easy to implement using a computer. Assume that there are a maximum of 100 channels, and programs can begin and end in time units that are multiples of 15 minutes. Design a numbering system using these criteria. What percentage of the total possible numbers are actually used?

3.4   Polynomial Congruences

In the previous sections, we studied the solution of linear and simultaneous linear congruences. A natural generalization is to consider the solution of polynomial congruences $f(x) \equiv 0 \pmod{m}$, where $f(x)$ is a polynomial with integer coefficients. Consider an example.

Example 3.4.1. Suppose we wish to find all integers n such that n and n^2 have the same last three digits in the decimal representation. The last three digits of n can be obtained by computing $n \bmod 1000$, so we have to solve the congruence $n^2 \equiv n \pmod{1000}$. This is an example of a polynomial congruence. We have to try a thousand values of n, so the solutions can be found easily by using a computer. These are $0, 1, 376$, and 625.

When m is small, we can solve the congruence $x^2 \equiv x \pmod{m}$ by trying all the values, as there are only m possible values of $x \bmod m$. This is not feasible for large m, and in any case, it does not give any insight to the behavior of the solutions. To understand the solution of polynomial congruences, we show that they can be reduced to congruences modulo primes. For small primes, these polynomial congruences can be solved using a computer. For quadratic congruences, we discuss the complete solution in Chapter 9, but the techniques for higher degree congruences are beyond the scope of this book.

Let $f(x)$ be a polynomial with integer coefficients. For every integer x, $f(x)$ is an integer, and hence $f(x) \bmod m$ is defined. We can view $f(x)$ modulo m as a polynomial with coefficients modulo m taking values modulo m. If $x \equiv y \pmod{m}$, then $x^k \equiv y^k \pmod{m}$; hence $f(x) \equiv f(y) \pmod{m}$.

Definition 3.4.2. A **solution** or **(root)** of a polynomial congruence $f(x) \equiv 0 \pmod{m}$ is an integer r such that $f(r) \equiv 0 \pmod{m}$.

If r is a solution of $f(x) \equiv 0 \pmod{m}$ and $r \equiv r' \pmod{m}$, then $f(r') \equiv 0 \pmod{m}$; hence we can view the roots as elements of a residue system modulo m. This implies that there can be at most m solutions that are not congruent to each other modulo m.

Example 3.4.3.   1. We can verify that $x^2 + 2 \equiv 0 \pmod{7}$ has no solutions by checking $x = 0, 1, 2, \ldots, 6$.

2. The equation $x^2 - 2 \equiv 0 \pmod 7$ has two solutions, $x \equiv 3 \pmod 7$ and $x \equiv 4 \pmod 7$.

3. The quadratic equation $x^2 \equiv 1 \pmod{12}$ has four solutions, $x = 1, 5, 7, 11$.

Exercise. Find the solutions to $x^2 - x + 3 \equiv 0 \pmod{11}$.

We now discuss polynomial congruences modulo composite numbers and their reduction to prime moduli.

If $m = p_1^{a_1} \cdots p_k^{a_k}$, then $f(x) \equiv 0 \pmod m$ implies that $f(x) \equiv 0 \pmod{p_i^{a_i}}$ for $1 \leq i \leq k$; that is, a solution modulo m is a solution modulo $p_i^{a_i}$ for each i. Conversely, if there is a root r_i of $f(x) \equiv 0 \pmod{p_i^{a_i}}$ for each i, then, by the Chinese Remainder Theorem, there is a unique r modulo m such that $r \equiv r_i \pmod{p_i^{a_i}}$; that is, $f(r) \equiv 0 \pmod{p_i^{a_i}}$, $1 \leq i \leq k$, and this implies $f(r) \equiv 0 \pmod m$. This reduces the solution polynomial congruences modulo composite numbers to those modulo prime powers, and then we can use the Chinese Remainder Theorem to construct a solution modulo m.

Example 3.4.4. Consider $x^2 \equiv 1 \pmod{105}$. Since $105 = 3 \cdot 5 \cdot 7$, we have three congruences:

$$x^2 \equiv 1 \pmod 3$$
$$x^2 \equiv 1 \pmod 5$$
$$x^2 \equiv 1 \pmod 7.$$

Each of these has two solutions, and any combination of solutions can be combined to give a solution modulo 105. For example, take 2, 4, and 1 as the solutions to the first, second, and third equations, respectively. Then we must solve

$$r \equiv 2 \pmod 3$$
$$r \equiv 4 \pmod 5$$
$$r \equiv 1 \pmod 7.$$

Applying the technique of the Chinese Remainder Theorem yields $r \equiv 29 \pmod{105}$.

Exercise. How many solutions does $x^2 \equiv 1 \pmod{105}$ have? Determine all the solutions.

Next, we describe the reduction of a polynomial congruence modulo prime powers to a congruence modulo primes. These congruences must then be solved by trial and error. We motivate the technique of reduction with an example.

Example 3.4.5. Solve $x^2 \equiv 2 \pmod{7^2}$.

Any solution x must satisfy $x^2 \equiv 2 \pmod 7$. This congruence has solutions $x \equiv 3, 4 \pmod 7$; hence solutions to $x^2 \equiv 2 \pmod{7^2}$ are of the form $x = 3 + 7k$ or $x = 4 + 7l$ for some k and l. Consider the first case, $x = 3 + 7k$. Substituting $x = 3 + 7k$ into $x^2 \equiv 2 \pmod 7^2$ yields:

$$9 + 42k + (7k)^2 \equiv 2 \pmod{7^2},$$
$$9 + 42k \equiv 2 \pmod{7^2},$$
$$42k \equiv -7 \pmod{7^2}.$$

Since every term is divisible by 7, we obtain the linear congruence

$$6k \equiv -1 \pmod 7,$$

which has the solution

$$k \equiv 1 \pmod 7.$$

Therefore, $x = 3 + 7 \cdot 1 \equiv 10 \pmod{7^2}$ is a solution (Verify!).

The second case, $x = 4 + 7l$, yields the equation $8l \equiv -2 \pmod 7$. Solving this linear congruence gives $l \equiv 5 \pmod 7$, so $x = 4 + 7 \cdot 5 \equiv 39 \pmod{7^2}$ is another solution. These are the only solutions to the congruence!

Exercise. Imitate this technique to solve $x^2 \equiv 5 \pmod{11^2}$.

The example shows that solving a quadratic congruence modulo 7^2 reduces to solving a quadratic congruence modulo 7 and a linear congruence modulo 7. This is the case in general, where we may have to solve a sequence of linear congruences and a polynomial congruence modulo p.

Suppose we wish to solve $f(x) \equiv 0 \pmod{p^2}$. A solution modulo p^2 is of the form $x_0 + kp$, where k is some integer and x_0 is a solution to $f(x) \equiv 0 \pmod p$. This is because if $f(x) \equiv 0 \pmod{p^2}$; then $f(x) \equiv 0 \pmod p$, hence $x \equiv x_0 \pmod p$, where x_0 is a solution modulo p. Our problem is to determine the possible values of k for a given value of x_0. To evaluate $f(x_0 + kp)$, we will use the Binomial Theorem.

Suppose $f(x)$ is a polynomial with integer coefficients, say

$$f(x) = a_n x^n + a_{n-1} x^{n-1} + \cdots + a_1 x + a_0.$$

Then the derivative of f is

$$f'(x) = n a_n x^{n-1} + (n-1) a_{n-1} x^{n-2} + \cdots + a_1.$$

Consider the evaluation of $f(x + y)$ using the Binomial Theorem. We can group the terms by the powers of y. The terms without any y in them are the same as those in $f(x)$. The terms with just y occurring are the following:

$$na_n x^{n-1} y + (n-1)a_{n-1} x^{n-2} y + \cdots + a_1 y,$$

which is precisely $y f'(x)$. We will not need the coefficients of higher powers of y. (They can be determined using Taylor's Theorem, if necessary). We can write,

$$f(x + y) = f(x) + y f'(x) + y^2 g(x, y) \tag{1}$$

for some polynomial $g(x, y)$. Using this formula, we can find the solutions to $f(x) \equiv 0 \pmod{p^2}$. We have

$$f(x_0 + kp) = f(x_0) + kp f'(x_0) + k^2 p^2 g(x_0, kp)$$
$$\equiv f(x_0) + kp f'(x_0) \pmod{p^2}.$$

Since $f(x_0) \equiv 0 \pmod{p}$ and $f(x_0 + kp) \equiv 0 \pmod{p^2}$, we can divide by p and obtain

$$k f'(x_0) \equiv -(f(x_0)/p) \pmod{p}.$$

The linear congruence can be solved to obtain the value of k and the desired solution to the polynomial congruence modulo p^2.

The same analysis applies to the solution of a congruence $f(x) \equiv 0 \pmod{p^{k+1}}$ if we know the solutions to the congruence $f(x) \equiv 0 \pmod{p^k}$.

Theorem 3.4.6. *Let $f(x)$ be a polynomial with integer coefficients and $f'(x)$ its derivative. Let x_0 be a solution of $f(x_0) \equiv 0 \pmod{p^k}$.*

(a) *If $p \nmid f'(x_0)$, then there is a unique solution $x = x_0 + p^k t$ to $f(x) \equiv 0 \pmod{p^{k+1}}$, where t is the unique solution to*

$$p^k t f'(x_0) \equiv -f(x) \pmod{p^{k+1}}.$$

(b) *If $p \mid f'(x_0)$ and $p^{k+1} \mid f(x_0)$, then $f(x) \equiv 0 \pmod{p^{k+1}}$ has p incongruent solutions given by $x \equiv x_0 + p^k t \pmod{p^{k+1}}$ for any value of $t \bmod p$.*

(c) *If $p \mid f'(x_0)$ and $p^{k+1} \nmid f(x_0)$, then there is no solution x to $f(x) \equiv 0 \pmod{p^{k+1}}$ such that $x \equiv x_0 \pmod{p^k}$.*

PROOF. Suppose x is a solution to $f(x) \equiv 0 \pmod{p^{k+1}}$ such that $x \equiv x_0 \pmod{p^k}$. Write $x = x_0 + p^k t$ for some t. Using Equation (1), we obtain

$$f(x) \equiv f(x_0 + p^k t)$$

$$\equiv f(x_0) + p^k t f'(x_0) + \left(p^k t\right)^2 g(x, p^k t) \pmod{p^{k+1}},$$

for some $g(x, p^k t)$. But the coefficient of $(p^k t)^2$ is divisible by p^{k+1}; hence

$$f(x) \equiv f(x_0) + p^k t f'(x_0) \pmod{p^{k+1}}.$$

Using this formula, we rewrite $f(x) \equiv 0 \pmod{p^{k+1}}$ (in terms of the quantity t) as

$$t p^k f'(x_0) \equiv -f(x_0) \pmod{p^{k+1}}. \tag{2}$$

The conclusions of this theorem follow from the conditions for the solvability of this congruence. Since $p^k \mid f(x_0)$, if $p \nmid f'(x_0)$, then (2) is equivalent to

$$t f'(x_0) \equiv -\frac{f(x_0)}{p^k} \pmod{p},$$

and this has a unique solution. If $p \mid f'(x_0)$ but $p^{k+1} \nmid f(x_0)$, then there is no solution to the equation. If $p \mid f'(x_0)$ and $p^{k+1} \mid f(x_0)$, then any value of t modulo p is a solution; hence there are p solutions to $x \equiv x_0 \pmod{p^k}$. ■

Example 3.4.7. Find all the solutions to $x^2 + 4x + 2 \equiv 0 \pmod{7^2}$.
     First, we solve $x^2 + 4x + 2 \equiv 0 \pmod 7$. We see that $x = 1$ and $x = 2$ are solutions by trying the seven numbers $x = 0, 1, 2, 3, 4, 5, 6$.
     If we let $f(x) = x^2 + 4x + 2$, then $f'(x) = 2x + 4$. Applying the technique of the theorem, we have the following cases.

(a) Let $x_0 = 1$. Since $7 \nmid f'(1)(= 6)$, there is a unique solution x such that $x \equiv 1 \pmod 7$. It is obtained by solving

$$
\begin{aligned}
7 t f'(1) &\equiv -f(1) & \pmod{7^2} \\
7 t \cdot 6 &\equiv -7 & \pmod{7^2} \\
6 t &\equiv -1 & \pmod 7 \\
t &\equiv 1 & \pmod 7;
\end{aligned}
$$

therefore, $x \equiv 1 + 7 \equiv 8 \pmod{7^2}$ is a root.

(b) Now take $x_0 = 2$. Since $7 \nmid f'(2)(= 8)$, there is a unique solution extending this root. Solving, we obtain

$$
\begin{aligned}
7 t f'(2) &\equiv -f(2) & \pmod{7^2} \\
7 t \cdot 8 &\equiv -14 & \pmod{7^2} \\
4 t &\equiv -1 & \pmod 7 \\
t &= 2 = 5 & \pmod 7;
\end{aligned}
$$

so $x \equiv 2 + 7 \cdot 5 \equiv 37 \pmod{7^2}$ is another solution.

Example 3.4.8. Let us solve the congruence $n^2 \equiv n$ (mod 1000) (see Example 3.4.1) by this technique. We have to solve the two congruences $n^2 \equiv n$ (mod 125) and $n^2 \equiv n$ (mod 8). It is easy to see that the solutions are $n \equiv 0, 1$ (mod 125) and $n \equiv 0, 1$ (mod 8), respectively. We combine them using the Chinese Remainder Theorem to obtain the four solutions modulo 1000: 0, 1, 376, and 625 .

Example 3.4.9. Find all the solutions to $x^3 + 8x^2 - x - 1 \equiv 0$ (mod 11^2).

If we let $f(x) = x^3 + 8x^2 - x - 1$, then $f'(x) = 3x^2 + 16x - 1$. We first solve $f(x) \equiv 0$ (mod 11). This has solutions $x \equiv 4$ (mod 11) and $x \equiv 5$ (mod 11).

  (a) First, consider $x_0 = 4$. Let $x = 4 + 11t$. Then $f(4) = 187$ and $f'(4) = 111 \equiv 1$ (mod 11). There is a unique solution x extending x_0. We solve as in the previous example to obtain $t \equiv 5$ (mod 11), and $x = 4 + 11 \cdot 5 = 59$ is a solution.

  (b) Next, consider $x_0 = 5$. Now, $f'(5) = 154$ is divisible by 11, but $f(5) = 319$ is not divisible by 11^2. Hence there is no solution x to $f(x) \equiv 0$ (mod 11^2) such that $x \equiv 5$ (mod 11).

The congruence $x^3 + 8x^2 - x - 1 \equiv 0$ (mod 11^2) has only one root, $x = 59$.

Example 3.4.10. The equation $x^2 \equiv 1$ (mod 8) has solutions $x \equiv 1, 3, 5, 7$ (mod 8). Is there a solution to $x^2 \equiv 1$ (mod 16) such that $x \equiv 3$ (mod 8)? Let $x_0 = 3$, and apply Theorem 3.4.6. Here, $f(x) = x^2 - 1$ and $f'(x) \equiv 2x$. Now $2 \mid f'(x_0) = 6$, but $16 \nmid f(x_0) = 9 - 1 = 8$; hence by part (c) of Theorem 3.4.6 there is no solution to $x^2 \equiv 1$ (mod 16) such that $x \equiv 3$ (mod 8).

Example 3.4.11. If $n = pq$ is a product of two distinct primes p and q, then the congruence $x^2 \equiv 1$ (mod n) has four solutions, because the congruences $x^2 \equiv 1$ (mod p) and $x^2 \equiv 1$ (mod q) have two solutions each. The trivial solution to the congruence $x^2 \equiv 1$ (mod n) is $x = \pm 1$. If we have a nontrivial solution x, then $x \not\equiv \pm 1$ (mod n) implies that $(x - 1, n)$ and $(x + 1, n)$ are proper factors of n. This fact has applications to the determination of prime factors of a product of two primes.

────────────────── **Exercises for Section 3.4** ──────────────────

  1. Find all solutions to the following equations.

      (a) $x^2 + 4x + 2 \equiv 0$ (mod 7^3)
      (b) $x^2 + 4x + 10 \equiv 0$ (mod 11^2)

      (c) $x^3 + 5x^2 + 2x - 1 \equiv 0 \pmod{7^2}$

2. Solve the following polynomial congruences.

      (a) $x^2 + 12x - 17 \equiv 0 \pmod{143}$

      (b) $x^3 + 30x^2 + 27x + 23 \equiv 0 \pmod{45}$

3. Determine, without using a computer, all integers x such that the last three digits of x^3 are the same as those of x.

4. Let $m = p_1^{a_1} \cdots p_k^{a_k}$. Suppose the congruence $f(x) \equiv 0 \pmod{p_i^{a_i}}$ has n_i distinct solutions for $1 \le i \le k$. Show that $f(x) \equiv 0 \pmod{m}$ has $n_1 n_2 \cdots n_k$ distinct solutions.

5. Let p be an odd prime. Show that $x^2 \equiv 1 \pmod{p^k}$ has exactly two solutions for any $k \ge 1$.

6. Let p be an odd prime. Show that $x^2 \equiv a \pmod{p^k}$, $k \ge 2$ has exactly two solutions if $x^2 \equiv a \pmod{p}$ has two solutions, and no solutions if $x^2 \equiv a \pmod{p}$ has no solutions.

7. This exercise determines the solutions to $x^2 \equiv 1 \pmod{2^k}$, $k \ge 1$.

      (a) Use a computer to determine the solutions for $k = 1, 2, \ldots, 10$. Do you see a general pattern?

      (b) Use the technique of Theorem 3.4.6 to construct the solutions to $x^2 \equiv 1 \pmod{16}$ from the set of solutions to $x^2 \equiv 1 \pmod 8$. Which solutions of $x^2 \equiv 1 \pmod 8$ give solutions to $x^2 \equiv 1 \pmod{16}$ and why?

      (c) Construct solutions to $x^2 \equiv 1 \pmod{32}$ from those of $x^2 \equiv 1 \pmod{16}$. Which solutions of $x^2 \equiv 1 \pmod{16}$ extend to solutions of $x^2 \equiv 1 \pmod{32}$? Do you see a pattern in these computations that will allow you to prove your guess from part (a)?

      (d) Determine all the solutions to $x^2 \equiv 1 \pmod{2^k}$, $k \ge 5$.

8. How many solutions does the congruence $x^2 \equiv 1 \pmod{m}$ have?

9. How many solutions does the congruence $x^2 \equiv x \pmod{m}$ have? Similarly, determine the number of solutions to $x^3 \equiv x \pmod{m}$.

10. Suppose $f(x) = ax^2 + bx + c$, and p is an odd prime, $p \nmid a$. Show that $4af(x) \equiv (2ax + b)^2 + 4ac - b^2 \pmod{p}$. Let $\Delta = b^2 - 4ac$ and $y = 2ax + b$.

      (a) Show that x_0 is a solution to $f(x) \equiv 0 \pmod{p}$ if and only if $y_0 = 2ax_0 + b$ is a solution to $y^2 \equiv \Delta \pmod{p}$.

      (b) The equation $y^2 \equiv \Delta \pmod{p}$ may have two, one, or no solutions. When does $y^2 \equiv \Delta \pmod{p}$ have exactly one solution? Use this to give a condition when $ax^2 + bx + c \equiv 0 \pmod{p}$ has exactly one root (of multiplicity two).

[This exercise shows that by completing the square we can reduce the solutions of quadratic congruences to those of the type $y^2 \equiv \Delta \pmod{p}$. We study this congruence in Chapter 9.]

3.5   Magic Squares

A magic square consists of a series of integers arranged in a square so that the sum of the numbers in any row or column is always the same. Magic squares have been known since ancient times, the first examples of which occurred in China around 2200 B.C. The construction of magic squares is a favorite topic in recreational mathematics. The construction of some magic squares is a pleasant application of the congruence arithmetic discussed in this chapter. The method described in this section is known as the uniform step method. The primary sources for our discussion are E. McClintock,[3] and W. W. Rouseball and H. S. M. Coxeter.[4]

Definition 3.5.1. An $n \times n$ square filled with n^2 integers is said to be a **magic square** if the sums of the entries in every row and column are the same. The **order** of an $n \times n$ magic square is n.

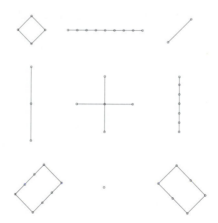

Figure 3.5.1 The Lo-Shu: An ancient Chinese magic square

Usually, it is also assumed that the main diagonals have the same sum. This condition is not a part of our definition. We are also assuming that a magic square is filled; that is, the n^2 entries occupy distinct cells. This is not

---

[3]E. McClintock. " On the Most Perfect Forms of Magic Squares, with Methods for Their Production." *American J. Math.*, 19(1897), 99-120.

[4]W. W. Rouseball and H. S. M. Coxeter. *Mathematical Recreations and Essays.* 12th ed. Buffalo, NY: University of Toronto Press, 1974.

necessary, but our goal is to consider the simplest magic squares. Our first method was described by De La Loubère in 1693. He learned the method in 1687 while serving as the envoy of Louis XIV to Siam. For reasons that will be clear later, the method is successful only for odd order squares.

Example 3.5.2. Consider a 5×5 square. We view the square as part of the larger plane such that the bottom row occurs again above the top row, the first column occurs again to the right of the last column, and so on. The method for constructing the magic squares uses this cyclic nature of the rows and columns. We place the digit 1 in the middle of the top row and the other digits are successively placed in the square using the following rules. We place the next digit by moving to the cell that is one square to the right and one square up from the current position. If this cell is occupied, then the next digit is placed in the cell directly below the current cell.

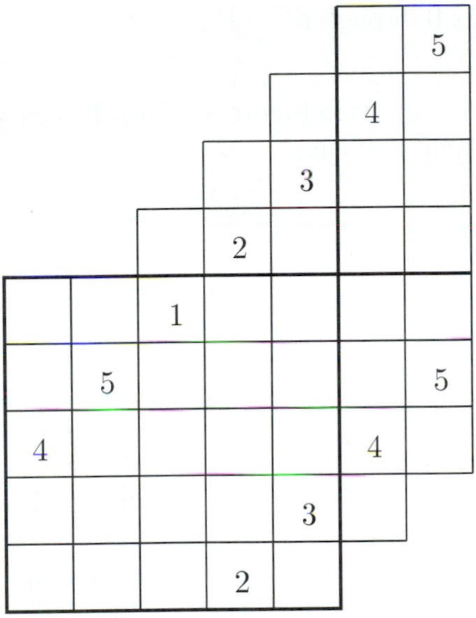

(a) Cyclic placement of integers consecutively

17	24	1	8	15
23	5	7	14	16
4	6	13	20	22
10	12	19	21	3
11	18	25	2	9

(b) 5×5 magic square

Figure 3.5.2 Construction of square in Example 3.5.2

Using these rules, 2 is placed one place to the right and above 1. This cell corresponds to the fourth cell of the bottom row. Similarly, 4 is placed in the middle cell of the first column (see Figure 3.5.2(a)). When we try to place 6, we notice that the next cell is occupied by 1, so we place 6 in the square directly below 5. Following this method, with the identification of every fifth

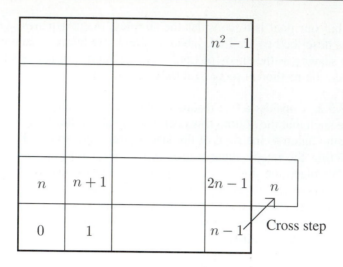

Figure 3.5.3 Integers 0 through $n^2 - 1$

row and every fifth column, we obtain the square in Figure 3.5.2(b). It is easy to check that this construction yields a magic square.

Exercise. Construct a 5×5 square using the same process but starting with 1 in the middle cell. Is it a magic square?

Exercise. Construct a 4×4 square using this method with 1 placed in any cell of your choice. Is it a magic square?

It is not at all clear that the method will fill every cell or that it yields a magic square. Our goal is to understand this method and derive conditions under which a magic square is produced. The cyclic nature of the method indicates that congruences play a role in the construction. In Figure 3.5.2(a) two cells are identified if their coordinates are congruent modulo 5. The use of the congruence notation leads to the uniform step method, a simple generalization of the method of Example 3.5.2. We shall describe the method after a few preliminaries regarding the choice of notation.

For ease of analysis, we will consider magic squares with entries from 0 to $n^2 - 1$. Any other magic square with consecutive entries can be obtained by adding an appropriate amount to every term of a magic square with entries from 0 to $n^2 - 1$. In addition, it is convenient to number the rows and columns from 0 to $n - 1$. Divide the entire plane into cells and identify cells which differ by n in the same row or column. Then, two cells with coordinates (x_1, y_1) and (x_2, y_2) are identified if $x_1 \equiv x_2 \pmod{n}$ and $y_1 \equiv y_2 \pmod{n}$.

Using the division algorithm, any number m, $0 \leq m \leq n^2 - 1$ can be written as $m = nq + r$, $0 \leq q \leq n - 1$, $0 \leq r \leq n - 1$. The quotient q is computed by the formula $q = \lfloor m/n \rfloor$. Conversely, any q and r satisfying

$0 \leq q \leq n - 1, 0 \leq r \leq n - 1$ gives a number m, $0 \leq m \leq n^2 - 1$ via the equation $m = nq + r$. If we think of r as the x-coordinate and q as the y-coordinate, the division algorithm identifies $m = nq + r$ with the cell (r, q). The n^2 numbers from 0 to $n^2 - 1$ are placed in the $n \times n$ square using this identification. (See Figure 3.5.3.)

Consider the placement of the integers 0 to $n^2 - 1$ using the De La Loubère's method. If r increases by one, that is, we move along a row, then we call it a direct step; when we move from one row to the next, we call it a cross-step. (In Figure 3.5.3 these are the steps from $n - 1$ to n, $2n - 1$ to $2n$, etc.) When we follow De La Loubère's method, a direct step increases x and y by one unit, and the cross-step decreases y by one unit.

A number r, $0 \leq r \leq n - 1$ with coordinates $(r, 0)$ ends up in the square with coordinates,

$$x_r \equiv x_0 + r \quad (\text{mod } n)$$
$$y_r \equiv y_0 + r \quad (\text{mod } n),$$

where (x_0, y_0) are the coordinates of the square where we have placed 0. The position of n is

$$x_n \equiv x_{n-1} \equiv x_0 + n - 1 \equiv x_0 - 1 \qquad (\text{mod } n)$$
$$y_n \equiv y_{n-1} - 1 \equiv y_0 + n - 1 - 1 \equiv y_0 - 2 \quad (\text{mod } n).$$

The position of $m = n+r$ is obtained by adding r to the coordinates (x_n, y_n), so we obtain

$$x_{n+r} \equiv x_0 + r - 1 \quad (\text{mod } n)$$
$$y_{n+r} \equiv y_o + r - 2 \quad (\text{mod } n).$$

The next cross-step yields the following coordinates for (x_{2n}, y_{2n}):

$$x_{2n} \equiv x_{2n-1} \equiv x_0 - 2 \qquad (\text{mod } n)$$
$$y_{2n} \equiv y_{2n-1} - 1 \equiv y_0 - 4 \quad (\text{mod } n).$$

Continuing in this way, we can prove by induction on q that the position of $m = nq + r$ is given by

$$x_m \equiv x_0 + r - q \quad (\text{mod } n)$$
$$y_m \equiv y_0 + r - 2q \quad (\text{mod } n).$$

These formulas lead to a simple generalization known as the **uniform step method**: select a cell (x_0, y_0) and integers a, b, c, and d; the position of $m = nq + r$ is determined by

$$x_m \equiv x_0 + ar + bq \quad (\text{mod } n)$$
$$y_m \equiv y_0 + cr + dq \quad (\text{mod } n).$$

The integers a, b, c, d can be written as a matrix $M = \begin{pmatrix} a & b \\ c & d \end{pmatrix}$. We say that M is the matrix for the uniform step method given by 1. A direct-step adds (a, c) to the coordinates of the current position A to move to the next cell B, while the cross-step adds $(a + b, c + d)$ to move from A to B.

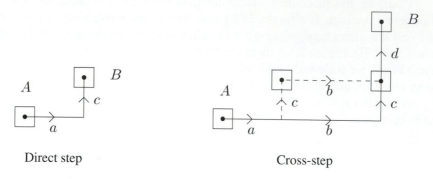

Direct step   Cross-step

Figure 3.5.4 The uniform step method

The method is called the uniform step method as each direct and cross-step is always transformed in the same way. In a direct-step, r increases by 1 and q is fixed, so after the transformation, x increases by a and y by c. In a cross-step, r and q increase by 1 each; hence x increases by $a + b$, and y by $c + d$. In Example 3.5.2 $a = 1$, $c = 1$, and $a + b = 0$, $c + d = -1$; hence $b = -1$ and $d = -2$.

Example 3.5.3. By an appropriate choice of parameters in the uniform step method, we can obtain a knight's step magic square. We construct a square where each step is a valid move of the knight on a chess board. In addition to

11	8	0	22	19
24	16	13	5	2
7	4	21	18	10
15	12	9	1	23
3	20	17	14	6

Figure 3.5.5 Knight's step magic square

the usual rules for a knight's jump, we are allowing a jump from the top row to the bottom row and from the rightmost column to the leftmost one, as in previous method. We take $a = 1$, $c = 2$ for the direct step, and $a + b = 2$, $c + d = 1$, that is, $b = 1$ and $d = -1$ for the cross-step.

A 5×5 square constructed with this matrix with 0 placed in the middle cell of the top row is given in Figure 3.5.5. The reader can verify that we have indeed constructed a magic square.

We will derive conditions under which the uniform step method will produce a filled square with the magic property, but first we give some examples to show that this is not always so.

Example 3.5.4. Consider the 4×4 square constructed by the uniform step method with matrix $\begin{pmatrix} 1 & -1 \\ 1 & -2 \end{pmatrix}$ and initial position $(1, 3)$.

14	0	6	8
3	5	11	13
4	10	12	2
9	15	1	7

The square is filled, but we do not get a magic square, as the sum of the entries in the first row is 28 and that of the second row is 32.

Example 3.5.5. We construct a 3×3 square using the uniform step method with matrix $\begin{pmatrix} 2 & 1 \\ -1 & 1 \end{pmatrix}$ and initial position $(1, 2)$. The square doesn't get filled, as 1 and 3 fall in the same cell:

	$0, 5, 8$	
$1, 3, 6$		
		$2, 4, 7$

This happens because $2 \equiv -1 \pmod 3$, and $-1 \equiv -1 \pmod 3$; that is, the cross-step gets is $(0, 0)$, so a number is placed in the same cell in the cross-step.

Proposition 3.5.6. *The uniform step method given by a matrix* $\begin{pmatrix} a & b \\ c & d \end{pmatrix}$ *fills an* $n \times n$ *square if and only if* $(ad - bc, n) = 1$.

Remark. We only need the implication that if $(ad - bc, n) = 1$, then the square is filled. The converse is interesting but not necessary for our purposes. Note the conclusion that the uniform step method fills an $n \times n$ square. The filled square need not be magic.

PROOF. We first show that if $(ad - bc, n) = 1$, then two distinct numbers $m_1, m_2, 0 \leq m_1, m_2 \leq n^2 - 1$ are not placed in the same cell.

Suppose $m_1 = nq_1 + r_1$ and $m_2 = nq_2 + r_2$ and both occupy the same cell; that is,

$$ar_1 + bq_1 \equiv ar_2 + bq_2 \pmod{n} \qquad (1)$$
$$cr_1 + dq_1 \equiv cr_2 + dq_2 \pmod{n}. \qquad (2)$$

Multiply (1) by d, (2) by b, and subtract the resulting congruences to obtain $(ad - bc)r_1 \equiv (ad - bc)r_2 \pmod{n}$. Since $ad - bc$ has an inverse modulo n, we see that $r_1 \equiv r_2 \pmod{n}$. Similarly, $q_1 \equiv q_2 \pmod{n}$, but $0 \leq r, q \leq n - 1$; hence $r_1 = r_2$ and $q_1 = q_2$, so $m_1 = m_2$. This proves that if $(ad - bc, n) = 1$, then the square gets filled.

Conversely, if $(ad - bc, n) = \delta \neq 1$ then we show that there exist two numbers which occupy the same cell. By Proposition 3.2.7, the congruence $(ad - bc)r \equiv 0 \pmod{n}$ has δ distinct solutions for r. We can assume that one of a, b, c, d is prime to n , say b. Choose r so that $(ad - bc)r \equiv 0 \pmod{n}$; then we can solve for q in the equation $ar + bq \equiv 0 \pmod{n}$. This implies that $cr + dq \equiv 0 \pmod{n}$ because

$$b(cr + dq) \equiv bcr + bdq \equiv adr + bdq \equiv d(ar + bq) \equiv 0 \pmod{n}.$$

Since $(b, n) = 1$, we must have $cr + dq \equiv 0 \pmod{n}$; hence there exists a nontrivial solution (r, q) to the congruences,

$$ar + bq \equiv 0 \pmod{n}$$
$$cr + dq \equiv 0 \pmod{n};$$

therefore, $m = nq + r \not\equiv 0 \pmod{n}$ and 0 occupy the same cell. ∎

Next, to derive conditions for obtaining a magic square, we must sum the entries in a row or column. The entries $m = nq + r$ in row i satisfy

$$i \equiv x_0 + ar + bq \pmod{n},$$

and the entries in column j satisfy

$$j \equiv y_0 + cr + dq \pmod{n}.$$

Our next result classifies the solutions to these linear congruences in two variables.

Lemma 3.5.7. *Suppose $(a, n) = (b, n) = 1$. Then the equation $ar + bq \equiv s$ (mod n) has exactly n distinct solutions for any s. Moreover, the solutions are such that every value of r and q from 0 to $n - 1$ occurs exactly once.*

PROOF. Fix a value of r, $0 \leq r \leq n - 1$. Since $(b, n) = 1$, the equation $bq \equiv s - ar \pmod{n}$ has a unique solution. For different r's, no two of these can be equal. Suppose r_1 and r_2 are such that $s - ar_1 \equiv s - ar_2 \pmod{n}$. Since $(a, n) = 1$, we must have $r_1 \equiv r_2 \pmod{n}$. Hence there are n distinct solutions, and every possible value of r and q occurs exactly once. ∎

Proposition 3.5.8. *Consider the uniform step method for an $n \times n$ square with matrix $\begin{pmatrix} a & b \\ c & d \end{pmatrix}$.*

(a) *If $(a, n) = (b, n) = 1$, then the sum of entries in any row is the same.*

(b) *If $(c, n) = (d, n) = 1$, then the sum of entries in any column is the same.*

(c) *If $(ad - bc, n) = (a, n) = (b, n) = (c, n) = (d, n) = 1$, then the procedure yields a magic square.*

PROOF.   (a) Fix a row, say i. Any element $nq + r$ in row i satisfies

$$i \equiv x_0 + ar + bq \pmod{n}.$$

From the lemma, this congruence has n distinct solutions, $m_k = nq_k + r_k$, for $1 \leq k \leq n$. If we add the entries in a row, then we obtain

$$\sum_k m_k = \sum_k (nq_k + r_k)$$
$$= \sum_k nq_k + \sum_k r_k,$$

but q_k and r_k take every value from 0 to $n - 1$ exactly once, so

$$\sum_k m_k = n \sum_{k=0}^{n-1} k + \sum_{k=0}^{n-1} k$$
$$= n \frac{(n-1)n}{2} + \frac{(n-1)n}{2}$$
$$= \frac{n(n^2 - 1)}{2}.$$

(b) Similar to the previous proof.

(c) This follows from Proposition 3.5.6 and the two previous parts. ∎

Example 3.5.9. Take $n = 7$, and $\begin{pmatrix} a & b \\ c & d \end{pmatrix} = \begin{pmatrix} 1 & 2 \\ 2 & -1 \end{pmatrix}$. The determinant $ad - bc = -5$, so all the conditions are satisfied. By the proposition, the uniform step method will yield a magic square. If $(x_0, y_0) = (3, 6)$, we obtain the following magic square,

26	15	11	0	45	41	30
48	37	33	22	18	7	3
14	10	6	44	40	29	25
36	32	21	17	13	2	47
9	5	43	39	28	24	20
31	27	16	12	1	46	35
4	42	38	34	23	19	8

The uniform step method fails to yield a magic square for n even. This is because $(a, n) = (b, n) = (c, n) = (d, n) = 1$ implies that a, b, c, d are all odd, but then $ad - bc$ is even; hence the square will not be filled. The interested reader should consult the references given at the beginning of the section for techniques of constructing magic squares of even orders.

———————————— **Exercises for Section 3.5** ————————————

1. The magic sum of an $n \times n$ magic square is the sum of the entries in any row or column. What is the value of the magic sum of an $n \times n$ magic square with entries from 1 to n^2?

2. Which of the following matrices give magic squares for the given values of n? Explain.

   (a) $\begin{pmatrix} 1 & 1 \\ 1 & -2 \end{pmatrix}$, $n = 3$.   (b) $\begin{pmatrix} 1 & 1 \\ 5 & 1 \end{pmatrix}$, $n = 6$.

   (c) $\begin{pmatrix} 3 & 2 \\ 2 & 3 \end{pmatrix}$, $n = 13$.   (d) $\begin{pmatrix} 7 & 1 \\ 1 & 2 \end{pmatrix}$, $n = 15$.

3. Show that the method of Example 3.5.2 always gives a filled square. When does it give a magic square?

4. Construct magic squares using the uniform step method and the given values.

   (a) $\begin{pmatrix} a & b \\ c & d \end{pmatrix} = \begin{pmatrix} 1 & 3 \\ 3 & -4 \end{pmatrix}$, $n = 7$, $(x_0, y_0) = (2, 3)$.

   (b) $\begin{pmatrix} a & b \\ c & d \end{pmatrix} = \begin{pmatrix} 2 & 3 \\ -1 & -1 \end{pmatrix}$, $n = 5$, $(x_0, y_0) = (1, 1)$.

5. Write a computer program to print a magic square using the uniform step method.

6. **Pandiagonal squares** A magic square with the additional property that the sum of entries in every diagonal in the same is called pan diagonal (or diabolic.) In addition to the main diagonals, we also consider the *broken diagonals*. The positive diagonals (see Figure 3.5.6) are given by the equation,

$$x + y \equiv k \pmod{n}.$$

The negative diagonals are defined by

$$y \equiv x + k \pmod{n}.$$

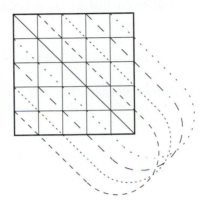

Figure 3.5.6 The positive diagonals

(a) Sketch the negative diagonals for a 5×5 square.

(b) Show that if $(a + c, n) = (b + d, n) = 1$, then the sum of entries in every positive diagonal is the same.

(c) Derive a condition so that the sum of entries in every negative diagonal is the same.

(d) Give conditions on a, b, c, d so that the uniform step method gives a pandiagonal square. Give examples of choices of a, b, c, and d, and construct at least one square of order 7.

7. Suppose n is odd; determine (x_0, y_0) so that the middle number is placed in the middle cell under the uniform step method.

8. **Symmetric Magic Squares:** The cells (i, j) and $(n - 1 - i, n - 1 - j)$ are said to be complementary. A square is symmetric if the sum of the entries of any two complementary cells is the same. If there is a middle cell, then that entry is counted twice.

(a) What is the sum of two complementary cells in a symmetric magic square with entries from 0 to $n^2 - 1$?

(b) Construct a symmetric magic square of order 5.

(c) Show that in the uniform step method, if (x_0, y_0) is chosen so that the middle entry is in the middle cell, then the square is symmetric.

9. **Magic Cubes:** A magic cube of nth order consists of n^3 consecutive integers (say, from 0 to $n^3 - 1$) arranged in the form of a cube such that the sum of every row, every column, and every file is the same.

   (a) What is the magic sum of a magic cube?

   (b) Generalize the uniform step method for constructing a magic cube of order n.

   (c) Construct a 3×3 magic cube using this technique.

   (d) Derive conditions so that the square is filled and magic.

Projects for Chapter 3

---

1. **Day of the week from the date.** The goal of this project is to derive a formula that gives the day of the week for any date in the Gregorian calendar. The Gregorian calendar was adopted in October 4, 1582 in many European countries, and in England and America, it was adopted on September 14, 1752. The Gregorian calendar accounts for the error in the 365-day year by designating the years divisible by 4 as leap years, except those divisible by 100 but not by 400.

   If d denotes the day of the week, we can write $d = 0$ for Monday, $d = 1$ for Tuesday, ..., and $d = 6$ for Sunday.

   Pick a reference date for which you know the day of the week (for example, January 1 of a nonleap year: January 1, 1995 was a Sunday)

   (a) Suppose the dates are numbered sequentially in the year, $1, \ldots, 365$. How would you determine the day of the week for any date on the reference year?

   (b) Now we want to figure out how to convert the information about months into days. Since January has 31 days, and $31 \equiv 3 \pmod{7}$, any date in February is 3 days ahead of the corresponding date in January. Suppose January 1 is the reference date and it is a Tuesday ($d = 1$). For January 17, as $17 - 1 \equiv 2 \pmod{7}$, we add 2 to 1 (for Tuesday) and obtain 3 (Thursday); therefore January 17 is a Thursday. February 17 is 3 days ahead; that is, February 17 is Sunday. Determine the number to add for each month so that you can determine the day for any date in the reference year.

   (c) Now, to go from one year to the next, we can use the fact that $365 \equiv 1 \pmod{7}$, so we can add 1 every time the year changes (except for leap years when we must add 2). Explain how to use this information to find the day of the week. Why does this show that it is better to use March 1 as the reference date? Repeat parts (a) and (b) by taking March 1 of a nonleap year as the reference date.

(d) Derive a formula for finding any day from the date from 1901 to 1999. Note that 1900 and 2000 are not leap years, so you will need to make an adjustment in your formula to find the day in another century.

2. **Card Shuffles:** Recall Example 3.1.9 about the standard riffle shuffle of a deck of playing cards. The example only applied to a deck of $2n$ cards. Suppose the deck has an odd number of cards. There are two ways to split the deck: one in which the top card lies in the half with n cards and another in which it lies in the half with $n + 1$ cards. Determine the position of the ith card in the deck after the shuffle in both cases. How many shuffles are needed to restore a deck of 53 cards to its original order?

Another way to shuffle a deck of $2n$ cards is the Monge shuffle. Cards are taken from the top of the deck. The second card is placed on the first, the third is placed below these, the fourth above them, and so on. (See figure for the shuffle on a deck of 8 cards.) Number the cards sequentially with

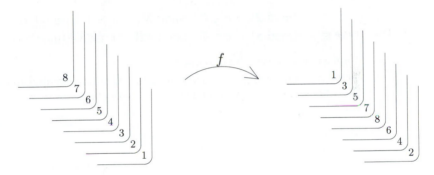

Figure 3.5.7 Monge shuffle

the first card at the bottom. Let f denote the position of the ith card after the shuffle.

(a) Show that $f(i) = i/2$ when i is even, and $f(i) = n - (i - 1)/2$ for i odd.

(b) Let f^{-1} be the inverse function; that is, $f(x) = y$ if and only if $f^{-1}(y) = x$. Show that $f^{-1} \equiv \pm 2i \pmod{4r + 1}$. Determine the values of i for which the positive sign is taken.

(c) Let f^k denote the result of the shuffle applied k times. Determine the position of the ith card after k shuffles.

(d) Show that $f^k = 1$ if and only if $f^{-k} = 1$, where f^{-k} represents the composition of f^{-1} with itself k times. Conclude that the number of shuffles required to return the deck to its original order is the smallest integer k such that $2^k \equiv \pm 1 \pmod{4n + 1}$.

(e) Determine the number of Monge shuffles needed to restore a deck to its original order if the deck has 32, 50, 52, or 64 cards.

Many card tricks are based on the properties of these shuffles. These tricks are studied in many books on magic and recreational mathematics. [5]

3. **Covering Congruences:** A set of congruences, $x \equiv a_i \pmod{n_i}$, where $n_1, \ldots, n_k$ are distinct integers larger than 1, is a covering congruence if every integer satisfies at least one of them.

   (a) Explain why it is sufficient to verify the covering property for integers modulo the LCM of $n_1, \ldots, n_k$.

   (b) Write a computer program to verify that a system of congruences is a covering congruence.

   (c) Show that if the congruences $x \equiv a_i \pmod{n_i}$ form a covering congruence, then $\displaystyle\sum_{i=1}^{k} \frac{1}{n_i} \geq 1$.

   (d) Show that the congruences,

   $$x \equiv 0 \pmod 2, \quad x \equiv 0 \pmod 3, \quad x \equiv 1 \pmod 4$$
   $$x \equiv 3 \pmod 8, \quad x \equiv 7 \pmod{12}, \quad x \equiv 23 \pmod{24},$$

   form a covering set of congruences.

   (e) In the previous example, notice that all the moduli n_i are divisors of 24. Explain how to construct a covering set of congruences with the moduli taken from a subset of divisors of 60.

4. **Binomial Coefficients modulo** p: The binomial coefficients modulo p satisfy many nice properties. The goal of this project is to observe the *fractal nature* of the coefficients modulo p, and derive a formula to explain the phenomena.

   The first nine rows of Pascal's triangle modulo 3 are given in Figure 3.5.8. The first three rows form the triangle:

   $$\triangle = \begin{matrix} & & 1 & & \\ & 1 & & 1 & \\ 1 & & 2 & & 1. \end{matrix}$$

   In fact, the first nine rows are of the form:

   $$\triangle_2 = \begin{matrix} & & & & 1 \cdot \triangle & & & & \\ & & 1 \cdot \triangle & & & & 1 \cdot \triangle & & \\ & 1 \cdot \triangle & & & 2 \cdot \triangle & & & & 1 \cdot \triangle \end{matrix}$$

   where the $\cdot$ indicates multiplication. The first twenty-seven rows are

   $$\begin{matrix} & & & & 1 \cdot \triangle_2 & & & & \\ & & 1 \cdot \triangle_2 & & & & 1 \cdot \triangle_2 & & \\ & 1 \cdot \triangle_2 & & & 2 \cdot \triangle_2 & & & & 1 \cdot \triangle_2. \end{matrix}$$

   Answer the following questions, and explain this phenomenon.

---

[5]See M. Gardner. *Mathematical Carnival*. New York: Knopf, 1975.

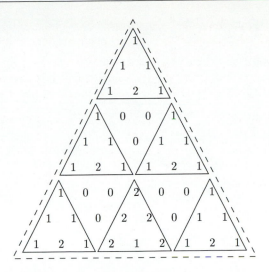

Figure 3.5.8 Pascal's triangle modulo 3

(a) Show that $\binom{p}{k} \equiv 0 \pmod{p}$ unless $k = 0$ or $k = p$.

(b) Show that $\binom{p+r}{k} \equiv \binom{r}{k \bmod p} \pmod{p}$ for $0 \le r \le p - 1$.

(c) Show that $\binom{2p}{p} \equiv 2 \pmod{p}$.

   Items (b) and (c) imply that the $2p$th row is

$$1\,0\,\ldots\,0\,2\,0\,\ldots\,0\,1.$$

(d) Show that

$$\binom{a_1 p + a_0}{b_1 p + b_0} \equiv \binom{a_1}{b_1}\binom{a_0}{b_0} \pmod{p}.$$

   (Hint: first prove the formula for $a_0 = 0$; then use a recursive argument to prove it in the general case.)

(e) Conclude the project by proving Lucas's Theorem. If

$$n = a_k p^k + a_{k-1} p^{k-1} + \cdots + a_1 p + a_0$$
$$m = b_k p^k + b_{k-1} p^{k-1} + \cdots + b_1 p + b_0,$$

   then

$$\binom{n}{m} \equiv \prod_{i=0}^{k} \binom{a_i}{b_i} \pmod{p}.$$

Chapter 4

Fundamental Theorems of Modular Arithmetic

4.1 Fermat's Theorem

Fermat wrote a letter to Mersenne in June 1640 asserting that if p is prime, then $2^p - 2$ is a multiple of $2p$, and that if q is a prime divisor of $2^p - 1$, then $q-1$ is a multiple of p. As with most of his results, Fermat never published his proof. This was a common practice at that time due to the lack of scholarly journals. Euler published a proof in 1730 using the binomial theorem, and a more algebraic proof of 1758. This allowed him to prove a generalization, now known as Euler's Theorem, for composite moduli.

The reader who has looked at Exercise 2.1.24 may have already guessed and proved the following theorem. We give two proofs, one based on the Binomial Theorem, and a second, more conceptual proof. This is the fundamental theorem in the arithmetic of congruences for prime moduli.

Theorem 4.1.1 (Fermat). *Let p be a prime. Then $a^p \equiv a \pmod{p}$ for all integers a. In particular, if $p \nmid a$, then $a^{p-1} \equiv 1 \pmod{p}$.*

FIRST PROOF. It suffices to prove the theorem for positive integers (why?). We prove the theorem by induction on a. The theorem is clearly true for $a = 0$ and $a = 1$. Assume the theorem is true for $a = n$. We must prove that it is true for $a = n + 1$. By the Binomial Theorem,

$$(n + 1)^p = n^p + \binom{p}{1}n^{p-1} + \binom{p}{2}n^{p-2} + \cdots + \binom{p}{p-1}n + 1.$$

For $1 \leq k \leq p - 1$, the binomial coefficient $\binom{p}{k}$ is divisible by p because $\binom{p}{k} = \frac{p!}{k!(p-k)!}$, and the factor of p in the numerator cannot be canceled by any term in the denominator. Therefore, $(n+1)^p \equiv n^p + 1 \pmod{p}$. By the induction hypothesis, $n^p \equiv n \pmod{p}$, hence $(n+1)^p \equiv n+1 \pmod{p}$, completing the proof.   ■

SECOND PROOF. Let $\{0, 1, 2, \ldots, p-1\}$ be a complete residue system modulo p. We show that $\{0, a, 2a, \ldots, a(p-1)\}$ is also a complete residue system when $p \nmid a$.

Indeed, if $ai \equiv aj \pmod{p}$, then $i \equiv j \pmod{p}$ (as $(a, p) = 1$, a is invertible modulo p). Since multiplication modulo p is independent of the choice of residue systems, we must have

$$1 \cdot 2 \cdot 3 \cdots (p-1) \equiv a(2a)(3a) \cdots ((p-1)a) \pmod{p},$$

or, by rearranging the terms,

$$1 \cdot 2 \cdot 3 \cdots (p-1) \equiv a^{p-1}(1 \cdot 2 \cdot 3 \cdots (p-1)) \pmod{p}.$$

Since $p \nmid 1 \cdot 2 \cdot 3 \cdots (p-1)$, we can cancel $(p-1)!$ on both sides (as it is invertible), to obtain $a^{p-1} \equiv 1 \pmod{p}$, or $a^p \equiv a \pmod{p}$.

If $p \mid a$, then $a \equiv 0 \pmod{p}$, so $a^p \equiv a \pmod{p}$.   ■

Exercise. Identify all the places in the proof where we use the fact that p is prime. What happens if we drop the condition that p is prime?

We give a few examples that show how the theorem simplifies computations modulo primes.

Example 4.1.2.   1. Show that $2^{50} + 3^{50}$ is divisible by 13.

By Fermat's Theorem, $2^{12} \equiv 1 \pmod{13}$. Since $50 = 4 \cdot 12 + 2$, we compute $2^{50} \equiv 2^{4 \cdot 12 + 2} \equiv \left(2^{12}\right)^4 2^2 \equiv 1 \cdot 4 \pmod{13}$. Also, $3^{12} \equiv 1 \pmod{13}$, so $3^{50} \equiv 3^{48} 3^2 \equiv 9 \pmod{13}$. Therefore, $2^{50} + 3^{50} \equiv 4 + 9 \equiv 13 \equiv 0 \pmod{13}$.

2. Determine the remainder when 3^{372} is divided by 37.

Since 37 is prime, $3^{36} \equiv 1 \pmod{37}$. Now, $372 = 10 \cdot 36 + 12$; therefore, $3^{372} = 3^{10 \cdot 36 + 12} = \left(3^{36}\right)^{10} 3^{12} \equiv 1 \cdot 3^{12} \pmod{37}$. As $3^4 \equiv 81 \equiv 7 \pmod{37}$, $3^{12} \equiv 7^3 \equiv 7 \cdot 49 \equiv 7 \cdot 12 \equiv 10 \pmod{37}$. Therefore, the remainder is 10.

3. Show that $7 \nmid n^2 + 1$ for any n.

If $n^2 + 1 \equiv 0 \pmod{7}$, then $n^2 \equiv -1 \pmod{7}$, which implies that $n^4 \equiv 1 \pmod{7}$. These two congruences imply that $n^6 \equiv n^4 n^2 \equiv 1(-1) \equiv -1 \pmod{7}$. On the other hand, Fermat's Theorem implies that $n^6 \equiv 1 \pmod{7}$, a contradiction.

Pierre de Fermat (1601–1665)

Fermat

Fermat was the son of a leather merchant and obtained his early education at home. He studied law and found employment at Toulouse. Working as a jurist, he devoted the bulk of his leisure time to mathematics. Fermat published little, but he established an extensive correspondence with leading mathematicians of his age. Fermat's most important contributions are in number theory. He seems to have been inspired by the Latin translation, by Bachet, of Diophantus's *Arithmetica*. He made many marginal statements in his copy of Bachet's translation, whose margin was "unfortunately too narrow" for Fermat to include proofs.

These statements of Fermat were published by his son in 1670. Fermat's contribution to number theory includes the method of infinite descent, results on divisibility, sums of squares, and numerous Diophantine equations.

Fermat's other contributions are to analytic geometry and infinitesimal calculus. He laid the foundations of analytic geometry in a letter to Roberval in 1636, a year before Descartes published his geometry. In calculus, Fermat computed the tangents to algebraic curves, and the area under these curves. In addition, Fermat was the first to state the first derivative test for maxima/minima, which is sometimes called Fermat's test in calculus texts. Fermat's correspondence with Pascal laid the foundations of the modern theory of probability.

Exercise. Determine the remainder when 2^{372} is divided by 37.

Exercise. Show that $11 \nmid n^2 + 1$ for any n.

Example 4.1.3. Recall the riffle shuffle of a deck of playing cards described in Example 3.1.9. In the example, we considered a deck of 52 cards. The same analysis applies to a deck of $2n$ cards. If $f(i)$ is the position of the ith card (numbered sequentially from the bottom), then $f(i) \equiv 2i \pmod{2n+1}$. Recall that the position of the ith card after k shuffles is $2^k i \bmod 2n + 1$. By Fermat's Theorem, $2^{2n} \equiv 1 \pmod{2n+1}$. If $2n + 1$ is prime, then a deck of $2n$ cards returns to its original order after $2n$ shuffles. Hence a deck of 12 cards returns to its original order after 12 riffle shuffles.

Example 4.1.4. If the converse of Fermat's Theorem were true, then we would have a simple test for primality because the exponent $a^{p-1} \bmod p$ can be computed fairly rapidly. Unfortunately, the converse does not hold. For example, $2^{340} \equiv 1 \pmod{341}$, but 341 is not prime. We say that 341 is a **pseudoprime** to base 2. In fact, there exist composite numbers n such that $a^{n-1} \equiv 1 \pmod{n}$ for all $(a, n) = 1$. Such numbers are called Carmichael numbers. We will explore these questions and their relation to primality tests in Chapter 6.

Fermat's Theorem is usually applied in combination with the following lemma.

Proposition 4.1.5. *Suppose $a^r \equiv 1 \pmod{p}$ and p is prime. If $(r, p-1) = d$, then $a^d \equiv 1 \pmod{p}$.*

PROOF. By Theorem 2.5.6, we can write $d = rx + (p-1)y$ for some x and y. This implies that

$$a^d \equiv a^{rx+(p-1)y} \pmod{p}$$
$$\equiv (a^r)^x \left(a^{p-1}\right)^y \pmod{p}$$
$$\equiv 1 \pmod{p}. \qquad \blacksquare$$

In particular, the last proposition implies that if r is the smallest positive integer such that $a^r \equiv 1 \pmod{p}$, then $r \mid p - 1$. The integer r satisfying this condition is called the **order of a modulo p.** Orders are studied in detail in Section 7.1. We give some applications of this proposition.

Example 4.1.6. We show that there are infinitely many primes of the form $8k + 1$. The technique is similar to Euclid's proof. In order to derive a contradiction, we let $p_1, \ldots, p_r$ be finitely many primes of this form. Let $N = (2p_1 \ldots p_r)^4 + 1$. Suppose p is a prime factor of N. Then $N \equiv 0 \pmod{p}$ implies that $(2p_1 \ldots p_r)^4 \equiv -1 \pmod{p}$, or $(2p_1 \ldots p_r)^8 \equiv 1 \pmod{p}$. Suppose $d = (p-1, 8)$. Then d can be 1, 2, 4, or 8. The previous proposition implies that $(2p_1 \ldots p_r)^d \equiv 1 \pmod{p}$, so d cannot be a divisor of 4; hence $d = 8$. Therefore, $8 \mid p - 1$, or $p \equiv 1 \pmod{8}$. But p cannot be one of $p_1, \ldots, p_r$, otherwise p will divide 1. Hence there are infinitely many primes of the form $8k + 1$.

A similar proof can be given for primes in the arithmetic progressions $16k + 1, \ldots, 2^r k + 1$ for any $r > 0$.

Example 4.1.7. Suppose p is prime, and q is a prime factor of the Mersenne number $2^p - 1$. Then $2^p \equiv 1 \pmod{q}$. But $2^{q-1} \equiv 1 \pmod{q}$, by Fermat's Theorem. The proposition implies that $p \mid q - 1$, or $q \equiv 1 \pmod{p}$. Since p and q are odd, we get $q \equiv 1 \pmod{2p}$.

—————————————— **Exercises for Section 4.1** ——————————————

1. Use Fermat's Theorem to compute the following quantities.

    (a) 31^{100} mod 19.

    (b) 2^{10000} mod 29.

    (c) 99^{999} mod 31.

2. Show that $11^{84} - 5^{84}$ is divisible by 7.

3. Show that if $n \equiv 2 \pmod 4$, then $9^n + 8^n$ is divisible by 5.

4. For which values of n is $3^n + 2^n$ divisible by 13? by 7?

5. Use Fermat's Theorem to show that $n^{13} - n$ is divisible by 2730 for all n.

6. Show that if $p > 3$ is prime, then $ab^p - ba^p$ is divisible by $6p$.

7. Show, using the Binomial Theorem, that if p is prime and a and b are integers, then $(a + b)^p \equiv a + b \pmod p$.

8. Show that no prime number of the form $4k + 3$ can divide a number of the form $n^2 + 1$.

9. Show that there are infinitely many primes of the form $16k + 1$. More generally, show that for any $r > 0$, there are infinitely many primes of the form $2^r k + 1$.

10. Let $n = r^4 + 1$. Show that 3, 5, and 7 cannot divide n. What is the smallest prime that can divide n? Determine the form of the prime divisors of n.

11. Show that any proper factor, whether prime or not, of a composite Mersenne number $2^p - 1$ is of the form $1 + 2pk$ for some k.

12. What can you say about the prime factors of a composite Fermat number $F_n = 2^{2^n} + 1$? Use Fermat's Theorem and Proposition 4.1.5 to find a factor of F_5, thereby disproving Fermat's statement that all the F_n are prime.

13. In 1909, Wiefrich proved that if p is prime and $x^p + y^p = z^p$ has integer solutions with $p \nmid xyz$, then p satisfies $2^{p-1} \equiv 1 \pmod{p^2}$. A prime p satisfying this latter congruence is called a Wiefrich prime. Determine the first two Wiefrich primes.

4.2 Euler's Phi Function

In the arithmetic of congruences, the number of invertible elements is an important quantity. It plays a role in the generalization of Fermat's Theorem to composite numbers.

Definition 4.2.1. The number of invertible elements in a complete residue system modulo m is denoted by $\phi(m)$. The function ϕ is called **Euler's Totient function** or **Euler's** *phi*-**function**.

The Totient function, $\phi(m)$, can be computed using any residue system. For example, we can take $S = \{0, 1, \ldots, m-1\}$ and determine the number of elements a in S such that $(a, m) = 1$. It can be easily verified that if $a \equiv b$ (mod m), then $a^{-1} \equiv b^{-1}$ (mod m), so the number of invertible elements does not depend on the choice of the residue system.

Example 4.2.2.   1. $\phi(8) = 4$, as the elements in $\{0, 1, 2, \ldots, 7\}$ relatively prime to 8 are 1, 3, 5, and 7.

2. $\phi(11) = 10$ because 11 is prime, and $(a, 11) = 1$ for $1 \le a \le 10$. For any prime p, we have $\phi(p) = p - 1$.

3. For a prime power p^r, if a, $1 \le a \le p^r$, is not relatively prime to p^r, then a is a multiple of p. Since the number of multiples of p that are less than p^r is $\left\lfloor \dfrac{p^r}{p} \right\rfloor = \lfloor p^{r-1} \rfloor = p^{r-1}$, we have

$$\phi(p^r) = p^r - p^{r-1} = p^{r-1}(p-1) = p^r \left(1 - \frac{1}{p} \right).$$

The last example evaluates ϕ for prime powers. There is a nice formula that allows the determination of $\phi(m)$ for composite m from the prime factorization of m. The formula is a corollary to the following fundamental property of the ϕ function.

Theorem 4.2.3. *If $(m, n) = 1$, then $\phi(mn) = \phi(m)\phi(n)$.*

PROOF. The proof is a consequence of the Chinese Remainder Theorem. We count the invertible elements modulo mn by relating them to pairs (a, b) where a and b are invertible modulo m and n, respectively. For an integer s, let U_s denote the set of invertible elements in the standard residue system modulo s.

Suppose $x \in U_{mn}$; then $(x, mn) = 1$ implies that $(x, m) = 1$, and $(x, n) = 1$. Consider the pair (a, b) with $a = x \bmod m$ and $b = x \bmod n$. It is clear that $a \in U_m$, and $b \in U_n$.

Conversely, if we have a pair (a, b) with $a \in U_m$ and $b \in U_n$, then the Chinese Remainder Theorem implies that there is a unique x modulo mn (because $(m, n) = 1$) such that $a = x \bmod m$ and $b = x \bmod n$. This establishes a bijection (a one-to-one correspondence) between U_{mn} and the set of pairs (a, b) with $a \in U_m$ and $b \in U_n$. There are $\phi(m)$ elements in U_m and $\phi(n)$ elements in U_n, so there are a total of $\phi(m)\phi(n)$ pairs. This completes the proof, as there are $\phi(mn)$ elements in U_{mn}. ∎

Corollary 4.2.4. *If $m = p_1^{a_1} \cdots p_k^{a_k}$ is the prime factorization of m, then*

$$\phi(m) = \prod_{i=1}^{k} p_i^{a_i-1}(p_i - 1)$$

$$= m \prod_{i=1}^{k} \left(1 - \frac{1}{p_i}\right).$$

PROOF. This follows from the formula $\phi(p^r) = p^r \left(1 - \frac{1}{p}\right)$ and the multiplicativity of ϕ.

$$\phi(p_1^{a_1} \cdots p_k^{a_k}) = \phi(p_1^{a_1}) \cdots \phi(p_k^{a_k})$$

$$= p_1^{a_1} \left(1 - \frac{1}{p_1}\right) \cdots p_k^{a_k} \left(1 - \frac{1}{p_k}\right)$$

$$= p_1^{a_1} \cdots p_k^{a_k} \left(1 - \frac{1}{p_1}\right) \cdots \left(1 - \frac{1}{p_k}\right)$$

$$= m \prod_{i=1}^{k} \left(1 - \frac{1}{p_i}\right). \qquad \blacksquare$$

Example 4.2.5. Compute $\phi(29 \cdot 5^2)$.

$$\phi(29 \cdot 5^2) = \phi(29)\phi(5^2)$$

$$= 28 \cdot 5^2 \left(1 - \frac{1}{5}\right)$$

$$= 28 \cdot 20$$

$$= 560.$$

Example 4.2.6. Find all n such that $\phi(n) = 6$. Consider the different possibilities for the prime factorization of n. Suppose n is prime. Then $\phi(n) = n - 1 = 6$ implies that $n = 7$. If $n = pq$, a product of two primes, then $\phi(n) = (p-1)(q-1) = 6$. There are two possibilities: $p - 1 = 2$ and $q - 1 = 3$, or $p - 1 = 1$ and $q - 1 = 6$. The first implies that $q = 4$, a composite number. In the second case, $p = 2$ and $q = 7$, so $n = 14$. Next, we can see that n cannot be the product of 3 distinct primes, because $\phi(n)$ has at most two factors when written as a product, so one of the three prime factors of n would have to be 2. It is easy to verify that this case is not possible.

If $n = p^2$, then $\phi(n) = p(p-1) = 6$. It is clear that the only possible value of p is 3, so $n = 9$. Similarly, if $n = p^2q$, then we see that $q = 2$ and $p = 3$; hence $n = 18$. We can also see in the same way that $n = p^2q^2$ is not possible.

It is not possible for a power of a prime higher than the second to divide n because if $p^3 \mid n$, then $p^2 \mid \phi(n)$, but no square divides 6. Therefore, we have considered all the possibilities for n and the only n such that $\phi(n) = 6$ are $7, 9, 14,$ and 18.

Example 4.2.7. The formula for the ϕ function can be used to give another proof of the infinitude of primes. Assume to the contrary that $2, 3, \ldots, p$ is the finite list of primes, and let $N = 2 \cdot 3 \cdot 5 \cdots p$. Then there is only one element prime to N in a residue system modulo N, namely 1, because every prime number is a factor of N. Hence $\phi(N) = 1$, but the formula for ϕ shows that $\phi(N) = 2(5-1) \cdots (p-1)$ is quite large. The contradiction proves the infinitude of primes.

In many cases, computation of ϕ is as difficult (time-consuming) as obtaining the prime factorization of a number. More precisely, if $\phi(n)$ can be computed for composite n that are products of two primes, then it is possible to find the prime factors of n. We will see in Chapter 5 that this is the basis for applications to cryptography.

─────────────── **Exercises for Section 4.2** ───────────────

1. Determine $\phi(120)$ and $\phi(225)$.

2. Use the program that finds the prime factorization of an integer by trial division to write another program that evaluates the Euler ϕ–function.

3. Show that $\phi(n) = n - 1$ if and only if n is prime. (This result has applications to proofs of primality.)

4. Determine all n for which $\phi(n) = n - 2$.

5. If d is a factor of n, then determine the number of integers a, $1 \le a \le n$, such that $(a, n) = d$.

6. Prove that the equation $\phi(n) = k$ has finitely many solutions for a given k. Determine the values of n such that $\phi(n) = 12$, and all n such that $\phi(n) = 18$ and $\phi(n) = 100$.

7. What is the smallest even number k such that there is no n satisfying $\phi(n) = k$? Use a computer to find all k, $1 \le k \le 100$ such that $\phi(n) = k$ has no solution. To search for the values of n, observe that any prime factor of a number n satisfying $\phi(n) = k$ is less than $k + 1$. Also, the largest exponents in the prime factorization of n can be at most one more than the largest exponent in the prime factorization of k.

8. Show that if $\phi(n) \mid n - 1$, then n is squarefree. (You might be able to conjecture a stronger result if you look at a table of numbers satisfying this property.)

9. Prove that there is no n satisfying $\phi(n) = 2 \cdot 7^k$ for any $k \geq 1$. Generalize this to show that if p is prime and $2^r p + 1$ is composite, then $\phi(n) = 2^r p$ has no solution.

10. Make a table of values of $\phi(n)$ using a computer. Search the table to characterize integers satisfying the following properties. Prove your conjectures.

   (a) For which n is $\phi(n) = n/2$?

   (b) What are the values of n such that $\phi(2n) = \phi(n)$?

   (c) When does $\phi(n) \mid n$?

   (d) Determine all n for which $\phi(n) = 2^k$.

   (e) Are there any n such that $\phi(n) = n/3$? Are there infinitely many? What about integers n such that $\phi(n) = n/4$?

11. (Schroeder[1]) An n-pointed star has n points lying equidistant on a circle joined by n straight lines. A point cannot be joined to a neighboring point on the circle. We require that the star be symmetric, that is, the angle between the two lines that meet at a point is the same for all points. Show that the number of n-pointed stars that can be drawn with n lines using consecutive strokes without lifting the pen is $(\phi(n)-2)/2$; hence a 6-pointed star cannot be drawn with 6 consecutive strokes.

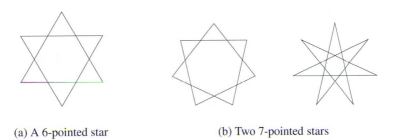

(a) A 6-pointed star   (b) Two 7-pointed stars

Figure 4.2.1

12. Investigate the average value of $\phi(n)$ by computing $\sum_{k=1}^{n} \phi(k)/n$ for increasing values of n.

13. Visible points in the plane were defined in Exercise 2.5.17. These are points (a, b) such that a and b are relatively prime. Show that the number of visible points in the square $1 \leq a, b \leq n$ is $2 \sum_{k=1}^{n} \phi(k)$.

14. Use the table of values of $\phi(n)$ to investigate the sum $\sum_{d \mid n} \phi(d)$ for different values of n. The sum is over all positive divisors of n, including 1 and n.

---

[1]M. R. Schroeder. *Number Theory in Science and Communication*. New York: Springer–Verlag, 1986.

Once you have formulated a conjecture regarding $\sum_{d|n} \phi(d)$, you can try to prove it using induction.

15. If P denotes the product of primes common to m and n, prove that $\phi(mn) = P\phi(m)\phi(n)/\phi(P)$. Hence, prove that if $(m,n) \neq 1$, then $\phi(mn) > \phi(m)\phi(n)$.

4.3   Euler's Theorem

Let m be a composite number. To find the analog of Fermat's Theorem in the case of a composite modulus, we must first restrict attention to the invertible elements modulo m. To see this, let a be noninvertible, that is, $(a,m) \neq 1$. In this case $a^k \not\equiv 1 \pmod{m}$ for any k because $m \mid a^k - 1$ implies that $(a,m) \mid 1$, a contradiction. It is this observation, that only the invertible elements should be considered, which allowed Euler to generalize Fermat's Theorem to composite moduli. The second proof of Fermat's Theorem (due to Euler) uses the fact that all the nonzero elements in a residue system have inverses, allowing the cancellation of the invertible factor $1 \cdot 2 \cdot 3 \cdots (p-1)$. This technique is also used in Euler's Theorem.

Theorem 4.3.1 (Euler). *If a and m are integers such that $(a,m) = 1$, then $a^{\phi(m)} \equiv 1 \pmod{m}$.*

PROOF. The proof is essentially the same as the second proof of Fermat's Theorem. Let $(a,m) = 1$, and let $r_1, \ldots, r_{\phi(m)}$ be the $\phi(m)$ invertible elements in a residue system. Then $ar_1, \ldots, ar_{\phi(m)}$ are all invertible, no two of which are congruent modulo m. (Because $ar_i \equiv ar_j \pmod{m}$ implies that $r_i \equiv r_j \pmod{m}$ as a is invertible modulo m.) Therefore, we have

$$(ar_1)(ar_2) \cdots (ar_{\phi(m)}) \equiv r_1 r_2 \cdots r_{\phi(m)} \pmod{m},$$

or, by rearranging the terms we obtain

$$a^{\phi(m)} r_1 r_2 \cdots r_{\phi(m)} \equiv r_1 r_2 \cdots r_{\phi(m)} \pmod{m},$$

which implies that $a^{\phi(m)} \equiv 1 \pmod{m}$. ∎

Example 4.3.2.   1. Determine the last two digits of 1993^{1993}.

We need to compute $1993^{1993} \bmod 100$. Since $\phi(100) = \phi(25)\phi(4) = 20 \cdot 2 = 40$, we divide the exponent 1993 by 40. Then $1993 = 40q + 33$ for some q. Therefore, $1993^{1993} \equiv 93^{1993} \equiv 93^{40k+33} \equiv \left(93^{40}\right)^k 93^{33} \equiv 93^{33} \pmod{100}$, where we used Euler's Theorem to write $93^{40} \equiv 1 \pmod{100}$.

We compute 93^{33} mod 100 by observing that $93 \equiv -7 \pmod{100}$; therefore, $93^{33} \equiv (-1)^{33} 7^{33} \pmod{100}$. Now, $7^2 = 49$, $7^3 \equiv 343 \equiv 43 \pmod{100}$, $7^4 \equiv 301 \equiv 1 \pmod{100}$, so then $7^{33} \equiv (7^4)^8 7 \equiv 7 \pmod{100}$. Therefore, $1993^{1993} \equiv -7 \equiv 93 \pmod{100}$; that is, the last two digits are 93.

2. It is also possible to prove Euler's Theorem by using the Binomial Theorem, Fermat's Theorem, and the Chinese Remainder Theorem. Actually, it is possible to prove a stronger statement than Euler's Theorem. For example, $\phi(100) = 40$, so $n^{40} \equiv 1 \pmod{100}$ for all $(n, 100) = 1$, but more is true. Since $\phi(25) = 20$ and $\phi(4) = 2$, we have $n^{20} \equiv 1 \pmod{25}$ and $n^2 \equiv 1 \pmod 4$ for all n prime to 5 and 2. Since $2 \mid 20$, we have $n^{20} \equiv 1 \pmod 4$, which, together with $n^{20} \equiv 1 \pmod{25}$, implies that $n^{20} \equiv 1 \pmod{100}$ by the Chinese Remainder Theorem (because the two numbers n^{20} and 1 are the same modulo 25 and 4).

3. Euler's Theorem can be used to compute the inverse of a modulo m. Since $a^{\phi(m)} \equiv a a^{\phi(m)-1} \equiv 1 \pmod m$; hence $a^{\phi(m)-1}$ is an inverse of a modulo m. For example, $2^{24} \equiv 1 \pmod{35}$, as $\phi(35) = 6 \cdot 4 = 24$; then the inverse of 2 must be 2^{23} mod 35, which can be computed to be 18. If the exponent modulo m can be computed rapidly, then this method is as efficient as the extended Euclidean Algorithm. We describe below an efficient technique for computing a^k mod m in approximately $\log_2(k)$ steps.

Exercise. Determine the remainder when 2^{720} is divided by 225.

Exercise. Determine the inverse of 3 modulo 40.

Applications of the theorems of Fermat and Euler involve computing exponents a^k mod m. We will use Fermat's Theorem to test primality, and Euler's Theorem has applications to cryptography. The following method computes a^k mod m by the method of successive squaring and multiplication. The technique is ancient; the ancient Egyptians did multiplication by a very similar method. To evaluate a^k, we compute successive squares of a, that is, a, a^2, a^4, a^8, and so on, and include only those that are necessary in the computation. For example, $a^{37} = a^{32} a^4 a^1$. (Is there any other way to write a^{37} as a product of powers of a with exponents that are powers of 2?)

Exercise. Write a^{77} as a product of powers of a where the exponents are powers of 2.

We need a systematic procedure to determine which powers of 2 need to be included as exponents in the calculation of a^k mod m. The method is to write k in base-2, and include the exponents of a to the powers of 2 that occur in the base-2 expansion of k. For example,

$$k = 37 = 1 \cdot 2^5 + 0 \cdot 2^4 + 0 \cdot 2^3 + 1 \cdot 2^2 + 0 \cdot 2 + 1$$
$$= 100101 \quad \text{in base 2.}$$

Therefore, $a^{37} = a^{32}a^4a^1$.

We describe an algorithm in which the base 2 expansion of k is implicit. The idea is to traverse the digits of k (in base 2) from right to left. Every time we encounter a 0, divide k by 2 and square a, but do not add this term to the result. If k is odd, we multiply the result by the current power of a and subtract 1 from k.

Algorithm 4.3.3 (Exponentiation modulo m). Given integers a, k, and m, this algorithm computes $a^k \bmod m$ for $k \geq 0$.

1. [Initialize] Set *result*=1.

2. [Check if done] If $k = 0$, return *result* and terminate.

3. [k is odd] If $k \bmod 2 = 1$, then let *result* $= (result \cdot a) \bmod m$, $k = k - 1$, and go to step 2.

4. [k is even] Let $a = a^2 \bmod m$, $k = k/2$, and go to step 2.

The following example gives the steps in the algorithm along with the binary expansion of k in the computation of a^{37}.

Example 4.3.4. Let $k = 37$; the values in the computation are as follows:

Iteration	a	*Result*	k base 10	k base 2
0	a	1	37	100101
1	a	a	36	10000
2	a^2	a	18	10010
3	a^4	a	9	1001
4	a^4	a^5	8	1000
5	a^8	a^5	4	100
6	a^{16}	a^5	2	10
7	a^{32}	a^5	1	1
8	a^{32}	a^{37}	0	0

Example 4.3.5. As an application of the method to compute $a^k \bmod m$, we describe a method to determine if a number n is a prime power. If $n = p^k$, then Fermat's theorem implies that $a^n \equiv a \pmod{p}$. Hence $d = (a^n - a, n)$ is divisible by p. If d is not prime, then it is a power of a prime. We can

Leonhard Euler (1707–1783)

Euler

Euler was born in Basel, Switzerland. His father was a Protestant minister who gave Euler his early education. At the age of 16 he obtained a Master's degree in Philosophy from the University of Basel and joined the theology department. But he soon abandoned this is favor of mathematics and began independent mathematical investigations at the age of 18. Euler moved to St. Petersburg in 1727 and then to Berlin in 1740 under the invitation of Frederick the Great. He returned to St. Petersburg in 1766 and remained there till his death.

Euler laid the foundations of number theory by giving rigorous proofs of many assertions that Fermat had left without proof. Euler introduced the ϕ-function, generalized Fermat's Theorem, and discovered the law of quadratic reciprocity. Euler introduced infinite series into number theory through the study of the ζ-function and the generating functions for partitions. He also proved that $x^3 + y^3 = z^3$ has no nontrivial integer solutions by a study of numbers of the form $a + b\sqrt{-3}$.

Euler made fundamental contributions to analysis, algebra, geometry, complex variables, hydromechanics, astronomy, and the calculus of variations. He devoted considerable attention to the problem of lunar motion and applications of mathematics to engineering. Euler authored numerous texts, which standardized mathematical notation. Euler used e to represent the base of the natural logarithm and i for $\sqrt{-1}$. He also introduced the letter "f" and parentheses to represent functions. The modern notation for trigonometric functions and $\sum$ for summation are also due to Euler.

Euler was the most prolific mathematician ever, and his collected works span over 70 volumes.

then check if d is a power of a prime, and if it is, then we can repeatedly divide n by this prime to see if n is a prime power. In the computation of d, it is sufficient to compute $a^n - a \bmod n$ instead of $a^n - a$. For example, let $n = 28561$ and $a = 2$. We can easily compute (using Algorithm 4.3.3) that $4810 = a^n - a \bmod n$. We compute $(4810, n) = 13$, hence n could be a power of 13. Repeatedly dividing n by 13, we see that $n = 13^4$.

——————————— **Exercises for Section 4.3** ———————————

1. Use Euler's Theorem to compute the following quantities.

    (a) $3^{340} \bmod 341$

    (b) $7^{8^9} \bmod 100$

    (c) $2^{10000} \bmod 121$

2. Implement the algorithm for exponentiation modulo m on a computer. Use this to check the result of Exercise 4.1.1.

3. Use Euler's Theorem (and the Chinese Remainder Theorem) to show that $n^{12} \equiv 1 \pmod{72}$ for all $(n, 72) = 1$.

4. What is the smallest positive integer λ such that $n^{\lambda} \equiv 1 \pmod{100}$ for all $(n, 100) = 1$?

5. Prove that $m^{\phi(n)} + n^{\phi(m)} \equiv 1 \pmod{mn}$ if $(m, n) = 1$.

6. Show that if $n = pq$, a product of distinct primes, then $a^{\phi(n)+1} \equiv a \pmod{n}$ for all a.

7. Suppose that $n = rs$ with $r > 2$, $s > 2$ and $(r, s) = 1$. Show that $a^{\frac{\phi(r)\phi(s)}{2}} \equiv 1 \pmod{n}$, that is, $a^{\frac{\phi(n)}{2}} \equiv 1 \pmod{n}$.

8. Suppose n is the product of two odd primes, p and q. Let

$$\lambda(n) = (p-1)(q-1)/(p-1, q-1).$$

Show that $a^{\lambda(n)} \equiv 1 \pmod{n}$ for all integers a, satisfying $(a, n) = 1$.

9. Use a computer to find the composite numbers $n \leq 2000$ such that $2^n \equiv 2 \pmod{n}$. Repeat the exercise to find composite n such that $3^n \equiv 3 \pmod{n}$.

10. Prove that $a^{560} \equiv 1 \pmod{561}$ for all $(a, 561) = 1$.

11. Suppose $a^x \equiv 1 \pmod{m}$ and $a^y \equiv 1 \pmod{m}$; show that $a^{(x,y)} \equiv 1 \pmod{m}$.

12. Determine whether the following integers are prime powers, and if so, find the prime power decomposition.

    (a) 24137569

    (b) 500246412961

    (c) 486695567

4.4 Lagrange's Theorem

We saw in Section 3.4 that a quadratic congruence modulo m can have no solutions, two solutions, or more than two solutions in a complete residue system modulo m. Recall that the fundamental theorem of algebra states that for the real number system, a polynomial of degree n has at most n roots, and over the complex number system, it has exactly n roots counting multiplicities. For a quadratic equation over the complex numbers, we can have two distinct roots, or one root occurring twice, that is, $f(x) = (x - r_1)(x - r_2)$ or $f(x) = (x - r)^2$. A similar theorem is true for prime moduli, but not if the modulus is composite. The proof depends on the fact that every nonzero element has an inverse, which is true in $\mathbb{Z}/p\mathbb{Z}$ (and in the real and complex number systems).

Remark. The proof of the theorem uses the division algorithm for polynomials. If $f(x)$ and $g(x)$ are polynomials with integer coefficients, and $g(x)$ has leading coefficient 1, then there exist unique $q(x)$ and $r(x)$ such that $f(x) = q(x)g(x) + r(x)$, and $\deg r(x) < \deg g(x)$, where $\deg p(x)$ denotes the degree of a polynomial. (See Exercise 7.)

Theorem 4.4.1 (Lagrange). *Let p be a prime number, and let $f(x)$ be a polynomial of degree $n \geq 1$, not all of whose coefficients are divisible by p. Then the congruence $f(x) \equiv 0 \pmod{p}$ has at most n solutions in a complete residue system modulo p.*

PROOF. We prove the theorem by induction on the degree of the polynomial. If f is of degree 1, then $f(x) = ax + b$ for some a and b. If $p \nmid a$, then $ax + b \equiv 0 \pmod{p}$ has a unique solution. If $p \mid a$ and $p \nmid b$, then there is no solution. In either case, there is at most one solution. Note that we cannot have both $p \mid a$ and $p \mid b$ by the assumption in the theorem.

Now, we assume that all polynomials of degree less than n, not all of whose coefficients are divisible by p, satisfy the conclusion of the theorem. Let $f(x)$ be a polynomial of degree n, and let a be a root. (If the polynomial has no roots, then the conclusion of the theorem is valid.) By the long division for polynomials, there exist $q(x)$ and $r(x)$ such that

$$f(x) = (x - a)q(x) + r(x) \quad \text{and} \quad \deg r(x) < \deg(x - a);$$

that is, $r(x)$ is an integer, also denoted by r.

Now, $f(a) \equiv 0 \pmod{p}$ implies that $r \equiv 0 \pmod{p}$, that is, $p \mid r$. Therefore, $f(x) \equiv (x - a)q(x) \pmod{p}$. But $q(x)$ is a polynomial of degree at most $n - 1$, and by the assumption on f, not all the coefficients of $q(x)$ are divisible by p; hence q has at most $n - 1$ roots by the induction hypothesis. Suppose a is not a root of $q(x)$, that is, $q(a) \not\equiv 0 \pmod{p}$. Then a root,

b, of f is either equal to a or it must be a root of $q(x)$; because $f(b) \equiv 0 \equiv (b-a)q(b) \pmod{p}$, and $(b-a)$ is invertible modulo p so $q(b) \equiv 0 \pmod{p}$. By the induction hypothesis, $q(x)$ has at most $n-1$ roots in a complete residue system, so f has at most n roots.

Next, we deal with the possibility that a is a root of q. If $q(a) \equiv 0 \pmod{p}$, then we can divide q by $(x-a)$ and write $q(x) \equiv (x-a)q_1(x)$ $\pmod{p}$, that is, $f(x) = (x-a)^2 q_1(x)$. Repeated application of this procedure shows that we can write $f(x) \equiv (x-a)^k q'(x) \pmod{p}$ where $q'(a) \not\equiv 0$ $\pmod{p}$ (here, k is the multiplicity of a as a root of f). This polynomial q' has degree $n-k$, hence at most $n-k$ roots by the induction hypothesis. If b is a root of f, $b \not\equiv a$, then $f(b) \equiv (b-a)^k q'(b) \equiv 0 \pmod{p}$. Since $b-a \not\equiv 0 \pmod{p}$, we must have $q(b) \equiv 0 \pmod{p}$. Therefore, every root of f different from a must be a root of q', and we have at most $n-k$ of these; hence f has at most n roots (counting a, k times). ∎

The theorem does not apply to composite moduli. For example, $x^2 \equiv 1$ $\pmod{8}$ has four solutions, $x = 1, 3, 5, 7$. The reader should study the proof to discover where the primality of p is used. Note that if p divides all the coefficients of f, then $f(x) \equiv 0 \pmod{p}$ for all x independently of the degree of f. Where do we use the condition that p does not divide all the coefficients of f in the proof of the theorem?

The theorem only describes the maximum number of possible roots. Usually, it is difficult to find the roots or to determine their number. The complete solution of quadratic congruences is dealt with in Chapter 9.

Example 4.4.2.   1. $x^3 \equiv 8 \pmod{13}$ has three solutions, $x \equiv 1, 5, 6$ $\pmod{13}$.

2. The quadratic congruence $x^2 + 3x + 4 \equiv 0 \pmod{7}$ has at most two solutions by Lagrange's theorem, but substituting $x = 0, 1, \dots, 6$ shows that only $x = 2$ is a solution to the congruence.

We can deduce a very useful corollary to the theorem by combining the theorems of Fermat and Lagrange.

Corollary 4.4.3. *Let p be a prime and $d \mid p-1$; then the congruence $x^d - 1 \equiv 0 \pmod{p}$ has exactly d solutions.*

PROOF. By Fermat's Theorem, $x^{p-1} \equiv 1 \pmod{p}$ for all $(x, p) = 1$; hence the congruence has $p-1$ solutions. Let $p - 1 = dk$; then

$$x^{p-1} - 1 = (x^d - 1)(x^{d(k-1)} + x^{d(k-2)} + \cdots + x^d + 1).$$

By Lagrange's Theorem, $x^d - 1 \equiv 0 \pmod{p}$ has at most d solutions, and $x^{d(k-1)} + x^{d(k-2)} + \cdots + x^d + 1$ has at most $d(k-1)$ solutions. Hence the right-hand side has at most $d(k-1) + d = dk = p - 1$ solutions, but the left-hand side has exactly $p - 1$ solutions. Therefore, each polynomial factor on the right-hand side must have the maximum number of solutions possible. Hence $x^d \equiv 1 \pmod p$ has d solutions when $d \mid p - 1$. ∎

Example 4.4.4. $x^4 \equiv 1 \pmod{13}$ has four solutions. Two solutions to the congruence are 1 and -1. Since $4 \mid 12$, the corollary implies that there are two more fourth roots of unity modulo 13.

Example 4.4.5. If $p \equiv 1 \pmod 4$, then the congruence $x^2 \equiv -1 \pmod p$ has a solution.

The proof is an application of the theorems of Fermat and Lagrange. Write $p = 4k + 1$. By Fermat's Theorem, the equation $x^{4k} \equiv 1 \pmod p$ has $4k$ solutions in a complete residue system. We can factor

$$x^{4k} - 1 \equiv (x^{2k} - 1)(x^{2k} + 1) \pmod p.$$

Every root of the left-hand side is either a root of $x^{2k} - 1 \equiv 0 \pmod p$ or a root of $x^{2k} + 1 \equiv 0 \pmod p$. By Corollary 4.4.3, the congruence $x^{2k} - 1 \equiv 0 \pmod p$ has exactly $2k$ solutions. Therefore, the congruence $x^{2k} + 1 \equiv 0 \pmod p$ has $2k$ solutions. If a is a root, then $(a^k)^2 \equiv -1 \pmod p$, so a^k is a solution to the congruence $x^2 \equiv -1 \pmod p$.

The assertion of this example has many consequences. It can be used to show that a prime p of the form $4k + 1$ can be written as a sum of 2 squares. (See Section 14.4 for a proof.)

──────── **Exercises for Section 4.4** ────────

1. Find the four solutions of $x^4 \equiv 1 \pmod{13}$.

2. Suppose p is prime and $d \nmid p - 1$. Show that $x^d - 1 \equiv 0 \pmod p$ has exactly $(d, p - 1)$ solutions. Conclude that if $4 \mid p - 1$, then $x^2 \equiv -1 \pmod p$ has two solutions, and if $4 \nmid p - 1$, then $x^2 \equiv -1 \pmod p$ has no solutions.

3. Suppose $p = 6k + 1$ is prime. How many solutions does the congruence $x^{3k} \equiv -1 \pmod p$ have?

4. Let p be prime.
   (a) If $a \not\equiv 1 \pmod p$, show that $1 + a + \cdots + a^{p-1} \equiv 1 \pmod p$.
   (b) If $a \not\equiv -1 \pmod p$, show that $1 - a + a^2 - a^3 + \cdots + a^{p-1} \equiv 1 \pmod p$.

5. Let p be an odd prime and let
$$f(x) = (x - 1)(x - 2) \cdots (x - (p - 1))$$
$$= x^{p-1} + a_{p-2}x^{p-2} + \cdots + a_1 x + a_0.$$

(a) Show that $x^{p-1} - 1 \equiv f(x) \pmod{p}$.

(b) Compare the coefficients of x on both sides and conclude that $a_0 \equiv -1 \pmod{p}$ and $a_i \equiv 0 \pmod{p}$ for $1 \leq i \leq p-2$.

(c) Show that a_1 is the numerator of $1 + \dfrac{1}{2} + \dfrac{1}{3} + \cdots + \dfrac{1}{p-1}$. Since $f(p) \equiv -1 \pmod{p}$, and $a_2, a_3 \ldots$ are divisible by p, show that a_1 is divisible by p^2. This result is known as Wolstenholme's Theorem.

6. Show that the polynomial congruence $f(x) \equiv 0 \pmod{p}$ of degree n has exactly n solutions if and only if $f(x)$ is a factor of $x^p - x \bmod p$; that is, there exists $q(x)$ such that $x^p - x \equiv f(x)q(x) \pmod{p}$.

7. Prove the division algorithm for polynomials with integer coefficients. For polynomials $f(x)$ and $g(x)$ with leading coefficient 1, show that there exist unique $q(x)$ and $r(x)$ with integer coefficients such that

$$f(x) = q(x)g(x) + r(x), \quad \deg r(x) < \deg g(x).$$

[Consider the polynomial of smallest degree among all combinations of the form $f(x) - q(x)g(x)$.]

8. Let p be an odd prime. Use the identity

$$x^p - 1 \equiv (x-1)(x^{p-1} + x^{p-2} + \cdots + x + 1) \pmod{p}$$

to show that if a prime q divides $x^{p-1} + x^{p-2} + \cdots + x + 1$, then $q = p$ or $q \equiv 1 \pmod{2p}$. Show that there are infinitely many primes in the sequence $2pk + 1$, $k = 0, 1, \ldots$.

Chapter 5

Cryptography

5.1 Classical Cryptosystems

Cryptography is the science of keeping communications private. Its goal is to secure communication in the form of a cipher between legitimate users, the sender and the receiver, using a key known only to them. Cryptography has a long and fascinating history.[1] For most of its history, cryptography has been used to conceal military strategies and sensitive diplomatic secrets. While the primary use of cryptography is still the concealment of data and communications, there are important new applications. Cryptography now plays an important role in Authentication and Identification, problems which have become important in an increasingly digital society.

First, let us introduce some terminology to talk about cryptography. The message or information to be secured is called **plaintext**. **Encipherment** (or encryption) is the process of converting the plaintext into **ciphertext**. The process of encipherment uses **keys**, which the legitimate recipient also uses to **decipher** the ciphertext into the original plaintext. **Cryptography** is the science of securing the ciphertext so as to make it difficult or computationally infeasible for anyone other than the intended recipient to recover the plaintext. **Cryptanalysis** is the science of decoding the ciphertext without knowledge of the key.

We discuss a few elementary cryptographic methods and their cryptanalysis. The methods presented belong to the class of **secret-key** systems, in contrast to the **public-key** systems discussed in the next section. In a secret-key system, the sender and the receiver share a key known only to them. Knowledge of the key allows the decipherment of the ciphertext, hence the need for secrecy. We will denote the enciphering algorithm by E_K and the deciphering algorithm by D_K, where K represents the key.

For simplicity, we assume that the plaintext is composed of the 26 letters

---

[1]See D. Kahn. *The Codebreakers*. New York: Macmillan, 1967.

Table 5.1.1 Table of English letters and their numeric values

letter:	A	B	C	D	E	F	G	H	I	J	K	L	M
value:	0	1	2	3	4	5	6	7	8	9	10	11	12
letter:	N	O	P	Q	R	S	T	U	V	W	X	Y	Z
value:	13	14	15	16	17	18	19	20	21	22	23	24	25

of the English alphabet. All punctuation and spaces are removed from the plaintext before converting it into ciphertext. It is easy to add appropriate punctuation once the message is recovered. The same methods apply to any other alphabet.

Shift Cipher: This is one of the simplest and oldest ciphers. Julius Caesar seems to be the first to have used it. In its original form, each letter in the plaintext is replaced by a letter 3 steps to the right. So A is replaced by D, B is replaced by E and so on. The table gives the transformation:

plain:	A	B	C	D	E	F	G	H	I	J	K	L	M
cipher:	D	E	F	G	H	I	J	K	L	M	N	O	P
plain:	N	O	P	Q	R	S	T	U	V	W	X	Y	Z
cipher:	Q	R	S	T	U	V	W	X	Y	Z	A	B	C

For example, the plaintext `retreat at once` becomes

$$\texttt{uhwuhdwdwrqfh.}$$

We use integers 0 through 25 to represent the letters A through Z, by letting A= 0, B= 1, ..., Z= 25. These are given in Table 5.1.1.

Suppose **p** represents the plaintext composed of the letters $p_1, \ldots, p_k$; then the ith letter, c_i, of Caesar's cipher is given by the formula

$$c_i = (p_i + 3) \bmod 26.$$

In this cipher, $c_i = E_K(p_i)$ where $E_K(x) = x+3 \bmod 26$. The key here is 3, and knowledge of this allows us to recover the plaintext using the decryption formula $p_i = D_K(c_i)$ where

$$D_K(y) = (y - 3) \bmod 26.$$

We don't have to be restricted to shifting by 3, so the general shift cipher on each letter of the plaintext is given by the enciphering function $E_k(x) = x + k \bmod 26$, for some key k, $0 \le k \le 25$. If k is known, then the plaintext is recovered using the decryption formula $D_k(y) = y - k \bmod 26$.

It is not necessary to know the key to recover the plaintext. Since there are only 26 possibilities for k, we can simply try all of them to see which one

Table 5.1.2 Frequency of occurrence of the 26 letters in a random sample of 100 characters

letter	frequency	letter	frequency
A	8.167	N	6.749
B	1.492	O	7.507
C	2.782	P	1.929
D	4.253	Q	0.095
E	12.702	R	5.987
F	2.228	S	6.327
G	2.015	T	9.056
H	6.094	U	2.758
I	6.966	V	0.978
J	0.153	W	2.360
K	0.772	X	1.974
L	4.025	Y	0.074
M	2.406	Z	0.074

yields a meaningful plaintext. Another approach is to use frequency analysis. It has been determined that e is the most frequently occurring letter in the English alphabet, followed by t, a, o, i, n, s, h, d, r, l, u. Table 5.1.2 gives the frequency of occurrence of each of the 26 letters in a random sample of 100 English characters.[2]

Suppose we have the following ciphertext, which has been enciphered using the shift cipher.

```
cptyqzcupxpyedcpbftcpoezszwoazdtetzy
```

The two most frequently occurring letters in the cipher are z and p, so it is likely that one of these is e. If z represents e, then the key is $25 - 4 = 21$. This means that t is represented by o because $19 + 21 = 40 \equiv 14 \bmod 26$. The letter o occurs twice in the cipher, so it is not likely to be t. If we assume that p represents e, then the key is 11. This means that t is represented by e. Using this key, we readily recover the plaintext (with appropriate punctuation added) to read

send reinforcements immediately.

Affine Ciphers: Our next cipher is a simple generalization of the shift cipher. It operates on each letter of the plaintext by converting them to numbers 0 through 25. If $0 \leq x \leq 25$, then the enciphering function is

---

[2]The table is from H. Beker and F. Piper. *Cipher Systems, The Protection of Communications.* New York: Wiley, 1982.

$E_K(x) = ax + b \bmod 26$, where a and b are integers. We write the key as $K = (a, b)$. It is necessary that $(a, 26) = 1$ for this system to work (why?). If $c_i = E_K(p_i)$, then p_i can be recovered from c_i using the function $D_K(y) = a^{-1}(y - b) \bmod 26$. For example, with $a = 3$ and $b = 7$, the plaintext attack at noon becomes hmmhnlhmuxxu.

The number of possible keys is small enough that it is possible to break the cipher by exhaustive search of the keys. Even without doing a complete search of keys, we can use frequency analysis judiciously to determine the keys. Suppose the following ciphertext is known to be enciphered using an affine cipher.

jatkkabwhwzbwzbwhrjwqkibwtzobatzibqunkibubzkhmlbw

The frequency of the letters is given in the table below.

letter:	A	B	C	D	E	F	G	H	I	J	K	L	M
frequency:	3	9	0	0	0	0	0	3	3	2	5	1	1
letter:	N	O	P	Q	R	S	T	U	V	W	X	Y	Z
frequency:	1	1	0	2	1	3	2	0	7	0	0	5	0

We guess that b represents e, but this information is insufficient to find the two elements a and b of the key. To guess the values of the other letters, it is convenient to look at the frequency of digrams (combinations of two letters). The 15 most common digrams in decreasing order are:

th   he   in   er   an   re   ed   on
es   sr   en   at   to   nt   ha.

Observe that in the ciphertext, the digram bw occurs 5 times, and ib occurs thrice. If b represents e, then w is likely to be either r, s or n. Since n occurs more frequently than r or s in English text, we assume that w represents n. This leads to the equations,

$$1 = 4a + b \quad \bmod 26$$
$$22 = 13a + b \quad \bmod 26.$$

Subtracting, we obtain the congruence $21 \equiv 9a \pmod{26}$. Since $9 \cdot 3 \equiv 1 \pmod{26}$, we have $a = 11$ and $b = 9$. The plaintext is then recovered to read

A little nonsense now and then
is relished by the best of men.

We were lucky in our analysis, as our first guess led to a solution. In general, it will be necessary to experiment with a few values before a meaningful plaintext is obtained. The use of frequency analysis of letters and digrams reduces the number of trials needed to recover the plaintext.

Substitution Ciphers: The two ciphers discussed above are examples of a general monoalphabetic substitution scheme in which each letter is replaced by another. In general, we can choose a permutation π of the 26 letters and apply it to each letter by the formula $E_K(x) = \pi(x)$. A permutation of the 26 letters is a one-to-one function on these letters with the same range. The key, k, in this cryptosystem is the permutation, π, of 26 letters.
Consider the following permutation.

plain:	A	B	C	D	E	F	G	H	I	J	K	L	M
cipher:	Z	X	C	V	B	N	M	A	S	D	F	G	H
plain:	N	O	P	Q	R	S	T	U	V	W	X	Y	Z
cipher:	J	K	L	Q	W	E	R	T	Y	U	I	O	P

The plaintext `attack now` becomes the cipher `zrrzcfjku`.
Frequency analysis can still be used to cryptanalyze this cipher in spite of the large number of possible keys. Of course, knowledge of one or two letters is not enough to recover the entire plaintext, but if a reasonably large piece of ciphertext is available, then enough letters can be determined to enable the recovery of the plaintext.
Consider the following cipher encrypted using a substitution cipher.

```
rabrabkwoknjthxbwesebelbcszggobjrsrgbvrkzeblzwzrb
aserkwokjzccktjrknrabmwbzrsjrbwberuascaazexbbjrzf
bjsjsrckjrsjktegorawktmarabcbjrtwsbenwkhrabrshbk
nlorazmkwzezesjrbwbereazwbvkjrabkjbbirwbhbxojbzwg
obybwojkrbvhzrabhzrscszjzjvkjrabkrabwbirwbhbxojth
bwktezhzrbtwezrrwzcrbvxojkkrabwlzwrknhzrabhzrsce
```

The frequency count is as follows.

letter:	A	B	C	D	E	F	G	H	I	J	K	L	M
frequency:	19	44	9	0	16	1	5	11	2	21	22	4	3
letter:	N	O	P	Q	R	S	T	U	V	W	X	Y	Z
frequency:	5	10	0	0	39	15	8	1	5	22	5	1	24

The frequency of digrams and trigrams (three letter sequences) is given below.

digram:	RA	AB	JR	ZR	BW	WB	RB	RS
frequency:	13	11	9	8	7	7	6	6

trigram:	RAB	HZR	BWB	ABK	BJR	KJR	XOJ	ZRS
frequency:	11	5	3	3	3	3	3	3

We guess that b represents e and r represents t. This is justified from the letter frequencies and from the fact that the is the most frequently occurring trigram in the English language, which we expect to correspond to the trigram rab in the cipher. The trigram bwb occurs thrice, and it is most likely to represent ere. If we fill in the letters we have obtained so far, it is clear that k must stand for o. The next two most frequently occurring letters in the cipher are z and l. Notice that the string zwzr occurs in the cipher. Since r and w represent consonants, z must be a vowel. From the frequency counts, it is likely to be a or i. The digram zr occurs 8 times, and, as at occurs more frequently than it, we assume that z represents a. It is also clear that o represents y. Let's fill in all this information into the first part of the cipher.

```
thetheoryo_____e____e__e__a_____t_t_e_toa_e_arate
rabrabkwoknjthxbwesebelbcszggobjrsrgbvrkzeblzwzrb

h__toryo_a__o__to_the_reat__tere_t_h__hha__ee_ta_
aserkwokjzccktjrknrabmwbzrsjrbwberuascaazexbbjrzf

e_____t_o_t__o___ythro__hthe_e_t_r_e__ro_thet__eo
bjsjsrckjrsjktegorawktmarabcbjrtwsbenwkhrabrshbk
```

Figure 5.1.1 Partial decipherment of a substitution cipher

In the last line, it is clear that the word through occurs, so that t is u and m is g. The letter j occurs very frequently and cannot represent a vowel, so it is likely to be n. We can safely assume that s represents i. This is sufficient to recover the following plaintext.[3]

> The theory of numbers is especially entitled to a separate history on account of the great interest which has been taken in it continuously through the centuries from the time of Pythagoras, an interest shared on the one extreme by nearly every noted mathematician and on the other extreme by numerous amateurs attracted by no other part of mathematics.

The defect of the substitution cipher is that it does not change the relative frequency of letters of the alphabet, permitting cryptanalysis by counting the frequency of occurrence of letters. This is in spite of the large number of possible keys, which prevents exhaustive search, unlike the shift and affine cipher. The moral is that it is not enough to have a large key space; the cryptosystem should make the distribution of letters in the cipher completely random. This cannot be achieved by monoalphabetic ciphers like the substitution ciphers. We must turn to the polyalphabetic ciphers to randomize

---

[3]L. E. Dickson. *History of the Theory of Numbers*. Vol. 1. New York: Chelsea, 1952.

the distribution of the letters, but as we shall see, this too turns out to be insufficient to guarantee security.

Vigenère Cipher: A simple example of a polyalphabetic cipher is the Vigenère cipher, named after Blaise de Vigenère, who lived in the sixteenth century. It operates on blocks of letters. The key is a word or phrase that is repeated as many times as necessary to encipher the plaintext. If the key has length m, then the plaintext is divided into blocks of length m, and the key is added to each block (modulo 26) to produce the cipher. Suppose the plaintext is i came i saw i conquered and the keyword is egypt; then the ciphertext is produced as follows.

```
icame isawi conqu ered
egypt egypt egypt egyp
miybx myylb gulfn ixcs
```

The plaintext can be easily recovered if the key word is known, but cryptanalysis is more difficult because the frequency distribution is now more uniform. For example, the three e's become three different letters. This cipher was known as *Le chiffre indéchiffrable* because of its apparent immunity to analysis. The first chink in its armor was found by Friedrich Kasiski in 1863. His idea is that a letter or a phrase gets encoded to the same value if it repeats at positions that differ by multiples of the key length. For example, if the keyword has length 7, and the word the occurs at positions 17 and 45, it results in the same cipher phrase. This is because the key word is used repeatedly, and $45 - 17 = 28$ is a multiple of 7. Kasiski's technique is to look for repetitions of phrases in the cipher and compute the differences in the position. The keyword length m is a divisor of these differences, so it can be determined by evaluating the GCD of the differences.

As an example, consider the following ciphertext.

```
ganvvnatozseumdvmulployelqwymtcmynaigbpcmj
fwgzpjlppminwzpiswgoddrcgvojrkzmqjkjomczxy
wzpjjtgccnizfyfvrkabjwykabyziuwllrvzlmcosh
mqnfieemynenszpiijkazolrlbszmiaulbmesbtjrt
gcwykisaavvvsttockgesdgylpptlrvxczzzgcdgcs
wmydruanqzvvfbhzeiwttvfcwbznivavnjviwkegcl
fbtgejzqqoswsbezrkawywvzfodolvgjuzgkavejjf
ucdolvjmtnxywepgpbfwhigrkmzaxywkszwjtwlmhj
wmyavfewyzeeytpolvoptoijiclmijhzzovlvmlihu
gutiekwapzrwjwxvrflppmxywjwvgbkyfvvvkaevru
gcensdwbsdrxdqvzxyaasvwysxazrvvqyolvutzphp
owcghfxkcttksvlgcjaaijns
```

To apply Kasiski's procedure we look for repeat occurrences of trigrams in the cipher. A few trigrams obtained from this analysis are given in Table 5.1.3.

Table 5.1.3 Trigrams and their positions in a Vigenère cipher

trigram	positions	differences in positions
myn	32, 134	102
lpp	49,193,397	144, 204
ppm	50, 398	348
xyw	83, 305, 323, 401	222, 18, 78
tgc	90, 168	78
kab	102, 108	6
lrv	117, 197	80
ens	137, 423	286
dol	279, 297	18
olv	280, 298, 352, 454	18, 54, 102

We observe that 6 divides all the numbers in the right hand column except for 80 and 286. So it seems likely that the keyword used for encryption has length 6. We can proceed with this information to deduce the keyword. Let **c** be the ciphertext with c_i denoting the letter at position i. If the keyword has length 6, then the string $c_1 c_7 c_{13} c_{19} \ldots$ is encrypted by one letter, the first letter in the keyword. This is just the shift cipher, and the shift can be found by frequency analysis. Similarly, $c_2 c_8 c_{14} \ldots$ is encrypted using the same letter, so again we can employ frequency analysis.

We write the cipher in columns with 6 rows, a part of which is shown below.

```
gaullcgflwggzowgfaawlmesklasgssglvgwa
atmpqmbwpzovmmzcybblmqmzabubcatepxcmn
nodlwypgppdoqcpcfjylcnypzsltwatspcdyq
vzvoynczmidjjzjnvwzrofniozbjyvodtzgdz
vsmymampisrrkxjiryivsieilmmrkvcglzcrv
neuetijjnwckjytzkkuzhenjrietivkyrzsuv
```

The ith row has been encrypted using the ith letter of the keyword. If a large enough piece of ciphertext is available, we take the most frequently occurring letter to be e. The following table gives the two most frequently occurring letters in each row and the corresponding keys when the letters are taken to be e.

row	most frequent letters	corresponding keys
1	w, g	s, c
2	m,b	i, x
3	p, y,l	l, u, h
4	z, o	v, k
5	r, v	n, r
6	j, v	f, r

This reduces the number of keywords we have to try. Now it is possible to try combinations of these letters to see that the word *silver* decrypts the cipher into the plaintext.[4]

> Oscar Wilde once said that until Dickens wrote his novels there were no fogs in London. The fogs were there of course, in quantity, but it needed a writer to quicken men's awareness so that their imagination could grasp a reality to which they had previously been indifferent. We are liable to see incorrectly until a shift of attention brings the object into focus. There is the well known case of the chessboard, seen from one angle the white squares protrude and dominate, seen from another the black squares standout. Something like this has happened in the cloudy world of cryptanalysis.

Kasiski's test in the cryptanalysis of Vigenère's cipher is usually used together with the evaluation of various indices of coincidence.[5]

Pohlig–Hellman Exponentiation Cipher: Before discussing the implementation of public-key cryptography, we look at a simple exponentiation cipher, invented by Pohlig and Hellman in 1976.

Let p be an odd prime. Let M represent the numeric equivalent of the plaintext, where each letter of the plaintext is replaced by its two digit equivalent. Subdivide the message M into blocks M_i so that $0 < M_i < p$. If all the blocks are not of the same size, then we add a few random letters at the end of the message so that all the blocks have the same number of characters. Let e be an integer $0 < e < p$ such that $(e, p-1) = 1$. Then the message blocks M_i are encrypted for transmission by

$$E(M_i) = M_i^e \bmod p.$$

How do we recover M from $E(M)$? Let d be the inverse of e modulo $p-1$, that is, $ed \equiv 1 \pmod{p-1}$. If we write $ed = (p-1)k + 1$ for some k, then

---

[4]R. Lewin. *The American Magic: Codes, Ciphers, and the Defeat of Japan.* New York: Farrar Straus Giroux, 1982.

[5]See H. Beker and F. Piper. *Cipher Systems, The Protection of Communications.* New York: Wiley, 1982.

Fermat's Theorem implies that $M^{ed} \equiv \left(M^{p-1}\right)^k M \equiv M \pmod{p}$ for all M. Hence we can use the decrypting function

$$D(M) = M^d \bmod p.$$

The encryption and decryption functions can be evaluated by the method of successive squaring and multiplication discussed in Section 4.3.

Example 5.1.1. We choose $p = 4578971$ and $e = 3317271$ so that $(e, p - 1) = 1$. We compute the inverse of e modulo $p - 1$, $d = 573651$.

Suppose our message text is BEGIN ATTACK; this is converted to

$$M = 01040608130019190021020.$$

We break the message into blocks M_i so that $0 < M_i < p$, where M_i is as large as possible. It is convenient to take all blocks to be of the same length. As p has seven digits, we divide M into blocks of six digits each. If the last block doesn't have three characters, we add a few random characters so that it has. In our case, we add the letter u at the end of the message to obtain

$$M = \underbrace{010406}_{M_1}\,\underbrace{081300}_{M_2}\,\underbrace{191900}_{M_3}\,\underbrace{021020}_{M_4}.$$

We compute $E(M_i) \equiv M_i^e \pmod{p}$ to obtain the cipher

$$4137884\ 0438421\ 3227477\ 233970.$$

Note that the encoded message is in blocks of seven digits each.

Exercise. Verify that $D(E(M)) = M$.

Additional Secret-key Systems: There are many secret-key systems that are widely used. Examples of these include the Data Encryption Standard (DES), and stream and block ciphers like SEAL and RC5. The description of these systems is technical and does not involve any number theory. Our goal is to study cryptosystems which use number theory, particularly the public-key systems discussed in the next section. The interested reader should consult one of the numerous well-written books for details on the implementation of the modern secret-key systems.[6] These are important in practice because they tend to be a lot faster than the public-key systems discussed in this book. The systems operate on the binary alphabet consisting of 0's and 1's, and are designed to prevent cryptanalysis by statistical methods. The methods

---

[6]A. J. Menezes, P. C. van Oorschot, and S. A. Vanstone. *Handbook of Applied Cryptography*. Boca Raton, FL: CRC Press, 1997.

B. Schneier, *Applied Cryptography*. 2nd ed. New York: Wiley, 1996.

are still vulnerable because they use a secret-key, which can be compromised during key exchange. For maximum security and efficiency, both secret-key and public-key systems should be used together.

—————————— **Exercises for Section 5.1** ——————————

1. You are given the cipher `rljvnrbjfrlxwzdnam`, which has been encrypted using the shift cipher. Decipher it.

2. Determine the number of keys in the affine cipher with an alphabet of 26 letters. What is the number of keys if the alphabet consists of m letters?

3. Show that the number of possible keys in the substitution cipher is 26!.

4. Write a computer program to encipher text using the shift and affine ciphers.

5. Write a computer program to encipher text using the substitution cipher.

6. Write a program to compute the frequency of occurrence of the letters, digrams, and trigrams in the ciphertext.

7. Decipher the following examples of the substitution cipher. Both are taken from the short story, *The Gold-Bug*, by Edgar Allan Poe. The story is one of the first literary treatments of the substitution ciphers.

   (a) `sxhqazcafaxhglfasxraarsxlvvglfafynfagcahkcshs`
      `xdhqanscfhbiafhsyxcadlcrfhqavlxdildaynhqagszq`
      `acnychqazcsxgszvafynfyvihsyxfynlcamzagslvvolf`
      `hqawycafswzvagszqacflcagyxgacxarrazaxryxlxrlc`
      `ajlcsareohqadaxsifynhqazlchsgivlcsrsywsxdaxac`
      `lvhqacasfxylvhacxlhsjaeihamzacswaxhrscaghareo`
      `zcyelesvshsafynajacohyxdiauxykxhyqswkqylhhawz`
      `hflfyvihsyxixhsvhqahciayxaglxealhhlsxar`

   (b) `rkqgtjonrswqtrmowgsatwonrmshzvahstvssntrmsaqz`
      `swqtrmowgtjtwqmcoiyzvqtnqyzonqqgtqowqzwtcwsjs`
      `vtmeziisnesisnqwtnaqsviontqoznwzuzqgsvhzvawms`
      `qkwvsusvuzvsftiymsqzqgsmtwqonwqtnesrkqznsonhg`
      `oegqgsezirontqoznuzvqcsorgqzeekvwnzqutvuvziqg`
      `ssna`

8. Write a computer program to perform Kasiski's test on a Vigenère cipher.

5.2 Public-Key Cryptography

In the cryptosystems we have considered until now, the sender and the receiver select a key K for the enciphering algorithm E_K. This determines the deciphering algorithm D_K. Since the same key is used for both operations, the two parties must take great care in exchanging the key so that an eaves-dropper does not obtain it. The problem of key exchange is one of the most difficult in classical cryptosystems. It is not possible to use the phone or mail system to exchange keys, as these channels are insecure. Couriers are usually the most secure and reliable channel for exchanging keys. For military and diplomatic communications, it is possible to use couriers, but it is expensive and impractical in a commercial setting. In a commercial situation, business transactions are conducted between two or more parties who may not know each other, and who may never do business together again, but still desire complete confidentiality and instant secure communication.

There are additional key management problems in classical cryptosys-tems. Consider a network of users who wish to communicate with each other, for example, a network of banks or a group of people exchanging e-mail. For complete security, each pair of users must be assigned a different key. For example, with 10 users, we need 45 different keys, and with 100 users, we would need 4950 keys. It is not easy to add a new user to this system, as each user's database of keys needs to be updated to include the new user. More-over, the large number of keys creates security problems because each user has to safeguard his or her key database.

Classical cryptosystems create problems of trust, as the same key is held by both users in any sender/receiver pair. Suppose Alice and Bob share a secret key. If Carol obtains the key, then she can impersonate Bob and send messages to Alice, who has no way of knowing that the messages she is re-ceiving are not from Bob. Carol can also intercept messages from Alice to Bob and modify them in transit, without Bob ever realizing it. It is also pos-sible for Bob to forge messages, that is, to send a message and later deny that he ever sent the message and claim that his keys were stolen. The problem of forgery is not serious in military/diplomatic settings, where the parties are more likely to have established trust, but it is one of the primary concerns in commercial situations. A bank's customer should not be able to deny making a transaction, but at the same time, the customer should be secure in the belief that the bank is not forging transactions.

We require that a cryptosystem satisfy the following requirements, in ad-dition to being cryptanalytically secure.

Authentication: The recipient of a message should be able to ascertain its origin.

Integrity: The recipient of a message should be able to determine that the message has not been modified in transit.

Non-Repudiability: A sender of a message should not be able to deny later that he sent the message.

These requirements are impossible to achieve in classical one-key cryp-tosystems. The breakthrough came in 1976 with the invention of Public-Key

Cryptography by Diffie and Hellman, and independently by Merkle. Public-key cryptography was invented to solve the problem of secret-key management. In this system there is no need to exchange keys. Each person uses two keys, an encrypting key and a decrypting key. The encrypting key is published (hence the name public-key) while the decrypting key is kept secret. Anyone wishing to send a message to a person uses that person's public key to encrypt a message. The point here is that the knowledge of the encrypting method does not allow decryption, ensuring that no exchange of secret keys is necessary. One can send confidential information using the published keys, confident that only the intended recipient will be able to decipher it.

It may help the reader to consider an analogy with the mail system. We can send mail to anyone knowing only their address, and once it is in the recipient's mailbox, only the recipient can read it (assuming that the mailbox can only be opened with the recipient's key). The public-key system is stronger in the sense that each message travels in its own safe box, which can only be opened by the legitimate owner. Even the sender cannot open the box to disclose its contents.

In a public-key cryptosystem, each user has two keys. This greatly reduces the number of keys needed to establish a communication network. Each user has a public key, which gives an encrypting function E, and a private key, which gives a decrypting function D. For the scheme to work correctly, we must have

$$D(E(m)) = m \text{ for any message } m.$$

The encrypting function E is known to everyone. Since E is known, an opponent can generate any amount of ciphertext. Hence any such system should be immune to a ciphertext-based attack in which the corresponding plaintext is also known. For the system to be practical, one should be able to compute $E(m)$ and $D(E(m))$ quickly, but for security, it should be impossible to determine D from E in any reasonable amount of time. It is easier to describe the requirements of a public-key system, than to construct the functions E and D, which meet these requirements.

Theoretically, a public-key system can be constructed by a special case of a **one-way function** known as a **trapdoor one-way function**. Intuitively, a one-way function f should be one-to-one, and $f(x)$ easy to compute for any input x, but f^{-1} should be hard to compute. That means that it should be difficult to solve the equation $f(x) = y$ when y is known. A trapdoor one-way function is one for which the equation $f(x) = y$ becomes easy to solve if additional information (the trapdoor) is available. Instead of giving a precise definition of what is meant by *easy to compute* and *hard to compute*, we consider a couple of examples. We should remark that no one has proved the existence of a one-way function. The following two functions are considered the most likely candidates for being one-way functions.

Example 5.2.1. Fix two integers g and n and consider $f(x) = g^x \bmod n$.

The function f is considered easy to compute because the number of steps needed to evaluate f is a polynomial function of the number of digits in the input x, g, and n. The exponential can be computed in at most $\log_2(x)$ squaring operations. The square y^2 of a number y can be computed in $(\log_2 y)^2$ steps. In any case, the number of steps is a polynomial function of the input size, the number of digits. The problem of determining x from the knowledge of g^x mod n and g and n is known as the **discrete-log problem**. Despite considerable research, no method is known to evaluate the general discrete-log problem in polynomial time in the number of digits. All known algorithms require time that is polynomial in the input g^x mod n, not polynomial in the input size. To settle this question, one needs to show that the discrete-log problem is difficult or find an algorithm to compute it in polynomial time in the input size.

Example 5.2.2. Consider the function $f(p,q) = pq$, where p and q are prime numbers. The function is easy to compute because multiplication takes time proportional to the product of the number of digits of p and q, but computing the inverse of f is an extremely difficult problem. There is no efficient method to determine p and q from $n = pq$. The method of factoring by trial division takes time proportional to $\sqrt{p}$, if p is the smaller prime. This is exponential in the number of digits. While trial division is rather crude and cannot factor numbers with more than 18 to 20 digits, modern factoring methods still cannot factor numbers with more than 120 digits. If p and q have 100 digits each, then for practical purposes, this is a one-way function. Again, there is no proof that it is a true one-way function.

The first application of one-way functions seems to have been to the protection of passwords of users in a multi-user computer system. A user needs a password known only to her to be able to access the system. If the computer stores the passwords in a file, then the password is susceptible to attack. The solution is to apply a one-way function to the password and store the result in the password file. To authenticate a user, the computer applies the one-way function to the typed password and compares it to the entry in the password file. If the two are the same, then the user is allowed access. The advantage of this system is that the actual passwords are not stored, and if the function used is one-way, then anyone seeing the password entries will not be able to determine the actual password.

Unlike the previous example, for encryption and decryption, a one-way function is not directly useful. What is needed is a special type of one-way function, the trapdoor one-way function. In Example 5.2.2, if $n = pq$ is known, then p and q can be determined if additional information is available, for example, $\phi(n)$ or $\sigma(n)$. This additional information is the trapdoor. If these functions are used in cryptography, the trapdoor is needed to determine the decrypting algorithm D from the knowledge of E.

The following are some of the public-key systems that have been proposed.

RSA: This method is described in the next section. Its security lies in the difficulty of the integer factoring problem.

ElGamal: This method is based on the presumed intractability of the discrete logarithm problem.

McEliece: The McEliece system is based on the theory of Goppa codes. The theory is not within the scope of this book.

Elliptic Curve: The Elliptic Curve Cryptosystems are generalizations of the ElGamal system and are based on the difficulty of computing discrete logarithms in Elliptic Curves. Elliptic curves are discussed in the last chapter.

The first application of public-key systems is to the problem of key exchange in secret-key systems. The following key exchange protocol was proposed by Diffie and Hellman:[7] Suppose that Alice and Bob wish to exchange a secret key. They choose numbers g and n, which are made public. The key exchange protocol is as follows.

1. Alice chooses a random number a, computes $u = g^a \bmod n$, and sends u to Bob.

2. Bob chooses a random number b, computes $v = g^b \bmod n$, and sends v to Alice.

3. Bob computes the key $k = u^b = (g^a)^b \pmod{n}$.

4. Alice computes the key $k = v^a = (g^b)^a \pmod{n}$.

Notice that Alice and Bob compute the same key because $(g^a)^b = (g^b)^a = g^{ab}$. If g and n are chosen appropriately, then an intruder cannot determine k from the knowledge of u and v. To compute k, the intruder needs to know either a or b, and there is no known algorithm to accomplish this in a reasonable amount of time. The discrete-log problem is studied in Section 7.3.

The Diffie–Hellman key exchange protocol is susceptible to a man-in-the-middle attack. It is possible for Carol to intercept u and v and substitute her own u' and v'. If she can intercept all communication between Alice and Bob, then she can substitute her own messages. This type of attack can be foiled by combining a key-exchange protocol with authentication and identification schemes.

---

[7]W. Diffie and M. E. Hellman. "New Directions in Cryptography." *IEEE Transactions in Information Theory*, 22(1976), 644-654.

An important application of public-key systems is user authentication and message integrity. As a simple example, suppose a public-key system satisfies $E(D(m)) = m$, in addition to $D(E(m)) = m$ (the latter is required for successful encryption and decryption). We denote by E_A and E_B the public encryption keys of Alice and Bob, respectively, and D_A and D_B, their respective private keys. The basic signature protocol is as follows.

1. Alice encrypts the message m with her private key D_A.

2. Alice sends $E_B(D_A(m))$ to Bob.

3. Bob recovers $D_A(m)$ with his private key and then recovers m using Alice's public key.

Since Bob is able to recover m using Alice's public key, he is sure that only Alice could have sent it because only she knows her private key. The security of the scheme derives from infeasibility of computing D_A from knowledge of E_A. The signature satisfies the following properties:

1. Bob can verify that Alice sent the message because she signed it with her private key.

2. The signature depends on the contents of the message; hence no one can use the signature with another document.

In practice, instead of signing the entire document, only a hash of the document is signed. The hash function or a message digest of the document is a value that depends on the document. It is usually of fixed length, but the function is such that it is difficult to find two messages with the same hash function. In signature scheme, the hash function of a document is signed with the private key, and it can be verified by anyone using the public key. Detailed signature protocols are discussed in a later section.

Public-key cryptography has found numerous applications:

1. Paper documents such as checks, stocks, lottery tickets can be authenticated using public-key techniques. A document is assigned a unique digital signature based on the fiber patterns of the paper and the contents of the document. The signature is encoded with a private key, and the resulting cipher is affixed to the document. Anyone can verify the authenticity of the document by using an embedded public key.

2. Digital signatures and authentication methods are used for network access control. By these methods, a user's password never needs to travel over the network.

3. Public-key techniques are now used in smart cards, digital cash, and other types of electronic banking and commerce.

4. Verification of some international treaties can be done by public-key techniques.[8]

Exercises for Section 5.2
---

1. Suppose that a secret-key system is installed in a community of n persons. How many keys are needed so that every person can securely communicate with another?

2. Extend the Diffie–Hellman key exchange protocol so that three people, Alice, Bob, and Carol, can exchange the same key.

5.3   The RSA Scheme

The RSA scheme is one of the most popular public-key cryptosystems. It was invented by R. L. Rivest, A. Shamir, and L. M. Adleman in 1978.[9] Recall the Pohlig–Hellman exponentiation method in which the enciphering function is $E(m) = m^e \bmod p$ for a prime number p and key e. The decrypting function can be easily computed from E by choosing d so that $ed \equiv 1 \pmod{p-1}$. Then $D(m) = m^d \bmod p$. The RSA method is a simple modification of the Pohlig–Hellman scheme, but it makes the determination of d very difficult. The idea is to use a product of two primes, $n = pq$, as the modulus. We choose an integer e such that $(e, \phi(n)) = 1$. Then the enciphering function is

$$E(m) = m^e \bmod n. \tag{1}$$

To determine the decryption function, we compute d such that

$$ed \equiv 1 \pmod{\phi(n)}. \tag{2}$$

Then the decryption operation is given

$$D(y) = y^d \bmod n. \tag{3}$$

The following proposition proves that the encryption and decryption functions are inverse operations.

Proposition 5.3.1. *Suppose $n = pq$ is a product of two distinct primes. If E and D are as given above, then*

$$D(E(m)) = m \text{ for all } m, \ 0 \le m < n.$$

---

[8]See G. J. Simmons. "How to Insure That Data Acquired to Verify Treaty Compliance Are Trustworthy." *Contemporary Cryptology: The Science of Information Integrity*, ed. G. J. Simmons, IEEE Press, 1992, 615–630.

[9]R. L. Rivest, A. Shamir, and L. M. Adleman. "A Method for Obtaining Digital Signatures and Public-Key Cryptosystems." *Communications of the ACM*, 21(1978), 120–126.

PROOF. We have to verify that $m^{ed} \equiv (m^e)^d \equiv m \pmod{n}$ for all m. Since $\phi(n) \mid ed - 1$, using Euler's Theorem, we see that $m^{ed-1} \equiv 1 \pmod{n}$ when $(m, n) = 1$. Therefore, $m^{ed} \equiv m \pmod{n}$ when $(m, n) = 1$.

To see that the assertion is true even when $(m, n) \neq 1$, take m with p or q as a factor. Suppose $m = pk$ with $(k, q) = 1$. Then $m^{ed} \equiv 0 \pmod{p}$. The value of $\phi(n)$ is $(p-1)(q-1)$, which implies that $(q-1)$ divides $ed - 1$. Therefore, $m^{ed} \equiv m^{ed-1}m \equiv m \pmod{q}$ because $(m, q) = 1$. The two congruences, $m^{ed} \equiv 0 \equiv m \pmod{p}$ and $m^{ed} \equiv m \pmod{q}$, imply that $m^{ed} \equiv m \pmod{n}$. The proof is the same when m is a multiple of q. ∎

The RSA system can also used for digital signatures because of the property $D(E(m)) = m$ and $E(D(m)) = m$. A system based on this property was described in the previous section.

To implement the RSA system, we convert text into numeric form. As usual, we remove all spaces and punctuation and assign numbers from 0 through 25 for the letters A through Z. The message M is divided into blocks M_i such that $0 \leq M_i < n$. It is necessary to have $M_i < n$, otherwise, M_i^{ed} mod n will not be equal to M_i. If n has k digits, then we can guarantee that $M_i < n$ by letting M_i have at most $k - 1$ digits.

Example 5.3.2. Suppose $n = 4067351 = 1733 \cdot 2347$. Both factors of n are prime, so $\phi(n) = 1732 \cdot 2346$. We can use $e = 31$ as the encrypting key because $(e, \phi(n)) = 1$. Then the encrypting function is $E(m) = m^{31}$ mod n. The phrase `public key cryptography` is divided into blocks of six digits each to yield the following plaintext stream.

M_1	M_2	M_3	M_4	M_5	M_6	M_7
152001	110802	100424	021724	151914	061700	150724

We encrypt each block using E and obtain the cipher:

$E(M_1)$	$E(M_2)$	$E(M_3)$	$E(M_4)$
2721372	3969831	2416419	1795753

$E(M_5)$	$E(M_6)$	$E(M_7)$
2110079	0242624	0889174.

The decryption exponent is computed by solving $31d \equiv 1 \pmod{\phi(n)}$. The Euclidean Algorithm yields the solution $d = 3145759$, and we see that applying $D(y) = y^d$ mod n to the cipher recovers the plaintext stream.

The security of the system is derived from the belief that the encrypting function $E(m) = m^e$ mod n is a one-way function. This is not a proven fact. It is believed that determining the decryption function D from E is equivalent to factoring n. While this equivalence has not been proved for the RSA system, some variations of the RSA scheme have been shown to be equivalent to factoring. We will show that computing $\phi(n)$ or d is equivalent to factoring n, but this still leaves out the issue of being able to recover m from $E(m)$ without knowing d or $\phi(n)$. There is also the possibility that someone will find an efficient method to factor arbitrarily large numbers.

Proposition 5.3.3. *Suppose $n = pq$ is a product of two distinct primes. Then determining $\phi(n)$ is equivalent to factoring n.*

PROOF. If we know the factorization of n, then $\phi(n) = (p-1)(q-1)$.

On the other hand, if n and $\phi(n)$ are known, then it is easy to compute the factors p and q. We write $q = n/p$ and substitute this in the formula for $\phi(n)$.

$$\phi(n) = (p-1)(q-1)$$
$$= (p-1)(n/p-1).$$

Simplifying this equation shows that p is one root of the quadratic equation

$$p^2 + (n+1-\phi(n))p + n = 0.$$

It is easy to see that q is also a root of the quadratic equation. Hence we have shown that determining $\phi(n)$ is equivalent to factoring n. ■

Next, we will show that if there is a method to compute the decrypting exponent d without knowing $\phi(n)$, then we will still be able to factor n. Here we are assuming that d satisfies the equation $ed \equiv 1 \pmod{\phi(n)}$. The algorithm below is probabilistic, in the sense that it depends on random inputs, and is not guaranteed to succeed every time. We will show that if the procedure is applied once, then the probability of success is at least $1/2$. Therefore, if the procedure is repeated k times, then the probability of k failures is at most $(1/2)^k$; that is, the probability that we are able to factor the number in k attempts is at least $1 - (1/2)^k$. In practice, we are guaranteed of finding a factor in a few trials.

The algorithm is based on the fact that the equation $x^2 \equiv 1 \pmod{n}$ has four solutions when n is a product of two distinct primes. Any solution α to $x^2 \equiv 1 \pmod{n}$ is such that $n \mid (\alpha-1)(\alpha+1)$. Since $\alpha \not\equiv \pm1 \pmod{n}$, we can compute a factor of n by computing either $(\alpha-1, n)$ or $(\alpha+1, n)$.

Suppose we know d such that $ed \equiv 1 \pmod{\phi(n)}$. Write $ed - 1 = 2^r s$ where s is odd. Choose a random number w such that $0 < w < n$. We compute the sequence

$$z_0 = w^s \bmod n, \ z_1 = z_0^2 \bmod n, \ldots, \ z_i = z_{i-1}^2 \bmod n, \ldots,$$
$$z_r = z_{r-1}^2 \bmod n.$$

Notice that $z_i = w^{2^i s} \bmod n$. Since $\phi(n) \mid ed - 1 = 2^r s$, Euler's Theorem implies that the last term of the sequence, z_r, is 1. If the first term of the sequence is not 1, then there must be a number $z_k \not\equiv 1 \pmod n$ such that $z_k^2 \equiv 1 \pmod n$. Our hope is that $z_k \not\equiv -1 \pmod n$. Suppose a z_k exists with these properties; then we have found a nontrivial square root of unity modulo n, and we can factor n.

Example 5.3.4. Consider the RSA public-key $n = 13289$ with encrypting exponent $e = 7849$. Suppose the decrypting exponent is $d = 2713$. We can use this to factor the number as follows. We write $ed - 1 = 2^r \cdot s$, with $r = 8$ and $s = 83181$. Take a random number w, say $w = 493$. Then $x_0 = w^s \bmod n = 5032$. If $x_i = x_{i-1}^2 \bmod n$, then $x_1 = 5479$, $x_2 = 12879$, and $x_3 = 8632$. Next, $x_4 = 1$; therefore, $8632^2 \equiv 1 \pmod n$; therefore, $(8632 - 1, n)$ is a proper factor of n. Evaluating this, we obtain 137, and the other factor of n is 97.

The procedure described above fails when $z_0 = 1$ or if there exists an index k such that $z_k = -1$. To estimate the probability of failure, we count the number of integers w, $0 \le w < n$ such that $z_0 = 1$ or some $z_k = -1$; that is, we are interested in the number of solutions to the congruences

$$w^s \equiv 1 \quad (\bmod\ n) \tag{4}$$

$$w^{2^i s} \equiv -1 \quad (\bmod\ n) \quad \text{for} \quad 0 \le i < r. \tag{5}$$

We show that the total number of solutions to these congruences is at most $n/2$. The crucial fact in the proof is Lagrange's Theorem, or, rather, its corollary that the number of solutions to the equation $x^d \equiv 1 \pmod p$ is $(p - 1, d)$ when p is prime.

Lemma 5.3.5. *Suppose p is a prime number and d an integer. Let $p - 1 = 2^r s$ and $d = 2^{r'} s'$ with s and s' odd integers.*

(a) *The congruence $w^d \equiv 1 \pmod p$ has exactly $(d, p - 1)$ solutions.*

(b) *If $r' < r$, then $w^d \equiv -1 \pmod p$ has exactly $(d, p - 1)$ solutions, and if $r' \ge r$, the congruence has no solutions.*

PROOF. From Corollary 4.4.3 to Lagrange's Theorem, if p is prime and $d \mid (p - 1)$, the equation $w^d \equiv 1 \pmod p$ has exactly d solutions.

If $d \nmid (p - 1)$, then we can write $(d, p - 1)$ as a linear combination of d and $(p - 1)$, say $(d, p - 1) = xd + y(p - 1)$. By Fermat's Theorem, $w^{p-1} \equiv 1 \pmod p$; if $w^d \equiv \pmod p$, then $(w^d)^x (w^{p-1})^y \equiv 1 \pmod p$, hence $w^{(d,p-1)} \equiv 1 \pmod p$. Therefore, every solution to $w^d \equiv 1 \pmod p$ is a solution to $w^{(d,p-1)} \equiv 1 \pmod p$, and vice versa. This shows that the number of solutions to $w^d \equiv 1 \pmod p$ is $(d, p - 1)$.

For the second part, if $w^d \equiv -1 \pmod{p}$, then $w^{2d} \equiv 1 \pmod{p}$. The latter has exactly $(2d, p-1)$ solutions. We can factor

$$w^{2d} - 1 \equiv (w^d - 1)(w^d + 1) \equiv 0 \pmod{p}.$$

Now $w^d - 1 \equiv 0 \pmod{p}$ has $(d, p-1)$ solutions. Therefore, the equation $w^d + 1 \equiv 0 \pmod{p}$ has a solution if

$$(2d, p-1) > (d, p-1). \tag{6}$$

By using the formula for the greatest common divisor in terms of the prime factorizations of the numbers, it is clear that

$$(2d, p-1) = 2^{\min(r'+1,r)}(s', s)$$

and

$$(d, p-1) = 2^{\min(r',r)}(s', s).$$

Inequality (6) translates into the condition $\min(r', r_1) < \min(r'+1, r)$. This is possible only when $r' < r$. If this is the case, then $(2d, p-1)$ is twice $(d, p-1)$; hence the equation $w^d \equiv -1 \pmod{p}$ has $(d, p-1)$ solutions. ∎

Consider an example.

Example 5.3.6. Let $n = pq$, with $p = 137$ and $q = 97$. We use the proposition to count the number of solutions to Equations (4) and (5). We have $p - 1 = 8 \cdot 17$ and $q - 1 = 32 \cdot 3$. Let $\phi(n) = ed - 1 = 2^r s$ where $s = 51$. The equation $w^s \equiv 1 \pmod{p}$ has $(s, p-1) = 17$, and $w^s \equiv 1 \pmod{q}$ has $(s, q-1) = 3$ solutions. Therefore, the equation $w^s \equiv 1 \pmod{n}$ has 51 solutions.

By the proposition, the congruences, $w^s \equiv -1 \pmod{p}$ and $w^s \equiv -1 \pmod{q}$ have 17 and 3 solutions, respectively. Therefore, $w^s \equiv -1 \pmod{n}$ has 51 solutions.

Similarly, the congruence $w^{2s} \equiv -1 \pmod{p}$ has $(2s, p-1) = 34$ solutions, and $w^{2s} \equiv -1 \pmod{q}$ has $(2s, q-1) = 6$ solutions. Hence $w^{2s} \equiv -1 \pmod{n}$ has $34 \cdot 6 = 204$ solutions.

The congruence $w^{4s} \equiv -1 \pmod{p}$ has $(4s, p-1) = 68$ solutions, and $w^{4s} \equiv -1 \pmod{q}$ has 12 solutions. Therefore, $w^{4s} \equiv -1 \pmod{n}$ has 816 solutions.

It is not possible to have $w^{8s} \equiv -1 \pmod{n}$, because $p - 1 \mid 8s$ implies that $w^{8s} \equiv 1 \pmod{p}$. Therefore, the total number of solutions to the congruences (4) and (5) is $51 + 51 + 204 + 814 = 1122$, whereas there are 13056 invertible elements modulo n.

Theorem 5.3.7. *Suppose $n = pq$, with p and q distinct primes, and suppose that $\phi(n) \mid ed - 1$ for some e and d. If $ed - 1 = 2^r s$, with s odd, the number of solutions to the congruences,*

$$w^s \equiv 1 \quad \pmod{n} \tag{7}$$

$$w^{2^i s} \equiv -1 \quad \pmod{n} \text{ for } 0 \le i < r, \tag{8}$$

is at most $n/2$.

PROOF. Let $ed - 1 = 2^r s$, $p - 1 = 2^{r_1} s_1$, and $q - 1 = 2^{r_2} s_2$, with s, s_1, and s_2 odd. Since $\phi(n) = (p-1)(q-1)$ divides $ed - 1$, we must have $2^{r_1 + r_2} \mid 2^r$ and $s_1 s_2 \mid s$.

Now, $w^d \equiv 1 \pmod{n}$ if and only if $w^d \equiv 1 \pmod{p}$ and $w^d \equiv 1 \pmod{q}$. The equation $w^s \equiv 1 \pmod{p}$ has $(s, p-1) = s_1$ solutions, and $w^s \equiv 1 \pmod{q}$ has $(s, q-1) = s_2$ solutions. By the Chinese Remainder Theorem, the number of solutions to $w^s \equiv 1 \pmod{pq}$ is $s_1 s_2$.

Next, we count the solutions to $w^{2^i s} \equiv -1 \pmod{n}$. This is equivalent to $w^{2^i s} \equiv -1 \bmod p$, and $w^{2^i s} \equiv -1 \bmod q$. By Fermat's Theorem, we must have $i < r_1$ and $i < r_2$, that is, $i < \min(r_1, r_2)$. By Lemma 5.3.5, the congruence $w^{2^i s} \equiv -1 \pmod{p}$ has $2^i s_1$ solutions, and $w^{2^i s} \equiv -1 \pmod{q}$ has $2^i s_2$ solutions. Therefore, the congruence $w^{2^i s} \equiv -1 \pmod{pq}$ has $2^{2i} s_1 s_2$ solutions.

We can assume that $r_1 \le r_2$. Adding the number of solutions to each of the congruences in the statement of the theorem, we obtain the sum

$$s_1 s_2 + \sum_{i=0}^{r_1 - 1} 2^{2i} s_1 s_2,$$

which simplifies to the sum

$$s_1 s_2 \left(1 + \frac{4^{r_1} - 1}{4 - 1} \right) = s_1 s_2 \frac{4^{r_1} + 2}{3}.$$

It is easy to verify that this sum is less than $\phi(n)/2$; hence it is less than $n/2$ and the theorem is proved. ■

There are additional weaknesses of some implementations of the system. These defects are known as **protocol failures**. They are not defects in the RSA scheme but rather in the specific implementation of it. Here are two examples of protocol failures.

Example 5.3.8. Common Modulus Protocol Failure. Suppose several users share the same modulus n in an RSA system (with different encrypting exponents). If the same message is sent to more than one person, then it can be recovered without knowing the decrypting key or the factorization of n. Suppose $m^{e_1} \bmod n$ and $m^{e_2} \bmod n$ are known. If e_1 and e_2 are relatively prime, then there exist x and y such that $e_1 x + e_2 y = 1$. Then m can be recovered using the formula $m = m^{e_1 x + e_2 y} = (m^{e_1})^x (m^{e_2})^y \bmod m$.

For instance, suppose $n = 5038301$ is the common modulus among several users Let $e_1 = 787$ and $e_2 = 6785$. Then the Euclidean Algorithm yields $888 e_1 - 103 e_2 = 1$. If we have $m^{e_1} \bmod n = 986536$ and $m^{e_2} \bmod n = 2294886$, then

$$m = (986536)^{888}(2294886)^{-103} \bmod n = 52013.$$

We recover the plaintext to read fun using our convention regarding the numeric values assigned to letters.

Example 5.3.9. Low Exponent Protocol Failure: Suppose that the encryption exponent is 3 among a group of users. If n_1, n_2, and n_3 are three different RSA moduli of three users, then the same message sent to the three users can be recovered. Suppose $y_i = m^3 \bmod n_i$ is known for $i = 1, 2, 3$. Since $m < n_i$ for each i, $m^3 < n_1 n_2 n_3$. We can use the Chinese Remainder Theorem to find a Y such that $Y \equiv y_i \pmod{n_i}$, $i = 1, 2, 3$. The solution to this simultaneous congruence is unique modulo $n_1 n_2 n_3$, (the n_i should be relatively prime, otherwise we can factor them by computing the greatest common divisors). Hence $Y = m^3$ because m^3 satisfies the three congruences. The equality is assured because $m^3 < n_1 n_2 n_3$. Therefore, we can recover m by taking the integer cube-root of Y.

──────────────── **Exercises for Section 5.3** ────────────────

1. The following cipher is obtained using an RSA scheme with modulus $n = 1146115723$ and encrypting key $e = 67$. Compute the deciphering key and recover the plaintext.

   474786165   509323344   171384806   982448832
   205150025   1122132935   1013717577 121618407

2. Suppose that $n = 10088821$ is a product of two distinct primes, and $\phi(n) = 10082272$. Determine the prime factors of n.

3. Show that the conclusion of Proposition 5.3.1 is valid whenever n is a product of distinct primes, but that the assertion is not true when n has a square divisor.

4. A message is said to be *fixed* in the RSA system if $E(m) = m$. Show that if n, e, and d are as described in this section, the number of messages fixed by E is

$$(e - 1, p - 1) \times (d - 1, q - 1).$$

5. The RSA public key for a secret agency is

$$n = 19108155832182962093640598524113899991$$

with encryption exponent

$$e = 16316342466904772882376009670880883193.$$

The private key d has been leaked to you and is

$$d = 16931978517931581023326224399166856057.$$

Determine the prime factors of n.

6. Modify the RSA system to design a multiparty cryptosystem, in which there are three keys. The goal is that a message encrypted with any one of the keys cannot be deciphered without the knowledge of the other two keys. The design of such a system could have applications to digital signature schemes where it is necessary to have the same document signed by more than one person.

7. The choice of the decrypting exponent d in the RSA system is not unique. Suppose
$$\lambda(n) = \frac{(p-1)(q-1)}{\gcd(p-1, q-1)}.$$

Show that if $ed \equiv 1 \pmod{\lambda(n)}$, then d can be used as a decrypting exponent in the RSA scheme with encrypting exponent e.

Chapter 6

Primality Testing and Factoring

6.1 Pseudoprimes and Carmichael Numbers

Applications of number theory to cryptography require a supply of large primes. The method of trial division is only practical for finding primes with at most 12 to 14 digits. We need better techniques to find much larger primes. Moreover, the methods have to be fast, as it is necessary to produce a large number of primes. The methods discussed in this section and the next are very efficient, but they are probabilistic. If one of these techniques returns that a number is composite, then it is certain that the number is composite, but if it returns that the number is prime, then there is a small probability that the number is actually composite. What this means is that occasionally the algorithms report composite numbers as primes. We will study the probability of failure of these algorithms, and it will be clear that the chances of failure of the strong pseudoprime test is negligible. The strong pseudoprime test is sufficient for all practical purposes.

To test if a number is prime, one would like to have a simple criterion that is computationally feasible. Unfortunately, the known criteria for primality fail to meet this requirement. For example, Wilson's Theorem (Exercise 3.2.6) states that $(n-1)! \equiv -1 \pmod{n}$ if and only if n is prime. Unfortunately, there is no way to compute $(n-1)!$ in a reasonable amount of time. A lot of so-called "formulas for primes" are based on Wilson's Theorem and are utterly useless.[1]

We appeal to Fermat's Theorem for salvation. The theorem states that for primes p, $a^p \equiv a \pmod{p}$ for all a. We investigate the extent to which the converse of this theorem holds, that is, does $a^n \equiv a \pmod{n}$ for all a imply

---

[1] For a detailed discussion, see U. Dudley. "Formulas for Primes." *Mathematics Magazine*, 56(1983), 17–22.

that n is prime? If it doesn't, for which n does it fail and how often? Note that this criterion would be computationally feasible, as $a^n \bmod n$ can be computed in $O(\log_2 n)$ steps by the technique of squaring and multiplication. (See Section 4.3.)

Definition 6.1.1. A composite number n is a **pseudoprime to base** 2 [or 2-pseudoprime or psp(2)] if $2^n \equiv 2 \pmod{n}$.

Example 6.1.2. Let's verify that 341 is a pseudoprime to base 2.

First, $341 = 11 \cdot 31$ is composite. To compute $2^{341} \bmod 341$, we compute $2^{341} \bmod 11$, and $2^{341} \bmod 31$. Notice that $2^{10} \equiv 1 \pmod{11}$ and $2^5 \equiv 1 \pmod{31}$; hence

$$2^{341} \equiv 2^{10 \cdot 34} 2^1 \equiv 2 \pmod{11}$$
$$2^{341} \equiv 2^{5 \cdot 68} 2^1 \equiv 2 \pmod{31}.$$

By Proposition 3.1.10, it follows that $2^{341} \equiv 2 \pmod{341}$, so 341 is a 2-pseudoprime.

In fact, 341 is the smallest 2-pseudoprime and gives a counterexample to the converse of Fermat's Theorem. It was discovered by Sarrus in 1819.

Exercise. Show that $561 = 3 \cdot 11 \cdot 17$ is a pseudoprime to base 2.

Note that if $2^n \not\equiv 2 \pmod{n}$, then n must be composite. This is an instance of a compositeness test, where we know a number is composite without knowing any of the factors. If $2^n \equiv 2 \pmod{n}$, we must check the compositeness by a different method, for example, trial division, to conclude that n is a pseudoprime.

We are interested in the number of 2-pseudoprimes and their distribution. If only a finite number of 2-pseudoprimes exist, then we could obtain a simple criterion for primality, at least for very large numbers. Unfortunately (but not unexpectedly), there are an infinite number of 2-pseudoprimes.

Proposition 6.1.3. *If n is a 2-pseudoprime, then $2^n - 1$ is a 2-pseudoprime. Therefore, there are infinitely many 2-pseudoprimes.*

PROOF. Let $n' = 2^n - 1$. We want to show that $2^{n'} \equiv 2 \pmod{n'}$. Now $n \mid n' - 1$ because $n' - 1 = 2^n - 2$, and n is a 2-pseudoprime. Let $n' - 1 = nk$ for some k; then using the formula for the sum of a geometric series, we obtain

$$2^{n'-1} - 1 = 2^{nk} - 1$$
$$= (2^n - 1)(2^{n(k-1)} + \cdots + 2^n + 1)$$
$$= n'(2^{n(k-1)} + \cdots + 2^n + 1).$$

This shows that $n' \mid 2^{n'-1} - 1$, or equivalently, $2^{n'-1} \equiv 1 \pmod{n'}$, or $2^{n'} \equiv 2 \pmod{n'}$.

The proof is not finished yet, because, for n' to be a 2-pseudoprime, we must also show that it is composite. Since n is composite, we can write $n = ab$ for some $a, b > 1$. Again, we use the formula for the sum of a geometric progression to factor $2^n - 1 = 2^{ab} - 1 = (2^a - 1)(2^{a(b-1)} + \cdots + 1)$; hence n' is composite, and so it is a 2-pseudoprime. ∎

The proof shows that there are infinitely many odd 2-pseudoprimes. The first even 2-pseudoprime was discovered by D. H. Lehmer in 1950 (see Exercise 3). It is also known that there are infinitely many even 2-pseudoprimes.

The definition of 2-pseudoprime can be modified to consider pseudoprimes to other bases.

Definition 6.1.4. A composite number n is a **pseudoprime to base** a ($\text{psp}(a)$ or a-pseudoprime) if $a^n \equiv a \pmod{n}$.

Note that if $(a, n) = 1$, then the condition is equivalent to $a^{n-1} \equiv 1 \pmod{n}$.

Example 6.1.5. Let's verify that 91 is a 3-pseudoprime.

A direct computation shows that $3^{91} \equiv 3 \pmod{91}$, and 91 is composite, as $91 = 7 \cdot 13$. We could also do the calculation as in Example 6.1.2 by using Fermat's Theorem and computing $3^{91} \bmod 7$ and $3^{91} \bmod 13$.

For any a, there are infinitely many a-pseudoprimes. (See Exercise 11.)

Definition 6.1.6. We say that a number n passes the **pseudoprime test to base** a if $a^n \equiv a \pmod{n}$.

The pseudoprime test does not require that n be composite. Prime numbers pass the pseudoprime test to any base a, and if n is composite and passes the pseudoprime test to base a, then it is an a-pseudoprime. If n fails the pseudoprime test, then it must be composite. To use this test as a test for primality, we need to know the number of a-pseudoprimes to determine the probability that a composite number is misidentified as prime.

There are $50,847,534$ primes less than 10^9, but only 5597 psp(2)'s. If a number n less than 10^9 satisfies $2^n \equiv 2 \pmod{n}$, then there is a very high probability $[= 1 - (5597)/(50847534)]$ that it is prime. Our chances are increased if we also test n with respect to other bases. For example, there are 685 numbers less than or equal to 10^9 that are pseudoprimes to bases 2, 3, and 5. We could construct a list of these and then obtain the following simple primality test.

Algorithm 6.1.7 (Pseudoprime Primality Test). This test applies to numbers less than 10^9. Make a table of the 685 pseudoprimes to bases 2, 3, and 5.

1. Does n pass the pseudoprime test to bases 2, 3, and 5? If not, n is composite.

2. Is n one of 685 precomputed pseudoprimes to bases 2, 3, and 5? If so, n is composite; otherwise, it is prime.

While the range of numbers to which this test is applicable is not as large as we would like, the test is much faster than trial division. We have three pseudoprime tests, which can be performed in time proportional to $\log n$. A search of 685 numbers can be done in less than 10 operations. (How?)

We will not actually use this algorithm, as there is a better method discussed in the next section. For large n, we might hope that if n is composite, its compositeness will be revealed in some pseudoprime test. Unfortunately, there are many composite numbers that are pseudoprimes to all bases. Such numbers were first studied by R. D. Carmichael in 1912.

Definition 6.1.8. A composite number n that satisfies $a^{n-1} \equiv 1 \pmod{n}$ for all a such that $(a, n) = 1$ is called a **Carmichael number**.

Example 6.1.9. 561 is a Carmichael number.

First notice that 561 is composite as $561 = 3 \cdot 11 \cdot 17$. To show that $a^{560} \equiv 1 \pmod{561}$ for all $(a, 561) = 1$, we use the Chinese Remainder Theorem and Fermat's Theorem.

$$a^2 \equiv 1 \pmod{3} \Rightarrow a^{560} \equiv \left(a^2\right)^{280} \equiv 1 \pmod{3}$$

$$a^{10} \equiv 1 \pmod{11} \Rightarrow a^{560} \equiv \left(a^{10}\right)^{56} \equiv 1 \pmod{11}$$

$$a^{16} \equiv 1 \pmod{17} \Rightarrow a^{560} \equiv \left(a^{16}\right)^{35} \equiv 1 \pmod{17}.$$

Hence $a^{560} \equiv 1 \pmod{561}$ for all a satisfying $(a, 561) = 1$.

Carmichael numbers are characterized by the following property.

Proposition 6.1.10. *A composite number n is a Carmichael number if and only if for every $p \mid n$, we have that $p - 1 \mid n - 1$.*

The proof of the necessity of the condition is similar to the computation of Example 6.1.9 (see Exercise 13). The sufficiency will be established in Exercise 7.2.17 after a study of primitive roots.

It was shown recently by Alford, Granville, and Pomerance[2] that there exist infinitely many Carmichael numbers.

---

[2]See A. Granville. "Primality Testing and Carmichael Numbers." *Notices of Amer. Math. Soc.*, 39(1992), 696–700.

—————————————— **Exercises for Section 6.1** ——————————————

1. Verify that 217 is a pseudoprime to base 5. Is 217 a Carmichael number?

2. Verify that 1105 is a pseudoprime to bases 2 and 3.

3. Write a computer program to find 2-pseudoprimes. (Be sure to check for compositeness using trial division.)

   (a) Which are the 2-pseudoprimes less than or equal to 2000?

   (b) Determine the smallest even 2-pseudoprime.

   (c) Let $P_2(x)$ be the number of 2-pseudoprimes that are less than or equal to x. Compute $P_2(k \times 10^4)$ and $P_2(k \times 10^5)$ for $1 \le k \le 10$. Draw a graph of $P_2(x)$.

4. Construct a table that contains all the numbers less than or equal to 10^9 that are pseudoprimes to bases 2, 3, and 5. Implement Algorithm 6.1.7, and determine which of the following are prime:

   (a) 622476791.   (b) 115921.
   (c) 203091001.   (d) 74251703.

5. Show that 341 is not a Carmichael number.

6. Show that if a number n is a pseudoprime to some base, then it is a pseudoprime to at least $\phi(n)/2$ bases.

7. Show that no prime power can be a Carmichael number.

8. Show that every composite Fermat number is a pseudoprime to base 2. (Recall that the kth Fermat number is $F_k = 2^{2^k} + 1$.)

9. Show that if p is prime and $2^p - 1$ is composite, then $2^p - 1$ is a 2-pseudoprime.

10. It is shown in Section 17.1 that if $q = 24m + 1$ is prime, then $2^{12m} \equiv 1 \pmod{q}$ and $3^{12m} \equiv 1 \pmod{q}$. Use this fact to show that $(12m + 1)(24m + 1)$ is a pseudoprime to bases 2 and 3 whenever both factors are prime.

11. This exercise shows that there are infinitely many pseudoprimes to any base a. Let p be an odd prime and consider

$$ n = \frac{a^p - 1}{a - 1}, \quad m = \frac{a^p + 1}{a + 1}, \quad N = nm. $$

   Note that $n = 1 + a + \cdots + a^{p-1}$ and $m = 1 - a + a^2 - \cdots + a^{p-1}$.

   (a) Use Exercise 4.4.4 to show that if $a \not\equiv 0, \pm 1 \pmod{p}$, then $n \equiv 1 \pmod{2p}$ and $m \equiv 1 \pmod{2p}$; hence $N \equiv 1 \pmod{2p}$.

(b) Show that $a^{2p} \equiv 1 \pmod{N}$, hence $a^{N-1} \equiv 1 \pmod{N}$, that is $a^N \equiv a \pmod{N}$ for $(a, N) = 1$, that is, N is psp(a).

12. Let $m > 0$ and suppose $6m + 1$, $12m + 1$, and $18m + 1$ are prime numbers. Show that $n = (6m + 1)(12m + 1)(18m + 1)$ is a Carmichael number. Use a computer to find at least five examples of Carmichael numbers of this type.

13. Suppose N is a product of distinct primes; $n = p_1 \cdots p_k$ and $p_i - 1 \mid n - 1$, for $1 \le i \le k$. Show that n is a Carmichael number. If we prove that the prime factorization of a Carmichael number consists of distinct factors, then the proof of the necessity in Proposition 6.1.10 will be complete. (See Exercise 7.2.17.)

14. Prove that all Carmichael numbers are odd.

15. Prove that a composite number is a Carmichael number if and only if $a^n \equiv a \pmod{n}$ for all a.

16. Write a program to verify if a number is Carmichael. Use this to determine the number of Carmichael numbers less than or equal to 10^6.

6.2 Strong Pseudoprimes and Probabilistic Primality Testing

An improvement in the pseudoprime test for identifying probable primes comes from the following observation. Suppose p is prime; then the equation $x^2 \equiv 1 \pmod{p}$ has two solutions, $x \equiv 1, -1 \pmod{p}$. The number of solutions to $x^2 \equiv 1 \pmod{n}$ is more than two for composite numbers. We want to exploit this fact to make the pseudoprime test reveal the compositeness of the number. Most of the results discussed here regarding strong pseudoprimes are due to Pomerance, Selfridge, and Wagstaff.[3]

We saw in the previous section that $a^{n-1} \equiv 1 \pmod{n}$, does not imply that n is prime. Suppose n is odd; then we can write $n - 1 = 2^r s$, with $r \ge 1$ and s odd. If n is prime, $\left(a^{(n-1)/2}\right)^2 \equiv 1 \pmod{n}$ implies that $a^{(n-1)/2} \equiv \pm 1 \pmod{n}$. If $a^{(n-1)/2} \equiv 1 \pmod{n}$ and $4 \mid n - 1$, then $a^{(n-1)/4} \equiv \pm 1 \pmod{n}$, and so on until $(n - 1)/2^r$ becomes odd. In each step, while solving $x^2 \equiv 1 \pmod{n}$, we should get 1 or -1 when n is prime. For composite n, we may obtain a number other than ± 1.

Since computing a square root is harder than computing a square, we start with a^s and compute $a^{2s}, a^{4s}, \ldots$, instead of computing the sequence

---

[3]C. Pomerance, J. L. Selfridge, and S. S. Wagstaff, Jr. " The Pseudoprimes to $25 \cdot 10^9$." *Mathematics of Computation*, 35(1980), 1003–1026.

$a^{n-1}, a^{(n-1)/2}, a^{(n-1)/4}, \ldots$. More precisely, computation of a^{n-1} is accomplished by the following sequence:

$$x_0 = a^s \bmod n, \quad x_1 = a^{2s} \bmod n, \quad x_2 = a^{4s} \bmod n,$$
$$\ldots, x_i = a^{2^i s} \bmod n, \ldots x_r = a^{2^r s} \bmod n.$$

Note that $x_r = a^{n-1} \bmod n$. Each term of the sequence $x_1, x_2, \ldots, x_r$ is computed by squaring the previous term in the sequence and taking the remainder modulo n.

For n prime, $x_r \equiv 1 \pmod{n}$ means $(x_{r-1})^2 \equiv x_r \equiv 1 \pmod{n}$, that is, $x_{r-1} \equiv \pm 1 \pmod{n}$. If $x_{r-1} \equiv 1 \pmod{n}$, then $x_{r-2} \equiv \pm 1 \pmod{n}$. Continuing in this manner, we have two possibilities: either $x_0 \equiv 1 \pmod{n}$ or there is an index i such that $x_i \equiv -1 \pmod{n}$. This means that for p prime, $\{x_0, \ldots, x_r\}$ can be of the form

$$\{1, 1, \ldots, 1\} \text{ or } \{*, \ldots, *, -1, 1, \ldots, 1\},$$

where $*$ represents some number different from 1 or -1, not important for our purposes.

If n is composite, the analysis fails, as $x^2 \equiv 1 \pmod{n}$ has more solutions than just ± 1. The sequence $\{x_0, \ldots, x_r\}$ can be of the form

$$\{*, \ldots, *, 1, \ldots, 1\} \text{ or } \{*, \ldots, *\} \text{ or } \{*, \ldots, *, -1, 1, \ldots, 1\}.$$

If we get a sequence of the type $\{*, \ldots, *, 1, \ldots, 1\}$ or $\{*, \ldots, *\}$, then n must be composite. (Why?)

Example 6.2.1. 1. We compute the sequence $x_0, x_1, \ldots, x_r$ with $a = 2$ for $n = 341$. (Recall that 341 is the smallest 2-pseudoprime.)

First, $340 = 4 \cdot 85 = 2^2 85$, so $r = 2$ and $s = 85$. Next,

$$x_0 \equiv 2^{85} \equiv 32 \qquad \pmod{341} \qquad (1)$$
$$x_1 \equiv x_0^2 \equiv 2^{170} \equiv 1 \quad \pmod{341} \qquad (2)$$
$$x_2 \equiv x_1^2 \equiv 1^2 \equiv 1 \quad \pmod{341}. \qquad (3)$$

The compositeness of 341 is revealed in (2) where $x_0^2 \equiv 1 \pmod{341}$, but $x_0 \equiv 32 \not\equiv \pm 1 \pmod{341}$.

In Example 3.4.11, we showed that if $x^2 \equiv 1 \pmod{n}$, but $x \not\equiv \pm 1 \pmod{n}$, then $(x - 1, n)$ and $(x + 1, n)$ are proper factors of n. Therefore, proper factors can be found from computing $(32 - 1, 341) = 31$ and $(32 + 1, 341) = 11$.

2. Consider 561, a Carmichael number, for which all pseudoprime tests fail to reveal its compositeness. First, factor $560 = 35 \cdot 2^4$, so $r = 4$ and $s = 35$. We compute the desired sequence with $a = 2$.

$$
\begin{aligned}
x_0 &\equiv 2^{35} &&\equiv 263 && (\text{mod } 561) \\
x_1 &\equiv x_0^2 &&\equiv 2^{70} \equiv 166 && (\text{mod } 561) \\
x_2 &\equiv x_1^2 &&\equiv 2^{140} \equiv 67 && (\text{mod } 561) \\
x_3 &\equiv 2^{280} &&\equiv 1 && (\text{mod } 561).
\end{aligned}
$$

The compositeness is revealed in step 4, where $67^2 \equiv 1 \pmod{561}$, but $67 \not\equiv \pm 1 \pmod{561}$.

3. Consider $n = 2047$. Take $a = 2$ and write $n - 1 = 2 \cdot 1023$. We can easily compute $x_0 \equiv 2^{1023} \equiv 1 \pmod{2047}$. But 2047 is not prime. ($2047 = 23 \cdot 89$.) This example shows that there are composite numbers masquerading as primes in the calculation of the sequence $x_0, x_1, \ldots, x_r$.

Definition 6.2.2. Suppose n is an integer and $n - 1 = 2^r s$. Then n is said to pass the **strong pseudoprime test to base** a if

1. $a^s \equiv 1 \pmod{n}$, or

2. $a^{s2^i} \equiv -1 \pmod{n}$ for some $0 \le i < r$.

Definition 6.2.3. An odd composite number n that passes the strong pseudoprime test to base a is called a **strong pseudoprime to base** a [or spsp(a)].

Given a base a, the pseudoprime test checks only the last component of the sequence $\{x_0, x_1, \ldots, x_r\}$, hence is a much weaker condition than the strong pseudoprime test. One should expect that there are fewer strong pseudoprimes to a given base. Nevertheless, there are infinitely many strong pseudoprimes.

Proposition 6.2.4. *If n is an odd pseudoprime to base 2, then $2^n - 1$ is a strong pseudoprime to base 2.*

PROOF. We proved in the previous section that $2^n - 1$ is a 2-pseudoprime. In particular, it is composite.

We have that $2^{n-1} \equiv 1 \pmod{n}$ as n is a 2-pseudoprime. Write $2^{n-1} - 1 = nk$, where k is necessarily odd. Let $n' = 2^n - 1$. We factor $n' - 1 = 2^n - 2 = 2(2^{n-1} - 1) = 2nk$; therefore, the odd part of $n' - 1$ is nk, and the sequence $\{x_0, \ldots, x_r\}$ has only two terms, 2^{nk} and 2^{2nk}.

Clearly, $2^n - 1 \equiv 0 \pmod{2^n - 1}$, or $2^n \equiv 1 \pmod{n'}$. This implies that $2^{nk} \equiv 1 \pmod{n'}$, satisfying condition (1) in the definition of spsp(2). This completes the proof, hence there are infinitely many strong pseudoprimes to base 2. ∎

The infinitude of spsp(a) for any base a was proven by Pomerance, Selfridge, and Wagstaff.

There are $14,884$ 2-pseudoprimes up to 10^{10}, but only 3291 strong pseudoprimes to base 2. The situation improves dramatically if we consider strong pseudoprimes to different bases. The smallest integer that is both spsp(2) and spsp(3) is $1,373,653 = 829 \cdot 1657$, whereas there are 66 integers less than or equal to 10^6 that are both psp(2) and psp(3). There are only 13 integers less than or equal to $25 \cdot 10^9$ that are strong pseudoprimes to bases 2, 3, and 5, while there are 2522 integers that are pseudoprimes to bases 2, 3, and 5. These strong pseudoprimes to bases 2, 3, and 5 are listed in Table 6.2.1 (along with the results of the pseudoprime tests to bases 7, 11, and 13). This leads to the following simple primality test for numbers up to $25 \cdot 10^9$.

Algorithm 6.2.5 (Simple primality test). Given $n \le 25 \cdot 10^9$, this algorithm determines if n is prime.

1. If n fails the strong pseudoprime test to base 2, then n is composite.

2. If n fails the strong pseudoprime test to base 3, then n is composite.

3. If n fails the strong pseudoprime test to base 5, then n is composite.

4. If n is one of the 13 numbers in Table 6.2.1, then n is composite; otherwise, n is prime.

In fact, from the table we see that no number less than or equal to $25 \cdot 10^9$ is a strong pseudoprime to bases $2, 3, 5, 7$, and 11; hence we can check for spsp(7) and spsp(11) and eliminate the table look-up. This gives a simple primality test that can even be performed on a pocket calculator.

Example 6.2.6. Let $n = 15790321$; then $n - 1$ has the factorization

$$2^4 \overbrace{986895}^{s}.$$

We compute the terms in the strong pseudoprime test with $a = 2$.

$$2^s \equiv 128 \pmod{n}$$
$$2^{2s} \equiv 16384 \pmod{n}$$
$$2^{4s} \equiv -1 \pmod{n}.$$

Table 6.2.1 Strong pseudoprimes to bases 2, 3, and 5 and the results of tests to bases 7, 11, and 13.

	base 7	base 11	base 13
25,326,001	no	no	no
161,304,001	no	spsp	no
960,946,321	no	no	no
1,157,839,381	no	no	no
3,215,031,751	spsp	psp	psp
3,697,278,427	no	no	no
5,764,643,587	no	no	spsp
6,770,862,367	no	no	no
14,386,156,093	psp	psp	psp
15,579,919,981	psp	spsp	no
18,459,366,157	no	no	no
19,887,974,881	psp	no	no
21,276,028,621	no	psp	spsp

This shows that n passes the strong pseudoprime test to base 2. Similarly, we verify that n passes the strong pseudoprime test to bases 3 and 5. Since n does not appear in the table, n is prime. This test requires approximately $O(\log_2 s)$ multiplications as opposed to $\sqrt{n}$ divisions for trial division.

For numbers less than 1,373,653 we only need to use the strong pseudoprime test for bases 2 and 3 and eliminate the table look-up step.

Example 6.2.7. Consider $n = 117371$. We factor $n - 1$ as $2 \cdot \overbrace{58685}^{s}$. Compute:

$$2^s \equiv -1 \pmod{n}$$
$$2^{2s} \equiv 1 \pmod{n}$$
$$3^s \equiv 1 \pmod{n}.$$

This implies that n is prime.

It is clear that strong pseudoprimes are much more useful than pseudoprimes to detect compositeness. There is another advantage of the strong pseudoprime tests: unlike pseudoprimes, there are no analogs of Carmichael numbers in this case. It was shown by Rabin[4] that no odd composite number n is a strong pseudoprime to more that $n/4$ bases a satisfying $(a, n) = 1$.

[4]M. O. Rabin. "Probabilistic Algorithms for Primality Testing." *Journal of Number Theory.* 12(1980), 128–128.

Suppose a number passes the strong pseudoprime test for some randomly selected integer $a < n$. If n is composite, the probability that we select an a that makes n pass the spsp test is less than $1/4$. Suppose we test this for more bases, assuming that each test is independent of the previous test; the probability that a composite n passes k tests is less than $(1/4)^k$. If we perform k successive tests for randomly chosen bases, the tests may not be independent, but it has been shown that for large n (about 10^{60}), it is indeed true that the probability that a composite number passes the strong pseudoprime test to k bases is less than $(1/4)^k$.

The proof of Rabin's Theorem that a composite number n is a strong pseudoprime to at most $n/4$ bases is too long to be presented here. But we can prove a weaker assertion.

Theorem 6.2.8. *A composite number n is a strong pseudoprime to at most $n/2$ bases.*

PROOF. Suppose there exists an integer a such that $a^{n-1} \not\equiv 1 \pmod{n}$; then $b^{n-1} \equiv 1 \pmod{n}$ implies that $(ab)^{n-1} \not\equiv 1 \pmod{n}$. Hence there are at least as many numbers x such that $x^{n-1} \not\equiv 1 \pmod{n}$ as there are numbers x such that $x^{n-1} \equiv 1 \pmod{n}$. Therefore, the strong pseudoprime test fails for at least $n/2$ bases.

If there is no a such that $a^{n-1} \not\equiv 1 \pmod{n}$, then n is Carmichael. We will show that a Carmichael number is squarefree. Suppose $p^r \mid n$ for some p and $r > 1$. Then, for all $(a, n) = 1$, $a^{n-1} \equiv 1 \pmod{p^r}$, and $a^{p^{r-1}(p-1)} \equiv 1 \pmod{p^r}$ implies that $a^d \equiv 1 \pmod{p^r}$, where $d = (n - 1, p^{r-1}(p - 1))$. Since $p \nmid n - 1$, d is a factor of $p - 1$. If $r > 1$, then it is impossible to have $a^{p-1} \equiv 1 \pmod{p^r}$ for all $(a, p) = 1$; to see this, take $a = p^{r-1} + 1$. Then by the Binomial Theorem,

$$(p^{r-1} + 1)^{p-1} \equiv 1 + (p - 1)(p^{r-1}) + \binom{p-1}{2} p^{2(r-1)} + \cdots \pmod{p^r},$$

where the terms other than the first two involve multiples of p^r; hence

$$(p^{r-1} + 1)^{p-1} \equiv 1 + p^r - p^{r-1} \pmod{p^r}$$
$$\equiv 1 - p^{r-1} \pmod{p^r}$$
$$\not\equiv 1 \pmod{p^r}.$$

This shows that n must be a product of distinct primes. When n is a product of two primes, we showed, in Theorem 5.3.7, that the number of bases to which n passes the strong pseudoprime test is at most $n/2$. The same proof applies when n has more than two distinct prime factors. ∎

Of course, if n fails a strong pseudoprime test, then it is composite. Suppose we check for 50 random bases; then the probability that a composite number successfully passes 50 random pseudoprime tests is less than $1/2^{50}$. In this case, the number is very likely to be prime. We call a number that passes this test a **probable prime**. Note that this does not prove that the number is prime. We will study how to provide a proof of primality in Section 7.4.

This discussion results in the following probabilistic algorithm for compositeness, or detecting probable primes.

Algorithm 6.2.9 (Rabin–Miller Probabilistic Primality Test). We use the strong pseudoprime test to several random bases to check if n is a probable prime. The test should only be applied after checking if the number has any small factors.

1. [Initialize] Let $s = n - 1, r = 0, k = 50$.

2. [Compute $n - 1 = 2^r s$] If $s \equiv 0 \pmod 2$, $s = s/2, r = r + 1$, and repeat step 2. Otherwise, go to step 3.

3. Choose a random integer $a < n$, and set $a = a^s \bmod n$ and $c = 0$. If $a = 1$, go to step 5; otherwise, go to step 4.

4. If $c > r - 2$ or $a = 1$, then n is composite. If $a = -1$, go to step 5. Otherwise, $a = a^2 \bmod n, c = c + 1$; repeat step 4.

5. Let $k = k - 1$. If $k > 0$, go to step 3; otherwise, report that n is a probable prime.

Example 6.2.10. Consider $n = 10^{16} - 10^8 + 1 = 9999999900000001$. We factor $n - 1 = 2^8 s$, with $s = 39062499609375$.

For any a, we have to compute the sequence $x_0 = a^s \bmod n, \ldots, x_8 = a^{n-1} \bmod n$. The following can be verified:

For $a = 2$, $x_2 \equiv -1 \pmod n$; hence n passes strong pseudoprime test to base 2.

For $a = 3$, $x_5 \equiv -1 \pmod n$; hence n passes the strong pseudoprime test to base 3.

For $a = 5$, $x_0 \equiv -1 \pmod n$; hence n passes the strong pseudoprime test to base 5.

One can similarly verify that n passes the strong pseudoprime test to numerous other bases, increasing the likelihood that n is prime.

The only way this test can prove the primality of a number is if we use the strong pseudoprime test to over $n/4$ bases, which is clearly impractical. There are composite numbers that will pass the strong pseudoprime test to an arbitrary number of bases.

There is another aspect that makes the strong pseudoprime test useful for testing primality. Miller[5] showed that by assuming a major unproved conjecture, the Generalized Riemann Hypothesis (GRH), the compositeness of a number can be revealed very quickly by strong pseudoprime tests. Miller showed that if n is composite, then there exists a, $1 < a < 2 (\log n)^2$ such that n is not a strong pseudoprime to base a.

If GRH is true, then we obtain a fast primality test that can be applied to very large integers. In practice, for composite n, we should be able to find a, $1 < a < 2 (\log n)^2$ such that n is not a strong pseudoprime to base a. For our purposes, given n, the lack of such an a only means that n is a probable prime since a proof of GRH has eluded the greatest mathematicians.

If one wants a deterministic primality test, a recent algorithm of Adleman, Pomerance, and Rumely is considered a breakthrough. Their idea is to use versions of the strong pseudoprimes in number systems obtained by adding complex roots of unity to the integers. A description of the test is beyond the techniques covered in this book.

Exercises for Section 6.2

1. Write a computer program that determines if a number is a strong pseudoprime to a given base a. Be careful to eliminate the genuine primes.

2. Verify that 2047 is the smallest $\text{spsp}(2)$, and 1373653 is the smallest $\text{spsp}(2)$ and $\text{spsp}(3)$.

3. Determine the number of $\text{spsp}(2)$'s that are less than 10^3, 10^4, 10^5, and 10^6. Compare with the corresponding number of $\text{psp}(2)$'s.

4. Determine the number of bases to which $n = 1729$ and $n = 294409$ are strong pseudoprimes.

5. Determine the smallest Carmichael number that is also a strong pseudoprime to base 2.

6. Implement the primality test described in Algorithm 6.2.5 for numbers up to $25 \cdot 10^9$.

7. Write a computer program to implement the algorithm for the probabilistic compositeness test.

8. Verify that 274177 divides $F_6 = 2^{64} + 1$. Check if $F_6/274177$ is likely to be a prime.

---

[5]G. L. Miller. "Riemann's Hypothesis and Tests for Primality." *Journal of Computer and System Sciences*, 13(1976), 300–317.

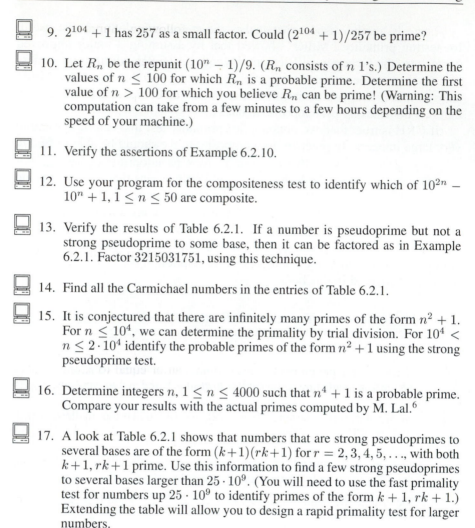

9. $2^{104} + 1$ has 257 as a small factor. Could $(2^{104} + 1)/257$ be prime?

10. Let R_n be the repunit $(10^n - 1)/9$. (R_n consists of n 1's.) Determine the values of $n \leq 100$ for which R_n is a probable prime. Determine the first value of $n > 100$ for which you believe R_n can be prime! (Warning: This computation can take from a few minutes to a few hours depending on the speed of your machine.)

11. Verify the assertions of Example 6.2.10.

12. Use your program for the compositeness test to identify which of $10^{2n} - 10^n + 1, 1 \leq n \leq 50$ are composite.

13. Verify the results of Table 6.2.1. If a number is pseudoprime but not a strong pseudoprime to some base, then it can be factored as in Example 6.2.1. Factor 3215031751, using this technique.

14. Find all the Carmichael numbers in the entries of Table 6.2.1.

15. It is conjectured that there are infinitely many primes of the form $n^2 + 1$. For $n \leq 10^4$, we can determine the primality by trial division. For $10^4 < n \leq 2 \cdot 10^4$ identify the probable primes of the form $n^2 + 1$ using the strong pseudoprime test.

16. Determine integers n, $1 \leq n \leq 4000$ such that $n^4 + 1$ is a probable prime. Compare your results with the actual primes computed by M. Lal.[6]

17. A look at Table 6.2.1 shows that numbers that are strong pseudoprimes to several bases are of the form $(k+1)(rk+1)$ for $r = 2, 3, 4, 5, \ldots$, with both $k+1, rk+1$ prime. Use this information to find a few strong pseudoprimes to several bases larger than $25 \cdot 10^9$. (You will need to use the fast primality test for numbers up $25 \cdot 10^9$ to identify primes of the form $k + 1, rk + 1$.) Extending the table will allow you to design a rapid primality test for larger numbers.

6.3 Pollard's $(p - 1)$-method

In the next two sections, we discuss two factorization methods developed by J. M. Pollard.

The $(p-1)$-method[7] seems to have been known to D. N. Lehmer but was never published. This is not a general purpose algorithm, that is, it will not work on all numbers; still its performance is impressive and give factors of some very large numbers.

---

[6]M. Lal. "Primes of the Form $n^4 + 1$." *Mathematics of Computation*, 21(1967), 245–247.

[7]J. M. Pollard. "Theorems on Factorization and Primality Testing." *Proc. Cambridge Phil. Soc.*, 76(1974), 521–528.

The idea behind the $(p-1)$-method is the following. Suppose n is the number to be factored, and say $p \mid n$, p prime. Now, $a^{p-1} \equiv 1 \pmod{p}$ for any $(a, p) = 1$. Suppose $p - 1$ divides a number M; then $a^M \equiv 1 \pmod{p}$, that is, $p \mid a^M - 1$. Since $p \mid n$ and $p \mid a^M - 1$, p will divide $(a^M - 1, n)$. Instead of computing $a^M - 1$, we can compute $a^M - 1 \bmod n$ and $(a^M - 1 \bmod n, n)$. If this GCD is not equal to n, then we would have found a nontrivial factor of n. This factor need not be p.

Example 6.3.1. Consider $n = 1073 = 29 \cdot 37$. If $p = 29$, $p - 1 = 28$. Let $a = 2$, then $2^{28} \equiv 900 \pmod{n}$ and $(900 - 1, n) = 29$. Similarly, $2^{36} \equiv 777 \pmod{n}$ and $(777 - 1, n) = 37$, the second factor.

Exercise. Let $n = 533 = 13 \cdot 41$. Compute $(2^{12} - 1 \bmod n, n)$ and $(2^{40} - 1 \bmod n, n)$ and verify that these give a factor of n.

In the previous example we know the prime factorization; hence we were able to choose the exponent M. If p is not known, then we have to choose M so that $p - 1 \mid M$. This will happen if the prime factors of $p - 1$ occur in that of M. The exponent M should be small enough so that we can compute a^M in a reasonable amount of time. Since we don't know the prime factorization of $p - 1$, M must include all the primes up to some upper bound B.

The simplest possibility is to take $M = k!$ with increasing values of k. If all the prime divisors of $p - 1$ are less than or equal to $k \leq B$, then $p - 1 \mid k!$, and we will have $p \mid (a^{k!} - 1 \bmod n)$ and $p \mid n$, and we might find a nontrivial factor using $(a^{k!} - 1 \bmod n, n)$.

Instead of computing $k!$, which has the small primes appearing to very large powers, we could also use $\mathrm{LCM}[1, 2, \ldots, k]$.

Example 6.3.2. Let $n = 2479$, we compute $2^{k!} \bmod n$ and the GCD $(2^{k!} - 1 \pmod{n}, n)$.

$$
\begin{array}{llll}
2^{1!} \equiv 2 & \pmod{n}, & (2 - 1, n) & = 1, \\
2^{2!} \equiv 4 & \pmod{n}, & (4 - 1, n) & = 1, \\
2^{3!} \equiv 64 & \pmod{n}, & (64 - 1, n) & = 1, \\
2^{4!} \equiv 1823 & \pmod{n}, & (1823 - 1, n) & = 1, \\
2^{5!} \equiv 618 & \pmod{n}, & (618 - 1, n) & = 1, \\
2^{6!} \equiv 223 & \pmod{n}, & (223, n) & = 37.
\end{array}
$$

The factor 37 is found after 6 steps. Note that $37 - 1 = 36 = 2^2 3^2$ and $36 \nmid 5!$, but $36 \mid 6!$, hence $p \mid 2^{6!} - 1$.

It may happen that $n \mid a^{k!} - 1$ before we find a nontrivial factor. In this case, one should try a different value of a. (Also, see the remarks at the end of this section.)

Since the GCD is 1 most of the time, it is not necessary to compute it in each step. Instead, we can accumulate the product of the numbers $a^{k!} - 1$ (mod n), $a^{(k+1)!} - 1$ (mod n), ... for a few values and then compute the GCD of n and this product.

Example 6.3.3. Let $n = 115943$; we compute $2^{1!}$ mod n, $2^{2!}$ mod n, and so on. Let $a_k \equiv 2^{k!}$ (mod n); then $a_{k+1} \equiv a_k^{k+1}$ (mod n). We accumulate the product $Q_1 = (a_1 - 1)(a_2 - 1) \cdots (a_{10} - 1)$ (mod n) and then compute (Q_1, n); then we accumulate the product $Q_2 = (a_{11}-1)(a_{12}-1)\cdots(a_{20}-1)$ (mod n) and compute (Q_2, n).

$$Q_1 \equiv 6645 \quad (\bmod\ n), \qquad (Q_1, n) = 1,$$
$$Q_2 \equiv 7988 \quad (\bmod\ n), \qquad (Q_2, n) = 1,$$
$$Q_3 \equiv 1692 \quad (\bmod\ n), \qquad (Q_3, n) = 1,$$
$$Q_4 \equiv 14453 \quad (\bmod\ n), \qquad (Q_4, n) = 149,$$

so 149 is a factor of n, and the other factor is 107. Notice that $149 - 1 = 14 \cdot 37$, which is why the factor appeared in Q_4, and $107 - 1 = 2 \cdot 53$, so this factor will appear after 149.

For larger numbers it may be better to check the GCD every 25 or 100 steps. It may happen that all the prime factors are found at once, that is, they all divide Q_k but not Q_{k-1}, in which case the GCD will go from being 1 to n; then one has to backtrack and compute the GCD in smaller intervals.

Example 6.3.4. Consider $n = 23489$ and let $a_k = 2^{k!}$ mod n and Q_k be as in the previous example. If we are checking the GCD every 10 steps, then we obtain:

$$Q_1 \equiv 21444 \quad (\bmod\ n), \qquad (Q_1, n) = 1,$$
$$Q_2 \equiv 12687 \quad (\bmod\ n), \qquad (Q_2, n) = 1,$$
$$Q_3 \equiv 1870 \quad (\bmod\ n), \qquad (Q_3, n) = 1,$$
$$Q_4 \equiv 1839 \quad (\bmod\ n), \qquad (Q_4, n) = 1,$$
$$Q_5 \equiv 0 \quad (\bmod\ n), \qquad (Q_5, n) = n.$$

Here, all the prime factors of n are such that $p - 1 \mid 50!$; hence they are all discovered at once. It is necessary to backtrack. Starting from $2^{40!}$, we compute successively $2^{41!}$ mod n, $2^{42!}$ mod n and so on, and we find that

$$2^{41!} \equiv 23074 \quad (\bmod\ n), \qquad (23074 - 1, n) = 83.$$

The other factor is 283, with $283 - 1 = 6 \cdot 47$.

Example 6.3.5. Let us consider a much larger example. Let $R_{17} = \frac{10^{17}-1}{9}$. It can be verified by a pseudoprime test that this number is composite. We compute $a_k = 2^{k!} \bmod n$ and evaluate the GCD every 100 steps. The Q's are computed as in the previous examples.

$$Q_1 \equiv 1857037302483324 \pmod{n}, \qquad (Q_1, n) = 1,$$
$$Q_2 \equiv 9411430890509582 \pmod{n}, \qquad (Q_2, n) = 1,$$
$$\vdots \qquad\qquad\qquad\qquad\qquad \vdots$$
$$Q_{11} \equiv 7398524334989123 \pmod{n}, \qquad (Q_{11}, n) = 2071723.$$

We find that 2071723 is a factor of n, and we can verify by the strong pseudoprime test given in the previous section that it is prime.

The quotient $R_{17}/2071723$ is 5363222357, which can also be verified by the same test to be prime. Note that $2071723 - 1 = 2 \cdot 3 \cdot 17 \cdot 19 \cdot 1069$, that is, $p - 1$ has all its factors less than 1100, which explains the success of this method.

Algorithm 6.3.6 (Pollard $p - 1$-method). Given n composite, this factorization algorithm computes $a^{k!} \bmod n$ successively for $k \leq B$, a prespecified bound. The GCD $(a^{k!} - 1 \bmod n, n)$ is computed every 25 steps by accumulating products.

1. [Initialize] Let $a = 2, m = 1, Q = 1, k \leftarrow 2$.

2. [Accumulate products] Let $a = a^k \bmod n$, $Q = Q(a - 1) \bmod n$, $k = k + 1, m = m + 1$. If $m = 25$, go to step 3; otherwise, repeat step 2.

3. [Compute GCD] Let $d = (Q, n)$; if $d \neq 1$ and $d \neq n$, return d as a factor of n; otherwise, go to step 4.

4. [Is it necessary to backtrack?] If $d = n$, then report that it is necessary to backtrack and compute the GCD more often. If $d = 1$ and $k < B$, $Q = 1, m = 1$, and go to step 2; otherwise, if $m \geq B$, terminate the algorithm, as n does not have a factor p with $p - 1$ consisting of small primes.

Remarks.   (1) Before factoring a number, one should apply the probabilistic compositeness test of the previous section.

(2) The algorithm should be tried only after removing the small factors of n by trial division.

(3) This is not a general purpose algorithm and will frequently fail to find any factor. But when it works, it is impressive and can find some very large factors.

(4) It seems reasonable to take $B \leq 10^6$ in the algorithm. The algorithm can be made more efficient by computing $a^{\text{LCM}[1,\dots,k]}$ instead of $a^{k!}$.

(5) It may happen in some rare cases that all the factors are discovered at once, that is, the GCD jumps from 1 to n in one step. Then a different value of a might reveal the factor. Examples of such numbers include 2047 and 536870911. The reader should find the factors of $p - 1$ for all prime factors of these numbers to see why this happens.

The ideas involved in the $(p-1)$-method have been used successfully in other methods, notably the elliptic curve factorization method. The elliptic curve factorization method is discussed in the last chapter.

──────────── **Exercises for Section 6.3** ────────────

1. Implement the Pollard $(p-1)$-method using Algorithm 6.3.6. Be sure to remove the small factors by a trial division step.

2. Modify your program to compute $a^{\text{LCM}[1,\dots,B]}$ instead of $a^{B!}$. For this you need to build a table of primes and prime powers less than or equal to B and use this to successively compute $a^{\text{LCM}(1,\dots,B)}$.

3. Obtain the probable factorization of the following integers using the $(p-1)$-method. For some large factors it may not be possible to prove primality, but we can check that they pass the strong pseudoprime test to several bases.

   (a) $\left(2^{62} - 2^{31} + 1\right)/3$
   (b) $10^{18} - 10^9 + 1$
   (c) 9999000099990001
   (d) $2^{81} - 2^{41} + 1$

4. Obtain the probable factorization of all repunits $R_n = \dfrac{10^n - 1}{9}$, $n \leq 50$.

5. Determine the probable prime factorization of the Mersenne numbers $2^p - 1$ using the Pollard $(p-1)$-method. These are known to be composite for $p \neq 2, 3, 5, 7, 13, 17, 19, 31, 61, 89$, $p \leq 100$. Why should you not use $a = 2$ in the implementation of Pollard's method for Mersenne numbers?

6.4   Pollard's ρ-method

The idea behind the ρ-method[8] of factoring an integer n is the following. Let $f(x)$ be a polynomial with integer coefficients. Starting with a number x_0, we

──────────────

[8]J. M. Pollard. "A Monte-Carlo Method for Factorization." *BIT*, 15(1975), 331–334.

construct a sequence of integers $x_1, x_2, x_3, \ldots$ by $x_{k+1} = f(x_k) \bmod n$. Let p be a prime divisor of n; we can consider the corresponding sequence $z_k = x_k \bmod p$, and the sequence z_k takes values in the set $\{0, 1, 2, \ldots, p-1\}$. If $p \leq \sqrt{n}$, then the number of possible values of z_k is much less than the number of possible values of the x_k; hence we can expect them to repeat much sooner.

Suppose $z_i = z_j$ for some $j > i$; then $x_i \equiv x_j \pmod{p}$, that is, $p \mid x_i - x_j$. Since $p \mid n$, we must have that $p \mid (x_i - x_j, n)$, and, if we are lucky, $(x_i - x_j, n)$ might be a nontrivial factor of n.

Let us look at an example to see how this method works.

Example 6.4.1. Let $n = 341 = 11 \cdot 31$. Take $f(x) = x^2 + 1$; we compute the sequence $x_i = f(x_{i-1}) \bmod 341$, starting with $x_0 = 3$:

x_0	x_1	x_2	x_3	x_4	x_5	x_6	x_7	x_8	x_9	x_{10}	$\cdots$
3	10	101	313	103	39	158	72	70	127	103	$\cdots$

We compute $z_i = x_i \bmod 11$ and obtain the sequence

z_0	z_1	z_2	z_3	z_4	z_5	z_6	z_7	z_8	z_9	z_{10}	$\cdots$
3	10	2	5	4	6	4	6	4	6	4	$\cdots$

The sequence of $\{x_i\}$ has $x_{10} = x_4$, that is, after the initial part $3, 10, 101, 313$, the terms $103, 39, 158, 72, 70, 127$ repeat. The sequence $\{z_i\}$ repeats earlier, so we have $z_6 = z_4$; this implies that $x_4 \equiv x_6 \pmod{11}$, so $103 - 158 = -55$ is divisible by 11 and $(x_4 - x_6, 3410) = 11$ is a factor of 341. If we compute $z_i' = x_i \bmod 31$, then

z_0'	z_1'	z_2'	z_3'	z_4'	z_5'	z_6'	$\cdots$
3	10	8	3	10	8		$\cdots$

Since $z_3' = z_0'$, we have that $31 \mid x_3 - x_0$ and $(x_3 - x_0, 341) = 31$.

Definition 6.4.2. Let $z_0, z_1, \ldots, z_i, \ldots$ be a sequence modulo n given by $z_i = f(z_{i-1})$ for some f. Then there exist i and j, $j > i$, such that $z_i = z_j$, hence the sequence is periodic. Let T be the smallest positive integer such that $z_{k+T} = z_k$ for some k. If k is the smallest such integer, then $z_0, \ldots, z_{k-1}$ is called the **preperiod** of $\{z_k\}$, and $z_k, \ldots, z_{k+T-1}$ is the **periodic part** and T is the **period**. The sequence $\{z_i\}$ is called an **eventually periodic** sequence.

Example 6.4.3. In Example 6.4.1, the preperiod of $\{x_i\}$ is $3, 10, 101, 313$; the periodic part is $103, 39, 158, 72, 70, 127$; and the period length is 6.

One can picture such a sequence as in Figure 6.4.1, which also explains the name of the method.

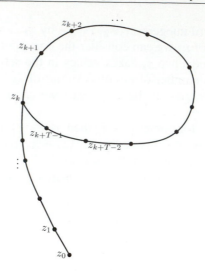

Figure 6.4.1 An eventually periodic sequence

For application to factorization, we cannot compute the sequence $\{z_i\}$ while computing the sequence $\{x_i\}$ because $p \mid n$ is not known. So we have to compute x_i and try to determine when z_i has a cycle. An additional difficulty is presented because it is not practical to keep track of all the x_i and compute all the GCD's $(x_i - x_j, n)$. The method that Pollard employed is a clever procedure to detect a cycle in a long list, called Floyd's cycle detecting algorithm. The method applies to any periodic sequence $\{x_i\}$ generated by a formula $x_{i+1} = f(x_i)$. Suppose $z_i = x_i \bmod p$. Since a term in the sequence depends only on the previous value, the period starts with the first repetition of a number. If $z_0, z_1, \ldots, z_{n_0}$ is the preperiod, then there is exactly one multiple of T in the interval $[n_0, n_0 + T]$. Let this multiple be $k = rT$ for some r. Then $z_k = z_{2k}$ because $2k - k$ is a multiple of T. Also, k is the smallest positive integer i such that $z_i = z_{2i}$.

In order to detect a cycle, we compute $x_k - x_{2k}$ for $k = 0, 1, 2, \ldots$. When k is some multiple of T, we will have $x_k - x_{2k}$ divisible by p (because $z_k = z_{2k}$). To find a factor, we compute $(x_k - x_{2k}, n)$ for successive values of k, and when $x_k \equiv x_{2k} \pmod{p}$, then $p \mid (x_k - x_{2k}, n)$.

Example 6.4.4. Let $n = 341$, $f(x) = x^2 + 1 \bmod n$, and $x_0 = 3$. Let $y_i = x_{2i}$. The values of x_i and y_i are:

i	0	1	2	3	4	5	6
x_i	3	10	101	313	103	39	158
y_i	3	101	103	158	70	103	158

Since $341 = 11 \cdot 31$, we reduce x_i and y_i modulo 31 to get

i	0	1	2	3	4	5	6
$x_i \bmod 31$	3	10	8	3	10	8	3
$y_i \bmod 31$	3	8	10	3	8	10	3

We see that $x_3 \equiv y_3 \pmod{31}$; hence the period of the sequence modulo 31 must be 3. We can find a factor by computing the greatest common divisor $(x_i - y_i, n)$. We see that the $(x_3 - y_3, 341) = 31$, a nontrivial factor of 341.

We have obtained a factor of 31 in three steps. Of course, to do this computation, we do not need to know in advance the factors 11 and 31.

To compute $y_i = x_{2i}$, do we need to compute $x_{i+1}, x_{i+2}, \ldots, x_{2i-1}$ until we get to x_{2i}? This would make the method impractical, but fortunately, there is a more efficient method to compute y_i. Consider the following:

$$y_1 = x_2 = f(x_1) = f(f(x_0)) = f(f(y_0)),$$
$$y_2 = x_4 = f(x_3) = f(f(x_2)) = f(f(y_1)),$$
$$\vdots$$
$$y_i = f(f(y_{i-1})).$$

In each step, we compute $x_i = f(x_{i-1}) \bmod n$ and $y_i = f(f(y_{i-1})) \bmod n$; that is, each step requires three evaluations of f. (Hence, it is beneficial to keep f as simple as possible.)

Example 6.4.5. Let $n = 15857$ and take $f(x) = x^2 + 1$, with initial value $x_0 = 2 = y_0$.

We tabulate the results for the sequences $\{x_i\}, \{y_i\}$, the differences $y_i - x_i$, and the GCD $(y_i - x_i, n)$.

i	x_i	y_i	$y_i - x_i$	$(y_i - x_i, n)$
0	2	2	0	-
1	5	26	21	1
2	26	14334	14308	1
3	677	5640	4963	1
4	14334	4541	-9793	1
5	5640	9408	3768	157

We have obtained a factor of 157 in six steps of the algorithm.

Exercise. In the previous example, what are the possible values of the period of the sequence $x_i \bmod 157$. Compute the sequence and determine the value of the period.

Exercise. Factor $n = 407$ using $f(x) = x^2 + 1$ and $x_0 = 3$. What happens if $x_0 = 2$?

Since the GCD is 1 most of the time, it is not necessary to compute it at each step. Instead, we accumulate the product $Q = \prod(y_i - x_i) \bmod n$ for approximately 20 or 25 steps and compute the GCD. If the GCD jumps to n in one step, then all the factors have been discovered at once, and it is necessary to backtrack and compute the GCD more often.

Algorithm 6.4.6 (Pollard's ρ-method). The algorithm uses $f(x) = x^2 + 1$ and $x_0 = 2$ in this implementation of Pollard's method. There is an upper limit of 10^6 iterations and the GCD is checked every 20 steps by accumulating the product.

1. [Initialize] Let $x = 2$, $y = 2$, $Q = 1$, $k = 1$, $c = 1$.

2. If $k > 10^6$, terminate algorithm; otherwise, $x = f(x) \bmod n$, $y = f(f(y)) \bmod n$, $k = k + 1$.

3. [Accumulate product] If $y = x$ terminate the algorithm; otherwise, $Q = Q \cdot (y - x)$ and $c = c + 1$. If $c = 20$, go to step 4; otherwise, go to step 2.

4. [Compute GCD] Let $d = (Q, n)$. If $d > 1$ go to step 5; otherwise, $Q = 1$, $c = 0$ and go to step 2.

5. If $d \neq n$, return d as a factor; otherwise, report that it is necessary to compute the GCD in shorter intervals.

Example 6.4.7. For $n = \left(11^{23} - 6^{23}\right)/5$, using $f(x) = x^2 + 1 \bmod n$ and $x_0 = 2$, the algorithm yields a factor $n_1 = 7345513129$ after 46,740 iterations. We can verify that this factor is prime. The cofactor $n_2 = n/n_1 = 24380310477967$ is a probable prime, passing the strong pseudoprime test to several bases. The primality of n_2 is proved in Example 7.4.10. A factorization of $n_1 - 1$ shows that it has a factor greater that $2 \cdot 10^5$; hence it would have required many more steps to obtain the same factorization by the $(p - 1)$-method.

Remarks.   1. Pollard discovered that a random sequence $\{x_i\} \bmod p$ is usually recurring after only $O(\sqrt{p})$ steps. No one has proved that the sequence produced by the polynomial $x^2 + 1$ is random or that any other simple function gives a random sequence of length about $\sqrt{p}$.

2. The method should only be used after determining if a number is composite by using the probabilistic primality test and eliminating the small prime factors by trial division.

3. The method is not guaranteed to succeed. The sequences modulo all prime factors may have very large periods, or in some rare circumstances, all the prime factors might have the same period, in which case, they are all detected at once. If this happens, then try a different polynomial or a different value of x_0.

──────────── **Exercises for Section 6.4** ────────────

1. Identify the preperiod and the period for the sequence

$$\{x_i = f(x_{i-1}) \bmod n\},$$

where x_0, f, and n are as follows.

   (a) $x_0 = 2$, $f(x) = x^2 + 1$, $n = 61 \cdot 73$.
   (b) $x_0 = 2$, $f(x) = x^2 - 1$, $n = 59 \cdot 67$.
   (c) $x_0 = 3$, $f(x) = x^2 - 1$, $n = 59 \cdot 67$.

2. Write a computer program to implement Pollard's ρ-method. Be sure to first eliminate small factors by trial division.

3. Factor the following integers. As in the previous section, the factorization is only probable, pending the proofs of the primality of the factors that pass the strong pseudoprime test to several bases.

   (a) $2^{58} + 1$.
   (b) $3^{53} + 2^{53}$.
   (c) $6^{29} - 5^{29}$.
   (d) $3^{61} - 2^{61}$.

Chapter 7

Primitive Roots

7.1 The Concept of Order

Further development of congruence arithmetic comes from the study of the concept of order, primitive roots, and index arithmetic. The ideas are fundamental and will be used for proofs of primality, solution of quadratic and higher congruences, random number generation, and numerous other applications.

Definition 7.1.1. Let a and n be integers such that $(a, n) = 1$. Then the **order of** a **modulo** n, denoted by $\mathrm{ord}_n(a)$ is the smallest positive integer k such that $a^k \equiv 1 \pmod{n}$.

Note that by Euler's Theorem, $a^{\phi(n)} \equiv 1 \pmod{n}$; hence a smallest positive integer k exists satisfying $a^k \equiv 1 \pmod{n}$. We have $\mathrm{ord}_n(a) \leq \phi(n)$ for all a satisfying $(a, n) = 1$.

Example 7.1.2. 1. The order of -1 is always 2 modulo any $n \neq 2$ because $(-1)^2 \equiv 1 \pmod{n}$ and $-1^1 \not\equiv 1 \pmod{n}$. Since $-1 \equiv 1 \pmod{2}$, $\mathrm{ord}_2(-1) = 1$.

2. The order of 2 modulo 31 is 5 because 5 is the smallest integer k such that $2^k \equiv 1 \pmod{31}$. This follows from the computations $2^5 \equiv 1 \pmod{31}$, $2^4 \equiv 16 \pmod{31}$, $2^3 \equiv 8 \pmod{31}$, $2^2 \equiv 4 \pmod{31}$, $2 \equiv 2 \pmod{31}$.

3. To determine $\mathrm{ord}_{31}(3)$, we compute 3^k for $1 \leq k \leq 30$ and see that the smallest integer k such that $3^k \equiv 1 \pmod{31}$ is 30.

4. Suppose $x^5 \equiv 1 \pmod{n}$ and $x \not\equiv 1 \pmod{n}$. What is the order of x modulo n? First, we cannot have $x^2 \equiv 1 \pmod{n}$; otherwise, $x^5 \equiv \left(x^2\right)^2 x \equiv x \not\equiv 1 \pmod{n}$, a contradiction. Similarly, $x^3 \equiv 1$

169

(mod n) and $x^4 \equiv 1$ (mod n) are impossible. Hence, $\mathrm{ord}_n(x) = 5$. A similar argument can be used to show that if p is prime and $x^p \equiv 1$ (mod n), then $\mathrm{ord}_n(x) = p$ unless $x \equiv 1$ (mod n).

5. From part 2, we have $2^5 \equiv 1$ (mod 31). What is the smallest integer $k > 5$ such that $2^k \equiv 1$ (mod 31)? We compute

$$
\begin{aligned}
2^6 &\equiv 2^5 2 &\equiv 2 &\not\equiv 1 &&(\text{mod } 31) \\
2^7 &\equiv 2^5 2^2 &\equiv 4 &\not\equiv 1 &&(\text{mod } 31) \\
2^8 &\equiv 2^5 2^3 &\equiv 8 &\not\equiv 1 &&(\text{mod } 31) \\
2^9 &\equiv 2^5 2^4 &\equiv 16 &\not\equiv 1 &&(\text{mod } 31) \\
2^{10} &\equiv 16 \cdot 2 \equiv 32 &\equiv 1 &&&(\text{mod } 31).
\end{aligned}
$$

Now, can you guess the smallest integer $k > 10$ such that $2^k \equiv 1$ (mod 31)?

Exercise. What is $\mathrm{ord}_{17}(3)$?

Exercise. What are all the integers k such that $2^k \equiv 1$ (mod 31)?

The last example reveals a fundamental property of orders.

Proposition 7.1.3. *Suppose $a^m \equiv 1$ (mod n); then $\mathrm{ord}_n(a) \mid m$.*

PROOF. Let $k = \mathrm{ord}_n(a)$. By the division algorithm, there exist a quotient q and a remainder r such that $m = kq + r$, $0 \le r < q$. If $\mathrm{ord}_n(a) \nmid m$, then $r \neq 0$.

Now, $a^m \equiv a^{kq+r} \equiv a^{kq}a^r \equiv a^r$ (mod n), that is, $a^r \equiv 1$ (mod n). Since $r < k$, this is not possible unless $r = 0$ (as k is the smallest positive integer satisfying this property); therefore, $r = 0$ and $\mathrm{ord}_n(a) \mid m$. ∎

Corollary 7.1.4. *Suppose $(a, n) = 1$ and $a^i \equiv a^j$ (mod n); then $i \equiv j$ (mod $\mathrm{ord}_n(a)$).*

PROOF. Since $(a, n) = 1$, a^j is invertible and its inverse is $(a^{-1})^j$. Therefore, $a^i a^{-j} \equiv a^j a^{-i} \equiv 1$ (mod n), that is, $a^{i-j} \equiv 1$ (mod n). This implies that $\mathrm{ord}_n(a) \mid i - j$, that is, $i - j \equiv 0$ (mod $\mathrm{ord}_n(a)$) or $i \equiv j$ (mod $\mathrm{ord}_n(a)$). ∎

Corollary 7.1.5. *If $(a, n) = 1$, then $\mathrm{ord}_n(a) \mid \phi(n)$. In particular, if p is prime and $(a, p) = 1$, then $\mathrm{ord}_p(a) \mid p - 1$.*

PROOF. This follows from Euler's Theorem and Proposition 7.1.3. ∎

Corollary 7.1.6. *If* $\operatorname{ord}_n(a) = k$, *then* $1, a, a^2, \ldots, a^{k-1}$ *are distinct modulo* n.

PROOF. This follows from Corollary 7.1.4, since $\{0, 1, 2, \ldots, k-1\}$ is a complete residue system modulo k. ∎

Example 7.1.7. The proposition simplifies the determination of the order. To find $\operatorname{ord}_{31}(3)$, we know that $\operatorname{ord}_{31}(3) \mid 30$, that is, $\operatorname{ord}_{31}(3)$ is one of 2, 3, 5, 6, 10, 15, or 30. We compute

$$3^2 \equiv 9 \quad (\bmod\ 31), \quad 3^3 \equiv 27 \quad (\bmod\ 31), \quad 3^5 \equiv 5 \quad (\bmod\ 31),$$
$$3^6 \equiv 15 \quad (\bmod\ 31), 3^{10} \equiv 8 \quad (\bmod\ 31), 3^{15} \equiv 16 \quad (\bmod\ 31).$$

Thus $\operatorname{ord}_{31}(3) = 30$.

From Corollary 7.1.6, the 30 elements $1, 3, 3^2, \ldots, 3^{29}$ are all distinct modulo 31. But there are only 30 nonzero elements modulo 31; hence

$$\{0, 1, 3, 3^2, \ldots, 3^{29}\}$$

is a complete residue system modulo 31.

Exercise. Is $\{0, 2, 2^2, \ldots, 2^{16}\}$ a complete residue system modulo 17?

Exercise. Find a number a so that the powers of a together with 0 can be used to construct a complete residue system modulo 23.

If we know the order of a, then we can determine the order of any power of a. The following identity is very useful is solving problems related to orders.

Lemma 7.1.8. *If* $(a, n) = 1$, *then* $\operatorname{ord}_n(a^k) = \dfrac{\operatorname{ord}_n(a)}{(k, \operatorname{ord}_n(a))}$.

PROOF. Let $x = \dfrac{\operatorname{ord}_n(a)}{(k, \operatorname{ord}_n(a))}$. On one hand,

$$\left(a^k\right)^x \equiv a^{kx} \equiv \left(a^{\operatorname{ord}_n(a)}\right)^{k/(k, \operatorname{ord}_n(a))} \equiv 1 \quad (\bmod\ n),$$

and this implies that $\operatorname{ord}_n(a^k) \mid x$. On the other hand, if $l = \operatorname{ord}_n(a^k)$, then $\left(a^k\right)^l \equiv a^{kl} \equiv 1 \pmod{n}$, so $\operatorname{ord}_n(a) \mid kl$ by Proposition 7.1.3. We can write $kl = \operatorname{ord}_n(a)c$ for some c. Dividing by the greatest common divisor of k and $\operatorname{ord}_n(a)$ on both sides, we see that $x \mid l$. Hence $\operatorname{ord}_n(a^k) = \dfrac{\operatorname{ord}_n(a)}{(k, \operatorname{ord}_n(a))}$. ∎

Example 7.1.9. Suppose $\operatorname{ord}_n(a) = 12$, then $\operatorname{ord}_n(a^8) = 12/(8,12) = 12/4 = 3$. Clearly, $\left(a^8\right)^2 \equiv a^{16} \equiv a^4 \not\equiv 1 \pmod{n}$, but $\left(a^8\right)^3 \equiv a^{24} \equiv \left(a^{12}\right)^2 \equiv 1 \pmod{n}$.

Example 7.1.10. Let $\operatorname{ord}_n(a) = 12$; then $1, a, a^2, \ldots, a^{11}$ are distinct modulo n. We want to know which of these have order 12. From Lemma 7.1.8, we must have $\operatorname{ord}_n(a^k) = \dfrac{12}{(k,12)} = 12$. This implies that $(k,12) = 1$. The values of k satisfying $(k,12) = 1$ are $k = 1, 2, 7, 11$, that is, a, a^5, a^7, a^{11} all have order 12.

Exercise. Suppose $\operatorname{ord}_n(a) = 28$. Identify the powers of a that have orders 4, 7, and 14. Can you guess which familiar function gives the number of powers of a with the same order as that of a?

An element with the maximal possible order, $\phi(n)$, has many nice properties. For example, any invertible element can be expressed as a power of a.

Definition 7.1.11. An integer a satisfying $\operatorname{ord}_n(a) = \phi(n)$ is called a **primitive root** modulo n.

Example 7.1.12.   1. Since $\operatorname{ord}_{31}(3) = 30$, 3 is a primitive root modulo 31. As $\operatorname{ord}_{31}(2) = 5$, 2 is not a primitive root modulo 31.

2. Since $a^2 \equiv 1 \pmod{8}$ for all a satisfying $(a,8) = 1$, there is no primitive root modulo 8 ($\phi(8) = 4$). Here, we have three elements of order 2.

Exercise. Is there a primitive root modulo 15? What about modulo 23?

Exercise. How many elements of order 2 exist modulo 16? Modulo 2^k, for $k > 4$?

The following two results show that for most composite numbers there is no primitive root.

Proposition 7.1.13. *There is no primitive root modulo 2^k for $k \geq 3$.*

PROOF. Suppose there exists an integer a that is a primitive root modulo 2^k. Since $\phi(2^k) = 2^{k-1}$, the order of a is 2^{k-1}. Thus the 2^{k-1} elements

$$\{a, a^2, a^3, \ldots, a^{2^{k-1}}\}$$

are distinct modulo 2^k and hence they must be all the invertible elements. This implies that there is only one element of order 2, because if $x = a^i$ has order 2, then Lemma 7.1.8 implies that $\operatorname{ord}_{2^k}(a^i) = 2^{k-1}/(i, 2^{k-1}) = 2$.

Hence $(i, 2^{k-1}) = 2^{k-2}$. The only possible value of $i < 2^{k-1}$ satisfying this condition is $i = 2^{k-2}$.

But by Exercise 3.4.7, for $k \geq 3$, there are four solutions to the equation $x^2 \equiv 1 \pmod{2^k}$, and so there are three elements of order 2. This contradiction proves the proposition. ∎

Proposition 7.1.14. *Suppose p is an odd prime. If $n \neq p^k$ or $2p^k$ for some k, or $n \neq 2, 4$, then there is no primitive root modulo n.*

PROOF. We determined the number of solutions of $x^2 \equiv 1 \pmod n$ in Exercise 3.4.8. Using that result, we can argue exactly as in the proof of the previous proposition. Here is another proof.

Suppose $n = rs$ with both r and s greater than 2 and $(r, s) = 1$. (The only integers not of this form are 2^k, p^b, and $2p^b$, for p an odd prime.) For such n, $\phi(n) = \phi(r)\phi(s)$; since $\phi(r)$ and $\phi(s)$ are even, by using Euler's Theorem, we obtain the following congruences.

$$a^{\frac{\phi(r)\phi(s)}{2}} \equiv \left(a^{\phi(r)}\right)^{\frac{\phi(s)}{2}} \equiv 1 \pmod r,$$

$$a^{\frac{\phi(r)\phi(s)}{2}} \equiv \left(a^{\phi(s)}\right)^{\frac{\phi(r)}{2}} \equiv 1 \pmod s.$$

Using the Chinese Remainder Theorem, we obtain

$$a^{\frac{\phi(r)\phi(s)}{2}} \equiv 1 \pmod n \quad \text{for all } (a, n) = 1$$

or $a^{\frac{\phi(n)}{2}} \equiv 1 \pmod n$. Hence there is no primitive root modulo n, because the maximum possible order is $\phi(n)/2$. This assertion, together with the previous proposition, completes the proof. ∎

Example 7.1.15. There is no element a such that $a^8 \equiv 1 \pmod{16}$; hence we must have $a^4 \equiv 1 \pmod{16}$ for all $(a, 2) = 1$ [because $a^k \equiv 1 \pmod{16}$ implies that $k \mid \phi(16) = 8$].

Since $x^2 \equiv 1 \pmod{16}$ has four solutions, there is one element of order 1, three of order 2, and four of order 4.

Suppose a is a primitive root modulo n, then all the invertible elements can be expressed as powers of a. The elements $a, a^2, a^3, \ldots, a^{\phi(n)-1}, a^{\phi(n)}$ are distinct. Writing all the invertible elements in this form simplifies congruence arithmetic by reducing multiplication to addition and exponentiation to multiplication (just like logarithms). We can use this to solve equations involving exponents.

Example 7.1.16. We have $3^4 \equiv 13 \pmod{17}$ and $3^8 \equiv -1 \pmod{17}$; therefore, 3 is a primitive root modulo 17. We tabulate the powers of 3 modulo 17.

k	0	1	2	3	4	5	6	7
$3^k \bmod n$	1	3	9	10	13	5	15	11
k	8	9	10	11	12	13	14	15
$3^k \bmod n$	16	14	8	7	4	12	2	6

Let us illustrate the applications of the exponential notation to multiplication and exponentiation.

Since $5 \equiv 3^5 \pmod{17}$ and $7 \equiv 3^{11} \pmod{17}$ then

$$5 \cdot 7 \equiv 3^5 3^{11} \equiv 3^{16} \equiv 1 \pmod{17}$$
$$5^3 \equiv \left(3^5\right)^3 \equiv 3^{15} \equiv 6 \pmod{17}.$$

Example 7.1.17. Find all the integers such that $7^x \equiv 4 \pmod{17}$.

From the previous example, we have $7 \equiv 3^{11} \pmod{17}$ and $4 \equiv 3^{12} \pmod{17}$; hence we have to solve $\left(3^{11}\right)^x \equiv 3^{12} \pmod{17}$, that is, $3^{11x} \equiv 3^{12} \pmod{17}$. From Corollary 7.1.4, this is equivalent to the congruence $11x \equiv 12 \pmod{16}$.

The inverse of 11 modulo 16 is 3, so

$$3 \cdot 11x \equiv 3 \cdot 12 \pmod{16}$$
$$x \equiv 4 \pmod{16}.$$

Then $x = 4, 20, \ldots$

(Compare with the solutions of the exponential equation $7^x = 4$ in the real number using logarithms, say, to base 3.)

Example 7.1.18. Solve $9^x \equiv 2 \pmod{17}$.

Since $9 \equiv 3^2 \pmod{17}$ and $2 \equiv 3^{14} \pmod{17}$, we have $3^{2x} \equiv 3^{14} \pmod{17}$ or $2x \equiv 14 \pmod{16}$, that is, $x \equiv 7 \pmod{8}$.

Exercise. What is the inverse of 13 modulo 17? [Use the fact that $3^4 \equiv 13 \pmod{17}$.]

─────────────── **Exercises for Section 7.1** ───────────────

1. Determine the order of the following elements.

   (a) 9 mod 17   (b) 11 mod 47   (c) 5 mod 31
   (d) 3 mod 76   (e) 11 mod 42   (f) 13 mod 54

2. Show that $2^n + 3^n$ is never divisible by 17.

3. For which of the following integers m is there a primitive root modulo m?

  (a) $m = 54$ (b) $m = 686$ (c) $m = 100$ (d) $m = 752$

4. If a has order k, how many of $1, a, a^2, \ldots, a^{k-1}$ have order k?

5. Determine a primitive root modulo each of the following integers, and use it to find all the primitive roots.

  (a) 19   (b) 54   (c) 37

6. Determine the number of elements of order 4 and order 8 mod 32. What is the smallest positive integer λ such that $a^\lambda \equiv 1 \pmod{32}$ for all $(a, 2) = 1$? Repeat the exercise for 2^k, $k > 5$.

7. Show that if m has a primitive root, then the equation $x^2 \equiv 1 \pmod{m}$ has only two solutions.

8. Suppose n has a primitive root g. For which values of a (in terms of the primitive root g) does the equation $x^2 \equiv a \pmod{n}$ have solutions?

9. If a is a primitive root modulo $m_1 m_2$, prove that a is a primitive root modulo both m_1 and m_2. Is the converse true?

10. Suppose there is an integer a such that $a^{\phi(n)/q} \not\equiv 1 \pmod{n}$ for any prime divisor q of $\phi(n)$. Show that a is a primitive root modulo n.

11. Suppose $\text{ord}_m(a) = k$ and $\text{ord}_n(a) = l$. If $(m, n) = 1$, show that the order of a modulo mn is $[k, l]$, the least common multiple of k and l.

12. (a) Suppose a has order k and b order l modulo n with $(k, l) = 1$. Show that $\text{ord}_n(ab) = kl$.

  (b) Investigate what happens if $(k, l) \neq 1$. What can you say about $\text{ord}_n(ab)$?

13. Suppose p is prime and $\text{ord}_p(a) = 3$. Show that $3 \mid p - 1$ and $1 + a + a^2 \equiv 0 \pmod{p}$. Conclude that $a + 1$ has order 6.

14. Let $F_k = 2^{2^k} + 1$ be the kth Fermat number and let $p \mid F_k$. What is $\text{ord}_p(2)$? What can we conclude about the prime divisors of F_k? Use this to simplify the search for prime factors of F_5.

15. Let $N = F_m F_{m-1} \cdots F_n$, with $m > n$. Recall (Exercise 2.5.15) that $(F_i, F_j) = 1$ for $i \neq j$. Use the previous exercise to determine the order of 2 modulo N. Show that $N - 1 = 2^{2^n} Q$ where Q is odd. Hence conclude that N is a 2-pseudoprime if and only if $m > 2^n$. (This gives another proof of the infinitude of 2-pseudoprimes.)

16. Show that if n is a pseudoprime to base a and $p \mid n$, then

$$n \equiv 1 \pmod{\text{ord}_p(a)},$$

and hence $n \equiv p \pmod{p\,\text{ord}_p(a)}$.

17. Solve the following equations.

    (a) $13^x \equiv 43 \pmod{54}$.

    (b) $7 \cdot 5^x \equiv 6 \pmod{19}$.

18. Make a table of the orders of the invertible elements modulo 2^k for $k = 2, 3, 4, \ldots$. What do you notice? Can you make a conjecture regarding the order of an odd integer a modulo 2^k?

7.2   The Primitive Root Theorem

In the previous section, we showed that there can be no primitive root modulo m unless $m = 2$, 4, p^k, or $2p^k$, for an odd prime p. The existence of the primitive root will be established in this section. The results in this section were first established by Gauss, who was led to the study of orders and primitive roots while studying the periods of decimal fractions. (See the projects at the end of this chapter.)

The existence of primitive roots for an odd prime can be proved by counting the elements of a given order. The idea is to use Lagrange's Theorem, which implies that $x^d \equiv 1 \pmod{p}$ has exactly d solutions when $d \mid p - 1$. We study a few special cases to see the general pattern.

Example 7.2.1.   1. Consider the elements of order 3 modulo p. If x has order 3, then $x^3 \equiv 1 \pmod{p}$. This equation has exactly three solutions (as $3 \mid p - 1$), of which one has order 1. The other two solutions must have order 3. (Why?)

2. Suppose $\text{ord}_p(x) = 4$. Then $4 \mid p - 1$, and $x^4 \equiv 1 \pmod{p}$ has four solutions. If $\text{ord}_p(a) = 2$, then $a^2 \equiv 1 \pmod{p}$ implies that $a^4 \equiv 1 \pmod{p}$. Among the four solutions of $x^4 \equiv 1 \pmod{p}$, one has order 1, one has order 2, and the other two must have order 4.

3. Consider the elements of $\mathbb{Z}/13\mathbb{Z}$. All nonzero elements satisfy $x^{12} - 1 \equiv 0 \pmod{13}$. The divisors of 12 are 1, 2, 3, 4, 6, and 12. From the previous examples, there are two elements of order 3, one of order 2, and two of order 4. Similarly, $x^6 - 1 \equiv 0 \pmod{p}$ has six solutions, of which only two have order 6. (Why?) We list the number of elements of various order modulo 12 in the following table.

k	1	2	3	4	6
number of elements of order k	1	1	2	2	2

This accounts for eight elements; hence the remaining four invertible elements must have order 12, that is, there are four primitive roots modulo 13.

Exercise. Determine the number of primitive roots modulo 11.

The reader may have noticed that in the example above, there are $\phi(k)$ elements of order k, hence $\phi(p-1)$ primitive roots. The existence of a primitive root along the lines of Example 3 can be given using the identity $\sum_{d|n} \phi(d) = n$. (See Exercise 5.)

We use a slightly different approach to show the existence of a primitive root, and then count the elements of various orders. This also gives another proof of the identity $\sum_{d|n} \phi(d) = n$.

Definition 7.2.2. An integer $\lambda > 0$ is called the **minimal universal exponent** if λ is the smallest integer such that $a^{\lambda} \equiv 1 \pmod{n}$ for all $(a, n) = 1$.

Compare this definition with the definition of order; what is the difference? We usually write $\lambda(n)$ for the minimal universal exponent modulo n to show the dependence on n. The function was introduced by R. D. Carmichael.

Example 7.2.3.   1. If there is a primitive root modulo n, then $\lambda(n) = \phi(n)$.

 2. If a has order k, then $k \mid \lambda(n)$, that is, all the orders divide $\lambda(n)$, so the least common multiple of all the orders divides $\lambda(n)$.

Example 7.2.4. We have $\lambda(2^k) < 2^{k-1}$ for $k \geq 3$ as there is no primitive root modulo 2^k; therefore, $\lambda(2^k) \mid 2^{k-2}$. Since the possible orders are 2^i, to find $\lambda(2^k)$ we must find an element of maximal order. Exercise 10 shows that the order of 3 is 2^{k-2}. This shows that $\lambda(2^k) = 2^{k-2}$ for $k \geq 3$.

The definition of $\lambda(n)$ does not require that there be an element of order $\lambda(n)$. We show the existence of an element of order $\lambda(n)$ by the following technique.

Proposition 7.2.5. *Suppose* $\mathrm{ord}_n(a) = k$ *and* $\mathrm{ord}_n(b) = l$; *then there exists an element g of order* $[k, l]$.

PROOF. Using the unique factorization for integers, we can write

$$k = xu, \quad l = yv \quad \text{with } (x, y) = 1 \text{ and } xy = [k, l].$$

(See Exercise 2.5.22.) Then a^u has order x and b^v has order y. (Why?) Since x and y are relatively prime, $a^u b^v$ has order xy, or if $g = a^u b^v$, then $\mathrm{ord}_n(g) = [k, l]$. ∎

Example 7.2.6. Take $n = 465$, $\text{ord}_{465}(2) = 20$, and $\text{ord}_{465}(7) = 30$. We construct an element of order $60(= [20, 30])$ using the procedure given above.

$$k = 20 = 2^2 5, \qquad l = 30 = 2 \cdot 3 \cdot 5.$$

Then $x = 2^2$, $u = 5$, $y = 15$, and $v = 2$; therefore, $g = 2^5 7^2 = 32 \cdot 49 = 173$ is an element of order 60.

Exercise. Let $n = 210$. You are given that $\text{ord}_n(11) = 6$ and $\text{ord}_n(83) = 4$. Construct an element of order 12.

Proposition 7.2.7. *Suppose M is the maximum possible order of all elements modulo n; then $M = \lambda(n)$, hence there exists an element of order $\lambda(n)$.*

PROOF. Suppose M is the maximum possible order; then $\text{ord}_n(x)$ divides M for all $(x, n) = 1$. Otherwise, $[M, \text{ord}_n(x)] > M$, and by Proposition 7.2.5, we can construct an element of order $[M, \text{ord}_n(x)]$, contradicting the maximality of M.

Since $\text{ord}_n(x) \mid M$ for all $(x, n) = 1$, then $x^M \equiv 1 \pmod{n}$ for all x, and this implies $\lambda \leq M$. On the other hand, $M \leq \lambda$, hence $\lambda = M$. ∎

Theorem 7.2.8. *Let p be a prime number; then there exists a primitive root modulo p.*

PROOF. Let λ be the minimal universal exponent of p. If $\lambda = p-1$, then there exists an element of order λ, hence a primitive root. Assume that $\lambda < p - 1$. Now, all $p - 1$ invertible elements satisfy the equation $x^\lambda \equiv 1 \pmod{p}$; contradicting Lagrange's Theorem that it has at most λ solutions. Hence $\lambda = p - 1$, that is, there exists a primitive root modulo p. ∎

Corollary 7.2.9. *Suppose p is prime and $d \mid p - 1$, $d > 0$. Then the number of elements of order d modulo p is $\phi(d)$.*

PROOF. Suppose g is a primitive root modulo p, then $\{g, g^2, \ldots, g^{p-1}\}$ are the invertible elements. We have $\text{ord}_p(g^i) = \dfrac{p-1}{(i, p-1)} = d$, then $(i, p-1) = (p-1)/d$, that is, $i = j(p-1)/d$. Suppose $p - 1 = dd'$; then, on one hand, $(i, p-1) = \big(j(p-1)/d, dd'\big) = (d'j, dd') = d'(j, d)$, and on the other hand $(i, p-1) = (p-1)/d = d'$. Then we must have $(j, d) = 1$ and also $j \leq d$ as $g^{j(p-1)/d} \equiv g^{j'(p-1)/d} \pmod{p}$ for $d'j \equiv d'j' \pmod{d'd}$, that is, $j \equiv j' \pmod{d}$. Then the distinct elements of order d are of the form $g^{j(p-1)/d}$, for $(j, d) = 1$, $1 \leq j \leq d$. This shows that there are $\phi(d)$ elements of order d. ∎

We can now prove the existence of primitive roots for $n = p^k$ and $n = 2p^k$, for p an odd prime. The reader should study the proof carefully while paying attention to the hypothesis. Where do we use the fact that p is an odd prime in the proof?

Theorem 7.2.10. *Let p be an odd prime.*

(a) *If g is a primitive root modulo p, then either g or $g + p$ is a primitive root modulo p^2.*

(b) *If g is a primitive root modulo p^2, then g is a primitive root modulo p^{k+1} for every $k \geq 2$.*

(c) *If g is odd and a primitive root modulo p^k for $k \geq 1$, then g is a primitive root modulo $2p^k$. Otherwise, if g is even, then $g + p^k$ is a primitive root modulo $2p^k$.*

PROOF.   (a) Since $\phi(p^2) = p(p-1)$, $\operatorname{ord}_{p^2}(g) \mid p(p-1)$. Let $k = \operatorname{ord}_{p^2}(g)$, then $g^k \equiv 1 \pmod{p^2}$ implies that $g^k \equiv 1 \pmod{p}$, so $p - 1 \mid k$; that is, $k = p - 1$ or $k = p(p-1)$. If $k = p(p-1)$, then g is a primitive root modulo p^2. If $k = p - 1$, then we claim that $g + p$ is a primitive root modulo p^2. Since $g + p \equiv g \pmod{p}$, then $\operatorname{ord}_{p^2}(g+p)$ is a multiple of $p - 1$, so it is $p - 1$ or $p(p-1)$. If $\operatorname{ord}_{p^2}(g+p) = p - 1$, then, $(g+p)^p \equiv (g+p)(g+p)^{p-1} \equiv g+p \pmod{p^2}$. By the Binomial Theorem, $(g + p)^p \equiv g^p + p^2 g^{p-1} \equiv g^p \pmod{p^2}$. We are assuming that $g^{p-1} \equiv 1 \pmod{p^2}$, which implies that $g^p \equiv g \pmod{p^2}$. We have shown that $(g + p)^p \equiv g + p \equiv g \pmod{p^2}$, which implies that $p \equiv 0 \pmod{p^2}$. This is a contradiction, hence $g + p$ is a primitive root modulo p^2.

(b) If g is a primitive root modulo p^2, then it is also a primitive root modulo p. Hence $g^{p-1} \equiv 1 \pmod{p}$ implies that $g^{p-1} = 1 + kp$ for some k. Now, $g^{p-1} \not\equiv 1 \pmod{p^2}$, so k is not a multiple of p. Using the Binomial Theorem, $g^{p^k(p-1)} \equiv 1 + kp \cdot p^k(p-1) \equiv 1 \pmod{p^{k+1}}$ and $g^{p^{k-1}(p-1)} \equiv 1 + kp \cdot p^{k-1}(p-1) \equiv 1 - kp^k \pmod{p^{k+1}}$. Therefore, $g^{p^{k-1}(p-1)} \not\equiv 1 \pmod{p^{k+1}}$. This implies that g is a primitive root modulo p^{k+1}.

(c) The proof is easy because $\phi(2p^k) = \phi(2)\phi(p^k) = \phi(p^k)$. We leave it as an exercise for the reader. ■

Example 7.2.11. It is not necessarily true that a primitive root for p is a primitive root for p^2. For example, 14 is a primitive root modulo 29, but 14 is not a primitive root modulo 29^2, because $\operatorname{ord}_{29^2}(14) = 28$. In this case $14 + 29 = 43$ is a primitive root modulo 29^2, and by the part (b) of the theorem, it is a primitive root modulo 29^k for any $k \geq 2$.

The existence of a primitive root shows that for an odd prime, $\lambda(p^i) = \phi(p^i) = p^{i-1}(p-1)$. We know that $\lambda(2) = 1$, $\lambda(4) = 2$, and $\lambda(2^k) = 2^{k-2}$ for $k \geq 2$. (See Exercise 10.) Using this, we can give a formula $\lambda(n)$.

Proposition 7.2.12. *Suppose* $n = 2^d p_1^{a_1} \cdots p_k^{a_k}$, *where* $p_1, \ldots, p_k$ *are distinct odd primes; then* $\lambda(n)$ *is the least common multiple*

$$\left[\lambda(2^d), \lambda(p_1^{a_1}), \ldots, \lambda(p_k^{a_k}) \right].$$

PROOF. Let $m = \text{LCM} \left[\lambda(2^d), \lambda(p_1^{a_1}), \ldots, \lambda(p_k^{a_k}) \right]$; then $x^m \equiv 1 \pmod{n}$ for all $(x, n) = 1$, as $x^m \equiv 1 \pmod{2^d}$, and $x^m \equiv 1 \pmod{p_i^{a_i}}$ for all $1 \leq i \leq k$; therefore, $\lambda \leq m$. On the other hand, let $x \equiv 3 \pmod{2^d}$ and $x \equiv g_i \pmod{p_i^{a_i}}$, where g_i is a primitive root modulo $p_i^{a_i}$. Then x has order m. Hence $m \leq \lambda$, that is, $\lambda = m$ and x has the maximal order. Here, we make use of Exercise 10. ■

Example 7.2.13. Let $n = 100 = 2^2 5^2$; then $\lambda(n) = \text{LCM}(\lambda(4), \lambda(25)) = LCM(2, 20) = 20$. Equivalently, $x^{20} \equiv 1 \pmod{100}$ for all $(x, 100) = 1$. [Euler's Theorem gives $x^{40} \equiv 1 \pmod{100}$ for all $(x, 100) = 1$.]

To find an element of order 20, we compute that 2 is a primitive root modulo 25. Take $x \equiv 3 \pmod{4}$ and $x \equiv 2 \pmod{25}$; which gives $x \equiv 27 \pmod{100}$ as an element of order 20.

A primitive root modulo an odd prime can be found quickly using the following criterion. For prime powers, we can use the theorem above to determine a primitive root using the primitive root modulo the corresponding prime.

Proposition 7.2.14. *Let* p *be an odd prime; suppose there exists an* a *such that* $a^{(p-1)/q} \not\equiv 1 \pmod{p}$ *for all prime divisors* q *of* $p - 1$; *then* a *is a primitive root modulo* p.

The proof is quite easy and is given in the next section. Notice that we need the prime factorization of $p - 1$.

Algorithm 7.2.15 (Primitive root modulo p). We write $p - 1 = q_1^{a_1} \cdots q_k^{a_k}$.

1. [Initialize] Let $a = 2$, $i = 1$.

2. If $i > k$, a primitive root has been found and return a; otherwise, go to step 3.

3. If $a^{(p-1)/q_i} \not\equiv 1 \pmod{p}$, let $i = i + 1$ and go to step 2; otherwise, let $i = 1$, $a = a + 1$ and repeat step 3.

If the prime factorization of $p - 1$ is not available, one has to search for a primitive root by starting with $2, 3, \ldots$, and computing the orders. The algorithm can be improved by Proposition 7.2.5. When two elements of small orders are found, an element of a larger order can be constructed. In practice, primitive roots tend to be small integers and can be quickly found. It should be noted that there exist primes with arbitrarily large smallest primitive roots.

———————————— **Exercises for Section 7.2** ————————————

1. Determine the primitive roots modulo the following integers.

   (a) 17   (b) 27   (c) 25   (d) 98

2. Show that 18 is a primitive root modulo 37. Is it a primitive root modulo 37^2? Determine primitive root modulo 37^k for every $k \geq 2$.

3. Suppose there exists a primitive root modulo n; show that there are $\phi(\phi(n))$ primitive roots.

4. Write a program to find a primitive root modulo a prime using Algorithm 7.2.15.

5. This exercise gives another proof of the existence of a primitive root modulo a prime p. We use the identity $\sum_{\delta \mid d} \phi(\delta) = d$. (See Exercise 4.2.14.) Suppose $f(d)$ is the number of elements of order d for $d \mid p - 1$. Show by induction that $f(d) = \phi(d)$.

6. Suppose $p \nmid a$. Show that the equation $x^2 \equiv a \pmod{p}$ has a solution if and only if $a^{(p-1)/2} \equiv 1 \pmod{p}$.

7. Suppose g is a primitive root of a prime p, $p \equiv 1 \pmod 4$; then show that $-g$ is also a primitive root. Is the assertion also true for primes p, $p \equiv 3 \pmod 4$? Prove or disprove.

8. Suppose $p \equiv 1 \pmod 4$ and $q = 2p + 1$ are both primes. If $2^p \equiv -1 \pmod q$, show that 2 is a primitive root modulo p.

9. Let p be an odd prime. Prove that there are exactly $\phi(p-1)$ primitive roots of p that are incongruent modulo p^2 and that are not primitive roots of p^2.

10. (a) Show that for g odd, $g^{2^{k-2}} \equiv 1 \pmod{2^k}$ for $k \geq 3$.

    (b) Show by induction that $3^{2^{k-3}} \not\equiv 1 \pmod{2^k}$ for $k \geq 3$.

11. Compute $\lambda(n)$ for the following numbers.

    (a) 800   (b) 2268   (c) 5733

12. Suppose that g is a primitive root modulo p^3, where p is an odd prime. Is it true that $g^{(p-1)/2} \equiv -1 \pmod p$? Explain.

13. Find the least positive residue of all solutions to $x^3 \equiv 1 \pmod{37}$.

14. What is the largest integer m such that $a^{12} \equiv 1 \pmod m$ for all $(a, m) = 1$?

15. Suppose N is the product of k distinct odd primes. Show that $\dfrac{\lambda(n)}{\phi(n)} \leq \dfrac{1}{2^{k-1}}$.

16. Suppose p is an odd prime, $p \equiv 1 \pmod 4$. Show that there exists an element x of order 4. What is $x^2 \bmod p$? Does $x^2 \equiv -1 \pmod p$ have a solution for $p \equiv 3 \pmod 4$?

17. Prove that if n is a Carmichael number and $p \mid n$, then $p - 1 \mid n - 1$. (This is the converse of Exercise 6.1.13.) Conclude that a Carmichael number is squarefree. This, together with Exercise 6.1.13, completes the proof of Proposition 6.1.10.

18. Incorporate the result of Proposition 7.2.5 to write a function that will find an element of maximal order for composite moduli.

7.3   The Discrete Logarithm

Suppose g is a primitive root modulo n; then $\{1, g, \ldots, g^{\phi(n)-1}\}$ are all distinct modulo n; hence they are all the invertible elements modulo n. Therefore, for any integer y such that $(y, n) = 1$, there exists an x with $0 \leq x \leq \phi(n) - 1$ such that $g^x \equiv y \pmod n$. Our goal in this section is to study the exponent x.

Definition 7.3.1. Suppose g is a primitive root modulo n. If $g^x \equiv y \pmod n$, then the **discrete logarithm** or **index** of y (to the base g) is

$$\text{ind}_g (y) = x \bmod \phi(n).$$

Example 7.3.2. First, 3 is a primitive root modulo 17. We have $3^8 \equiv -1 \pmod{17}$ and $3^{12} \equiv 4 \pmod{17}$, so $\text{ind}_3 (-1) = 8$ and $\text{ind}_3 (4) = 12$. Another primitive root modulo 17 is 7, and $7^4 \equiv 4 \pmod{17}$, so $\text{ind}_7 (4) = 4$.

The discrete logarithm has many of the same properties as the standard logarithm. Note that $\text{ind}_g (y)$ depends on the primitive root g and the modulus n. The modulus n is not included in the notation, as it is usually clear.

Proposition 7.3.3. *Let g be a primitive root modulo n. Then*

(a) $\text{ind}_g (1) = 0.$

(b) $\text{ind}_g (g) = 1.$

(c) $\text{ind}_g (y_1 y_2) = [\text{ind}_g (y_1) + \text{ind}_g (y_2)] \bmod \phi(n).$

(d) $\text{ind}_g (y^a) = a\,\text{ind}_g (y) \bmod \phi(n).$

(e) $a \equiv b \pmod{n}$ *if and only if* $\text{ind}_g(a) = \text{ind}_g(b)$.

PROOF.   (a) Since $g^0 \equiv 1 \pmod{n}$, we have $\text{ind}_g(1) = 0$.

(b) It is clear from the definition.

(c) Let $x_1 = \text{ind}_g(y_1)$ and $x_2 = \text{ind}_g(y_2)$. Then $g^{x_1} \equiv y_1 \pmod{n}$ and $g^{x_2} \equiv y_2 \pmod{n}$ and $g^{x_1+x_2} \equiv g^{x_1}g^{x_2} \equiv y_1y_2 \pmod{n}$. This shows that

$$\text{ind}_g(y_1y_2) = (x_1 + x_2) \bmod \phi(n)$$
$$= [\text{ind}_g(y_1) + \text{ind}_g(y_2)] \bmod \phi(n).$$

(d) If $g^x \equiv y \pmod{n}$, then $y^a \equiv g^{ax} \pmod{n}$. Hence

$$\text{ind}_g(y^a) = ax \bmod \phi(n)$$
$$= a\,\text{ind}_g(y) \bmod \phi(n).$$

(e) If $\text{ind}_g(a) = \text{ind}_g(b)$, then $g^{\text{ind}_g(a)} \equiv g^{\text{ind}_g(b)} \pmod{n}$, hence $a \equiv b$ $\pmod{n}$. Conversely, if $a \equiv b \pmod{n}$, then $g^{\text{ind}_g(a)} \equiv g^{\text{ind}_g(b)}$ $\pmod{n}$. This implies that $\text{ind}_g(a) - \text{ind}_g(b)$ is a multiple of $\phi(n)$. Since both numbers are smaller than $\phi(n)$, they must be equal.   ■

Example 7.3.4. We can use the properties of the discrete logarithm to solve equations. Let us solve $7^x \equiv 4 \pmod{17}$ using the discrete logarithm. Since 3 is a primitive root modulo 17, we can take the index to the base 3

$$\text{ind}_3(7^x) = \text{ind}_3(4)$$
$$x\,\text{ind}_3(7) \equiv \text{ind}_3(4) \quad (\bmod\ \phi(17)).$$

Since $\text{ind}_3(7) = 11$ and $\text{ind}_3(4) = 12$, solving the equation is equivalent to solving

$$11x \equiv 12 \pmod{16}.$$

Hence the solution is $x \equiv 4 \pmod{16}$.

Example 7.3.5. Solve $8k^5 \equiv 3 \pmod{17}$.
   Again we can use 3 as the primitive root. We have $\text{ind}_3(8) = 10$. Then

$$\text{ind}_3(8k^5) = \text{ind}_3(3)$$
$$[\text{ind}_3(8) + 5\,\text{ind}_3(k)] \bmod 16 \quad = 1$$
$$[10 + 5\,\text{ind}_3(k)] \bmod 16 \quad = 1$$
$$5\,\text{ind}_3(k) \bmod 16 \quad = -9.$$

The inverse of 5 modulo 16 is -3, and we multiply both sides by this to obtain $\text{ind}_3(k) = 11$ and $k \equiv 3^{11} \equiv -7 \pmod{17}$.

Most of the time we will be interested in the discrete logarithm modulo prime numbers. From now on p will denote a prime.

One way to compute the discrete logarithm is to tabulate $g^x \bmod p$ for $1 \leq x \leq p - 1$. We can arrange the table in increasing values of $g^x \bmod p$ so that x can be easily found. For example, consider the following table in which $p = 11$ and $g = 2$:

x	0	1	2	3	4	5	6	7	8	9	10
$g^x \bmod p$	1	2	4	8	5	10	9	7	3	6	1

This table can be rewritten as:

y	1	2	3	4	5	6	7	8	9	10
$\mathrm{ind}_g (y)$	0	1	8	2	4	9	7	3	6	5

The number of steps in the computation is proportional to p and so are the storage requirements. This is not practical for large p. The following algorithm, due to Shanks, reduces the time and storage to the order of $\sqrt{p} \log p$, increasing the range of p for which it is possible to compute the discrete logarithm.

Suppose $\mathrm{ind}_g (y) = x$; then, for any m, we can write $x = mq + r$, where $q = \lfloor x/m \rfloor \leq \lfloor (p-1)/m \rfloor$ and $r = x \bmod m$. Then $g^{mq+r} \equiv y \pmod{p}$, and this implies

$$g^{mq} \equiv y g^{-r} \pmod{p}. \tag{1}$$

Shanks's method consists of making a table of values of g^{mq} for $1 \leq q \leq (p-1)/m$ and the values $y g^{-r}$ for $0 \leq r \leq m - 1$, sorting the tables, and finding the values of q and r that satisfy (1). The two tables are of equal size when $m = \lceil \sqrt{p-1} \rceil$. For this value of m, the first table consists of the values $g^{mq} \bmod p$, $0 \leq q \leq m - 1$, and the second table consists of the values $y g^{-r} \bmod p$, $0 \leq r \leq m - 1$.

Example 7.3.6. Let $p = 37$. Since $2^{36} \equiv 1 \pmod{37}$, $2^{18} \equiv -1 \pmod{37}$, and $2^{12} \equiv 26 \pmod{37}$, we see that 2 is a primitive root modulo 37. Let us solve $2^x \equiv 22 \pmod{37}$ using Shanks's algorithm.

1. Let $m = \lceil \sqrt{37 - 1} \rceil = 6$.

2. Tabulate $2^{6q} \bmod 37$ for $0 \leq q \leq 5$ and arrange it in increasing order of $2^{6q} \bmod 36$.

$2^{6q} \bmod 37$	1	10	11	26	27	36
q	0	4	5	2	1	3

3. Tabulate $22 \cdot 2^{-r} \bmod 37$ for $0 \leq r \leq 5$. Note that $2^{-1} \equiv 19$ (mod 37), so we have to compute $22 \cdot 19^r \bmod 37$ and arrange the values in increasing order.

$22 \cdot 19^r \bmod 37$	3	6	11	12	22	26
r	5	4	1	3	0	22

We see that 11 occurs in the first row of both tables, with $q = 5$ and $r = 1$. Therefore, $x = 5 \cdot 6 + 1 = 31$.

Recall that an exponential $a^k \bmod p$, $1 \leq k \leq p - 1$ can be computed in approximately $\log p$ multiplications. Each of the tables can then be constructed in $\sqrt{p} \log p$ steps. While simple sorting methods will sort a list of k elements in k^2 steps, we can do much better by using a sorting method like quicksort or heapsort. These methods require $\sqrt{p} \log p$ steps. Therefore, the total time complexity of Shanks's method is approximately $O(\sqrt{p} \log p)$, and the storage complexity is $O(\sqrt{p})$.

——————————— **Exercises for Section 7.3** ———————————

1. Suppose p is prime and g is a primitive root modulo p. Determine the index of -1.

2. Solve the following equations:

    (a) $2^x \equiv 35 \pmod{37}$
    (b) $59^x \equiv 63 \pmod{71}$

3. Suppose g and h are primitive roots modulo n. Show that

$$\mathrm{ind}_h (y) = \mathrm{ind}_h (g) \, \mathrm{ind}_g (y) \bmod \phi(n).$$

4. Implement Shanks's Algorithm for the computation of the discrete logarithm.

5. Solve the following equations:

    (a) $3^x \equiv 141 \pmod{331}$
    (b) $2^x \equiv 318 \pmod{509}$
    (c) $7^x \equiv 30 \pmod{919}$

6. Explain how to modify Shanks's method to find the order of an integer q modulo p.

7. Solve $9k^7 \equiv 61 \pmod{67}$.

7.4 Primality Testing

> *To tell if a given number of 15 to 20 digits*
> *is prime or not, all time would not suffice for the test,*
> *whatever use is made of what is already known.*

> — M. MERSENNE, 1644

The probabilistic algorithms to determine compositeness described in Chapter 6 are quite sufficient to construct large primes. How does one establish rigorously that a number is prime? A demonstration of primality should be short and easy to verify. This rules out a demonstration by trial division to show that there are no factors, because even if we were able to carry out the trial division in the first place, we would have to do the computation all over again to verify its correctness. The techniques described in this section give simple proofs of primality, which are easy to verify.

The idea of the proof is to use the fact that $\phi(n) = n - 1$ if and only if n is prime. Of course, we know of no method to compute $\phi(n)$ without knowing the factorization of n, so we must find another way to show that $\phi(n) = n - 1$. If we can find an integer a whose order modulo n is $n - 1$, then $n - 1 \mid \phi(n)$ because $\operatorname{ord}_n(a) \mid \phi(n)$. On the other hand, $\phi(n) \leq n - 1$, hence $\phi(n) = n - 1$. This is the basis of the following test devised by Lucas in 1876 and first published by D. H. Lehmer in 1927.[1]

Theorem 7.4.1 (Lucas–Lehmer). *Suppose there exists an integer a such that* $a^{n-1} \equiv 1 \pmod{n}$*, but for each prime q dividing $n - 1$, $a^{(n-1)/q} \not\equiv 1$* *(mod n), then n is prime.*

PROOF. We wish to demonstrate that $\operatorname{ord}_n(a) = n - 1$.

The congruence $a^{n-1} \equiv 1 \pmod{n}$ implies that $\operatorname{ord}_n(a) \mid n - 1$. Let $n - 1 = \operatorname{ord}_n(a)k$ for some k. We want to show that $k = 1$, so suppose $k > 1$ and a prime q divides k. Then $q \mid n - 1$, and we can write

$$a^{(n-1)/q} \equiv a^{\operatorname{ord}_n(a)k/q} \equiv 1 \pmod{n}.$$

This contradicts the hypothesis of the theorem, so $k = 1$ and $\operatorname{ord}_n(a) = n-1$. As $\operatorname{ord}_n(a) \mid \phi(n)$, we must have $\phi(n) \geq n-1$, but $\phi(n) \leq n-1$; therefore, $\phi(n) = n - 1$ and n is prime. ∎

---

[1] D. H. Lehmer. "Tests for Primality by the Converse of Fermat's Theorem." *Bulletin of the American Mathematical Society,* 33(1927), 327–340.

Example 7.4.2. Consider $n = 29$; $n - 1$ has the prime factorization $28 = 2^2 7$. Take $a = 2$ and compute the quantities in the theorem.

$$2^{28} \equiv 1 \quad (\bmod\ 29)$$

$$2^{28/2} \equiv 28 \quad (\bmod\ 29)$$

$$2^{28/7} \equiv 16 \quad (\bmod\ 29).$$

The hypotheses of Theorem 7.4.1 are satisfied with $a = 2$, so 29 is prime.

Example 7.4.3. Let $n = 2071723$, a number that appeared in Example 6.3.4 as a factor of the repunit R_{17}. We factor (by trial division):

$$n - 1 = 2 \cdot 3 \cdot 17 \cdot 19 \cdot 1069,$$

and taking $a = 2$ in the theorem, we have:

$$2^{n-1} \equiv 1 \qquad (\bmod\ n)$$

$$2^{(n-1)/2} \equiv -1 \qquad (\bmod\ n)$$

$$2^{(n-1)/3} \equiv 321129 \quad (\bmod\ n)$$

$$2^{(n-1)/17} \equiv 100000 \quad (\bmod\ n)$$

$$2^{(n-1)/19} \equiv 71064 \quad (\bmod\ n)$$

$$2^{(n-1)/1069} \equiv 1573595 \quad (\bmod\ n).$$

Again, all the conditions are satisfied, so n is prime.

Exercise. Verify that 31 is prime using the Lucas–Lehmer test.

It is important to verify that $a^{n-1} \equiv 1 \pmod{n}$. Without this condition, the rest of the proof will not work.

This integer a satisfying the condition of the theorem is a primitive root. In most cases, a primitive root can be found quite quickly by trying consecutive integers, $2, 3, 4 \ldots$. In some cases the smallest primitive root tends to be very large, so finding a number a to satisfy the requirements of the theorem may take a long time. The following improvement in the conditions of the theorem shows that it is not necessary to find a primitive root to prove that $\phi(n) = n - 1$.

If a number a doesn't satisfy the congruence condition of the Lucas–Lehmer test, then there exists q such that $a^{(n-1)/q} \equiv 1 \pmod{n}$. It is possible for many other primes q' that $a^{(n-1)/q'} \not\equiv 1 \pmod{n}$. We would like to use this information instead of discarding it and moving to a new base a. The following theorem shows how to use this information to prove that $\phi(n) = n - 1$. The improvement is possible because we don't need to find an element of order $n - 1$, but only need to show that it exists (using Proposition 7.2.5).

Theorem 7.4.4. *Suppose $n - 1$ has the prime factorization $q_1^{e_1} q_2^{e_2} \cdots q_k^{e_k}$. Suppose that for each i, $1 \leq i \leq k$, there exists an integer a_i such that $a_i^{n-1} \equiv 1 \pmod{n}$ and $a_i^{(n-1)/q_i} \not\equiv 1 \pmod{n}$; then n is prime.*

PROOF. We will show that each $q_i^{e_i}$ divides $\phi(n)$, and hence the product divides $\phi(n)$.

Fix an index i; if a_i satisfies $a_i^{n-1} \equiv 1 \pmod{n}$, then $\text{ord}_n(a_i) \mid n - 1$. Let $n - 1 = \text{ord}_n(a_i)k$; we claim that $q_i \nmid k$. Otherwise,

$$a_i^{(n-1)/q_i} \equiv a_i^{\text{ord}_n(a_i)k/q_i} \equiv 1 \pmod{n},$$

a contradiction. Since $(q_i, k) = 1$, and $q_i^{e_i}$ occurs in the prime factorization of $n - 1$, we must have $q_i^{e_i} \mid \text{ord}_n(a)$. Hence $n - 1 \mid \phi(n)$, implying the primality of n. ∎

Example 7.4.5. Let $n = 911$, $n - 1 = 2 \cdot 5 \cdot 7 \cdot 13$.
  Computations show that

$$7^{n-1} \equiv 1 \pmod{n} \qquad 7^{(n-1)/2} \equiv -1 \pmod{n}$$
$$3^{n-1} \equiv 1 \pmod{n} \qquad 3^{(n-1)/5} \equiv 482 \pmod{n}$$
$$2^{n-1} \equiv 1 \pmod{n} \qquad 2^{(n-1)/7} \equiv 568 \pmod{n}$$
$$2^{n-1} \equiv 1 \pmod{n} \qquad 2^{(n-1)/13} \equiv 577 \pmod{n}.$$

Hence 911 is prime by Theorem 7.4.4. A search for a primitive root shows that 17 is the smallest primitive root. Our computation is more efficient than searching for a primitive root, as we do not discard the previously computed values.

A proof of the primality of a number n using this theorem involves the following:

1. Give the factorization of $n - 1$.

2. For each factor q of $n - 1$ give an a satisfying $a^{n-1} \equiv 1 \pmod{n}$ and $a^{(n-1)/q} \not\equiv 1 \pmod{n}$.

In applying the previous methods, it is also necessary to prove the primality of numbers appearing in the factorization of $n - 1$. In Example 7.4.3, a complete proof will also include a proof of the primality of 1069 and the smaller primes until we get to 2. Such a proof is called a primality certificate.[2] The virtue of this method lies in its ease of verification. One can write a program to generate a certificate of primality and another program to verify a certificate. Since this involves only exponentiation operations, it will be very fast. (See Exercise 11.)

---

[2]V. R. Pratt. "Every Prime Has a Succinct Certificate." *SIAM Journal of Computing*, 4(1975), 214-220.

Exercise. What are the certificates for $3, 5$, and 911?

The primary obstacle in applying this technique is obtaining the prime factorization of $n - 1$. This can be very difficult for large numbers. The following criterion is sometimes useful when only a partial factorization is available.

Theorem 7.4.6 (Pocklington–Lehmer). *Suppose that $n - 1 = FR$, where all the prime factors of F are known and $(F, R) = 1$. Suppose that, for some a, we have $a^{n-1} \equiv 1 \pmod{n}$, and $\left(a^{(n-1)/q} - 1, n\right) = 1$ for all primes $q \mid F$; then for each prime factor p of n, we have $F \mid p - 1$. This implies that if $F > R$, then n is prime.*

PROOF. Let p be a prime divisor of n. We wish to show that $F \mid p - 1$.

Now, $a^{n-1} \equiv 1 \pmod{n}$ implies that $a^{n-1} \equiv 1 \pmod{p}$; hence $\mathrm{ord}_p(a)$ divides $n - 1$. Write $n - 1 = \mathrm{ord}_p(a)k$.

Let $q \mid F$, if $q \mid k$; then we can write

$$a^{(n-1)/q} \equiv a^{\mathrm{ord}_p(a)k/q} \equiv 1 \pmod{p};$$

that is, $p \mid a^{(n-1)/q} - 1$, and $p \mid \left(a^{(n-1)/q} - 1, n\right)$. But by the assumption on a, this gcd is 1; hence we obtain a contradiction to $q \mid k$. So $(q, k) = 1$, that is, all the prime factors of F occur in $\mathrm{ord}_p(a)$, that is, $F \mid \mathrm{ord}_p(a)$. Since $\mathrm{ord}_p(a) \mid p - 1$, we have $F \mid p - 1$, that is, $p \equiv 1 \pmod{F}$.

If $F > R$, then $F \geq \sqrt{n}$ (why?), hence $p \geq \sqrt{n}$, and this implies that n is prime. ∎

Example 7.4.7. Let $n = 10998989$, $n - 1 = FR$, with $F = 4004 = 4 \cdot 7 \cdot 11 \cdot 13$, and $R = 2747$. We take $a = 3$ and compute

$$3^{n-1} \equiv 1 \qquad (\bmod\ n)$$

$$3^{(n-1)/2} \equiv -1 \qquad (\bmod\ n) \qquad \left(3^{(n-1)/2} - 1, n\right) = 1$$

$$3^{(n-1)/7} \equiv 7756665 \quad (\bmod\ n) \qquad (7756665 - 1, n) = 1$$

$$3^{(n-1)/11} \equiv 6406997 \quad (\bmod\ n) \qquad (6406997 - 1, n) = 1$$

$$3^{(n-1)/13} \equiv 7502214 \quad (\bmod\ n) \qquad (7502214 - 1, n) = 1.$$

This guarantees that n is prime. We do not need a prime factorization of $n-1$.

The criterion of Pocklington–Lehmer is especially useful to discover prime numbers of special forms. As an application, consider:

Theorem 7.4.8 (Proth). *Let $n = k2^m + 1$, where $k < 2^m$ and odd. Suppose that there exists an integer a satisfying $a^{(n-1)/2} \equiv -1 \pmod{n}$; then n is prime.*

PROOF. We use the Pocklington–Lehmer criterion. The factored part F of $n - 1$ is 2^m, and the unfactored part, R, is k. If $a^{(n-1)/2} \equiv -1 \pmod{n}$, then $\left(a^{(n-1)/2} - 1, n\right) = 1$ because only 2 can be a factor of -2 and n, but this is not possible as n is odd. Also $k < 2^m$ and the conditions of Theorem 7.4.6 are satisfied, proving the primality of n. ∎

The virtue of the test is that only one exponentiation modulo n needs to be performed for each a.

Example 7.4.9. Let $n = 5 \cdot 2^{75} + 1$. It is easy to compute that $3^{(n-1)/2} \equiv -1$ $\pmod{n}$ (requiring 74 squaring operations), and by Proth's criterion, n is prime.

A number of very large primes have been determined by this method. For illustration, consider an example of a complete factorization.

Example 7.4.10. Let $n = (11^{63} - 6^{23})/5$. By the Pollard ρ-method (in Example 6.4.7), we obtained a factor $n_1 = 7345513129$ of n. We can verify that n_1 is a probable prime and that the cofactor $n_2 = n/n_1 = 24380310477967$ is also a probable prime.

To prove that n_1 is prime, factor $n_1 - 1 = 2^3 \cdot 3 \cdot 23 \cdot 61 \cdot 218149$ by trial division. We find that $a = 7$ satisfies the hypothesis of Theorem 7.4.1, proving the primality of n_1 if the large factor occurring in $n_1 - 1$, $n_3 = 218149$, is prime. We have to show that n_3 is prime.

Factor $n_3 - 1$ as $2^2 \cdot 3 \cdot 7 \cdot 53$ and compute

$$2^{n_3-1} \equiv 1 \pmod{n_3} \qquad 2^{(n_3-1)/2} \not\equiv 1 \pmod{n_3}$$

$$3^{n_3-1} \equiv 1 \pmod{n_3} \qquad 3^{(n_3-1)/3} \not\equiv 1 \pmod{n_3}$$

$$3^{(n_3-1)/7} \not\equiv 1 \pmod{n_3} \qquad 3^{(n_3-1)/53} \not\equiv 1 \pmod{n_3}.$$

By Theorem 7.4.4, n_3 is prime.

To show the primality of n_2, factor $n_2 - 1$ by trial division,

$$n_2 - 1 = 2 \cdot 3 \cdot 23 \cdot 113 \cdot \overbrace{1563441739}^{n_4}.$$

Again, we compute $2^{n_2-1} \equiv 1 \pmod{n_2}$, and $2^{(n_2-1)/q} \not\equiv 1 \pmod{n_2}$ for $q = 3$, 23, 113, and n_4. For $q = 2$, we find that $3^{n_2-1} \equiv 1 \pmod{n}$, and $3^{(n_2-1)/2} \not\equiv 1 \pmod{n}$; hence by Theorem 7.4.4, n_2 is prime, provided that n_4 is prime.

Finally, $n_4 - 1 = 2 \cdot 3 \cdot 103 \cdot 277 \cdot 9133$ and $a = 2$ satisfies the hypothesis of Theorem 7.4.1, hence n_4 is prime. This shows that the probable factorization obtained in Example 6.4.7 is indeed the prime factorization.

In applications to cryptography, it is necessary to construct large primes. In practice it is sufficient to construct probable primes using the strong pseudoprime tests to several bases. If it is necessary to have a large prime, together with a certificate of primality, then the criterion of Pocklington–Lehmer can be modified[3] to generate larger primes from known primes. There is also a recent method of Maurer[4] to generate large primes for applications to cryptography.

─────────────── **Exercises for Section 7.4** ───────────────

1. Show that the following integers are prime by finding an a satisfying the conditions of Theorem 7.4.1.

   (a) 271   (b) 21649   (c) 757
   (d) 62921   (e) 1321   (f) 2906161

2. Write a computer program that will output a proof of the primality of n if a factorization of $n - 1$ is available. Your program should first check if the number passes the strong pseudoprime test to several bases, and then obtain the factorization of $n - 1$. The output should consist of an integer a, $a^{n-1} \bmod n$ and $a^{(n-1)/q} \bmod n$ for each $q \mid n - 1$.

3. Prove that the following integers are prime.

   (a) $10^{12} - 10^6 + 1$ (b) $R_{19} = (10^{19} - 1)/9$

   (c) $10^{16} - 10^8 + 1$ (d) $R_{23} = (10^{23} - 1)/9$

4. Suppose that for each prime $p \mid n - 1$, there exists an a such that $a^{(n-1)/2} \equiv -1 \pmod{n}$, but for $p > 2$, $a^{(n-1)/(2p)} \not\equiv -1 \pmod{n}$. Show that n is prime.

5. Let p be a prime. If $n = 2p + 1$, then $n - 1 = 2p$, so the size of the problem of proving that n is prime has come down by a factor of 2. Show that if $2^{n-1} \equiv 1 \pmod{n}$ and $3 \nmid n$, then n is prime.

   One can use this to construct **Cunningham chains** of primes with each member one more than twice the previous one. Write a computer program to find Cunningham chains of length 5 and length 6.

[3] See P. Ribenboim. "Selling Primes." *Mathematics Magazine*, 68(1995), 175–182.
[4] U. M. Maurer. "Fast Generation of Prime Numbers and Secure Public-Key Cryptographic Parameters." *Journal of Cryptology*, 8(1995), 123–155.

6. The sixth Fermat number F_6 has the factorization

$$F_6 = 274177 \cdot \overbrace{67280421310721}^{p_1}.$$

Prove the primality of both factors. (Be sure to prove also the primality of any large factors appearing in $274177 - 1$ and $p_1 - 1$.)

7. The expression $n! + 1$ appeared in Euclid's proof of the infinitude of primes. Apply the strong pseudoprime test to identify n such that $N = n! + 1$ is likely to be a prime for $n \leq 100$. Use your program from Exercise 2 to identify the primes.

8. Use Proth's Theorem to determine as many primes of the forms $3 \cdot 2^m + 1$ and $5 \cdot 2^m + 1$ as possible.

9. Recall Exercise 6.2.16 on probable primes of the form $n^4 + 1$. Use your computer program of Exercise 2 to prove the primality of these for $2000 \leq n \leq 2100$.

10. Let $n - 1 = mp$, where p is an odd prime such that $2p + 1 > \sqrt{n}$. Show that if there exists an integer a for which $a^{(n-1)/2} \equiv -1 \pmod{n}$ but $a^{m/2} \not\equiv -1 \pmod{n}$, then n is prime.

11. We would like to generate the certificate for primality of a number. This requires proving the primality of any intermediate factors. Hence the certificate can be generated recursively until 2 is reached. Decide on a format for a certificate and write a program to produce a certificate. You should write another program that takes a certificate as input and verifies that the number passes the certificate.

--- **Projects for Chapter 7** ---

1. A delightful application of the concept of order is to the periods of decimal fractions. This was one of the first arithmetical problems considered by Gauss, who was led to many of his results on congruences through the study of this problem.

   The problem is to understand the relation between an integer m and the period of the decimal fraction expansion of $1/m$. Recall that any rational number p/q has either a finite or a recurring decimal expansion. If $0 < p/q < 1$ and it has an infinite decimal expansion, then we can write $p/q = 0.a_1 a_2 \ldots a_r \overline{b_1 b_2 \ldots b_s}$, where the periodic part of the decimal is $b_1 b_2 \ldots b_s$. Here s is the period length of the decimal fraction expansion of p/q. The bar on top of a string of numbers indicates the recurrence of this string in the decimal expansion. For example, $1/7 = 0.\overline{142857}$ and $1/11 = 0.\overline{09}$. Why does $1/7$ have period length 6 and $1/11$ only 2? To understand the question better, answer the following questions.

(a) Show that if $(m, 10) \neq 1$, then either $1/m$ is a terminating decimal or the periodic part does not start immediately after the decimal place.

(b) Assume that $(m, 10) = 1$. The following example for the computation of $1/7$ illustrates the use of the division algorithm to obtain the terms in the decimal expansion.

$$10 \cdot 1 = 7 \cdot 1 + 3$$
$$10 \cdot 3 = 7 \cdot 4 + 2$$
$$10 \cdot 2 = 7 \cdot 2 + 6$$
$$10 \cdot 6 = 7 \cdot 8 + 4$$
$$10 \cdot 4 = 7 \cdot 5 + 5$$
$$10 \cdot 5 = 7 \cdot 7 + 1.$$

Explain how the terms of the decimal fraction are obtained in the computation. Explain the procedure in general for the fraction $1/m$.

(c) In the computation of the previous part, do you see how to determine the period of the fraction? Make a suitable conjecture and prove it.

(d) Show that if $(m, 10) = 1$, then the periodic part of $1/m$ starts right after the decimal place.

(e) Describe the period length in terms of congruence notation and orders. Compute the period using this characterization for some values of m.

(f) Describe the period length of $1/m$ when $(m, 10) \neq 1$.

2. This project analyzes "the most complicated card trick ever." The trick due to Charles Sanders Pierce is described in the article "Mathematical Games" by Martin Gardner, *Scientific American*, July, 1978, pages 18–26. First, understand the trick and perform it; then, answer the following questions to explain why the trick works.

We number the cards with Ace $= 1$ through K $= 13$.

(a) After the first rearrangement of the black and red decks (12 dealings of the "black" deck discarding one card and replacing with a red card), we end up with a completely red deck and a completely black deck. Answer the following questions to explain this phenomenon.

   (i) Describe the function $f(i)$ that gives the *position* (from the bottom) of the ith card in the "black" deck after one dealing. (This is only with the position of the cards and has nothing to do with the value of the cards. This permutation on the 12 cards is the inverse of the standard interlacing shuffle.)

   (ii) Show that $f^{-1}(j) \equiv 2j \pmod{13}$. (Here f^{-1} is the inverse function, not the inverse modulo 13 of its value. Recall that if $f(x) = y$, then $f^{-1}(y) = x$.)

   (iii) Prove that $f^{(12)} = Id$, the identity function. Here $f^{(k)}$ means that f is composed with itself k times.

(b) After the two decks are arranged, they satisfy a reciprocity property. Prove this by answering the following questions.

   (i) Show that the face values of the black deck are obtained by taking the powers of 2 modulo 13.

   (ii) Let a denote the number of cards after A spade in the black deck and $b(i)$ be the face value of the ith card in the black deck, counting from the top and $r(i)$, the value of the ith card in the red deck. Prove that $b(i) = 2^{i'}$, where $i' \equiv i + a \pmod{12}$ and $r(i) = \text{ind}(i) - a \pmod{12}$. [Recall that $\text{ind}(i)$ is defined as the smallest positive integer k such that $2^k \equiv i \pmod{13}$.]

   (iii) Show that $b(i)$ and $r(i)$ are inverses of each other.

(d) Prove that doing a k-shuffle in the red deck is equivalent to arranging the 13 cards in a circle, counting off k positions cyclically, and extracting that card until all the cards have been exhausted.

(e) Show that after putting the King of hearts at the bottom of the red deck, the shuffled deck is

$$r(k), r(2k \bmod 13), \ldots, r(13k \bmod 13).$$

(f) Use the above description to show that we only need to cut the black deck once to preserve the reciprocity of the pointers between the red and black decks, and that this cut is precisely putting the Ace of spades in the position $r(k)$.

(g) Complete the analysis of the trick by explaining the result of performing several k_i-shuffles in sequence.

Chapter 8

Applications

8.1 The ElGamal System

The ElGamal system, like the Diffie–Hellman key exchange, is based on the presumed difficulty of the discrete logarithm problem. Recall that if $g^x \equiv y \pmod{p}$, then x is the discrete logarithm of y modulo p. There is no known method to compute x efficiently with the knowledge of g, y, and p, where p is an appropriately chosen large prime. The system was introduced by T. ElGamal in 1985. [1]

The ElGamal system can be used for both encryption and signatures. Like the RSA system, ElGamal is a public-key method. Suppose Alice wishes to receive encrypted communication. She chooses a prime p and a primitive root g modulo p. She selects a private key a, $0 < a < p - 1$, and computes $b = g^a \bmod p$. Alice's public key is $K = (g, b, p)$.

To encrypt a message m, $0 < m < p$, in this system, the procedure is the following:

1. Choose a random number r, $1 < r < p - 1$.

2. Compute $y_1 = g^r \bmod p$, and $y_2 = mb^r \bmod p$.

3. The ciphertext is $E_K(m) = (y_1, y_2)$.

Note that the cipher includes the choice of a random number r. It is necessary to include this for true security. Randomization increases the possible plaintexts and decreases the likelihood of any attacks based on statistical analysis of the ciphertext. Note also that the ciphertext is twice as long as the plaintext. As in the RSA system, if the plaintext is longer than p, then it is broken into blocks that are smaller than p, and each block is encrypted.

---

[1] T. ElGamal. "A Public Key Cryptosystem and a Signature Scheme Based on Discrete Logarithms." *Advances in Cryptology–Proceedings of CRYPTO 84 (LNCS 196)*, 1985, 10–18.

Alice can recover m using the decryption function

$$D_K(y_1, y_2) = y_2 (y_1)^{-a} \bmod p.$$

This is valid because

$$
\begin{aligned}
D_k(y_1, y_2) &= y_2 (y_1)^{-a} && \bmod p \\
&= y_2 \cdot g^{-ar} && \bmod p \\
&= y_2 (g^a)^{-r} && \bmod p \\
&= m \cdot b^r \cdot b^{-r} && \bmod p \\
&= m && \bmod p
\end{aligned}
$$

Example 8.1.1. Let $p = 37$, $g = 2$, and $a = 31$. Then $b = 2^{31} \bmod 37 = 22$. Suppose the plaintext is $m = 19$. Bob chooses a random number, say $r = 7$; then

$$
\begin{aligned}
y_1 &= 2^7 && \bmod 37 \\
&= 17, \\
y_2 &= 19 \cdot 22^7 \bmod 37 \\
&= 19 \cdot 2 && \bmod 37 \\
&= 1.
\end{aligned}
$$

Then Bob sends $E_K(m) = (17, 1)$ to Alice. To decipher this, Alice computes $D_K(17, 1) = 17^{-31} \bmod 37 = 19$.

Now we discuss some aspects of the security of the system. The computation of the private key a from the public-key parameters is the discrete log problem, which is presumably difficult. Is it possible to recover the plaintext m without knowledge of a? If it were, then we would be able to compute $m y_2^{-1} = b^{-r} \bmod p = y_1^{-a} \bmod p$, that is, we can calculate either b^{-r} or y_1^{-a} without knowledge of a and r. This is the same situation as in the Diffie–Hellman key exchange protocol described in Section 5.2. Recall that in the key exchange protocol, Alice and Bob exchange $g^a \bmod p$ and $g^b \bmod p$, where a and b are kept secret. The key is $g^{ab} \bmod p$, which both can compute. An intruder must be able to compute $g^{ab} = (g^a)^b = (g^b)^a$ without knowledge of a or b. It has been shown by Bert den Boer[2] that if $\phi(p-1)$ consists of small prime factors, then computing g^{ab} from g^a or g^b is equivalent to the discrete log problem. Hence in this case, when $\phi(p-1)$ has small prime factors, the security of the ElGamal scheme is equivalent to the discrete log problem.

---

[2]B. den Boer. "Diffie–Hellman Is As Strong As Discrete Log for Certain Primes." *Advances in Cryptology–CRYPTO '88 (LNCS 403)*, 530–539, 1990.

Another feature of the system that is important for its security is the random number r. The same value of r should not be used with different plaintexts. If the same r is used for different plaintexts, then it will be possible to recover all the plaintexts from the knowledge of one. Suppose that the same r is used to encrypt m_1 and m_2; the corresponding ciphertexts are $E_K(m_1) = (y_1, y_2)$ and $E_K(m_2) = (z_1, z_2)$. Then, $y_2/z_2 = m_1/m_2$, and m_2 can be recovered from the knowledge of m_1 (without knowing the private key).

Next, we study the ElGamal signature scheme. As before, let p be a prime number and g a primitive root modulo p; let $b = g^a \bmod p$. Again the public key is (g, b, p), and the private key is a. If m is the message to be signed, then the signature scheme is applied to a hash of the message, $H(m)$. For simplicity, we also denote this by m. The signature scheme is the following:

1. Select a random number r such that $(r, p - 1) = 1$, and compute $y = g^r \bmod p$.

2. Compute $s = (m - ay)r^{-1} \bmod (p - 1)$.

3. The signature is (y, s).

The signature depends on the random number r. Anyone possessing the public key and the message m can verify the signature by computing $v_1 = y^s b^y \bmod p$ and $v_2 = g^m \bmod p$. The signature is valid if $v_1 = v_2$. The justification of the verification step is based on the following computation:

$$
\begin{aligned}
y^s &= y^{(m-ay)r^{-1}} && \bmod p \\
&= \left(y^{r^{-1}}\right)^{m-ay} && \bmod p \\
&= g^{m-ay} && \bmod p \\
&= g^m g^{-ay} && \bmod p \\
&= g^m b^{-y} && \bmod p,
\end{aligned}
$$

where r^{-1} is the inverse of r modulo $p - 1$.

Example 8.1.2. Let $p = 16563$, $g = 2$, and $a = 3457$; then $b = g^a \bmod p = 12758$. Let $m = 2019$ be the message to be signed. We choose a random number $r = 7841$; then $r^{-1} \bmod p - 1 = 15101$. Next, we compute $y = g^r \bmod p = 7037$, and $s = (m - ay)r^{-1} \bmod p - 1 = 13714$. To verify the signature, we compute $y^s g^y \bmod p = 1037^{13714} 12758^{7037} \bmod p = 15057$ and $g^m = 2^{2019} \bmod p = 15057$.

Regarding the security of the signature scheme, let us consider the problem of determining y and s to satisfy $y^s b^y = g^m \bmod p$. If we choose y, then we have to solve for s in $y^s \equiv g^m b^{-y} \pmod{p}$, which is the discrete log problem. If s is selected first, then we obtain the equation in y, $y^s b^y \equiv g^n \pmod{p}$. This is not exactly the discrete logarithm problem, but it is an equation about which very little is known and the standard techniques do not apply to it. There might be a way to compute y and s simultaneously to satisfy $y^s b^y \equiv g^m \pmod{p}$; no one has shown that this is not possible. Also, if y and s are known, then solving for m is equivalent to the discrete log problem.

———————————————— **Exercises for Section 8.1** ————————————————

1. Suppose the random number r used in the ElGamal signature scheme is revealed. Explain how to compute the private key a.

2. Write a computer program to implement the ElGamal cryptosystem.

3. Write a computer program to implement the ElGamal signature scheme.

4. Suppose the same random number r is chosen to sign two different messages in the ElGamal system. Suppose $s_1 = (m_1 - ay)r^{-1} \bmod (p-1)$ and $s_2 = (m_2 - ay)r^{-1} \bmod (p-1)$, where $y = g^r \bmod p$. Show that r can be computed by solving $(s_1 - s_2)r \equiv (m_1 - m_2) \pmod{p-1}$. Explain how to determine the private key a using the knowledge of r.

5. Consider the following variation of the ElGamal signature scheme in which $y = g^r \bmod p$ and $s = (m + ay)r^{-1} \bmod (p-1)$. What is the verification function in this scheme? Similarly, determine the verification scheme if the signature is given by $s = am + ry \bmod (p-1)$.

8.2 Signature Schemes

In this section, we continue our study of digital signature schemes. A digital signature on an electronic document plays the role of the handwritten signature on paper documents. Its purpose is to establish the identity of the document's signer. A signature on paper documents is needed for most legal and financial transactions. Digital signatures are designed to serve the same purpose for electronic transactions and should have the same legal standing.

An important feature of the conventional signature on paper documents is that the signature cannot be separated from the document, and usually the contents of the document cannot be changed, so the signature serves to verify the contents of the message. In the same way, a digital signature is based on the contents of the document so that the signature cannot be separated from the document. It serves the purpose of verifying the authenticity of the document.

Conventional signatures are used to verify the identity of the signer by comparing the signature to an original, authentic version. Since it is relatively easy to forge a conventional signature, this verification can lead to many problems. It would be easy for the signer of a document to deny that he signed the document. This is also true for the two digital signature schemes that we have discussed so far: the RSA method and the ElGamal method. In both methods, a private key is used to sign the message. It is possible for a signer to deny ever signing the message by claiming that the private key has been compromised. Hence there is a need for a scheme where it is not possible for the signer to deny ever signing a message.

An undeniable digital signature requires the cooperation of the signer to verify the validity of the scheme. The schemes allow the signer to prove that a signature is a forgery; if a signer does not cooperate, it can be assumed that the signature is valid. Moreover, it is not computationally possible for the signer to prove that a valid signature is a forgery.

We discuss the undeniable signature scheme of Chaum–van Antwerpen.[3] As its name suggests, it has the important feature that a valid signature cannot be denied by the signer. The verification of the scheme requires the cooperation of the signer. The scheme derives its security from the difficulty of the discrete logarithm problem.

Suppose Alice wants to sign documents using the Chaum–van Antwerpen scheme. Then she is assigned a public key (p, g, b), where p is prime, g has a large order modulo p, and $b = g^a \bmod p$, where a is Alice's private key. In this scheme we assume that $p = 2q + 1$, where q is prime and g has order q. We also assume that a is invertible modulo q. Let G be the set of powers of g, that is, $G = \{1, g, g^2, \ldots, q^{q-1}\} \bmod q$. Suppose $x \in G$ is the message to be signed. Then the digital signature is

$$s(x) = x^a \bmod p.$$

This signature is simpler than the ElGamal method, but the verification step is more complicated. The verification step is a *challenge-response* protocol. Suppose Bob wants Alice to verify her signature $y = s(x)$. The verification protocol consists of the following steps.

1. Bob selects two numbers k_1 and k_2 at random.

2. Bob computes $z = y^{k_1} b^{k_2} \bmod p$ and sends z to Alice.

3. Alice returns $c = \left(z^{a^{-1} \bmod q} \right) \bmod p$.

4. Bob verifies that $c = x^{k_1} g^{k_2} \bmod p$.

---

[3]D. Chaum and H. van Antwerpen. "Undeniable Signatures." *Advances in Cryptology-CRYPTO '89 (LNCS 435)*, 212–216, 1990.

Let us check the validity of the verification step:

$$c \equiv z^{a^{-1} \bmod q} \pmod{p}$$
$$\equiv y^{k_1 a^{-1} \bmod q} b^{k_2 a^{-1} \bmod q} \pmod{p}$$
$$\equiv x^{k_1 a a^{-1} \bmod q} g^{k_2 a a^{-1} \bmod q} \pmod{p}$$
$$\equiv x^{k_1} g^{k_2} \pmod{p}.$$

The above shows that y is a valid signature for x. Suppose Eve does not possess the private key a. Can she compute c to satisfy the verification step without knowing k_1 and k_2? The answer is no because Eve would have to solve the discrete logarithm problem.

Example 8.2.1. Suppose $p = 83$, where $q = 41$. Since 2 is a primitive root modulo 83, the order of 2^2 is 41. Let $g = 4$ and $a = 57$; then $b = 77$. Let the message x to be signed be $x = 29$. The signature $29^{57} \bmod 83$ is equal to 10. Suppose Bob selects random numbers $k_1 = 8$ and $k_2 = 33$. Then $z = 69$ in step 2 of the protocol. Alice computes the inverse $a^{-1} \bmod q$ to equal 28, and from this $c = z^{28} \bmod p$ equals 21. Bob can verify that $x^{k_1} g^{k_2} \equiv 29^{10} 4^{33} \equiv 21 \pmod{p}$.

Another issue is the following: Can Alice fool Bob into accepting an invalid signature as valid? That is, if $y \not\equiv x^a \pmod{p}$, then what are the chances that Alice can pick a c to satisfy the verification step? Since $c = x^{k_1} g^{k_2} \bmod p$, it lies in G. The equation $c^a \equiv z \pmod{p}$ implies that z also lies in G, and hence y must lie in G.

Suppose $x = g^{r_1} \bmod p$ and $y = g^{r_2} \bmod p$ for some r_1 and r_2, which are uniquely determined modulo q, the order of g. If c satisfies the verification step, then $c^a \equiv z \pmod{p}$. The number of solutions to the congruence can be determined by computing $c^a z^{-1} \bmod p$:

$$c^a z^{-1} \equiv x^{a k_1} g^{a k_2} y^{-k_1} b^{-k_2} \pmod{p}.$$

Since $g^a \equiv b \pmod{p}$, we obtain,

$$c^a z^{-1} \equiv g^{a r_1 k_1} g^{-r_2 k_1} \pmod{p}$$
$$\equiv \left(g^{(a r_1 - r_2)} \right)^{k_1} \pmod{p}.$$

We are assuming that $y \not\equiv x^a \pmod{p}$. This implies that $g^{a r_1 - r_2} \not\equiv 1 \pmod{p}$. If r denotes $a r_1 - r_2$, then the order of g^r modulo p is q because q is prime. This implies that $c^a z^{-1} \equiv g^{r k_1} \pmod{p}$ can take all the q different values in G. Hence among all these possibilities, the probability of correctly choosing c, such that $c^a z^{-1} \equiv 1 \pmod{q}$ is $1/q$.

An important feature of this scheme is that Alice can prove that an invalid signature is a forgery. The proof consists of the following protocol, known as the *disavowal protocol*.

1. Bob selects two random numbers k_1 and k_2.

2. Bob computes $z_1 = y^{k_1} b^{k_2} \bmod p$.

3. Alice computes $c_1 = z_1^{a^{-1} \bmod q} \bmod p$.

4. Bob verifies that $c_1 \not\equiv x^{k_1} g^{k_2} \bmod p$.

5. Bob selects two random numbers l_1 and l_2.

6. Bob computes $z_2 = y^{l_1} b^{l_2} \bmod p$.

7. Alice computes $c_2 = z_2^{a^{-1} \bmod q} \bmod p$.

8. Bob verifies that $c_2 \not\equiv x^{l_1} g^{l_2} \bmod p$.

9. Bob is convinced that the signature is a forgery by verifying that

$$(c_1 g^{-k_2})^{l_1} \equiv (c_2 g^{-l_2})^{k_1} \pmod{p}.$$

Observe that the disavowal protocol consists of two rounds of the verification protocol. The validity of the disavowal protocol is based on the following computation. Let a^{-1} denote the inverse of a modulo q, and suppose $y \not\equiv x^a \pmod{p}$, so y is not a valid signature. Then the two rounds of the verification protocol fail, so

$$c_1 \not\equiv x^{k_1} g^{k_2} \pmod{p}$$
$$c_2 \not\equiv x^{l_1} g^{l_2} \pmod{p}.$$

But

$$c_1 \equiv z_1^{a^{-1} \bmod q} \pmod{p}$$
$$\equiv y^{a^{-1} k_1} g^{k_2} \pmod{p}$$

and

$$c_2 \equiv z_2^{a^{-1} \bmod q} \pmod{p}$$
$$\equiv y^{a^{-1} l_1} g^{l_2} \pmod{p}.$$

This implies that

$$c_1 g^{-k_2} \equiv y^{-a^{-1}k_1} \quad (\text{mod } p) \tag{1}$$

$$c_2 g^{-l_2} \equiv y^{-a^{-1}l_1} \quad (\text{mod } p). \tag{2}$$

If we raise the terms in (1) to the power l_1 and the terms in (2) to the power k_1, then we see that

$$(c_1 g^{-k_2})^{l_1} \equiv y^{-a^{-1}k_1 l_1} \quad (\text{mod } p)$$

$$(c_2 g^{-l_2})^{k_1} \equiv y^{-a^{-1}k_1 l_1} \quad (\text{mod } p),$$

which is the equality we desire to prove that the signature is a forgery.

Is it possible for Alice to substitute values of c_1 and c_2 so that she can prove that a valid signature is a forgery? We show in the following proposition that the probability of this happening is $1/q$, and if q is very large, then the successful disavowal protocol guarantees that the signature is a forgery.

Proposition 8.2.2. *Suppose y is a valid signature for x, that is, $y = x^a$ mod p. Suppose Bob carries out the disavowal protocol and*

$$c_1 \not\equiv x^{k_1} g^{k_2} \quad (\text{mod } p)$$

and

$$c_2 \not\equiv x^{l_1} g^{l_2} \quad (\text{mod } p);$$

then the probability that $(c_1 g^{-k_2})^{l_1} \not\equiv (c_2 g^{-l_2})^{k_1} \ (\text{mod } p)$ is $1 - 1/q$.

PROOF. Suppose $y = x^a$ mod p and

$$(c_1 g^{-k_2})^{l_1} \not\equiv (c_2 g^{-l_2})^{k_1}.$$

We solve for c_2 to obtain

$$c_2 g^{-l_2} \equiv c_1^{l_1 k_1^{-1}} g^{-k_2 l_1 k_1^{-1}} \quad (\text{mod } p)$$

$$\equiv c_1^{l_1 k_1^{-1}} g^{-k_2 l_1 k_1^{-1}} g^{l_2} \quad (\text{mod } p).$$

Here k_1^{-1} in the exponent is the inverse of k_1 mod q. If we write $x_0 = c_1^{k_1^{-1}} g^{-k_2 k_1^{-1}}$ mod p, then we have $c_2 \equiv x_0^{l_1} g^{l_2} \ (\text{mod } p)$. This is the verification step for the signature $y = x_0^a \ (\text{mod } p)$, that is, y is a valid signature for x_0. We are assuming that y is a valid signature for x. We showed before that the probability that Alice can substitute an invalid signature that passes the verification step is $1/q$. Hence, with a probability of $1 - 1/q$, we can conclude that $x \equiv x_0 \pmod{p}$.

But the first part of the disavowal protocol implies that

$$c_1 \not\equiv x^{k_1} g^{k_2} \pmod{p}.$$

If we solve for x, then we see that

$$x \not\equiv c_1^{k_1^{-1}} g^{-k_2 k_1^{-1}} \pmod{p},$$

that is,

$$x \not\equiv x_0 \pmod{p}.$$

We obtain a contradiction, so if the disavowal protocol is successful, the probability that the signature is a forgery is $1 - 1/q$. ∎

──────────── **Exercises for Section 8.2** ────────────

1. **Blind signatures** are useful in some circumstances where it is not necessary for a signer to see the contents of a document (e.g., signatures of a notary public). Here is an example of a blind signature scheme using the RSA signature scheme. The public notary has an RSA-modulus n, public key e, and private key d.

   (a) Alice chooses a random number r and blinds m by computing $m' = mr^e \bmod n$.

   (b) The notary signs with his private key $s' = (m')^d \bmod n$.

   (c) Alice computes the actual signature $s = s'r^{-1} \bmod n$.

   Verify that s is a valid RSA signature for m.

2. **Fail-stop** signatures are another signature mechanisms that allow a signer A to prove that a signature purportedly signed by A is a forgery (if A did not actually sign it). In this scheme, if a forger succeeds in constructing a valid signature, then it is possible (with a high probability) to prove that it is a forgery. The van Heyst–Pedersen fail-stop signature scheme is as follows. A trusted authority selects a prime p, a large prime $q \mid p - 1$, an element g of order q modulo p, a number a, and $b = g^a \bmod p$. If Alice wants to sign a message, then she selects four random numbers k_1, k_2, l_1, l_2 and computes $c_1 = g^{k_1} b^{k_2}$ and $c_2 = g^{l_1} b^{l_2} \bmod p$. Alice's public key is (c_1, c_2, p, q, g, b) and her private key is (k_1, k_2, l_1, l_2). Only the trusted authority knows a, the logarithm of b to base g. To sign a message m, Alice computes $s_1 = k_1 + ml_1 \bmod q$ and $s_2 = k_2 + ml_2 \bmod q$. Her signature is (s_1, s_2).

   Anyone wishing to verify the signature will compute $v_1 = c_1 c_2^m \bmod p$ and $v_2 = g^{s_1} b^{s_2} \bmod p$. The signature is accepted as valid if and only if $v_1 = v_2$.

(a) Show that the verification procedure correctly identifies a valid signature.

(b) Show that there are q^2 values of quadruples (k_1, k_2, l_1, l_2) that yield the same public-key components (c_1, c_2).

(c) Fix a choice of c_1 and c_2. Show that for each message m, $0 < m < q - 1$, there are exactly q choices of quadruples (k_1, k_2, l_1, l_2) that give the same signature.

(d) Show that if a forger creates a signature for a message m using Alice's public key, then the probability that it is the same signature as Alice's is $1/q$.

(e) Alice can prove that a valid signature she did not sign is a forgery. Suppose (s_1', s_2') is a valid signature for a message m and Alice's signature is $(s_1, s_2) \neq (s_1', s_2')$. Explain how Alice can use this to compute a, which is only known to the trusted authority. Since Alice is able to solve an instance of the discrete logarithm problem, this proves that the signature is a forgery.

8.3 Pseudorandom Number Generators-I

In many applications it is necessary to have a set of "randomly chosen numbers." A sequence of random integers $x_0, x_1, \ldots, x_n$ lying between m_1 and m_2 is such that every integer in the range has as equal a chance of occurring as any of the x_i's, independent of all the other numbers in the sequence. For example, a random sequence of 1's and 0's can be constructed by tossing a coin. Random sequences have many applications.

1. Random numbers are used to test if a number is composite or "probably prime" in Chapter 6. The construction of large primes is important in applications to cryptography. A source of random numbers is necessary in many public-key encryption and signature schemes.

2. A random sequence of 1's and 0's can be used as a one-time pad to construct secure ciphers. A plaintext $m = m_1, \ldots, m_k$ consisting of 1's and 0's is added to a random bit sequence $r_1, \ldots, r_k$. The ciphertext is $c = c_1, \ldots, c_k$, where $c_i = m_i + r_i \bmod 2$. The plaintext can be recovered from the knowledge of the random sequence by $m_i = c_i - r_i \bmod 2$. The randomness of r prevents cryptanalysis of c based on statistical properties.

3. Tossing coins, rolling dice, and spinning a roulette wheel all generate random numbers. Their simulation requires a source of random numbers. Many computer programs require random numbers to vary the response of the system to user input.

4. Random numbers are important in numerical analysis, where "Monte-Carlo methods" are employed in a large number of problems. The term "Monte-Carlo method" is applied to any algorithm that uses a source of random numbers. For example, integrals over an interval $[a, b]$ can be estimated by computing a random sequence $x_0, x_1, \ldots x_n$ in $[a, b]$ and evaluating the corresponding Riemann sum.

5. A famous example of the use of random numbers is the "Buffon needle experiment," where the value of π can be estimated by randomly dropping needles of equal length on a sheet marked with parallel and equidistant lines.[4]

Random sequences can be constructed by tossing coins, rolling dice, or spinning a roulette wheel. Many natural phenomena exhibit randomness that can be used to generate random numbers. A computer can generate a few random numbers using events occurring within the computer, like timing interrupts or by using the clock. But the deterministic nature of computers prevents us from using these methods to generate a large sequence of random numbers. Instead, we appeal to arithmetical methods. Naturally, the numbers generated by a formula are not truly random, but by a careful choice of methods, the numbers that are generated can be shown to pass a variety of statistical tests for randomness. We call a sequence generated by arithmetical operations a **pseudorandom sequence**.

The most popular pseudorandom number generator seems to be the linear congruential method invented by D. H. Lehmer in 1948. The method has numerous applications, although its predictability precludes its use in cryptography.

We fix a positive integer m (the modulus). The method uses a seed x_0, a multiplier a, and an increment c. Random numbers x_i are computed in sequence using the formula,

$$x_{n+1} = (ax_n + c) \bmod m. \qquad (1)$$

This formula gives a sequence of integers $x_0, x_1, \ldots, x_n, \ldots$ between 0 and $m - 1$. If we want to get a random sequence in the interval $[0, 1]$, then we divide each term of the sequence $\{x_i\}$ by m. A pseudorandom bit sequence can be obtained from this generator by $z_n = x_n \bmod 2$.

If any value occurs twice in the sequence, then the sequence must repeat from the second occurrence. Suppose $x_i = x_j$ with $j > i$; then $x_{i+1} = x_{j+1}, \ldots, x_{i+k} = x_{j+k}$ for any $k > 0$. Since there are only m possible values for each x_i, the sequence must repeat after at most m terms. The **period** of a sequence is the smallest positive integer T such that $x_{i+T} = x_i$ for all $i \geq i_0$ for some i_0.

---

[4]H. Dorrie. *100 Great Problems of Elementary Mathematics. Their History and Solution.* Translated by D. Antin. New York: Dover, 1965.

Example 8.3.1. Let us take $c = 0$ for simplicity. Then $x_1 = ax_0 \bmod m$, $x_2 = ax_1 \bmod m = a^2 x_0 \bmod m$, and by induction, we see that $x_n = a^n x_0 \bmod m$. If $(a, m) = 1$, then $a^k \equiv 1 \pmod m$ when k is a multiple of $\mathrm{ord}_m(a)$. Since $a, a^2, \ldots, a^{\mathrm{ord}_m(a)}$ are distinct modulo m, the sequence $\{x_i\}$ repeats after $\mathrm{ord}_m(a)$ terms.

Example 8.3.2. Suppose m is prime, and a is a primitive root modulo m. Then the period of the sequence is $\mathrm{ord}_m(a) = m - 1$. For example, let $m = 31$ and $a = 5$. We take the seed x_0 to be 5. Then we obtain the sequence,

$$5, 3, 8, 11, 19, 30, 18, 17, 4, 21, 25, 15, 9, 24, 2,$$
$$26, 28, 23, 20, 12, 1, 13, 14, 27, 10, 6, 16, 22, 7, 29.$$

For a pseudorandom number generator to be useful, it must have a long period. In addition to having a long period, the sequence must also satisfy statistical tests for randomness. A comprehensive analysis of the statistical properties of the linear congruential generator is given by Knuth.[5] Our goal is to study the period length of the sequence.

To study the period length when $c \neq 0$, we derive a formula for x_k in terms of x_0. Consider the following equations:

$$x_1 = (ax_0 + c) \bmod m$$
$$x_2 = (ax_1 + c) \bmod m$$
$$= (a^2 x_0 + (1 + a)c) \bmod m$$
$$x_3 = (a^3 x_0 + (1 + a + a^2)c) \bmod m.$$

The pattern is clear, and it is not hard to prove that

$$x_k = a^k x_0 + (1 + a + a^2 + \cdots + a^{k-1})c \bmod m. \qquad (2)$$

It will be convenient to abbreviate $y_k = 1 + a + a^2 + \cdots + a^{k-1} \bmod m$. Define y_0 to be 0. Using the formula for the sum of a geometric progression, we see that

$$y_k(a - 1) \equiv (a - 1)(1 + a + \cdots + a^{k-1}) \pmod m$$
$$\equiv a^k - 1 \pmod m.$$

Using y_k, we can rewrite Equation (2) as

$$x_k = (a^k - 1)x_0 + x_0 + y_k c \bmod m$$
$$= (a - 1)y_k x_0 + y_k c + x_0 \bmod m$$
$$= [(a - 1)x_0 + c]y_k + x_0 \bmod m.$$

[5]D. E. Knuth. *The Art of Computer Programming.* Vol. 2. Reading, MA: Addison-Wesley, 1980.

If we let $a' = (a - 1)x_0 + c$, then

$$x_k = a'y_k + x_0 \bmod m.$$

Lemma 8.3.3. *If $c \neq 0$, then the period of the sequence x_k is the same as the period of y_k mod $[m/(a', m)]$.*

PROOF. Suppose $x_{k+T} = x_k$; then $a'y_{k+T} \equiv a'y_k \bmod m$, which implies that $y_{k+T} \equiv y_k \pmod{m/(a', m)}$. Similarly, we observe that $y_{i+T} \equiv y_i \pmod{m/(a', m)}$ implies that $x_{i+T} = x_i$. ∎

Hence it is sufficient to determine the period of the sequence $y_i = 1 + a + a^2 + \cdots + a^{i-1} \pmod{m}$. The sequence $\{y_i\}$ satisfies the following simple properties.

Lemma 8.3.4. *Let $y_i = 1 + a + a^2 + \cdots + a^{i-1} \bmod m$ for some integers a and m.*

(a) *If $y_t \equiv y_s \pmod{m}$ for some $s < t$, then $a^s y_{t-s} \equiv 0 \pmod{m}$.*

(b) *If $t = rs$, then $y_t \equiv y_r(1 + a^r + a^{2r} + \cdots + a^{(s-1)r}) \pmod{m}$.*

PROOF.   (a) If $y_s = y_t$ such that $s < t$, then canceling common terms on both sides of $1 + a + \cdots + a^{s-1} \equiv 1 + a + \cdots + a^{t-1} \pmod{p^r}$, we obtain $a^s + a^{s+1} + \cdots + a^{t-1} \equiv 0 \pmod{p^r}$. Factoring out a^s from every term gives the desired result.

(b) We see, by grouping every consecutive set of r terms, that

$$1 + a + \cdots + a^{rs-1}$$
$$\equiv 1 + a + \cdots + a^{r-1} + a^r(1 + a + \cdots + a^{r-1})$$
$$+ \cdots + a^{(s-1)r}(1 + a + \cdots + a^{r-1}) \pmod{m}$$
$$\equiv y_r(1 + a^r + \cdots + a^{(s-1)r}) \pmod{m}. \quad ∎$$

We determine the period of the sequence y_i modulo m by computing the period modulo the prime powers occurring in the factorization of m.

Lemma 8.3.5. *Suppose $x_0, x_1, \ldots$ is an eventually periodic sequence of integers modulo m and m has the prime factorization $m = p_1^{r_1} \ldots p_k^{r_k}$.*

(a) *If S is an integer such that $x_{i+S} \equiv x_i \pmod{m}$ for all i, then S is a multiple of the period.*

(b) *The period of the sequence x_n mod m is the LCM of the periods of the sequence modulo $p_i^{r_i}$, $i = 1, \ldots, k$.*

PROOF. Let $T(n)$ denote the period of the sequence $\{x_i\}$ modulo n. If there exists an integer S such that $x_{i+S} \equiv x_i \pmod{n}$ for all i, then $T(n) \mid S$. Otherwise, we can use the division algorithm to write $S = qT(n) + r$, such that $r < T(n)$. Then $x_{i+S} \equiv x_{i+qT(n)+r} \equiv x_{i+r} \pmod{n}$ for all i; hence $x_{i+r} = x_i \pmod{n}$ for all i. If $r \neq 0$, then this cannot happen because the period $T(n)$ is the smallest integer with this property. Therefore, $r = 0$ and $T(n) \mid S$.

The congruence $x_{i+T(m)} \equiv x_i \pmod{p_j^{r_j}}$ for all $i \geq i_0$ and $j = 1, \ldots, k$ implies that

$$T(p_j^{r_j}) \mid T(m).$$

Therefore, the LCM of the $T(p_j^{r_j})$ divides $T(m)$. Conversely, the congruences

$$x_{i+[T(p_1^{r_1}), \ldots, T(p_k^{r_k})]} \equiv x_i \pmod{p_j^{r_j}} \text{ for all } j, \text{ such that } 1 \leq j \leq k,$$

imply that $x_{i+[T(p_1^{r_1}), \ldots, T(p_k^{r_k})]} \equiv x_i \pmod{m}$, that is,

$$[T(p_1^{r_1}), \ldots, T(p_k^{r_k})] \mid T(m). \qquad \blacksquare$$

Using the lemma, it suffices to determine the period of the sequence y_k modulo prime powers. First, we determine the period length for odd primes.

Theorem 8.3.6. *Let p be an odd prime, a and r integers with $r > 0$, and $y_k = 1 + a + a^2 + \cdots + a^{k-1} \pmod{p^r}$.*

(a) *If $p \mid a$, then the period of y_k is 1.*

(b) *If $a \not\equiv 1 \pmod{p}$, then the period of y_k is $\mathrm{ord}_{p^r}(a)$.*

(c) *If $a \equiv 1 \pmod{p}$, then the period of y_k is p^r.*

PROOF.   (a) If $p \mid a$, then $p^r \mid a^r$, then we have

$$y_r = y_{r-1} \bmod p^r,$$
$$y_{r+1} = y_{r-1} \bmod p^r,$$

and so on. The sequence remains constant after the first $r - 1$ terms, so the period is 1.

(b) We first show that the period starts with the first term $y_0 = 0$. If $y_t = y_s$, then $a^s y_{t-s} \equiv 0 \pmod{p^r}$. Since $p \nmid a$, we see that $y_{t-s} \equiv 0 \pmod{p^r}$. Since $t - s < t$, we have shown that there is no preperiod.

If $a \not\equiv 1 \pmod{p}$, then $a - 1$ has an inverse modulo p^r. Using the formula for the sum of a geometric series, we can write

$$y_t \equiv 1 + a + a^2 + \cdots + a^{t-1} \equiv (a-1)^{-1}(a^t - 1) \pmod{p^r}.$$

Therefore, $y_t = 0$ if and only if $a^t - 1 \equiv 0 \pmod{p^r}$. The smallest possible value of t is $\operatorname{ord}_{p^r}(a)$. This proves that when $a \not\equiv 1 \pmod p$, the period is $\operatorname{ord}_{p^r}(a)$.

(c) The case $a \equiv 1 \pmod p$ is more difficult and is outlined in the exercises. The simplest occurs when $r = 1$. Then $a = 1$, and the sequence produced, $0, 1, 2, \ldots$, clearly has period length p. ∎

Next, we determine the period length of the sequence modulo powers of 2. For example, modulo 8, if $a \equiv 1, 5 \pmod 8$, then the period length is 8, whereas the period length is 4 for $a \equiv 3 \pmod 7$ and 2 for $a \equiv 7 \pmod 8$. Computing the sequence for 2^4, we see that if $a \equiv 1 \pmod 4$, then the period length is 16. This is also the case in general.

Theorem 8.3.7. *Given an integer a, let $y_k = 1 + a + a^2 + \cdots + a^{k-1} \bmod 2^r$. The period length of the sequence y_k is 2^r if and only if $a \equiv 1 \pmod 4$.*

PROOF. Let $a \equiv 1 \pmod 4$. We prove the theorem by induction on r. We assume by induction that the period length is 2^r for $r = k - 1$. Since the period begins with the first term y_0, we have $y_{2^{k-1}} \equiv 0 \pmod{2^{k-1}}$ but $y_{2^{k-2}} \not\equiv 0 \pmod{2^{k-2}}$. We wish to show that $y_{2^k} \equiv 0 \pmod{2^k}$, but $y_{2^{k-1}} \not\equiv 0 \pmod{2^k}$. For this, observe that

$$y_{2^k} \equiv y_{2^{k-1}}(1 + a^{2^{k-1}}) \pmod{2^k}.$$

We know that $a^{2^{k-1}} \equiv 1 \pmod{2^k}$, hence

$$y_{2^k} \equiv 2y_{2^{k-1}} \pmod{2^k}$$
$$\equiv 0 \pmod{2^k}.$$

But

$$y_{2^{k-1}} \equiv y_{2^{k-2}}(1 + a^{2^{k-2}}) \pmod{2^k}.$$

We also know that $a^{2^{k-2}} \equiv 1 \pmod{2^k}$ because $\lambda(2^k) = 2^{k-2}$, hence

$$y_{2^{k-1}} \equiv 2y_{2^{k-2}} \pmod{2^k}$$
$$\not\equiv 0 \pmod{2^k}.$$

This completes the induction step. We used the condition $a \equiv 1 \pmod 4$ in the first step of the induction, for $r = 3$.

If $a \not\equiv 1 \pmod 4$, then the equation

$$a^s - 1 \equiv (a - 1)y_s \pmod{2^{r+1}}$$

implies that $y_s \equiv 0 \pmod{2^r}$ if and only if $\mathrm{ord}_{2^{r+1}}(a) \mid s$. Hence the maximal period length in this case is the maximum possible order modulo 2^{r+1}, which is 2^{r-1}. This proves that the maximal period length 2^r is achieved if and only if $a \equiv 1 \pmod 4$. ∎

Example 8.3.8. A popular random number generator uses $a = 69069$, $c = 1$, and $m = 2^{32}$. Since $a \equiv 1 \pmod 4$, the period length of the sequence is maximal. Since a is odd and $x_n = ax_{n-1} + 1 \bmod 2^{32}$, the sequence alternates between even and odd values. The corresponding bit generator using the lower order bit $x_n \bmod 2$ is completely predictable.

Example 8.3.9. Another popular random number generator (RANDU) uses $a = 65539$, $c = 0$, and $m = 2^{31}$. Since $c = 0$ and x_0 is odd, the period is the order of a. But there is no primitive root modulo 2^{31}; hence the maximum possible order is 2^{29}. It can be verified that the order of a is 2^{29} and hence the period length is 536870912. If the initial term x_0 is even, then the period can be much smaller, so an even seed should never be chosen with this generator.

Example 8.3.10. Park and Miller[6] suggest that a good choice (based on a variety of statistical tests) of parameters in the linear congruential generator is $a = 7^5 (= 16807)$ and $m = 2^{31} - 1$. Here m is prime and a is a primitive root, so the period length is $m - 1$.

──────── **Exercises for Section 8.3** ────────

1. Determine the period length of the sequence of pseudorandom numbers generated by linear congruential generator using the following parameters.

   (a) $m = 169, a = 3, c = 7, x_0 = 23$.

   (b) $m = 169, a = 27, c = 30, x_0 = 6$.

   (c) $m = 9163, a = 1698, c = 3544, x_0 = 5193$.

2. The middle square method is a random number generator invented by Von Neumann. Start with an integer x_0, with $2a$ digits in its decimal expansion. If x_n is known, then the next number x_{n+1} is formed by taking the middle $2a$ digits of x_n^2. (If x_n^2 doesn't have $4a$ digits, then we add 0's to the front to make it have $4a$ digits.) Show that

$$x_{n+1} = \left\lfloor \frac{x_n^2}{10^a} \right\rfloor - \left\lfloor \frac{x_n^2}{10^{3a}} \right\rfloor 10^{2a}.$$

3. This exercise proves part (c) of Theorem 8.3.6.

──────────────

[6]S. K. Park and K. W. Miller. "Random Number Generators: Good Ones Are Hard to Find." *Comm. ACM*, 31(1988), 1192–1201.

(a) Suppose $a \equiv 1 \pmod{p}$. Show by induction on r that $a^{p^{r-1}} \equiv 1 \pmod{p^r}$ and $1 + a + \cdots + a^{p^{r-1}} \not\equiv 0 \pmod{p^r}$. For the first assertion, write $a^{p^{r-2}} = 1 + fp^{r-1}$ (induction hypothesis) and use the Binomial Theorem to compute $a^{p^{r-1}} \bmod p^r$. For the second assertion, use the identity of Lemma 8.3.4.

(b) If $y_t = y_s$, show that $t - s = kp^{r-1}$ for some k. Use the results of the first part to conclude that the period length is p^r.

4. Determine the period length of the sequence $x_n = ax_{n-1} + c \bmod m$, where $m = 2^{31}$, $a = 214013$, and $c = 2531011$.

5. Suppose that $m = 2^n$ and $c = 0$. Show that the period length of x_n is maximum when $a \equiv 3, 5 \pmod 8$.

6. Suppose $m = 10^n$ and $c = 0$. Determine conditions on a so that the period length is as large as possible.

Chapter 9

Quadratic Congruences

Quand on a étudier un grand nombre, il faut commencer par en déterminer quelques résidus quadratiques.[1]

— M. KRAITCHIK

9.1 Quadratic Residues

Our next topic of study is the solution of quadratic congruences $f(x) \equiv 0$ (mod m), where $f(x)$ is a quadratic polynomial. By completing the square, the solution of this congruence is equivalent to solving the congruence $x^2 \equiv D$ (mod m), where $-D$ is the discriminant of $f(x)$. This study culminates in the Quadratic Reciprocity Law, one of the most beautiful theorems in mathematics. The law was first discovered by Euler in 1738 and independently by Legendre in 1785 and by Gauss in 1795, who gave the first complete proof of the theorem in 1796, at the age of 19. Legendre published an incomplete proof of the theorem in 1785. In this chapter, we study the solution of quadratic congruences, which does not require the use of reciprocity, and in Chapter 17, we discuss the Quadratic Reciprocity Law and its applications. An important application of quadratic reciprocity is to understand the prime divisors of binary quadratic forms. This topic is studied in greater detail in Chapter 18.

Suppose we wish to solve the congruence $ax^2 + bx + c \equiv 0$ (mod m), where a, b, and c are integers, and $(2a, m) = 1$. Multiplying by $4a$, we have $4a^2x^2 + 4abx + 4ac \equiv (2ax + b)^2 + 4ac - b^2 \equiv 0$ (mod m). Hence we have to solve the equation $(2ax + b)^2 \equiv b^2 - 4ac$ (mod m). If we let $y = 2ax + b$, then we have reduced the problem to solving the congruence

---

[1]When we study a large number, we must begin by determining some quadratic residues.

$y^2 \equiv b^2 - 4ac \pmod{m}$. If y is known, then x can be found, and vice versa because $(2a, m) = 1$. The analysis shows that it is sufficient to focus on the simpler congruences of the type $y^2 \equiv d \pmod{m}$. The situation is similar when $(2a, m) \neq 1$. (See Section 9.3.) Also, it is enough to find the solutions in any complete residue system. From now on, solutions to congruences will mean solutions in some fixed complete residue system.

Definition 9.1.1. Let a, m be integers such that $(a, m) = 1$. If the congruence $x^2 \equiv a \pmod{m}$ has an integer solution, then a is a **quadratic residue modulo** m; otherwise, it is a **quadratic nonresidue modulo** m.

Example 9.1.2.   1. Let $m = 7$; then 1, 2, and 4 are quadratic residues, and 3, 5, and 6 are nonresidues. This follows from the equations

$$1^2 \equiv 6^2 \equiv 1 \pmod{7}$$
$$2^2 \equiv 5^2 \equiv 4 \pmod{7}$$
$$3^2 \equiv 4^2 \equiv 2 \pmod{7}.$$

Here 3, 5, and 6 are nonresidues, as we have squared all the invertible elements and obtained 1, 2, and 4. Note that $a \equiv b \pmod{m}$ implies that $a^2 \equiv b^2 \pmod{m}$; hence it is enough to consider elements in a complete residue system to determine the quadratic residues modulo m.

2. Let $m = 15$; there are 8 [$= \phi(15)$] invertible elements, but only two quadratic residues,

$$2^2 \equiv 7^2 \equiv 8^2 \equiv 13^2 \equiv 4 \pmod{15}$$
$$1^2 \equiv 4^2 \equiv 11^2 \equiv 14^2 \equiv 1 \pmod{15}.$$

Exercise. Verify that the quadratic residues modulo 23 are 1, 2, 3, 4, 6, 8, 9, 12, 13, 16, and 18.

It can be observed that the product of any two quadratic residues is a quadratic residue, and the product of a quadratic nonresidue and a quadratic residue is a nonresidue.

Exercise. Is the product of two quadratic nonresidues a nonresidue? Is it a quadratic residue?

Given integers a and m, to determine if $x^2 \equiv a \pmod{m}$ is solvable, it is not feasible to enumerate all the quadratic residues to see if a occurs in it. This makes the problem appear computationally difficult, and the existence of an elegant solution all the more appealing.

Suppose m has the prime factorization $m = p_1^{e_1} \cdots p_k^{e_k}$. Then $x^2 \equiv a$ (mod m) implies that $x^2 \equiv a$ (mod $p_i^{e_i}$) for $1 \leq i \leq k$. Conversely, if we solve the congruence $x^2 \equiv a$ (mod $p_i^{e_i}$) to find a solution x_i, $1 \leq i \leq k$, then any x that is a solution to the simultaneous congruences $x \equiv x_i$ (mod $p_i^{e_i}$), $1 \leq i \leq k$, is a solution to the congruence $x^2 \equiv a$ (mod m). This last assertion is a consequence of the Chinese Remainder Theorem. We conclude that it suffices to solve quadratic congruences modulo prime powers.

From Theorem 3.4.6, when p is an odd prime, we can solve $x^2 \equiv a$ (mod p^e) if we can solve $x^2 \equiv a$ (mod p) (see Section 9.3 for additional examples); therefore, we restrict our attention to the case of the prime modulus. Example 9.1.2 above shows that for 7, exactly half of the invertible elements are quadratic residues, and for a given residue a, there are exactly two solutions to $x^2 \equiv a$ (mod 7). This happens because 7 is a prime.

Lemma 9.1.3. *Let p be an odd prime and a an integer such that $(a, p) = 1$; then*

(a) *the equation $x^2 \equiv a$ (mod p) has either no solutions or exactly two solutions. If x_0 is a solution, then $-x_0 \equiv p - x_0$ (mod p) is the other solution.*

(b) *There are exactly $(p-1)/2$ quadratic residues modulo p, and hence exactly $(p-1)/2$ quadratic nonresidues modulo p.*

PROOF.   (a) If x and y are solutions to $x^2 \equiv a$ (mod p), then $x^2 \equiv y^2$ (mod p), that is, $p \mid x^2 - y^2$. Since $x^2 - y^2 = (x - y)(x + y)$, we must have $p \mid x - y$ or $p \mid x + y$, that is, $x \equiv \pm y$ (mod p). Hence any two distinct solutions modulo p differ by a factor of -1.

(b) Consider the $p - 1$ equations

$$x^2 \equiv 1 \quad (\text{mod } p)$$
$$x^2 \equiv 2 \quad (\text{mod } p)$$
$$\vdots$$
$$x^2 \equiv p - 1 \quad (\text{mod } p).$$

Every invertible element is a solution to exactly one of these equations; hence the total number of solutions is $p - 1$. On the other hand, an equation has either no solution or exactly two solutions; thus exactly half the equations must have solutions; that is, there are $(p - 1)/2$ quadratic residues modulo p.   ∎

The last result can also be proved by using the existence of a primitive root modulo primes.

Lemma 9.1.4. *Let p be a prime and g a primitive root modulo p; then the quadratic residues are the even powers $g^2, g^4, \ldots, g^{p-1}$, and the quadratic nonresidues are the odd powers $g, g^3, \ldots, g^{p-2}$.*

PROOF. If g is a primitive root, then $g, g^2, g^3, \ldots, g^{p-1}$ are the invertible elements. Clearly, $g^{2r} \equiv (g^r)^2$ is a quadratic residue. Suppose $x^2 \equiv g^i$ (mod p) has a solution, say $x = g^k$; then $g^{2k} \equiv g^i$ (mod p) implies that $2k \equiv i$ (mod $p-1$) by Corollary 7.1.4. Since $p-1$ is even, this is only possible if i is even; therefore, the elements $g^2, g^4, \ldots, g^{p-1}$ are the quadratic residues, and all the odd powers of g are the nonresidues. ∎

The study of quadratic residues is simplified by the introduction of the Legendre symbol $\left(\dfrac{a}{p}\right)$. This is another example (like $\equiv$), where the choice of a good notation aids in the understanding and elucidation of the underlying concepts.

Definition 9.1.5. Let p be an odd prime and a an integer; the **Legendre symbol** $\left(\dfrac{a}{p}\right)$ is defined to be 1 if a is a quadratic residue modulo p; -1 if a is a quadratic nonresidue modulo p; and 0 if $p \mid a$.

Example 9.1.6.   1. From Example 9.1.2,

$$\left(\frac{1}{7}\right) = \left(\frac{2}{7}\right) = \left(\frac{4}{7}\right) = 1$$
$$\left(\frac{3}{7}\right) = \left(\frac{5}{7}\right) = \left(\frac{6}{7}\right) = -1.$$

2. It is clear that $\left(\dfrac{1}{p}\right) = 1$ for all p.

3. The Legendre symbol is only defined for prime moduli; hence $\left(\dfrac{4}{15}\right)$ is not defined. For composite moduli, we define the Jacobi symbol in Section 17.3.

4. The Legendre symbol is defined for odd primes; hence $\left(\dfrac{a}{2}\right)$ has no meaning. (Every odd integer is a quadratic residue modulo 2.)

The Legendre symbol satisfies the following wonderful properties.

Proposition 9.1.7. *Let p be an odd prime, and a, b two integers such that $(p, ab) = 1$. Then the following are true:*

(a) $\left(\dfrac{a^2}{p}\right) = 1.$

(b) *If $a \equiv b \pmod{p}$, then* $\left(\dfrac{a}{p}\right) = \left(\dfrac{b}{p}\right).$

(c) $\left(\dfrac{ab}{p}\right) = \left(\dfrac{a}{p}\right)\left(\dfrac{b}{p}\right).$

PROOF. (a) and (b) are immediate from the definition of the Legendre symbol. Regarding (c), we use Lemma 9.1.4 describing the residues and non-residues in terms of a primitive root g.

Observe that if $a \equiv g^i \pmod{p}$, where g is a primitive root modulo p, then $\left(\dfrac{a}{p}\right) = (-1)^i$. (Gauss used the expression on the right in his *Disquisitiones Arithmeticae* to denote the residues and nonresidues.) If $a \equiv g^i \pmod{p}$ and $b \equiv g^j \pmod{p}$, then $ab \equiv g^{i+j} \pmod{p}$. The result follows by noting that $i + j$ is even if i and j are both even or both odd; that is, ab is a quadratic residue if both a and b are quadratic residues or both are quadratic nonresidues. And $i + j$ is odd if one of i or j is even and the other is odd; that is, ab is a nonresidue if either a or b is a residue and the other a nonresidue. Finally, if $\left(\dfrac{ab}{p}\right) = 0$, then $ab \equiv 0 \pmod{p}$; hence $a \equiv 0 \pmod{p}$ or $b \equiv 0 \pmod{p}$, so $\left(\dfrac{a}{p}\right) = 0$ or $\left(\dfrac{b}{p}\right) = 0.$ ∎

Example 9.1.8.   1. From property (b) above, to determine $\left(\dfrac{a}{p}\right)$, we can first determine the remainder upon division of a by p; that is, if $r = a \bmod p$, then $\left(\dfrac{a}{p}\right) = \left(\dfrac{r}{p}\right)$. For example, $\left(\dfrac{74}{7}\right) = \left(\dfrac{4}{7}\right) = \left(\dfrac{2^2}{7}\right) = 1.$

2. Using property (c), $\left(\dfrac{18}{29}\right) = \left(\dfrac{2 \cdot 9}{29}\right) = \left(\dfrac{2}{29}\right)\left(\dfrac{9}{29}\right)$. By property (a), $\left(\dfrac{9}{29}\right) = 1$; therefore, $\left(\dfrac{18}{29}\right) = \left(\dfrac{2}{29}\right).$

3. Suppose $a > 0$ such that $(a, p) = 1$, and a has the prime factorization $a = q_i^{e_1} \cdots q_k^{e_k}$; then repeatedly using property (c), we obtain

$$\left(\frac{a}{p}\right) = \left(\frac{q_1}{p}\right)^{e_1} \cdots \left(\frac{q_k}{p}\right)^{e_k}.$$

Adrien-Marie Legendre (1752-1833)

Legendre was born into a well-to-do family in Paris. He received a strong education, especially in mathematics. In 1770, at the age of 18, he received his doctorate in mathematics and physics. Legendre's personal wealth was sufficient to allow him to devote himself entirely to research. The revolution in 1789 destroyed his small fortune, and Legendre found work in a series of academic and administrative positions, in one of which he contributed to the creation of the metric system.

Legendre

Legendre's favorite topics for research were celestial mechanics, number theory, and elliptic functions. In number theory, Legendre was the first to establish the Three Squares Theorem. He gave an incorrect proof of the Quadratic Reciprocity Law in 1785, but made fundamental contributions to the study of quadratic forms. Legendre discovered the method of least squares (also independently discovered by Gauss) and made contributions to the problem of distinguishing between maxima and minima in the calculus of variations. At the age of 75, along with Dirichlet, Legendre settled the case $n = 5$ of Fermat's Last Theorem.

As the Legendre symbols are ± 1, if the exponent e_i is even, then the term corresponding to q_i is 1.

Exercise. Determine $\left(\dfrac{83}{17}\right)$.

The last example shows that the study of the Legendre symbol $\left(\dfrac{a}{p}\right)$ is reduced to the determination of the symbols $\left(\dfrac{-1}{p}\right)$, $\left(\dfrac{2}{p}\right)$, and $\left(\dfrac{q}{p}\right)$ for odd primes q. [$\left(\dfrac{-1}{p}\right)$ is needed to determine the Legendre symbol of negative integers.]

The computation of the Legendre symbol can be accomplished by the following.

Proposition 9.1.9 (Euler's Criterion). *Let p be an odd prime and a an integer such that $(a,p) = 1$; then*

$$a^{\frac{p-1}{2}} \equiv \left(\frac{a}{p}\right) \quad (\text{mod } p).$$

PROOF. Let g be a primitive root modulo p. If a is a quadratic residue, then $a \equiv g^{2r} \pmod{p}$ for some integer r. Hence,

$$a^{\frac{p-1}{2}} \equiv g^{2r\frac{p-1}{2}} \equiv \left(g^{p-1}\right)^r \equiv 1 \equiv \left(\frac{a}{p}\right) \quad (\text{mod } p).$$

If a is not a quadratic residue, then $a \equiv g^{2r+1} \pmod{p}$ for some r, and

$$a^{\frac{p-1}{2}} \equiv g^{(2r+1)\frac{p-1}{2}} \equiv g^{(p-1)r} g^{\frac{p-1}{2}}$$

$$\equiv g^{\frac{p-1}{2}} \equiv -1 \equiv \left(\frac{a}{p}\right) \quad (\text{mod } p). \qquad \blacksquare$$

Exercise. If g is a primitive root, why is $g^{\frac{p-1}{2}} \equiv -1 \pmod{p}$?

From a computational point of view, Euler's Criterion is enough to determine if a number is a quadratic residue or not. Recall from Section 4.3 that exponents in modular arithmetic can be computed efficiently. Euler's Criterion, together with the method given in the next section, is sufficient to solve quadratic congruences.

In this chapter, we are studying solutions to $x^2 \equiv a \pmod{p}$, when p is fixed. The dual question is the following: For which primes p is a a quadratic residue? The solution to this question is more difficult and is answered by the Quadratic Reciprocity Law in Chapter 17.

Example 9.1.10.   1. Does there exist an integer n such that $73 \mid n^2 - 3$?

We are looking for a solution to $n^2 - 3 \equiv 0 \pmod{73}$ or $n^2 \equiv 3 \pmod{73}$, so we want to know if 3 is a quadratic residue. Since 73 is prime, we can use Euler's Criterion to compute the desired Legendre symbol. $\left(\frac{3}{73}\right) \equiv 3^{\frac{73-1}{2}} \equiv 3^{36} \pmod{73}$. The last term can easily be computed to be 1; hence 3 is a quadratic residue modulo 73. So there exists an integer n such that $73 \mid n^2 - 3$.

2. We determine $\left(\frac{31}{67}\right)$ using Euler's Criterion. We have $\left(\frac{31}{67}\right) \equiv 31^{33}$ (mod 67); a quick computation shows that $\left(\frac{31}{67}\right) = -1$, that is, 31 is not a quadratic residue modulo 67.

Exercise. What is the smallest positive integer n such that $73 \mid n^2 - 3$?

Exercise. Does $x^2 \equiv 8 \pmod{31}$ have a solution?

A simple application of Euler's Criterion determines the symbol $\left(\dfrac{-1}{p}\right)$. This symbol was also determined in Example 4.4.5 as an application of Lagrange's Theorem.

Proposition 9.1.11. *Let p be an odd prime; then*

$$\left(\frac{-1}{p}\right) = \begin{cases} 1 & \text{if } p \equiv 1 \pmod 4, \\ -1 & \text{if } p \equiv 3 \pmod 4. \end{cases}$$

PROOF. By Euler's Criterion, $\left(\dfrac{-1}{p}\right) \equiv (-1)^{\frac{p-1}{2}} \pmod p$; The exponent $(-1)^{\frac{p-1}{2}}$ is 1 if and only if $(p-1)/2$ is even; therefore, -1 is a quadratic residue modulo p if and only if $4 \mid p - 1$ or $p \equiv 1 \pmod 4$. ∎

Example 9.1.12.

Let us compute $\left(\dfrac{70}{73}\right)$. The computation is simplified if we notice that by property (2) of the Legendre symbol, $\left(\dfrac{70}{73}\right) = \left(\dfrac{-3}{73}\right)$. Now, $\left(\dfrac{-3}{73}\right) = \left(\dfrac{-1}{73}\right)\left(\dfrac{3}{73}\right)$ by the third property of the Legendre symbol. As $73 \equiv 1 \pmod 4$, $\left(\dfrac{-1}{73}\right) = 1$, and we showed in Example 9.1.10 (1) above, that $\left(\dfrac{3}{73}\right) = 1$, hence $\left(\dfrac{70}{73}\right) = 1$.

Example 9.1.13. Suppose an odd prime p is the sum of two squares, $x^2 + y^2 = p$; then $x^2 + y^2 \equiv 0 \pmod p$, that is, $x^2 \equiv -y^2 \pmod p$. As y is invertible, we multiply by $\left(y^{-1}\right)^2$ to obtain $\left(xy^{-1}\right)^2 \equiv -1 \pmod p$, that is, -1 is a quadratic residue modulo p, hence $p \equiv 1 \pmod 4$ by Proposition 9.1.11.

This shows that $p \equiv 3 \pmod 4$ cannot be expressed as a sum of two squares, a result that was established earlier in Exercise 2.1.26. Next, we prove that a prime p, $p \equiv 1 \pmod 4$, is a sum of two squares. There are many proofs of this important result. More general techniques for answering question of this type are in Section 14.4 and Chapter 18.

Example 9.1.14. Another application is to show that a prime $p \equiv 1 \pmod{4}$ is a sum of two squares. Suppose $0 < x < p$ is a solution to $x^2 \equiv -1 \pmod{p}$. We claim that there exist a and b, $0 \le a, b < \sqrt{p}$ such that $bx \equiv a \pmod{p}$. There are $\lfloor \sqrt{p} \rfloor + 1$ choices for each of a and b, hence there are $(\lfloor \sqrt{p} \rfloor + 1)^2$ numbers of the form $bx - a$, $0 \le a, b < \sqrt{p}$. Since we have more than p numbers of the form $bx - a$, two of these are congruent modulo p. If $b_1 x - a_1 \equiv b_2 x - a_2 \pmod{p}$, then $(b_1 - b_2)x \equiv (a_1 - a_2) \pmod{p}$. Since $b_1 \ne b_2$ or $a_1 \ne a_2$, we see that both are nonzero. Let $b = b_1 - b_2$ and $a = a_1 - a_2$.

Since $x^2 \equiv -1 \pmod{p}$, we see that $b^2 x^2 \equiv a^2 \equiv -b^2 \pmod{p}$ or $p \mid a^2 + b^2$. Since $0 < a, b < \sqrt{p}$, we must have $p = a^2 + b^2$.

An important feature of Proposition 9.1.11 is worth noticing. The proposition determines all primes for which -1 is a quadratic residue. Similarly, it is possible to find all primes for which 2 is a quadratic residue, those for which 3 is a quadratic residue, and so on.

———————————— **Exercises for Section 9.1** ————————————

1. Find all quadratic residues and quadratic nonresidues in a complete residue system modulo each of the following integers.

   (a) 11   (b) 29
   (c) 19   (d) 33

2. Evaluate the following Legendre symbols.

   (a) $\left(\dfrac{11}{29} \right)$ (b) $\left(\dfrac{23}{61} \right)$

   (c) $\left(\dfrac{7}{31} \right)$ (d) $\left(\dfrac{60}{79} \right)$

3. Write a computer program to evaluate the Legendre symbol $\left(\dfrac{a}{p} \right)$ using Euler's Criterion.

4. Does there exist an integer n such that $29 \mid n^2 - 5$? If so, find the smallest such integer.

5. Determine all prime numbers p such that $p \mid n^2 + 1$ for some integer n.

6. Determine the number of quadratic residues modulo p^n, where p is an odd prime.

7. Show that there are infinitely many primes of the form $4k + 1$ using the properties of $\left(\dfrac{-1}{p}\right)$. (Recall that the infinitude of primes of the form $4k+3$ was proved in Exercise 2.2.6, but that method does not apply to primes of the form $4k + 1$.)

8. Use Euler's Criterion to give another proof of the property

$$\left(\frac{ab}{p}\right) = \left(\frac{a}{p}\right)\left(\frac{b}{p}\right).$$

9. Show that at least one of 2, 5, or 10 is a quadratic residue modulo p.

10. Show that for $p > 5$, there are always consecutive integers that are quadratic residues mod p, and consecutive integers that are quadratic nonresidues modulo p.

11. What can you say about the product of all the quadratic residues a, $(a, p) = 1$, in a residue system modulo p? Similarly, investigate the product of all the quadratic nonresidues in a residue system modulo p.

12. Show that a primitive root modulo p is never a quadratic residue. Determine all primes p for which every quadratic nonresidue is also a primitive root.

13. Show that the smallest positive quadratic nonresidue modulo p is always a prime.

14. Suppose $p = 4n + 3$ is prime and a is a quadratic residue modulo p. Show that $x = a^{n+1}$ satisfies $x^2 \equiv a \pmod{p}$.

15. **Definition.** Let p be a prime number. A number a is said to be a **cubic residue mod** p if the congruence $x^3 \equiv a \pmod{p}$ has a solution.

   (a) Determine the cubic residues modulo 11 and modulo 19.

   (b) Show that if $p \equiv 2 \pmod{3}$, then every invertible element is a cubic residue.

   (c) Show that if $p \equiv 1 \pmod{3}$, exactly a third of the invertible elements are cubic residues.

   (d) Show that if the congruence $x^3 \equiv -1 \pmod{p}$ has a solution $x \not\equiv -1 \pmod{p}$, then $6 \mid p - 1$. Conversely, if $6 \mid p - 1$, show that $x^3 \equiv -1 \pmod{p}$ has a solution $x \not\equiv -1 \pmod{p}$, using the identity

$$x^{p-1} - 1 = (x^6 - 1)g(x) = (x^3 - 1)(x^3 + 1)g(x)$$

   and Corollary 4.4.3. [Here $g(x)$ is some polynomial of degree $p - 7$.]

   (e) Use part (d) to show that there exist infinitely many primes of the form $6k + 1$.

16. State and prove an analog of Euler's Criterion to determine if an integer is a cubic residue modulo p.

17. For the study of cubic residues (see Exercise 15), one would like to define a cubic symbol $\left(\dfrac{a}{p}\right)_3$ so that it has nice properties like the Legendre symbol. How should $\left(\dfrac{a}{p}\right)_3$ be defined if it is to have the following properties?

(a) $\left(\dfrac{a^3}{p}\right)_3 = 1$ (b) $\left(\dfrac{ab}{p}\right)_3 = \left(\dfrac{a}{p}\right)_3 \left(\dfrac{b}{p}\right)_3$.

18. **Definition.** A number a is said to be a **biquadratic residue modulo** p if the congruence $x^4 \equiv a \pmod{p}$ has a solution.

   (a) Determine the biquadratic residues modulo 17 and modulo 23.
   (b) Show that -1 is a biquadratic residue modulo p if and only if $p = 2$ or $p \equiv 1 \pmod 8$.
   (c) Use the identity
$$x^4 + 4 = \left[(x+1)^2 + 1\right]\left[(x-1)^2 + 1\right]$$
   to show that -4 is a biquadratic residue modulo p if and only if $p \equiv 1 \pmod 4$.
   (d) Use part (b) to show the infinitude of primes of the form $8k + 1$.

19. Determine the sum $\displaystyle\sum_a \left(\dfrac{a}{p}\right)$ where a runs over the elements of a complete residue system modulo p.

9.2   Computing Square Roots mod p

Once we know that the congruence $x^2 \equiv a \pmod p$ has a solution, how do we find it? Enumerating the values of x until a solution is discovered is clearly impractical. To appreciate our technique, the reader may wish to solve $x^2 \equiv 5 \pmod{19}$ and $x^2 \equiv 2 \pmod{41}$ before reading further. We describe an algorithm that is extremely fast in its search for a solution. The procedure given here uses many of the properties of order and primitive roots discussed in Sections 7.1 and 7.2. The procedure to find solutions to quadratic congruences modulo p is used in the quadratic sieve factoring method of Section 12.2.

If $p \equiv 3 \pmod 4$, then the solution is simple. If $p = 4n + 3$ and a is quadratic residue mod p, then $x = a^{n+1} = a^{(p+1)/4}$ is a solution to $x^2 \equiv a \pmod p$. We verify this by $x^2 \equiv a^{2n+2} \equiv a^{2n+1}a \equiv a^{(p-1)/2}a \equiv 1 \cdot a \pmod p$, where we used $a^{(p-1)/2} \equiv 1 \pmod p$ for a quadratic residue a.

For the case $p \equiv 1 \pmod 4$, we can have $p \equiv 1, 5 \pmod 8$. If $p \equiv 5 \pmod 8$, we can solve the congruence in the following way. We will show in Section 17.1 that 2 is a quadratic nonresidue modulo p. From Euler's Criterion, we have $a^{(p-1)/2} \equiv 1 \pmod{p}$, which implies that $a^{(p-1)/4} \equiv \pm 1 \pmod{p}$. If $a^{(p-1)/4} \equiv 1 \pmod{p}$, then $x \equiv a^{(p+3)/8} \pmod{p}$ is a solution; otherwise $x \equiv 2a(4a)^{(p-5)/8} \pmod{p}$ is a solution. The verification of the latter requires the fact that 2 is a quadratic nonresidue modulo p.

Exercise. Verify the assertions above.

Exercise. Solve the congruences $x^2 \equiv 12 \pmod{59}$ and $x^2 \equiv 6 \pmod{29}$.

The difficulty occurs when $p \equiv 1 \pmod{8}$. One can write down increasingly complex formulas for special cases. But this exercise is futile, as there exists an efficient algorithm to solve the problem that does not depend on any special properties.

We motivate the general square root algorithm by looking at the case $p = 4n + 3$. We write $p - 1 = 2s$ with s odd. Then $x = a^{n+1} = a^{(s+1)/2}$ is the solution to $x^2 \equiv a \pmod{p}$ because $\left(a^{(s+1)/2}\right)^2 \equiv a^s a$ and $a^s \equiv 1 \pmod{p}$.

We cannot repeat this for $p \not\equiv 3 \pmod 4$, as $(p-1)/2$ is even; that is, if we write $p - 1 = 2s$, then $(s+1)/2$ is not an integer. We salvage this by isolating the even part of $p - 1$; that is, we write $p - 1 = 2^r s$, with s odd.

Let us try $x = a^{(s+1)/2}$ and see how much it differs from a solution.

$$x^2 \equiv a^{s+1} \equiv a^s a \pmod{p}.$$

Now, both a and x^2 are quadratic residues, so a^s is also a quadratic residue. Suppose we know an element z such that $z^2 \equiv a^s \pmod{p}$; then we can write

$$x^2 \equiv z^2 a \pmod{p}.$$

Multiplying by the inverse of z^2, we get

$$\left(xz^{-1}\right)^2 \equiv a \pmod{p};$$

that is, we can solve $x^2 \equiv a \pmod{p}$ if we can solve $z^2 \equiv a^s \pmod{p}$. This may not seem helpful at first, but it simplifies the computation necessary to solve the problem. To search for a solution to $z^2 \equiv a^s \pmod{p}$, we have to look at fewer possibilities, because $\mathrm{ord}_p(a^s) \mid 2^{r-1}$, and, if z is a solution, then $\mathrm{ord}_p(z) \mid 2^r$. By Lagrange's Theorem, there are 2^r elements of order dividing 2^r, and z must be one of these. Since 2^r can be much smaller than p, the search is simplified.

The idea of the method is to compare a^s with the elements of order 2^r and obtain the solution. We give an example and then describe the elements of order 2^r.

Example 9.2.1. Solve $x^2 \equiv 2 \pmod{41}$.

We write $41 - 1 = 8 \cdot 5$, so $r = 3$ and $s = 5$. To search for elements of order 8, the most natural idea is to look for a primitive root g (which has order 40); then g^5 and all its powers will have orders dividing 8. Now, 7 is a primitive root mod 41, and $7^5 [\equiv 38 \pmod{41}]$ has order 8. Let

$$S = \{38, 38^2, 38^3, 38^4, 38^5, 38^6, 38^7, 38^8\}.$$

All the elements of S have order dividing 8, and S includes all the elements of orders that divide 8, as there are eight elements of order dividing 8. We reduce S modulo 41 to get

$$\tilde{S} = \{38, 9, 14, 40, 3, 32, 27, 11\}.$$

Now, $2^5 \equiv 32 \equiv 38^6 \pmod{41}$ is in this set, and hence our solution is

$$
\begin{aligned}
x &\equiv 2^{(5+1)/2} 38^{-6/2} \\
&\equiv 2^3 38^{-3} \qquad \equiv 2^3 38^5 \\
&\equiv 8 \cdot 3 \qquad\quad \equiv 24 \pmod{41}.
\end{aligned}
$$

Notice $38^{-3} \equiv 38^5 \pmod{41}$, as 38 has order 8 modulo 41.

Definition 9.2.2. Let $p - 1 = 2^r s$, s odd. If $m \le r$, we denote by S_{2^m} all the elements $a \bmod p$ such that $\mathrm{ord}_p(a) \mid 2^m$.

The reader familiar with groups will notice that S_{2^r} is the 2–Sylow subgroup of the group of invertible elements modulo p. We describe a method to compute the elements of the set S_{2^m} using any quadratic nonresidue. This set can also be characterized using a primitive root, but in practice it is easier to find a quadratic nonresidue than a primitive root. As half the invertible elements are quadratic nonresidues, finding a quadratic nonresidue is usually faster than searching for a primitive root. In addition, finding a primitive root requires the factorization of $p - 1$, whereas, the Legendre symbol can be computed using Euler's Criterion.

Lemma 9.2.3. *Suppose n is a quadratic nonresidue modulo p, with $p - 1 = 2^r s$, s odd. Then $\mathrm{ord}_p(n^s) = 2^r$, and*

$$S_{2^r} = \left\{ n^s, n^{2s}, n^{3s}, \dots, n^{2^r s} \right\}.$$

PROOF. By Euler's Criterion, $n^{(p-1)/2} = n^{2^{r-1}s} \equiv -1 \pmod{p}$. This implies that $\mathrm{ord}_p(n^s) = 2^r$ because $(n^s)^{2^{r-1}} \equiv -1 \pmod{p}$. Hence the elements of the set S_{2^r} are distinct, and since there are exactly 2^r elements of order dividing 2^r, the proof is complete. ∎

Example 9.2.4. In Example 9.2.1, we used the primitive root 7 to find the set S_8. Taking a quadratic nonresidue, say 3, we compute

$$\{3^5, 3^{10}, 3^{15}, 3^{20}, 3^{25}, 3^{30}, 3^{35}, 3^{40}\} \bmod 41$$
$$= \{38, 9, 14, 40, 3, 32, 27, 1\}.$$

If we use a different quadratic nonresidue, we must get the same set. For example, with 17, we obtain

$$\{17^5, 17^{10}, 17^{15}, 17^{20}, 17^{25}, 17^{30}, 17^{35}, 17^{40}\} \bmod 41$$
$$= \{27, 32, 3, 40, 14, 9, 38, 1\}.$$

Exercise. What is the set S_8 in the residue system modulo 47?

Let a be a quadratic residue, then $\mathrm{ord}_p(a^s) \mid 2^r$ because $(a^s)^{2^r} \equiv a^{p-1} \equiv 1 \pmod{p}$. In fact, Euler's Criterion implies that $\mathrm{ord}_p(a^s) \mid 2^{r-1}$. Let n be a quadratic nonresidue, and $z = n^s$; then

$$S_{2^r} = \left\{z, z^2, \ldots, z^{2^r}\right\}.$$

Now, $a^s \in S_{2^r}$, so $a^s = z^k$ for some k. Since $\mathrm{ord}_p(z^k) = 2^r/(k, 2^r) = \mathrm{ord}_p(a^s)$ implies that k is even, we can compute $x_0 = a^{(s+1)/2} z^{-k/2}$ as a solution.

The search for the value of k can take at most 2^r steps. If 2^r is not too large, then one can search the set and find the solution. There is a procedure due to Tonelli and Shanks that will solve the problem in r steps without enumerating all the 2^r elements. The method is not too complicated, so read on. The idea of this method is that we only want to compute $a^{(s+1)/2} z^{-k/2}$ and don't need the actual value of k.

Let $x_0 = a^{(s+1)/2} \bmod p$ be our first guess for a solution to $x^2 \equiv a \pmod{p}$. If $b_0 = a^s \bmod p$, then $\mathrm{ord}_p(b_0) \mid 2^{r-1}$, and $x_0^2 \equiv a^{s+1} \equiv a^s a \pmod{p}$ implies that $x_0^2 \equiv b_0 a \pmod{p}$. We want to successively compute $b_0, b_1, \ldots$ and $x_0, x_1, \ldots$ such that $ab_i \equiv x_i^2 \pmod{p}$ and $\mathrm{ord}_p(b_0) > \mathrm{ord}_p(b_1) > \cdots > \mathrm{ord}_p(b_i) > \cdots$. Since the order of a number is a positive integer, this sequence will stop, eventually reaching $b_k = 1$. As $\mathrm{ord}_p(b_0) \mid 2^{r-1}$, the process will take at most $r-1$ steps. Once $b_k = 1$, we have found the desired solution $x_k^2 \equiv a \pmod{p}$.

Lemma 9.2.5. *Let z be an element of order 2^r modulo p, with $p - 1 = 2^r s$, s odd. Suppose b_i has order 2^m, $m \leq r - 1$; then $b_{i+1} = b_i z^{2^{r-m}} \bmod p$ has order dividing 2^{m-1}.*

PROOF. Since z has order 2^r, we obtain $b_{i+1}^{2^{m-1}} \equiv b_i^{2^{m-1}} z^{2^{r-1}} \equiv (-1)(-1) \equiv 1 \pmod{p}$. Hence the order of b_{i+1} is a divisor of 2^{m-1}. (Notice that $b_i^{2^{m-1}} \equiv -1 \pmod{p}$ because $\mathrm{ord}_p(b_i) = 2^m$.) ∎

Using this result, we can construct an algorithm by letting $b_0 = a^s \bmod p$ and taking $b_{i+1} = b_i z^{2^{r-m}} \bmod p$. We solve for the values of $x_1, x_2, \ldots$ to satisfy the relation $x_i^2 \equiv ab_i \pmod{p}$. First $b_1 = b_0 z^{2^{r-m}}$; hence $ab_1 = ab_0 z^{2^{r-m}} \equiv x_0^2 z^{2^{r-m}} \equiv x_0^2 \left(z^{2^{r-m-1}} \right)^2 \pmod{p}$. Therefore, to satisfy the desired relation, we should take $x_1 = x_0 z^{2^{r-m-1}}$.

Inductively, it is easy to verify that the sequences $b_0, b_1, \ldots$ and $x_0, x_1, \ldots$ are given by

$$b_0 \equiv a^s \pmod{p} \qquad b_{i+1} \equiv b_i z^{2^{r-m}} \pmod{p}$$
$$x_0 \equiv a^{(s+1)/2} \pmod{p} \qquad x_{i+1} \equiv x_i z^{2^{r-m-1}} \pmod{p},$$

where $\mathrm{ord}_p(b_i) = 2^m$. We stop when $b_i = 1$ and then x_i is the desired solution.

Let's look at a few examples that illustrate the procedure.

Example 9.2.6. Solve $x^2 \equiv 2 \pmod{41}$.

Factor $41 - 1 = 2^3 \cdot 5$ and take $n = 3$ as a quadratic nonresidue. Then

$$z \equiv 3^5 \equiv 38 \qquad \pmod{41}$$
$$b_0 \equiv 2^5 \equiv 32 \qquad \pmod{41}$$
$$x_0 \equiv 2^{(5+1)/2} \equiv 2^3 \equiv 8 \quad \pmod{41}.$$

Here b_0 has order 2^2 because $32^2 \equiv 40 \equiv -1 \pmod{41}$. We compute

$$b_1 \equiv 32 \cdot 38^{2^{3-2}} \equiv 32 \cdot 38^2 \equiv 1 \quad \pmod{41}$$
$$x_1 \equiv 8 \cdot 38^{2^{3-2-1}} \equiv 8 \cdot 38^{2^0} \equiv 8 \cdot 38 \equiv 17 \quad \pmod{41}.$$

The solution $x_1 = 17$ is obtained in one iteration.

Example 9.2.7. Solve $x^2 \equiv 7 \pmod{113}$.

Factor $113 - 1 = 2^4 \cdot 7$, and take $n = 3$ as a quadratic nonresidue modulo 113. Then

$$z \equiv 3^7 \equiv 40 \qquad \pmod{113}$$
$$b_0 \equiv 7^7 \equiv 112 \qquad \pmod{113}$$
$$x_0 \equiv 7^4 \equiv 28 \qquad \pmod{113}.$$

We see that b_0 has order 2 because $b_0 \equiv -1 \pmod{113}$. Hence

$$b_1 \equiv 112 \cdot 40^3 \equiv 1 \quad \pmod{113}$$
$$x_1 \equiv 28 \cdot 40^4 \equiv 32 \quad \pmod{113}.$$

So $x_1 = 32$ is the desired solution.

The set S_{2^4} generated by $z = 40$ is

$$\{40, 18, 42, 98, 78, 69, 48, 112, 73, 95, 71, 15, 35, 44, 65, 1\}.$$

A straight search of this set would have required eight steps, as opposed to our computation, which required only one step.

Example 9.2.8. Solve $x^2 \equiv 103 \pmod{641}$.

Factor $641 - 1 = 2^7 \cdot 5$ and choose a nonresidue at random, say $n = 522$. Then

$$z \equiv 522^5 \equiv 488 \pmod{641}$$
$$b_0 \equiv 103^5 \equiv 625 \pmod{641}$$
$$x_0 \equiv 103^3 \equiv 463 \pmod{641}.$$

Here is the result of the first four iterations of our method.

i	b_i	$\mathrm{ord}_p(b_i)$	x_i
0	625	2^4	463
1	385	2^3	71
2	487	2^2	495
3	1	1	443

The solution $x = 443$ is found in the fourth step.

Exercise. Verify that the results in the table are correct.

In the last example, the following powers of z are computed:

$$\text{for } b_1, \quad z^{2^{7-4}} = z^{2^3};$$
$$\text{for } b_2, \quad z^{2^{7-3}} = z^{2^4};$$
$$\text{for } b_3, \quad z^{2^{7-2}} = z^{2^5}.$$

To implement the method efficiently, we can compute z^{2^4} from z^{2^3} by squaring, similarly for z^{2^5} from z^{2^4}, etc. For this purpose we also compute a sequence z_i as follows. Let $z_0 = n^s$. If b_i has order 2^m and z_i has order 2^r, then

$$z_{i+1} = z_i^{2^{r-m}},$$
$$b_{i+1} = b_i z_i^{2^{r-m}} = b_i z_{i+1},$$
$$x_{i+1} = x_i z_0^{2^{r-m-1}}.$$

Exercise. What are the values z_0, z_1, z_2, z_3 in the last example?

The above discussion is summarized in the following algorithm.

Algorithm 9.2.9 (Square root modulo p). This algorithm determines a solution to the congruence $x^2 \equiv a \pmod{p}$ if it exists. Otherwise, it will output that a is not a quadratic residue.

1. [Find r and s] Let $r = 0$, $s = p - 1$. While s is even, $r = r + 1$, $s = s/2$.

2. [Get nonresidue] Choose a random integer n such that $\left(\dfrac{n}{p}\right) = -1$.

3. [Initialize] Let $z = n^s \bmod p$, $x = a^{(s+1)/2} \bmod p$, $b = a^s \bmod p$.

4. [Are we done?] If $b = 1$, then return x as the solution.

5. [Find the order of b] Let $m = 1$, $y = b^2 \bmod p$. While $y \neq 1$; let $y = y^2 \bmod p$, $m = m + 1$.

6. [Nonresidue?] If $r = m$, then a is a quadratic nonresidue, terminate algorithm.

7. [Iterate] Let $x = xz^{2^{r-m-1}} \bmod p$, $b = bz^{2^{r-m}} \bmod p$, and $z = z^{2^{r-m}} \bmod p$. Go to step 4.

Remarks.   1. On average, the algorithm requires about $r^2/4$ multiplications. Since r is close to $\log p$, the expected running time of the algorithm is $O(\log^4 p)$.

2. The algorithm is probabilistic in the sense that the number of steps cannot be predicted in advance. This comes from choosing random numbers to find a nonresidue. One can make the algorithm deterministic by trying $n = 2, 3, 5, \ldots$ to find a nonresidue, that is, we find the smallest quadratic nonresidue.

——————————— **Exercises for Section 9.2** ———————————

1. Determine the set S_{2^r} with $p - 1 = 2^r s$, s odd, for the following primes.

    (a) 17   (b) 57

2. Verify the assertion at the beginning of the section regarding the solution to the congruence $x^2 \equiv a \pmod{p}$ for p of the form $8n + 5$.

3. Take $p \equiv 3 \pmod 4$ in Algorithm 9.2.9. What is the solution obtained for $x^2 \equiv a \pmod p$ when a is a quadratic residue?

4. Work through Algorithm 9.2.9 in the case $p = 8n + 5$. In this case you can use the fact that 2 is a quadratic nonresidue. Compare with the result of Exercise 2.

5. Suppose g is a primitive root modulo p, $p = 8n + 1$ a prime. Show that $x = g^n \pm g^{7n}$ is a solution to the congruence $x^2 \equiv \pm 2 \pmod p$.

6. Work through Algorithm 9.2.9 with a calculator or computer to solve the following equations. (In each case determine the set S_{2^r} and compare with the number of steps needed for a direct solution.)

   (a) $x^2 \equiv 11 \pmod{79}$ (b) $x^2 \equiv 5 \pmod{241}$
   (c) $x^2 \equiv 3 \pmod{37}$ (d) $x^2 \equiv 37 \pmod{73}$

7. Write a computer program to implement Algorithm 9.2.9.

8. (a) Suppose $2^m \mid p - 1$. Show that z is an element of order 2^m, then

$$S_{2^m} = \left\{ z, z^2, z^3, \ldots, z^{2^m} \right\}.$$

(b) Let g be a primitive root modulo p, with $p - 1 = 2^r s$, s odd, let $z = g^s$. Show that $\mathrm{ord}_p(z) = 2^r$ and

$$S_{2^r} = \left\{ z, z^2, z^3, \ldots, z^{2^r} \right\}.$$

9.3   Complete Solution of Quadratic Congruences

We can solve the congruence $x^2 \equiv a \pmod m$ for composite moduli using the technique of the previous section, the solution of a linear congruence, and the Chinese Remainder Theorem. We deal with the odd primes first.

Proposition 9.3.1. *Let p be an odd prime and a an integer not divisible by p. The congruence $x^2 \equiv a \pmod{p^k}$ has exactly two solutions if $x^2 \equiv a \pmod p$ has two solutions, and no solutions if $x^2 \equiv a \pmod p$ has no solutions.*

PROOF. If $x^2 \equiv a \pmod{p^k}$ has no solutions, then $x^2 \equiv a \pmod p$ has no solutions. Suppose that $x^2 \equiv a \pmod{p^k}$ has a solution x_0. Then any other solution x is such that $p^k \mid x^2 - x_0^2$, that is, $p^k \mid (x - x_0)(x + x_0)$. This means that $p^k \mid x - x_0$ or $p^k \mid x + x_0$, because $p \mid x - x_0$ and $p \mid x + x_0$ implies that $p \mid x_0$, but a is not divisible by p. Hence $x \equiv \pm x_0 \pmod{p^k}$. Since $-x_0$ is always a solution when x_0 is a solution, we see that the congruence has exactly two solutions. ∎

The congruences modulo prime powers are solved by first solving the congruence modulo a prime. The following example illustrates the technique.

Example 9.3.2. Solve $x^2 \equiv 47 \pmod{121}$.

The congruence $x^2 \equiv 47 \equiv 3 \pmod{11}$ has the solutions $x \equiv \pm 5$ (mod 11). Hence any solution to $x^2 \equiv 47 \pmod{121}$ is of the form $x = 11k + 5$ or $11k - 5$ for suitable values of k. We substitute this in the congruence and solve for k. We have $(11k+5)^2 \equiv 121k^2 + 110k + 25 \equiv 110k + 25 \equiv 47 \pmod{121}$. Subtracting 25, we obtain $110k \equiv 22 \pmod{121}$. Dividing by 11, we have $10k \equiv 2 \pmod{11}$. Since 2 is relatively prime to 11, we can cancel 2 on both sides. It remains to solve $5k \equiv 1 \pmod{11}$. This linear congruence can be readily solved and yields $k \equiv 9 \pmod{11}$. Therefore, a solution to the congruence is $x = 11 \cdot 9 + 5 = 104$. Another solution is $-104 \equiv 17 \pmod{121}$.

The procedure of the example is repeated to solve congruences modulo higher prime powers.

Exercise. Solve the congruence $x^2 \equiv 47 \pmod{11^3}$.

We study the solution of quadratic congruences modulo powers of 2 before doing examples of composite moduli and general quadratic congruences.

Proposition 9.3.3. *Let a be an odd integer; then*

(a) $x^2 \equiv a \pmod{2}$ *has exactly one solution for all a.*

(b) $x^2 \equiv a \pmod{4}$ *has two distinct solutions if and only if $a \equiv 1$ (mod 4) and no solutions otherwise.*

(c) $x^2 \equiv a \pmod{2^n}$ *with $n \geq 3$ has a solution if and only if $a \equiv 1$ (mod 8). In this case there are four solutions. If x_0 is a solution, then $\pm x_0$ and $\pm x_0 + 2^{n-1}$ are all the solutions.*

PROOF. The first two assertions are clear. Regarding (c), the congruence $x^2 \equiv a \pmod{8}$ is solvable only for $a \equiv 1 \pmod{8}$ and has 4 solutions 1, 3, 5, and 7. Hence there exists the necessity of the condition $a \equiv 1 \pmod{8}$ for $n \geq 3$.

We proceed by induction to prove the assertion for an arbitrary $n \geq 3$. Assume that $a \equiv 1 \pmod{8}$ and assertion (c) is true for k, that is, there are four solutions $\pm x_0$ and $\pm x_0 + 2^{k-1}$ to the congruence $x^2 \equiv a \pmod{2^k}$, where $0 < x_0 < 2^k$. Suppose $x^2 \equiv a \pmod{2^{k+1}}$ has a solution x_1 such that $0 < x_1 < 2^{k+1}$. Then it is easy to verify that $\pm x_1$ and $\pm x_1 + 2^k$ are distinct solutions in a complete residue system. Moreover, $x_1^2 \equiv a \pmod{2^k}$, so there are eight possibilities for x_1: $\pm x_0, \pm x_0 + 2^k, \pm x_0 + 2^{k-1}$ or $\pm x_0 + 2^{k-1} + 2^k$. To select four of these, it suffices to test if x_0 or $x_0 + 2^{k-1}$ is a solution to the congruence $x^2 \equiv a \pmod{2^{k+1}}$.

We claim that either $2^{k+1} \mid x_0^2 - a$ or $2^{k+1} \mid \left(x_0 + 2^{k-1}\right)^2 - a$. This will show that only one of x_0 or $x_0 + 2^{k-1}$ is a solution to $x^2 \equiv a \pmod{2^{k+1}}$. Indeed,

$$
\left(x_0 + 2^{k-1}\right)^2 - a \equiv x_0^2 - a + 2^k x_0 + 2^{2k-2}
$$
$$
\equiv x_0^2 - a + 2^k x_0 \pmod{2^{k+1}}.
$$

Since $x_0^2 - a \equiv 0 \pmod{2^k}$, we can write $x_0^2 - a = 2^k t$ for some t. Hence $\left(x_0 + 2^{k-1}\right)^2 - a \equiv 2^k(t + x_0) \pmod{2^{k+1}}$.

Now x_0 must be odd as a is odd. If t is odd, then $2^{k+1} \mid \left(x_0 + 2^{k-1}\right)^2 - a$; otherwise, $2^{k+1} \mid x_0^2 - a$. This completes the proof of the assertion. ∎

Example 9.3.4.   1. The congruence $x^2 \equiv 55 \pmod{256}$ has no solution because $55 \equiv 7 \pmod 8$.

2. Let's solve $x^2 \equiv 41 \pmod{128}$. Take $x \equiv 1 \pmod 8$; hence one of 1 or $1+4$ is a solution to $x^2 \equiv 41 \pmod{16}$. Since $41 \equiv 9 \pmod{16}$ and $5^2 \equiv 9 \pmod{16}$, we can take $x = 5$ as a solution. Next, one of 5 or $5 + 8$ is a solution to $x^2 \equiv 41 \pmod{32}$. Clearly, 5 is not a solution, so $x = 13$ satisfies $x^2 \equiv 41 \pmod{32}$. Continuing in this way, we see that $x = 13$ is a solution to $x^2 \equiv 41 \pmod{128}$. Then the other solutions are $x = -13$, $13 + 64$, and $-13 + 64$.

Exercise. Solve the congruence $x^2 \equiv 33 \pmod{512}$.

Next, we deal with composite moduli. The technique combines the above results with the Chinese Remainder Theorem.

Example 9.3.5. Solve $x^2 \equiv 410 \pmod{847}$. Since $847 = 121 \cdot 7$, we will first solve $x^2 \equiv 410 \equiv 47 \pmod{121}$ and $x^2 \equiv 410 \equiv 4 \pmod 7$. The first congruence has solutions $x \equiv \pm 17 \pmod{121}$ (from Example 9.3.2) and the second congruence has solutions $x \equiv \pm 2 \pmod 7$. If x satisfies $x^2 \equiv 410 \pmod{847}$, then $x \equiv \pm 17 \pmod{121}$ and $x \equiv \pm 2 \pmod 7$. There are four possible combinations of the signs, giving four solutions to the congruence. Using the Chinese Remainder Theorem, we see that these are $x \equiv \pm 380 \pmod{847}$ and $x \equiv \pm 138 \pmod{847}$.

Example 9.3.6. Solve $x^2 \equiv 2225 \pmod{3872}$. Since $3872 = 32 \cdot 121$, we first solve $x^2 \equiv 2225 \equiv 17 \pmod{32}$ and $x^2 \equiv 2225 \equiv 47 \pmod{121}$. The first equation can be solved by the technique of Theorem 9.3.3 to yield the solutions $x = 7, 9, 23, 25$. Therefore, any solution to the congruence $x^2 \equiv 2225 \pmod{3872}$ satisfies $x \equiv \pm 7, \pm 9 \pmod{32}$ and $x \equiv \pm 17$

(mod 121). There are eight possible combinations, hence eight distinct solutions to the congruence in a complete residue system modulo 3872. By applying the Chinese Remainder Theorem, the solutions are ±743, ±1193, ±2887, and ±2921.

The solution of a general quadratic congruence of the form $ax^2 + bx + c \equiv 0 \pmod{m}$ can be reduced to the solution of $x^2 \equiv d \pmod{m}$ using the method of completing the square. See the discussion at the beginning of Section 9.1 for the case $(2a, m) = 1$. When m is even and $(a, m) = 1$, then we can still complete the square, but solutions to $ax^2 + bx + c \equiv 0 \pmod{m}$ are a subset of the solutions to $(2ax + b)^2 \equiv b^2 - 4ac \equiv 0 \pmod{m}$. This happens because we multiply by $4a$ in the first step, and $4a$ is not invertible modulo m. In this case, it is best to multiply the modulus by 4, that is, solve the congruence $4a(ax^2 + bx + c) \equiv 0 \pmod{4m}$. This is because $4a(ax^2 + bx + c) \equiv 0 \pmod{4m}$ implies that $ax^2 + bx + c \equiv 0 \pmod{m}$ and conversely. A similar discussion applies when a is not relatively prime to m.

Example 9.3.7. Solve $5x^2 + 3x + 8 \equiv 0 \pmod{34}$. To complete the square, we multiply the equation by $20 = 4 \cdot 5$. Since $(20, 34) = 2$, we multiply the modulus by 4. We have

$$100x^2 + 60x + 160 \equiv (10x + 3)^2 + 160 - 9 \pmod{136}$$
$$\equiv (10x + 3)^2 + 15 \pmod{136}.$$

Let $y = 10x + 3$. We solve $y^2 \equiv -15 \pmod{136}$ by solving $y^2 \equiv -15 \pmod{17}$ and $y^2 \equiv -15 \pmod{8}$. The first has two solutions, $y \equiv \pm6 \pmod{17}$ and the second, four solutions $y \equiv \pm1, \pm3 \pmod{8}$. Combining these, we obtain the solutions $y \equiv \pm57, \pm91, \pm125, \pm23 \pmod{136}$. Next, we solve for x by solving the linear congruences, $10x + 3 \equiv y \equiv \pm57, \pm91, \pm125, \pm23 \pmod{136}$. Since $(10, 136) = 2$, there is a solution to each congruence modulo 68. We obtain the eight solutions

$$x \equiv 19, 62, 36, 45, 53, 28, 2, 11 \pmod{68}.$$

These are equivalent to $x \equiv 2, 11, 19, 28 \pmod{34}$.

The only remaining case regarding the solution of quadratic congruences $x^2 \equiv a \pmod{m}$ occurs when $(a, m) \neq 1$. Using the techniques of Section 3.4, the solution of a quadratic congruence can be reduced to the case of prime powers. If $m = p^k$ and $p \mid a$, then there is only one solution modulo p, namely $x = 0$. This solution can be extended to obtain solutions modulo p^k by writing $a = sp^r$ and dividing by powers of p (see Exercise 7).

———————————— **Exercises for Section 9.3** ————————————

1. Determine if each of the following congruences has a solution. If the congruence has solutions, then find all the roots.

   (a) $x^2 \equiv 61 \pmod{169}$

   (b) $x^2 \equiv 869 \pmod{961}$

   (c) $x^2 \equiv 191 \pmod{529}$

   (d) $x^2 \equiv 281 \pmod{512}$

2. Solve the following quadratic congruences.

   (a) $x^2 \equiv 696 \pmod{943}$

   (b) $x^2 \equiv 153 \pmod{236}$

   (c) $x^2 \equiv 1225 \pmod{1552}$

3. Solve the following quadratic congruences. (Find *all* the solutions.)

   (a) $x^2 \equiv 21 \pmod{629}$

   (b) $x^2 \equiv 5915 \pmod{12854}$

4. Determine if each of the following congruences has a solution and find all the roots when the equation is solvable.

   (a) $7x^2 + 13x + 26 \equiv 0 \pmod{97}$

   (b) $5x^2 + 7x + 78 \equiv 0 \pmod{136}$

   (c) $6x^2 + 14x + 8 \equiv 0 \pmod{21}$

5. Suppose $m = 2^k p_1^{e_1} \cdots p_r^{e_r}$ and $(a, m) = 1$. Show that $x^2 \equiv a \pmod{m}$ is solvable if and only if $x^2 \equiv a \pmod{2^k}$ and $x^2 \equiv a \pmod{p_i^{e_i}}$ is solvable. If $x^2 \equiv a \pmod{m}$ is solvable, how many solutions does it have?

6. Determine the number of invertible elements that are quadratic residues in a complete residue system modulo m.

7. Let a be an integer and p a prime number such that $(a, p) \neq 1$. Write $a = sp^r$, where $(s, p) = 1$. If $k > r$, then show that congruence $x^2 \equiv a \pmod{p^k}$ has a solution if and only if r is even and s is a quadratic residue modulo p^{k-r}.

8. Solve the congruence $ax^2 + bx + c \equiv 0 \pmod{2^k}$ in the case when a is even.

Chapter 10

Applications

10.1 Identification Schemes

In a previous section, we studied public-key signature schemes. A signature attached to a message can be used to verify the sender's identity. A document's authenticity can be verified using a signature that includes information about the contents of the document. A signature is message or document dependent and cannot be used in a different context. In many applications, it is necessary to verify the identity of the sender before a message is sent. For example, we use a Personal Identification Number, or PIN, to conduct many transactions. Automatic Teller Machines, telephone calling cards, and credit cards all use a PIN to validate transactions. In addition, passwords are needed to access many services over a computer network.

Suppose Alice is using a PIN as a secret key to access some services. Someone looking over her shoulder or eavesdropping electronically can recover her PIN and impersonate Alice. Identification schemes are designed to protect against security problems that arise when a secret key is compromised. Instead of revealing the secret key to verify Alice's identity, these schemes provide a mechanism for Alice to prove that she knows the secret key. A proof of identity is based on some computation involving this key, and the intermediate results of the computation are different for each identification session. Someone listening to the exchange will not be able to use the data to impersonate Alice because in a well-designed system, the computations will be different in the next session. The computations are simple enough that they can be performed by embedded electronic circuitry in "smart cards" or by client software in a computer network.

We describe two identification schemes: one based on the difficulty of factoring integers, and another based on the discrete logarithm problem.

The identification scheme of Feige, Fiat, and Shamir is based on the difficulty of factoring a large number.[1] Let $n = pq$, where p and q are large primes. Recall that solving quadratic congruences modulo n is equivalent to computing the prime factorization of n. Suppose a is a quadratic residue modulo n. If we know the prime factorization of n, then we can solve the congruence $x^2 \equiv a \pmod{n}$ by solving $x^2 \equiv a \pmod{p}$ and $x^2 \equiv a \pmod{q}$. Conversely, if we can compute the four roots of $x^2 \equiv a \pmod{n}$, then we can factor n. If the roots are $\pm x_0$ and $\pm y_0$, then $x_0^2 \equiv y_0^2 \pmod{n}$, but $x_0 \not\equiv \pm y_0 \pmod{n}$, so $(x_0 - y_0, n)$ is a proper factor of n.

In the Feige–Fiat–Shamir scheme, the number $n = pq$ can be shared among a group of users, without their knowing p and q. A trusted central authority can assign n and public and private keys based on n. Alice's public key is a number v such that v is a quadratic residue modulo n. Naturally, we require $(v, n) \neq 1$. Alice's private key is a number s, $0 < s < n$, such that $s^2 \equiv v^{-1} \pmod{n}$. The protocol for Alice to prove her identity to Bob consists of the following steps. Bob knows the number n and Alice's public key v.

1. Alice selects a random number r, $0 < r < n$ and sends $x = r^2 \bmod n$ to Bob.

2. Bob selects a random bit b, $b = 0$ or 1, and sends b to Alice.

3. If $b = 0$, the Alice returns r to Bob; otherwise, Alice returns $y = (rs) \bmod n$.

4. If $b = 0$, then Bob verifies that $x = r^2 \bmod n$ and if $b = 1$, then Bob verifies that $y^2 \equiv xv^{-1} \pmod{n}$.

Alice's identity is verified because she is proving her knowledge of the private key s, without revealing it. This protocol is repeated many times to validate Alice's identity. The number of times the protocol needs to be repeated will be clear after a discussion of its security. Why is this scheme secure? If Eve is trying to impersonate Alice without knowing s satisfying $s^2 \equiv v^{-1} \pmod{n}$, then Eve cannot satisfy both the conditions in step 4. If she sends x such that $x = r^2 \pmod{n}$, then she cannot find y such that $y^2 \equiv xv^{-1} \pmod{n}$ without knowing the prime factors of n. If she chooses x and y such that $y^2 \equiv xv^{-1} \pmod{n}$, then she will be able to satisfy the query when $b = 1$, but not when $b = 0$. If b is a randomly chosen bit, the probability that Eve can guess it correctly in advance is $1/2$. If the protocol is repeated k times, the probability that Eve has guessed each bit correctly is $1/2^k$. So if k is chosen to be large value, say $k \geq 50$, then the chances of someone impersonating Alice are slim.

---

[1]U. Feige, A. Fiat, and A. Shamir. "Zero Knowledge Proofs of Identity." *Journal of Cryptology*, 1(1988), 77–94.

The second identification scheme we discuss is due to Schnorr. Its security is based on the discrete logarithm problem. Let p be a large prime and q a factor of $p-1$. We have seen how to construct such primes in an earlier section. Select a number a such that $a^q \equiv 1 \pmod{p}$. A number a with order q can be computed by taking $a = g^{\frac{p-1}{q}} \bmod p$, where g is any primitive root modulo p. Recall that it is not difficult to find a primitive root modulo p if the prime factorization of $p-1$ is known.

The numbers a, q, and p are known publicly. Alice's public key is a number v modulo p and the private key is s satisfying $a^{-s} \equiv v \pmod{p}$. The Schnorr identification scheme uses the following protocol.

1. Alice selects a random number r and sends $x = a^r \pmod{n}$ to Bob.

2. Bob sends a number e, $0 < e < 2^t - 1$ to Alice.

3. Alice sends $y = r + se \bmod q$ to Bob.

4. Bob verifies that $x = a^y v^e \bmod p$.

To see that the verification step is correct, consider

$$
\begin{aligned}
a^y v^e &\equiv a^{r+se} v^e \pmod{p} \\
&\equiv a^r (a^s)^e v^e \pmod{p} \\
&\equiv x(v^{-1})^e v^e \pmod{p} \\
&\equiv x \pmod{p}.
\end{aligned}
$$

The security of the scheme is derived from the difficulty of solving the discrete logarithm problem. If Eve tries to impersonate Alice without knowing s, then she has to compute a value of y in the third step to satisfy $x \equiv a^y v^e \pmod{p}$. This is equivalent to solving the discrete logarithm problem $a^y \equiv xv^{-e} \pmod{p}$. Since we are assuming that the discrete logarithm problem is intractable, the Schnorr identification scheme is secure.

10.2 Pseudorandom Number Generators-II

In Section 8.3, we studied the generation of a sequence of pseudorandom numbers using the linear congruential generator. The linear congruential generator has been very well tested for statistical randomness properties and has numerous applications. What it lacks is the unpredictability needed for cryptographic applications. If $x_{n+1} = ax_n + b \bmod m$ is a linear congruential generator, then it is possible to compute a, b, and m by knowing a few terms of the sequence $x_0, x_1, \ldots$ Then all the terms of the sequence can be computed, so it is not suitable for many cryptographic applications.

In 1986, L. Blum, M. Blum, and M. Shub proposed a secure random number generator for cryptographic applications.[2] This generator is such that in the sequence $x_0, x_1, \ldots, x_k$, the knowledge of any number of terms x_i does not allow us to compute the previous terms x_{i-1} in a reasonable amount of time. This difficulty in computing x_{i-1} from x_i is based on the difficulty of factoring large integers.

Suppose $n = pq$ is a product of two primes such that $p \equiv q \equiv 3$ (mod 4). Given an integer x_0, the Blum–Blum–Shub generator outputs a sequence $x_0, x_1, \ldots, x_i, \ldots$ where

$$x_{i+1} = x_i{}^2 \bmod n.$$

We will abbreviate this generator as the BBS generator with seed x_0. The BBS bit generator is the sequence $b_0, \ldots, b_i, \ldots$ given by

$$b_i = x_i \bmod 2.$$

The BBS generator can be defined for any n, but for applications it will be important that $n = pq$, for $p \equiv q \equiv 3 \pmod 4$.

Example 10.2.1. Let $n = 7 \cdot 11$ and $x_0 = 5$. The sequence generated by the BBS generator is

$$5, 25, 9, 4, 16, 25, 9, 4, \ldots.$$

This sequence has a period length of 4 with period 25, 9, 4, 16.

If $x_0 = 63$, then we obtain the sequence

$$63, 42, 70, 49, 14, 42, 70, \ldots$$

of period length 4.

Our first task is to determine the period length of the sequence. If x_0 is not a quadratic residue, then it doesn't occur in the sequence again and the period will not start with the first element. In the example above, the period started with the second term and this will be the case whenever x_0 is not a quadratic residue and n is of the prescribed form. For the general BBS generator (arbitrary n), we can determine the start of the period and give a procedure to compute the period length. This depends on the properties of orders and minimal universal exponents. Recall that the order of x modulo n is the smallest integer $\mathrm{ord}_n(x)$ such that $x^{\mathrm{ord}_n(x)} \equiv 1 \pmod n$. The important property of order that we need is if $x^k \equiv 1 \pmod n$, then k is a multiple of the order.

Proposition 10.2.2. *Let $p(x_0)$ be the period of the BBS generator modulo n with seed x_0. If $\mathrm{ord}_n(x) = 2^r s$ with s odd, then $p(x_0) = \mathrm{ord}_s(2)$.*

[2]L. Blum, M. Blum, and M. Shub. "A Simple Unpredictable Pseudorandom Number Generator." *SIAM Journal of Computing*, 15(1986), 364–383.

PROOF. Let x_i be the $(i+1)$st term of the sequence given by $x_i = x^{2^i}$ mod n. Suppose that $x_i = x_j$ for some $i < j$. Then $x^{2^i} \equiv x^{2^j} \pmod{n}$ or $x^{2^j - 2^i} \equiv 1 \pmod{n}$. Then the properties of order imply that $2^j - 2^i \equiv 0 \pmod{\mathrm{ord}_n(x)}$. This implies that $2^r \mid 2^j - 2^i$, so $i \geq r$. After dividing by 2^r, we obtain the condition

$$2^{j-r} \equiv 2^{i-r} \pmod{s}$$

or

$$j - r \equiv i - r \pmod{\mathrm{ord}_s(2)}.$$

This implies that $i \equiv j \pmod{\mathrm{ord}_s(2)}$, so $\mathrm{ord}_2(s)$ divides the period length.

Conversely, if $i \equiv j \pmod{\mathrm{ord}_s(2)}$ and $i \geq r$, then we reverse the steps to see that

$$2^{i-r} \equiv 2^{j-r} \pmod{s}$$

and hence $x_i = x_j$. ∎

Example 10.2.3. In the example above with $n = 77$, we can easily compute $\mathrm{ord}_{77}(5) = 30$. Since $30 = 2 \cdot 15$, the period length is $\mathrm{ord}_{15}(2) = 4$. This is actually the longest possible period for any sequence produced by the BBS generator modulo 77. This is because $\lambda(77) = 30$, so 30 is the largest possible order modulo 77, thus, any value of s is going to divide 15, and so the length of 4 is the best possible. Observe that $r = 1$, so the period starts with the second term x_1.

Recall that the minimal universal exponent $\lambda(n)$ is defined as the smallest positive integer such that $a^{\lambda(n)} \equiv 1 \pmod{n}$ for all $(a, n) = 1$. The minimal universal exponent has the property that the order of every element divides it, and moreover, there is an element whose order is the minimal universal exponent. The minimal universal exponent $\lambda(n)$ is usually much smaller than $\phi(n)$ and can be computed in terms of the prime factorization of n. (See Proposition 7.2.12.)

Example 10.2.4. Suppose $n = pq$ with $p = 1471$ and $q = 6763$. We compute $\lambda(n) = \mathrm{LCM}[p-1, q-1] = 33810$. If x_0 is a quadratic residue, then its order can be at most $\lambda(n)/2 = 16905$. We can find the order of an element by computing $a^{\lambda(n)/t} \bmod n$ for all prime divisors t of $\lambda(n)$. By this method, it is easy to verify that $a = 564$ has order 16905, and by computing the Legendre symbol of 564 modulo p and q, we see that it is a quadratic residue. The period length of the sequence with seed $x_0 = 564$ is the order of 2 modulo 16905. Since $16905 = 3 \cdot 5 \cdot 7^2 \cdot 23$ has many prime factors, this order will be quite small. It is easy to compute that $\mathrm{ord}_{16905}(2) = 924$, so the period length of the BBS generator, with $n = pq$ and seed x_0, is 924.

As the example shows, the period length depends on prime factors of $\lambda(n)$, and if n prime factors, then the period length will be small. So for maximal period length, it is best to choose n so that $\lambda(n)$ has as few prime factors as possible.

For application to cryptography, we show that the BBS sequence with $n = pq$ of the prescribed form is unpredictable to the left. This means that if x_i is known, then computing x_{i-1} is equivalent to knowing the prime factorization of n. Since $x_{i-1}^2 \equiv x_i \pmod{n}$, it is possible for several different x_{i-1} to give the same x_i. But if we impose the restriction that x_{i-1} be a quadratic residue (which it is for $i \geq 1$), then x_{i-1} is uniquely determined by x_i.

Proposition 10.2.5. *Suppose x is a quadratic residue modulo $n = pq$, with p and q primes satisfying $p \equiv q \equiv 3 \pmod{4}$. Then there is a unique y such that $y^2 \equiv x \pmod{n}$ and y is a quadratic residue modulo n.*

PROOF. Since n is a product of two odd primes, there are four solutions to the congruence $y^2 \equiv x \pmod{n}$. If y_0 is a solution, then the four solutions are

$$
\begin{array}{ll}
y \equiv y_0 \pmod{p} & y \equiv y_0 \pmod{p} \\
y \equiv y_0 \pmod{q}, & y \equiv -y_0 \pmod{q},
\end{array}
$$

$$
\begin{array}{ll}
y \equiv -y_0 \pmod{p} & y \equiv -y_0 \pmod{p} \\
y \equiv y_0 \pmod{q}, & y \equiv -y_0 \pmod{q}.
\end{array}
$$

The four possible solutions give rise to four possible combinations of the Legendre symbols $\left(\dfrac{y}{p}\right)$ and $\left(\dfrac{y}{q}\right)$. The four pairs are:

$$
\left\{\left(\frac{y_0}{p}\right),\left(\frac{y_0}{q}\right)\right\}, \quad \left\{\left(\frac{y_0}{p}\right),\left(\frac{-y_0}{q}\right)\right\},
$$

$$
\left\{\left(\frac{-y_0}{p}\right),\left(\frac{y_0}{q}\right)\right\}, \quad \left\{\left(\frac{-y_0}{p}\right),\left(\frac{-y_0}{q}\right)\right\}.
$$

Since $p \equiv q \equiv 3 \pmod{4}$, the Legendre symbols $\left(\dfrac{-1}{p}\right)$ and $\left(\dfrac{-1}{q}\right)$ are both -1, and each of the four possible combinations of signs, $\{1, 1\}$, $\{1, -1\}$, $\{-1, 1\}$, and $\{-1, -1\}$ must occur once; irrespective of the values of $\left(\dfrac{y_0}{p}\right)$ and $\left(\dfrac{y_0}{q}\right)$. Hence, there is only one y such that $\left(\dfrac{y}{p}\right) = 1$ and $\left(\dfrac{y}{q}\right) = 1$; that is, there is only one solution to $y^2 \equiv x \pmod{n}$ such that y is a quadratic residue modulo n. ∎

Example 10.2.6. Consider $n = pq$ with $p = 331$ and 431. We select $x = 11535$ as a quadratic residue modulo n. Using the techniques of the previous chapter, we can solve $y^2 \equiv x \pmod{n}$, to obtain solutions $y \equiv \pm 127060, \pm 58962 \pmod{n}$. Of these four, only one, 127060 is a quadratic residue modulo n. The reader should also verify that all four possible combinations of signs of the Legendre symbols occur.

An application of the BBS pseudorandom sequence is to the construction of one-time pads in a public-key system. A one-time pad is a string $s_0, s_1, \ldots, s_k$ that is added to the plaintext $p_0, \ldots, p_k$ to produce the ciphertext $c_0, c_1, \ldots, c_k$. This operation can be represented by $c_i = p_i + s_i \bmod 2$.

Suppose Alice has set a public key $n = pq$, where p and q satisfy the condition $p \equiv q \equiv 3 \pmod 4$. If Bob wishes to send a coded message to Alice, then he chooses a number x_0 and computes the BBS sequence $x_0, x_1, \ldots, x_k, x_{k+1}$. Then Bob computes the bits $s_i = x_i \bmod 2$. The random bit stream $s_1, \ldots, s_k$ is used as a one-time pad to the plaintext $p_1, \ldots, p_k$. Bob sends the ciphertext $c_0, \ldots, c_k$ with $c_i = p_i + s_i \bmod 2$ and the integer x_{k+1}. Alice uses x_{k+1} to recover the numbers $x_k, x_{k-1}, \ldots, x_1$ and from this, the one-time pad. It is easy to recover the plaintext from the knowledge of the ciphertext and the one-time pad.

The scheme works because there is only one square root of each term that is a quadratic residue. The security of the scheme lies in the fact that computing y satisfying $y^2 \equiv x \pmod{n}$ is equivalent to factoring n.

──────────── **Exercises for Section 10.2** ────────────

1. Determine the number of seeds x_0 modulo $n = 77$ so that the BBS generator gives a sequence of period length 4.

2. Determine the maximum possible period length of a BBS sequence modulo $n = 23 \cdot 43$. Find a value of x_0 that gives this period length.

3. Show that the maximal possible period length in the BBS generator modulo n is $\lambda\left(\frac{\lambda(n)}{2}\right)$.

4. Determine the unique quadratic residue y that is a solution to the congruence $y^2 \equiv 198506 \pmod{225413}$.

5. (Blum–Blum–Shub) Let $n = pq$ where $p = 2p_1 + 1$, $p_1 = 2p_2 + 1$, $q = 2q_1 + 1$ and $q_1 = 2q_2 + 1$ and p_1, p_2, q_1, q_2 are all primes greater than 3. Show that if 2 is a quadratic residue modulo at most one of p_1, q_1, then the maximal possible period length is achieved in the BBS generator with a seed of order $\lambda(n)/2$.

Chapter 11

Continued Fractions

$$\cfrac{1}{1+\cfrac{e^{-2\pi}}{1+\cfrac{e^{-4\pi}}{1+\cfrac{e^{-6\pi}}{1+\cdots}}}} = \left\{ \sqrt{\frac{5+\sqrt{5}}{2}} - \frac{\sqrt{5}+1}{2} \right\} e^{2\pi/5}.$$

— L. Rogers and S. Ramanujan

11.1 Introduction

We usually think of a real or rational number in terms of its decimal expansion. This familiar representation is very convenient for computation. There is another representation of a real number that expresses certain geometric properties in a wonderful way. This is the simple continued fraction expansion of a real number. The representation occurs naturally when one tries to approximate real numbers by rational numbers and has applications to the solution of many Diophantine equations. The simplest examples of continued fractions, the finite ones, are essentially based on the Euclidean algorithm and hence date from ancient times. The continued fractions in the modern sense first appear in the work of Rafael Bombelli (1526–1573). The techniques were used by Huygens (1629–1695) in the construction of a mechanical planetarium. Similar methods were used by the Indian astronomer, Aryabhata, in the fifth century. The basic properties of continued fractions were developed by Euler and Lagrange.[1]

---

[1] For a historical discussion, see C. Brezinski. *History of Continued Fractions and Pade' Approximants*. New York: Springer–Verlag, 1991.

We begin with two examples to motivate the continued fraction representation.

Example 11.1.1. Consider the steps in the computation of $(181, 101)$ using the Euclidean algorithm.

$$181 = 1 \cdot 101 + 80$$
$$101 = 1 \cdot 80 + 21$$
$$80 = 3 \cdot 21 + 17$$
$$21 = 1 \cdot 17 + 4$$
$$17 = 4 \cdot 4 + 1.$$

We can rewrite these as fractions, and by combining them, write $\dfrac{181}{101}$ in terms of the successive quotients:

$$\frac{181}{101} = 1 + \frac{80}{101} \tag{1}$$

$$\frac{101}{80} = 1 + \frac{21}{80} \tag{2}$$

$$\frac{80}{21} = 3 + \frac{17}{21} \tag{3}$$

$$\frac{21}{17} = 1 + \frac{4}{17} \tag{4}$$

$$\frac{17}{4} = 4 + \frac{1}{4}. \tag{5}$$

Combining these equations yields the formula

$$\frac{181}{101} = 1 + \cfrac{1}{1 + \cfrac{1}{3 + \cfrac{1}{1 + \cfrac{1}{4 + \cfrac{1}{4}}}}}.$$

where in (1) we have replaced $\dfrac{80}{101}$ with $\cfrac{1}{1 + \cfrac{21}{80}}$ from (2), replace $\dfrac{21}{80}$ by

$\cfrac{1}{3 + \cfrac{17}{21}}$, and so on.

This representation of $181/101$ is known as a simple continued fraction, and the terms occurring in it are the successive quotients in the Euclidean algorithm.

Example 11.1.2. Consider the following approach to solve $x^2 - x - 1 = 0$. We write $x^2 = x + 1$, and dividing by x, obtain $x = 1 + \dfrac{1}{x}$. We can replace x in the denominator by the expression $1 + \dfrac{1}{x}$, so we must have

$$x = 1 + \cfrac{1}{1 + \cfrac{1}{x}}.$$

We can repeat this process any number of times to get

$$x = 1 + \cfrac{1}{1 + \cfrac{1}{1 + \cfrac{}{\ddots + \cfrac{1}{x}}}}.$$

The expression on the right involves x, and to get a formula for x independent of x, we bravely let the process repeat to infinity and write

$$x = 1 + \cfrac{1}{1 + \cfrac{1}{1 + \cdots}}$$

as the solution to the quadratic equation. Naturally, we must justify this step. This will follow from a study of the infinite continued fractions in Section 11.3.

Definition 11.1.3. An expression of the form

$$a_0 + \cfrac{b_1}{a_1 + \cfrac{b_2}{a_2 + \cfrac{b_3}{a_3 + \cfrac{}{\ddots + \cfrac{b_{n-1}}{a_{n-1} + \cfrac{b_n}{a_n}}}}}}$$

is called a **continued fraction**. It is called a **simple continued fraction** if all the b_i's are 1 and the a_i's are integers satisfying $a_1, a_2, \cdots \geq 1$. To avoid additional terminology, we will sometimes allow the last term, a_n, to be a positive irrational number and still call it a simple continued fraction.

We focus on the simple continued fractions in our study. Properties of other types of continued fractions are developed in some of the exercises and projects.

We denote simple continued fractions as $[a_0, a_1, a_2, \ldots, a_n]$, that is,

$$[a_0, a_1, a_2, \ldots, a_n] = a_0 + \cfrac{1}{a_1 + \cfrac{1}{a_2 + \cfrac{\ddots}{\ddots + \cfrac{1}{a_n}}}}.$$

In this notation,

$$\frac{181}{101} = [1, 1, 3, 1, 4, 4].$$

Any rational number can be expressed as a simple continued fraction. To express a/b as a simple continued fraction, we apply the Euclidean Algorithm to a and b.

Exercise. Express $256/101$ as a simple continued fraction.

Proposition 11.1.4. *A real number α can be expressed as a finite simple continued fraction if and only if α is rational.*

PROOF. Suppose $\alpha = [a_0, a_1, \ldots, a_n]$, where the a_i's are integers. Then the expression on the right-hand side involves a finite combination of additions, multiplications, and division of integers; hence it is rational.

We can also prove the assertion by induction on the length of the continued fraction, n. Clearly, if $n = 1$, $\alpha = [a_0] = a_0$ is an integer, hence rational. Assume that all finite continued fractions with positive *rational* terms a_i of length less than or equal to n are rational numbers. We take a simple continued fraction of length $n + 1$ with integral entries and rewrite it as a simple continued fraction of length n (with the last term rational).

$$\alpha = [a_0, a_1, \ldots, a_n]$$

$$= a_0 + \cfrac{1}{a_1 + \cfrac{\ddots}{\ddots + \cfrac{1}{a_{n-1} + \cfrac{1}{a_n}}}}$$

$$= \left[a_0, a_1, \ldots, a_{n-1} + \frac{1}{a_n}\right].$$

The continued fraction has length n, and hence it is rational by the induction hypothesis.

Conversely, if $\alpha = m/n$ is rational, the Euclidean algorithm applied to m and n terminates in a finite number of steps, so the simple continued fraction has a finite number of terms. ■

The continued fraction expansion is most useful for irrational numbers. From the proposition, an irrational number cannot have a finite simple continued fraction expansion. We will define an infinite continued fraction as a limit of finite ones. This requires some properties of finite continued fractions, which are developed in Section 11.2. Assuming that a real number can be expressed as a simple continued fraction, we can write down an algorithm to compute it. This is possible because the definition of the simple continued fraction implies that there is only one simple continued fraction expansion of a real number. To see this, suppose $\alpha = a_0 + \cfrac{1}{a_1 + \cfrac{1}{a_2 + \cdots}}$. Now, $a_1, a_2, \ldots$

are positive integers, hence $a_1 + \cfrac{1}{a_2 + \cdot}$ is larger than 1, or $\alpha - a_0$ is less than 1 and nonnegative. We must have $a_0 = \lfloor \alpha \rfloor$ because $0 < \alpha - a_0 < 1$. Using this process, we can determine a_1, a_2, and so on.

Let $\alpha = \underbrace{\lfloor \alpha \rfloor}_{a_0} + [[\alpha]]$, where $[[\alpha]]$ denotes the fractional part of α. Let $x_1 = 1/[[\alpha]]$. Then $\alpha = a_0 + \frac{1}{x_1}$. Let $a_1 = \lfloor x_1 \rfloor$, and x_2 be the reciprocal of the fractional part of x_1. (Notice that $a_1 \geq 1$.) Inductively, we compute a_i as the integer part of x_i and let x_{i+1} be the reciprocal of the fractional part of x_i, that is $x_i = a_i + \cfrac{1}{x_{i+1}}$. At any step, $\alpha = [a_0, a_1, \ldots, a_{i-1}, x_i]$.

Let's look at an example that illustrates this process.

Example 11.1.5. Determine the continued fraction expansion of $\sqrt{5}$.
Since $\lfloor \sqrt{5} \rfloor = 2$, the fractional part of $\sqrt{5}$ is $\sqrt{5} - 2$. We write

$$\sqrt{5} = 2 + (\sqrt{5} - 2)$$
$$= 2 + \cfrac{1}{1/(\sqrt{5} - 2)}.$$

We compute this by rationalizing the denominator:

$$\frac{1}{\sqrt{5} - 2} = \frac{\sqrt{5} + 2}{(\sqrt{5} - 2)(\sqrt{5} + 2)} = \frac{\sqrt{5} + 2}{5 - 4} = \sqrt{5} + 2,$$

which gives $\sqrt{5} = 2 + \cfrac{1}{\sqrt{5} + 2}$. Now, $\lfloor \sqrt{5} + 2 \rfloor = 4$ and $\sqrt{5} + 2 =$

$4 + (\sqrt{5} - 2)$. So

$$\sqrt{5} = 2 + \cfrac{1}{4 + (\sqrt{5} - 2)} = 2 + \cfrac{1}{4 + \cfrac{1}{\sqrt{5} + 2}}.$$

Now the computation repeats because the term $1/(\sqrt{5} + 2)$ occurs again and hence we can write

$$\sqrt{5} = 2 + \cfrac{1}{4 + \cfrac{1}{4 + \cfrac{}{\ddots \, + \cfrac{1}{\sqrt{5} + 2}}}}$$

for any number of terms, and we expect that $\sqrt{5}$ has the infinite simple continued fraction

$$[2, 4, 4, \ldots, 4, \ldots].$$

Exercise. Imitate the process in the above example to find the simple continued fraction expansion of $\sqrt{7}$.

The algorithm for the continued fraction expansion of an irrational number shows that the continued fraction expansion is unique. If the number is rational, then there are two possibilities for the last term of the continued fraction expansion.

──────────────── **Exercises for Section 11.1** ────────────────

1. Determine the continued fraction expansion of the following rational numbers.

    (a) $\dfrac{81}{19}$ (b) $\dfrac{290}{176}$

    (c) $\dfrac{387}{52}$ (d) $\dfrac{1729}{625}$

 2. Write a computer program that outputs the simple continued fraction expansion of a rational number using the Euclidean algorithm.

3. Is the simple continued fraction expansion of a rational number unique? If not, what are the possible continued fraction expansions of a rational number?

4. Write a computer program that determines the rational number from the continued fraction expansion. This can be accomplished by a procedure using the identity

$$[a_0, a_1, \ldots, a_k] = a_0 + \frac{1}{[a_1, a_2, \ldots, a_k]}.$$

5. Use the technique of Example 11.1.5 to determine the simple continued fraction expansions of $\sqrt{2}$, $\sqrt{3}$, and $\sqrt{11}$.

6. Use the following approximations of π and e to determine the first few terms in the simple continued fraction expansion of these numbers.

$$\pi = 3.1415926535897932384626434\ldots$$
$$e = 2.718281828459045235360287\ldots.$$

7. The kth convergent of a continued fraction $[a_0, \ldots, a_n]$ is the fraction $C_k = [a_0, \ldots, a_k]$. Compute the convergents $C_1, C_2, \ldots, C_6$ of π using the continued fraction in the previous exercise. To how many places is C_6 accurate as an approximation of π?

8. Show that if $p > q$ and $p/q = [a_0, a_1, \ldots, a_n]$, then $q/p = [0, a_0, \ldots, a_n]$.

9. Determine the first few terms in the simple continued fraction expansions of $e - 1$ and $(e - 1)/(e + 1)$. Make a conjecture regarding the infinite simple continued fractions of these numbers. (To obtain enough terms to see a pattern, you may need to use a decimal expansion to more than 20–25 places.)

10. Take the number e^2 to a hundred decimal places and use the program that computes the continued fraction expansion of a rational number to guess the continued fraction expansion of e^2.

11. Make a table of the first twenty terms of the continued fraction of $\sqrt{n}$ for n in the range 2 to 50. What conclusions can you reach about the continued fraction expansion of $\sqrt{n}$ from these examples?

11.2 Convergents

Further development of the theory of continued fractions comes from the properties of convergents.

Definition 11.2.1. We define the kth **convergent** C_k of the simple continued fraction $[a_0, a_1, \ldots, a_n]$ to be $C_k = [a_0, a_1, \ldots, a_k]$ for $k \leq n$.

Example 11.2.2. The convergents of $181/101 = [1, 1, 3, 1, 4, 4]$ are

$$C_0 = [1] = 1$$

$$C_1 = [1, 1] = 1 + \frac{1}{1} = 2$$

$$C_2 = [1, 1, 3] = 1 + \cfrac{1}{1 + \cfrac{1}{3}} = \frac{7}{4}$$

$$C_3 = [1, 1, 3, 1] = \frac{9}{5}$$

$$C_4 = [1, 1, 3, 1, 4] = \frac{43}{24}$$

$$C_5 = [1, 1, 3, 1, 4, 4] = \frac{181}{101}.$$

The convergents of a simple continued fraction are obtained by taking the terms from the beginning of the continued fraction. This process is analogous to the partial sums of a series. An infinite series is defined as the limit of its partial sums, and we will define an infinite continued fraction as the limit of its convergents. Before doing this, we develop some simple, but fundamental properties of the convergents.

The convergent C_k can be computed using the formula

$$C_k = a_0 + \frac{1}{[a_1, a_2, \ldots, a_k]}$$
$$= \frac{a_0[a_1, \ldots, a_k] + 1}{[a_1, \ldots, a_k]}.$$

Although this procedure is suitable for a computer, it is hard for us to evaluate a long continued fraction this way. Moreover, $[a_1, \ldots, a_k]$ is not a convergent of the continued fraction. It is necessary to develop a different formula for the convergents in order to see their basic properties. Let us derive a formula for the convergents which depends only the previous convergents of the continued fraction. We simplify the expressions for the first few convergents to see if there is a pattern.

$$C_0 = a_0$$

$$C_1 = a_0 + \frac{1}{a_1} = \frac{a_0 a_1 + 1}{a_1}$$

$$C_2 = a_0 + \cfrac{1}{a_1 + \cfrac{1}{a_2}} = a_0 + \cfrac{1}{\cfrac{a_1 a_2 + 1}{a_2}} = a_0 + \frac{a_2}{a_1 a_2 + 1}$$

$$= \frac{a_0 a_1 a_2 + a_2 + a_0}{a_1 a_2 + 1}$$

$$= \frac{a_2(a_0 a_1 + 1) + a_0}{a_2(a_1) + 1}.$$

Note that the terms in the numerator of C_2 are those of the numerators of C_1 and C_0, and the same holds for the denominators.

Exercise. Verify that

$$C_3 = \frac{a_3(a_2(a_0 a_1 + 1) + a_0) + a_0 a_1 + 1}{a_3(a_1 a_2 + 1) + a_1}.$$

If we write $C_k = p_k/q_k$, where p_k and q_k are integers, then these computations indicate that

$$p_k = a_k p_{k-1} + p_{k-2}$$

and

$$q_k = a_k q_{k-1} + q_{k-2}.$$

Proposition 11.2.3. *The numerator p_k and denominator q_k of the kth convergent of the continued fraction $[a_0, a_1, \ldots, a_n]$ satisfy the recurrence relation*

$$p_k = a_k p_{k-1} + p_{k-2} \tag{1}$$
$$q_k = a_k q_{k-1} + q_{k-2}, \tag{2}$$

with initial values

$$p_0 = a_0, \quad p_1 = a_1 a_0 + 1,$$
$$q_0 = 1, \quad q_1 = a_1.$$

PROOF. The proposition is proved by induction. For $k = 2$ the assertion was verified before the statement of the proposition.

We assume that the equations are true for all continued fractions of length less than or equal to $k - 1$. Consider

$$C_k = [a_0, a_1, \ldots, a_k]$$

$$= \left[a_0, a_1, a_2, \ldots, a_{k-1} + \frac{1}{a_k}\right].$$

C_k is the $(k-1)$st convergent of $[a_0, a_1, a_2, \ldots, a_{k-1} + 1/a_k]$. Let C_i' be the ith convergent of the continued fraction $[a_0, a_1, \ldots, a_{k-1} + 1/a_k]$. Note that $C_2' = C_2$, $C_3' = C_3, \ldots$ but $C_{k-1}' \neq C_{k-1}$. By definition, $C_k = p_k/q_k = p_{k-1}'/q_{k-1}'$. Applying the induction hypothesis to this continued fraction, we obtain

$$C_k = \frac{p_{k-1}'}{q_{k-1}'} = \frac{(a_{k-1} + 1/a_k)p_{k-2}' + p_{k-3}'}{(a_{k-1} + 1/a_k)q_{k-2}' + q_{k-3}'}.$$

We can drop the superscript on all the terms on the right because C_i and C_i' are the same for $i < k - 1$. Simplify the right-hand side to obtain

$$\frac{p_k}{q_k} = C_k = \frac{a_k(a_{k-1}p_{k-2} + p_{k-3}) + p_{k-2}}{a_k(a_{k-1}q_{k-2} + q_{k-3}) + q_{k-2}}.$$

Using the induction hypothesis, notice that the terms in the parentheses in the numerator and denominator of the above fraction are p_{k-1} and q_{k-1}, respectively, hence

$$\frac{p_k}{q_k} = C_k = \frac{a_k p_{k-1} + p_{k-2}}{a_k q_{k-1} + q_{k-2}}$$

and the proposition is proved.   ∎

The numerators and denominators of the convergents satisfy many nice relations that are the basis of their utility. The following two are the most important.

Proposition 11.2.4. *Let $C_k = p_k/q_k$ be the kth convergent of $[a_0, a_1, \ldots, a_k]$. Then*

$$p_k q_{k-1} - q_k p_{k-1} = (-1)^{k-1} \tag{3}$$

$$p_k q_{k-2} - q_k p_{k-2} = (-1)^k a_k. \tag{4}$$

PROOF. We prove (3) (by induction) and leave the proof of (4) to the reader.
For $k = 1$, $p_1 q_0 - q_1 p_0 = (a_1 a_0 + 1)1 - a_1 a_0 = 1 = (-1)^{1-1}$.
Assume that the formula is true for $C_1, \ldots, C_{k-1}$. Using the recurrence relations proved in Proposition 11.2.3, we have

$$\begin{aligned}
p_k q_{k-1} - q_k p_{k-1} &= (a_k p_{k-1} + p_{k-2})q_{k-1} - (a_k q_{k-1} + q_{k-2})p_{k-1} \\
&= a_k p_{k-1} q_{k-1} + p_{k-2} q_{k-1} - a_k q_{k-1} p_{k-1} - q_{k-2} p_{k-1} \\
&= p_{k-2} q_{k-1} - q_{k-2} p_{k-1} \\
&= -(p_{k-1} q_{k-2} - p_{k-2} q_{k-1}) \\
&= -(-1)^{k-2} \quad \text{by the induction hypothesis} \\
&= (-1)^{k-1}.
\end{aligned}$$

∎

The proposition shows that our choice of p_k and q_k are in lowest terms, that is, relatively prime. In addition, we adopt the convention that if $C_k < 0$, then $p_k < 0$ and $q_k > 0$; otherwise, both p_k and q_k are positive.

Since p_k and q_k are relatively prime, there exist x and y such that $p_k x + q_k y = 1$. We can solve for x and y by using the continued fraction expansion of p_k/q_k and computing the penultimate convergent. This is natural because the continued fraction method is just the Euclidean algorithm, which is the method by which we solve linear Diophantine equations.

Example 11.2.5. Let us solve $257x - 97y = 1$. The equation has a solution because $(257, 97) = 1$. The continued fraction expansion is $\frac{257}{97} = [2, 1, 1, 1, 5, 1, 4]$ and the convergents are $2, 3, \frac{5}{2}, \frac{8}{3}, \frac{45}{17}, \frac{53}{20}, \frac{257}{97}$. We compute $257 \cdot 20 - 97 \cdot 53 = -1$. The reason for the negative sign is because the continued fraction has an odd number of terms. To get a positive sign, we can write another continued fraction with an even number of terms by writing the last term in two parts, obtaining $\frac{257}{97} = [2, 1, 1, 1, 5, 1, 3, 1]$. Now the penultimate convergent is $\frac{204}{77}$ and we see that $257 \cdot 77 - 204 \cdot 97 = 1$.

The procedure of the previous example can be used to solve any linear Diophantine equation $ax - by = 1$ with $(a, b) = 1$. Let $b > 0$ and expand $\frac{a}{b} = [a_0, \ldots, a_n]$. If n is odd (the continued fraction has an even number of terms), then $x = q_{n-1}$, $y = p_{n-1}$ is a solution. If n is even, then write $\frac{a}{b} = [a_0, \ldots, a_n - 1, 1]$ and the penultimate convergent gives the solution. The solution obtained satisfies the condition $0 < x < b$, when $b > 0$.

There is another way to write equations (1) and (2) that is convenient for proving properties of continued fractions. For $k \geq 2$, using 2×2 matrices, we can write

$$\begin{pmatrix} p_k & q_k \\ p_{k-1} & q_{k-1} \end{pmatrix} = \begin{pmatrix} a_k & 1 \\ 1 & 0 \end{pmatrix} \begin{pmatrix} p_{k-1} & q_{k-1} \\ p_{k-2} & q_{k-2} \end{pmatrix}. \tag{5}$$

We can use the matrix representation for p_k and q_k to give a simpler proof of Equation (3). For a 2×2 matrix $g = \begin{pmatrix} a & b \\ c & d \end{pmatrix}$, the determinant is $ad - bc$, denoted $\det g$.

Use Equation (5) to observe that

$$\det \begin{pmatrix} p_k & q_k \\ p_{k-1} & q_{k-1} \end{pmatrix} = \det \begin{pmatrix} a_k & 1 \\ 1 & 0 \end{pmatrix} \det \begin{pmatrix} p_{k-1} & q_{k-1} \\ p_{k-2} & q_{k-2} \end{pmatrix}$$

or

$$p_k q_{k-1} - q_k p_{k-1} = (-1)(p_{k-1} q_{k-2} - q_{k-1} p_{k-2})$$

and since $p_1 q_0 - q_1 p_0 = 1$, the result is easily proved by induction.

The identities of Proposition 11.2.4 imply the following relations among the convergents of the continued fraction expansion. These relations are essential to justifying the existence of the continued fraction expansion of a real number.

Corollary 11.2.6. *If* $C_i = p_i/q_i$ *is the ith convergent of* $[a_0, \ldots, a_n]$, *then for any* k, $0 \le k \le n$, *we have*

$$C_k - C_{k-1} = \frac{(-1)^{k-1}}{q_k q_{k-1}} \tag{6}$$

$$C_k - C_{k-2} = \frac{a_k(-1)^k}{q_k q_{k-2}}. \tag{7}$$

PROOF. These are obtained from the identities of Proposition 11.2.4 after dividing by $q_k q_{k-1}$. ∎

Lemma 11.2.7. *Let* $C_k = p_k/q_k$ *be the kth convergent of* $[a_0, a_1, \ldots, a_n]$ *with* $a_i \ge 1$ *for* $i \ge 1$. *Then* q_k *form an increasing sequence of positive integers.*

The reader should have no trouble giving a proof of this using induction.

The positivity of the q_k implies the following inequalities among the convergents of the simple continued fractions. Equation (6) in the corollary implies that

$$
\begin{array}{ll}
C_1 - C_0 > 0 \qquad\qquad & C_1 > C_0 \\
C_2 - C_1 < 0 & C_2 < C_1 \\
C_3 - C_2 > 0 & C_3 > C_2 \\
C_4 - C_3 < 0 & C_4 < C_3 \\
\qquad\quad\vdots &
\end{array}
$$

and Equation (7) implies that

$$
\begin{array}{ll}
C_2 - C_0 > 0 \qquad\qquad & C_2 > C_0 \\
C_3 - C_1 < 0 & C_3 < C_1 \\
C_4 - C_2 > 0 & C_4 > C_2,
\end{array}
$$

and so on. If $\alpha = [a_0, a_1, \ldots, a_n]$, then $C_n = \alpha$ and we have the following diagram.

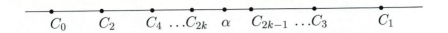

$C_0 \qquad C_2 \qquad C_4 \ldots C_{2k} \qquad \alpha \qquad C_{2k-1} \ldots C_3 \qquad\qquad C_1$

Corollary 11.2.8. *The odd convergents form a decreasing sequence, and the even convergents form an increasing sequence.*

Each convergent lies between the two preceding convergents.

PROOF. The assertion is clear from the inequalities given above. ∎

In the next section, this monotonicity property of even and odd convergents will allow us to define infinite continued fractions as a limit of the convergents.

─────────────── **Exercises for Section 11.2** ───────────────

1. Compute the convergents of the following continued fractions.

   (a) $[-1, 2, 3, 7, 1]$
   (b) $[1, 3, 6, 11]$
   (c) $[1, 3, 6, 3]$

2. Write a program to find the convergents of a simple continued fraction using the recurrence relations of Theorem 11.2.3.

3. Given positive integers a, b, c, and d, show that $[a, c] < [a, d]$ for $c > d$, but $[a, b, c] > [a, b, d]$.

4. Let $a_1, a_2, \ldots, a_n, x$ be positive real numbers. Determine the values of n for which $[a_0, a_1, \ldots, a_n] > [a_0, a_1, \ldots, a_n + x]$ holds.

5. Solve the following Diophantine equations.

   (a) $-718x + 123y = 1$
   (b) $417x - 172y = 1$
   (c) $1163x - 815y = 1$

6. Show that the recurrence relations of Proposition 11.2.3 are valid for $k = 0, 1$ by taking $p_{-1} = 1$, $q_{-1} = 0$, $p_{-2} = 0$, and $q_{-2} = 1$. Show that

$$\begin{pmatrix} p_k & q_k \\ p_{k-1} & q_{k-1} \end{pmatrix} = \begin{pmatrix} a_k & 1 \\ 1 & 0 \end{pmatrix} \begin{pmatrix} a_{k-1} & 1 \\ 1 & 0 \end{pmatrix} \cdots \begin{pmatrix} a_0 & 1 \\ 1 & 0 \end{pmatrix} \begin{pmatrix} 1 & 0 \\ 0 & 1 \end{pmatrix}$$

$$= \begin{pmatrix} a_k & 1 \\ 1 & 0 \end{pmatrix} \begin{pmatrix} a_{k-1} & 1 \\ 1 & 0 \end{pmatrix} \cdots \begin{pmatrix} a_0 & 1 \\ 1 & 0 \end{pmatrix}.$$

7. Show that the denominator q_k of the kth convergent of $[a_0, a_1, \ldots, a_n]$ satisfies $q_k \geq 2^{k/2}$ for $k \geq 2$, if $a_i \geq 1$ for all $i \geq 1$.

8. Prove Equation (4).

9. Show that for any real number $x > 0$,

$$[a_0, a_1, \ldots, a_{k-1}, x] = \frac{xp_{k-1} + p_{k-2}}{xq_{k-1} + q_{k-2}},$$

where p_i/q_i are the convergents of $[a_0, a_1, \ldots, a_{k-1}]$.

10. Suppose $a_0 \geq 1$. Determine all rational numbers p/q such that the first two convergents of p/q are $C_0 = a_0$ and $C_1 = a_0 + 1$.

11. Let $a/b = [a_0, a_1, \ldots, a_n]$ and let $C_k = p_k/q_k$ be the kth convergent of a/b. Suppose all the $a_i > 0$, then show that p_k form an increasing sequence and

$$\frac{p_n}{p_{n-1}} = [a_n, \ldots, a_0]$$

and

$$\frac{q_n}{q_{n-1}} = [a_n, \ldots, a_1].$$

Is the assertion valid if a_0 is not positive?

12. If p_k/q_k are the convergents of a simple continued fraction, show that

$$(p_k^2 - q_k^2)(p_{k-1}^2 - q_{k-1}^2) = (p_k p_{k-1} - q_k q_{k-1})^2 - 1.$$

13. Let $a/b = [a_0, \ldots, a_n, a_n, \ldots, a_0] = p_{2n+1}/q_{2n+1}$. Show that $p_{2n+1} = p_n^2 + p_{n-1}^2$ and $q_{2n+1} = p_n q_n + q_{n-1}p_{n-1}$.

14. (a) Suppose $a/b = [a_0, \ldots, a_n, a_n, \ldots, a_0]$ is a symmetric continued fraction with an even number of terms with $a, b > 1$. Show that $p_{2n} = b$ and $a \mid b^2 + 1$.

   (b) If $a/b = [a_0, \ldots, a_{n-1}, a_n, a_{n-1}, \ldots, a_0]$ is a symmetric continued fraction with an odd number of terms, show that $a \mid b^2 - 1$.

15. (a) Let f_n be the nth Fibonacci number, with $f_1 = 1$, $f_2 = 1$, $f_n = f_{n-1} + f_{n-2}$. Show that the continued fraction expansion of f_{n+1}/f_n consists of n 1's, that is,

$$\frac{f_{n+1}}{f_n} = \underbrace{[1, 1, \ldots, 1]}_{n \text{ ones}}.$$

   (b) What identities can you derive among the Fibonacci numbers using Proposition 11.2.4 and Exercises 13 and 14?

   (c) Conclude that any three consecutive Fibonacci numbers have no common factor.

16. Let p_k/q_k be the kth convergent of $[a_0, \ldots, a_n]$. For $s \leq k$ define $x_s = [a_s, \ldots, a_k]$, and let $x_s = A_{s,k}/B_{s,k}$, where $A_{s,k}$ and $B_{s,k}$ are coprime and positive.

   (a) Show that for $0 \leq s \leq k$,
   $$p_k q_s - q_s p_k = (-1)^{k-s} B_{s+1,k}.$$
   This generalizes the identities of Proposition 11.2.4.

   (b) Apply this identity to the continued fraction of Exercise 15 to derive a relation among the Fibonacci numbers.

   (c) Use the identity of (b) to determine (f_m, f_n). (The reader may have guessed this value based on computations done in Exercise 2.5.21.)

17. Show that if $[a_0, a_1, \ldots, a_n] = [b_0, b_1, \ldots, b_m]$ with all the a_i's positive and $a_n > 1$, and all the b_i's positive and $b_m > 1$, then $n = m$ and $a_0 = b_0$, $a_1 = b_1, \ldots, a_n = b_m$. (This shows that the continued fraction expansion of a rational number is unique if the last term is taken larger than 1.)

18. Let C_k be the kth convergent of $a/b = [a_0, a_1, \ldots, a_n]$. Show that each convergent is nearer to a/b than the preceding one.

19. The result of Exercise 11 can be used to prove the important theorem that a prime $p \equiv 1 \pmod 4$ is representable as a sum of two squares. This proof due to H. J. Smith is outlined in the following exercises.

   (a) Let S be the set of simple continued fraction expansions of the fractions p/q, $2 \leq q \leq \frac{p-1}{2}$. If $[a_0, a_1, a_2, \ldots a_k] \in S$, then show that the reversed fraction $[a_k, a_{k-1}, \ldots a_1, a_0]$ is also in S.

   (b) Let $f : S \to S$ be the function defined by $f([a_0, a_1, \ldots, a_k]) = [a_k, \ldots, a_0]$. Show that $f \circ f$ (the composition of f with itself) is the identity and that f leaves an element x of S fixed, that is, $f(x) = x$ for some x.

   (c) Suppose $f(x) = x$. Show that x is a symmetric continued fraction. Explain why x must have an even number of terms. Conclude that p is representable as a sum of two squares.

   (d) For $p = 13$ and $p = 29$, compute S, x, and the corresponding representation of p as a sum of two squares.

11.3 Infinite Continued Fractions

Our goal in this section is to define an infinite simple continued fraction and develop some of its basic properties. Recall that an infinite series, $\sum_{n=1}^{\infty} a_n$, is defined as a limit of the finite sums $s_N = \sum_{n=1}^{N} a_n$, when such a limit exists. Similarly, we would like to define an infinite continued fraction as a limit of finite continued fractions. The major difficulty for infinite series is to establish the existence of the limit. Fortunately, for infinite simple continued fractions, convergence can be established easily using the properties of the convergents proved in the previous section. The situation is more complicated for nonsimple continued fractions. Some special cases are explored in the projects at the end of the chapter.

We begin with a lemma that will be useful throughout this section.

Lemma 11.3.1. *Suppose $a_1, a_2, \ldots$ is a sequence of real numbers such that $a_i \geq 1$. If $q_0 = 1$, $q_1 = a_1$ and $q_i = a_i q_{i-1} + q_{i-2}$ for $i > 1$, then $q_{i-1} < q_i$ for all i and $\lim_{i \to \infty} q_i = \infty$.*

PROOF. The assertion follows from the fact that $q_i - q_{i-1} = (a_i - 1)q_{i-1} + q_{i-2} > q_{i-2} > 1$. ∎

Proposition 11.3.2. *Let $a_0, a_1, \ldots, a_k, \ldots$ be a sequence of integers with $a_i > 0$ for $i \geq 1$. Let $C_k = [a_0, a_1, \ldots, a_k]$. Then $\lim_{k \to \infty} C_k$ exists.*

PROOF. If $C_0, C_1, \ldots, C_N$ denote the convergents of the simple continued fraction $[a_0, a_1, \ldots, a_N]$, then they satisfy the monotonicity properties of Corollary 11.2.8, that is, the even convergents are increasing and the odd convergents are decreasing. Moreover, any even convergent is less than an odd convergent, and any odd convergent is greater than an even convergent. Since N is arbitrary, we have

$$C_0 < C_2 < C_4 < \cdots < C_{2k} < \cdots < C_1$$
$$C_0 < \cdots < C_{2k+1} < \cdots < C_5 < C_3 < C_1.$$

A fundamental property of the real number system is that a bounded monotone (only increasing or only decreasing) sequence has a limit.[2] Hence the even convergents have a limit, say α, and the odd convergents have a limit β, that is,

$$\lim_{k \to \infty} C_{2k} = \alpha \quad \text{and} \quad \lim_{k \to \infty} C_{2k+1} = \beta.$$

To see that $\alpha = \beta$, recall that $C_k - C_{k-1} = \dfrac{(-1)^{k-1}}{q_k q_{k-1}}$, and since q_k are an increasing sequence of integers, $q_k \to \infty$ as $k \to \infty$. This implies that the distance between consecutive convergents shrinks as $k \to \infty$; that is, the limit of the odd and even convergents must be the same, so the limit of the convergents exists. ∎

Remark. The proof of the proposition above only uses the fact that $a_i \geq 1$ for the q_i's to form an unbounded increasing sequence. It is not necessary that the a_i be integers for the convergence of the continued fraction.

---

[2]See any book on advanced calculus or real analysis.

Definition 11.3.3. Let $a_0, a_1, \ldots$ be integers such that $a_i \geq 1$ for $i \geq 1$. The **infinite simple continued fraction**, $[a_0, a_1, \ldots, a_k, \ldots]$, is defined as the limit of the convergents $C_k = [a_0, a_1, \ldots, a_k]$.

Recall the method of Section 11.1 to find the simple continued fraction expansion of a real number x. Denote x by x_0 and write $x_0 = a_0 + [[x_0]]$, where $a_0 = \lfloor x_0 \rfloor$ and $[[x_0]]$ is the fractional part. Write $x_1 = 1/[[x_0]]$, so $x_0 = a_0 + \dfrac{1}{x_1}$. Since $0 \leq [[x_0]] < 1$, we have $x_1 > 1$. We repeat this step and inductively write

$$x_k = a_k + [[x_k]], \qquad x_{k+1} = \frac{1}{[[x_k]]},$$

or

$$x_k = a_k + \frac{1}{x_{k+1}}.$$

The x_i's are real numbers and are called the **complete quotients** of x. The a_i's are called the **partial quotients** of x.

We will show that the simple continued fraction expansion of a real number is unique, that the entries are the partial quotients, and that for any positive integer k, $x = [a_0, a_1, \ldots, a_k, x_{k+1}]$.

Proposition 11.3.4.   (a) *Any real number x can be expressed as the continued fraction $x = [a_0, a_1, \ldots]$, where the a_i's are the partial quotients of x.*

(b) *If $x = [a_0, a_1, \ldots]$ is a simple continued fraction, then the a_i's are the partial quotients of x and the complete quotients of x are $x_{k+1} = [a_{k+1}, a_{k+2}, \ldots]$.*

PROOF.   (a) If x_{k+1} is the $(k+1)$st complete quotient, then for any positive integer k, $x = [a_0, a_1, \ldots, a_k, x_{k+1}]$. For this continued fraction, $C_{k+1} = x$ and $|C_k - C_{k+1}| = 1/(q_k q_{k+1})$. We use the fact that $a_i \geq 1$, hence $q_k \to \infty$. Then $|C_k - C_{k+1}| = |C_k - x|$ can be made as small as desired by taking k sufficiently large. Hence $\lim_{k \to \infty} C_k = x$, that is, $x = [a_0, a_1, a_2, \ldots]$.

(b) Let $y_{k+1} = [a_{k+1}, a_{k+2}, \ldots]$. Then for any positive integer k, we can show that $x = [a_0, a_1, \ldots, a_k, y_{k+1}]$ (see Exercise 1). Using this expression for x, we will prove by induction on k that the a_k's are the partial quotients and the y_k's are the complete quotients of x.

If $k = 1$, then $x = [a_0, y_1] = a_0 + \dfrac{1}{y_1}$, hence $y_1 = x_1$.

Suppose that $y_i = x_i$ for $1 \le i \le k$; then $x_k = y_k = a_k + \dfrac{1}{y_{k+1}}$. Since $y_{k+1} \ge 1$, $a_k = \lfloor x_k \rfloor$ and $x_{k+1} = y_{k+1}$. This proves that the a_k's are the partial quotients. ∎

The proposition proves the uniqueness of the simple continued fraction expansion of a real number and gives an algorithm to compute it.

All the properties of convergents proved in Section 11.2 are valid for the finite continued fraction $x = [a_0, a_1, \ldots, a_k, x_{k+1}]$ for any k, hence for the convergents of the infinite continued fraction. We restate them here for the reader's convenience.

Let $C_k = p_k/q_k$ denote the kth convergent of $[a_0, \ldots, a_n]$; then

$$p_k = a_k p_{k-1} + p_{k-2}$$
$$q_k = a_k q_{k-1} + q_{k-2},$$

with initial values

$$p_0 = a_0, \quad p_1 = a_1 a_0 + 1,$$
$$q_0 = 1, \quad\; q_1 = a_1,$$

We also have

$$x = \frac{x_{k+1} p_k + p_{k-1}}{x_{k+1} q_k + q_{k-1}} \quad \text{for any } k, \tag{1}$$

and

$$C_k - C_{k-1} = \frac{(-1)^{k-1}}{q_k q_{k-1}}, \quad C_k - C_{k-2} = \frac{a_k (-1)^k}{q_k q_{k-2}}.$$

In practice, computing the simple continued fraction of a real number requires a rational approximation of the number. Then the algorithm of Section 11.1 can be applied to this rational number. This is the best that we can do for a number like π, whose simple continued fraction has no discernible pattern. For e, a pattern can be detected by computation (Exercise 11.1.6). The proof of the conjectured expansion of e was given by Euler using the hypergeometric function. We present this proof in Section 11.6.

An exception to this rule (of requiring a decimal approximation) in computing the simple continued fraction of an irrational number occurs when the number is the solution to a polynomial equation with integer coefficients. The method is particularly simple if the irrational number under consideration is the only real positive root of the polynomial. Among these numbers, the quadratic irrationals are studied in more detail in the next two sections.

Let $f(x)$ be a polynomial (with integer coefficients) with a unique real positive root α. Assume that the leading coefficient of f is positive; then the first partial quotient a_0 is the largest integer n such that $f(n) < 0$. (Why?)

If $g(x) = f(x + a_0)$, then $g(\alpha - a_0) = f(\alpha - a_0 + a_0) = f(\alpha) = 0$. We are going to construct a polynomial with a root at $x_1 = 1/(\alpha - a_0)$, the first complete quotient. Let $f_1(x) = -x^d g(1/x)$, where d is the degree of f. Then $f_1(1/(\alpha - a_0)) = 0$ and f_1 is a polynomial of degree d with positive leading coefficient and a unique real positive root. Thus, a_1 is the largest integer n such that $f_1(n) < 0$. Inductively, we can find a polynomial $f_i(x)$ of degree d with a unique positive real root at x_i.

Note that it is important that f have only one positive real root so that a_0 can be found by the characterization above, and then f_1 also has the same property.

We can write the algorithm as follows.

Algorithm 11.3.5. Let $f(x)$ be a polynomial with integer coefficients, only one real positive root α, and a positive leading coefficient.

1. [Initialize] Let $n = 0$, $f_0 = f(x)$, and $d = \deg f(x)$.

2. [Find Partial Quotient] Find a_n such that a_n is the largest integer for which $f_n(a_n) < 0$.

3. [Transform the polynomial] Let $g(x) = f_n(x + a_n)$ and $f_{n+1}(x) = -x^d g(1/x)$.

4. [Repeat] Let $n = n + 1$ and go to step 2.

Remarks.   1. The algorithm uses only integer arithmetic and can be used to compute as many terms as necessary in the continued fraction.

2. Step 2 can be done very rapidly using the binary search method. Find a large integer n such that $f(n) > 0$. Set a to 0 and b to n. If $f([a + b]/2) > 0$, then let $b = (a + b)/2$; otherwise, let $a = (a + b)/2$. We stop when $b - a < 1$, and the desired integer is $\lfloor a \rfloor$.

3. Step 3 can be performed efficiently by writing down a formula for the coefficients of f_{n+1} using the Binomial Theorem (see Exercise 12).

Example 11.3.6. Compute the continued fraction expansion of $\sqrt[3]{3}$.

The polynomial $f(x) = x^3 - 3$ has one real positive root at $\sqrt[3]{3}$, hence we can use the algorithm.

Since $f(1) = -2$ and $f(k) > 0$ for $k \geq 2$, then $a_0 = 1$ and

$$g(x) = f(x + 1) = (x + 1)^3 - 3 = x^3 + 3x^2 + 3x - 2$$

$$f_1(x) = -x^3 \left(\frac{1}{x^3} + \frac{3}{x^2} + \frac{3}{x} - 2 \right) = 2x^3 - 3x^2 - 3x - 1.$$

Now $f_1(2) = -3$ and $f_1(k) > 0$ if $k \geq 3$, so $a_1 = 2$ and

$$g(x) = f_1(x+2) = 2x^3 + 9x^2 + 9x - 3$$
$$f_2(x) = 3x^3 - 9x^2 - 9x - 2.$$

In the next step, $f_2(3) = -29$ and $f_2(k) > 0$ for $k \geq 4$, so $a_2 = 3$. Continuing in this way, we obtain $\sqrt[3]{3} = [1, 2, 3, 1, 4, 1, 5, 1, 1, 6, 2, \cdots]$.

─────────────── **Exercises for Section 11.3** ───────────────

1. Let $x = [a_0, a_1, \ldots, a_k, \ldots]$ be an infinite simple continued fraction. Show that
$$x = [a_0, a_1, \ldots, a_k, [a_{k+1}, \ldots]].$$

2. Show that $|q_n \alpha - p_n| < 1/q_{n+1}$ where p_n/q_n is the nth convergent of the simple continued fraction expansion of an irrational number α.

3. Show that
$$\alpha - \frac{p_k}{q_k} = \frac{(-1)^k}{q_k(x_{k+1}q_k + q_{k-1})}.$$

4. If p/q is a convergent of the simple continued fraction expansion of an irrational number α, show that
$$\left| \alpha - \frac{p}{q} \right| < \frac{1}{q^2}.$$

5. Show that
$$\alpha = \lfloor \alpha \rfloor + \frac{1}{q_0 q_1} - \frac{1}{q_1 q_2} + \frac{1}{q_2 q_3} - \cdots,$$
where q_i is the denominator of the ith convergent C_i of α.

6. Suppose that two irrational numbers x and y have identical convergents C_0, $C_1, \ldots, C_n$. Prove that the first $(n+1)$ terms of their continued fractions are identical.

7. Let $x < z$ be two irrational numbers. Suppose that the continued fractions of x and z have the same first $n+1$ convergents $C_0, C_1, \ldots, C_n$. Prove that the continued fraction of any y with $x < y < z$ has the same convergents $C_0, C_1, \ldots, C_n$.

8. Two real numbers x and z are **equivalent** ($x \sim z$) if there exists integers $a, b, c,$ and d such that $ad - bc = \pm 1$ and
$$x = \frac{az + b}{cz + d}.$$

    (a) Show that if $x = [a_0, a_1, a_2, \ldots, a_k, z]$, then x is equivalent to z.

    (b) Show that $x \sim y$ if and only if $y \sim x$.

    (c) Show that if $x \sim y$ and $y \sim z$, then $x \sim z$.

    (d) Use these facts to show that if x and z have continued fractions with identical tail ends, then $x \sim z$. (The tail ends of the continued fractions are identical if there exists a y such that $x = [a_0, \ldots, a_k, y]$ and $z = [b_0, \ldots, b_l, y]$ for some k and l).

9. This exercise proves special cases for the converse of part (d) of the previous exercise.

    (a) Suppose $y = \frac{7x+3}{5x+2}$. Show that $7/5 = [1, 2, 2]$ and $y = [1, 2, 2, x]$, so x and y have continued fraction expansions with the same tail ends.

    (b) Suppose $y = \frac{13x+8}{5x+3}$. Write y as a continued fraction with tail end x.

    (c) For the general case, suppose A, B, C, and D are integers such that $C > D > 0$ and $AD - BC = 1$. If $y = \dfrac{Ax + B}{Cx + D}$ and $A/C = [a_0, \ldots, a_n]$. If n is even, then write $A/C = [a_0, a_1, \ldots, a_{n-1}, 1]$. With this choice of the continued fraction expansion, if $C_k = p_k/q_k$ are the convergents of $A/C = p_n/q_n$ then show that $D = q_{n-1}$ and $B = p_{n-1}$.

    (d) The case $y = -1/x$ is not covered by the previous exercise. Suppose $x = [a_0, a_1, \ldots]$, then $x = \frac{p_k x_{k+1} + p_{k-1}}{q_k x_{k+1} + q_{k-1}}$. Express $-1/x$ in terms of the complete quotient x_{k+1} of x and show that for some k, the result of part (c) can be applied to the expression. Hence y and x have the same tail ends.

10. Show that the polynomial f_1 constructed in Algorithm 11.3.5 is of degree d with positive leading term and only one real positive root.

11. Using Algorithm 11.3.5, find the first three terms of the continued fraction expansion of $\sqrt[3]{2}$.

12. Suppose $f_n(x) = c_d x^d + c_{d-1} x^{d-1} + \cdots + c_1 x + c_0$. Determine the jth coefficient of $f_{n+1}(x)$.

13. Implement Algorithm 11.3.5 to find the continued fraction expansion of an irrational number that is the only positive real root of a polynomial with integer coefficients. Use it to compute the first forty terms of the continued fraction expansion of the numbers $\sqrt[3]{2}$, $\sqrt[3]{3}$, $\sqrt[3]{4}$, $\sqrt[3]{5}$ and the real root of $x^5 - x - 1 = 0$.

14. Let $f_0(x) = x^2 - n$ with $n > 0$. Show that if $\lfloor \sqrt{n} \rfloor = a_0$, then $f_1(x) = -1 - 2a_0 x + (n - a_0^2)x^2$. Conclude that if x_1 and $\overline{x_1}$ are the roots of f_1 with $x_1 > 1$, then $-1 < \overline{x_1} < 0$.

15. Let $f(x) = ax^2 + bx + c$ have two real roots α and $\overline{\alpha}$ satisfying $\alpha > 1$ and $-1 < \overline{\alpha} < 0$. Show that $-1 < c/a < 0$ and $b/a > 0$.

    Let $a_0 = \lfloor \alpha \rfloor$, and conclude that the roots β and $\overline{\beta}$ of $f_1(x)$, where $f_1(x) = -x^2 f(1/[x + a_0])$, also satisfy $\beta > 1$ and $-1 < \overline{\beta} < 0$.

11.4 Quadratic Irrationals

The algorithm for continued fractions described at the end of Section 11.3 is particularly simple for a quadratic polynomial. Since there is a simple formula for the roots of a quadratic equation, we can find a simple method to describe the partial and complete quotients in the continued fraction of $\sqrt{n}$. This description shows that the continued fraction of a quadratic irrational is eventually periodic. Conversely, every eventually periodic continued fraction converges to a quadratic irrational of the form $a + b\sqrt{n}$, with a and b rational. Additional properties of the convergents of $\sqrt{n}$ have applications to the factorization of integers and the solution of some Diophantine problems.

The roots of a quadratic polynomial $f(x) = ax^2 + bx + c$ of positive discriminant can be written as $(A_0 \pm \sqrt{n})/B_0$. We can write each complete quotient x_k in the form $(A_k + \sqrt{n})/B_k$ by rationalizing the denominator. The following theorem gives the method to compute A_k and B_k in terms of the previous values.

Theorem 11.4.1. *The complete quotients in the continued fraction of* $(A_0 + \sqrt{n})/B_0$ *are* $x_k = (A_k + \sqrt{n})/B_k$ *with* $B_k \neq 0$. *If* $a_k = \lfloor x_k \rfloor$, *then for all* $k \geq 1$

$$A_{k+1} = a_k B_k - A_k, \qquad B_{k+1} = \frac{n - A_{k+1}^2}{B_k}.$$

Moreover, if $B_0 \mid n - A_0^2$, *then* $B_k \mid n - A_k^2$ *for all* k, *and all* A_k, B_k *are integers.*

PROOF. We denote the kth partial quotient by a_k.

For $k \geq 0$ we have,

$$x_{k+1} = \frac{1}{x_k - a_k}$$

$$= \frac{1}{\dfrac{A_k + \sqrt{n}}{B_k} - a_k}$$

$$= \frac{B_k}{A_k - a_k B_k + \sqrt{n}}.$$

Rationalizing the denominator, we obtain

$$x_{k+1} = \frac{B_k(A_k - a_k B_k - \sqrt{n})}{(A_k - a_k B_k)^2 - n}$$

$$= \frac{a_k B_k - A_k + \sqrt{n}}{\left(\dfrac{n - (a_k B_k - A_k)^2}{B_k}\right)}.$$

When $k = 0$, we can take $A_1 = a_0 B_0 - A_0$ and $B_1 = (n - A_1^2)/B_0$. Then by induction, the formula for x_k is valid for all k.

In addition, we wish to show that if $B_0 \mid n - A_0^2$, then the fraction in the denominator of x_{k+1} is an integer for all k. We can assume by induction that A_k and $B_k = (n - A_k^2)/B_{k-1}$ are integers. This implies that $A_{k+1} = a_k B_k - A_k$ is an integer. To show that B_{k+1} is an integer and divides $n - A_{k+1}^2$, write $n - A_{k+1}^2 = n - (a_k B_k - A_k)^2 = n - A_k^2 + B_k(2a_k A_k - B_k a_k^2)$. By the induction hypothesis, B_k divides the first term, and it is a factor of the second term. This implies that $(n - A_{k+1}^2)$ is divisible by B_k, hence B_{k+1} is an integer. This also shows that $B_{k+1} \mid n - A_{k+1}^2$. This completes the proof of the theorem. ∎

Remarks.   1. At each step we determine a_k by

$$a_k = \lfloor x_k \rfloor = \left\lfloor \frac{A_k + \sqrt{n}}{B_k} \right\rfloor.$$

2. In the important case of $\sqrt{n}$, $A_0 = 0$ and $B_0 = 1$. Hence all the A_i and B_i are integers.

3. If $(A_0 + \sqrt{n})/B_0$ is a root of $ax^2 + bx + c = 0$ such that $n = b^2 - 4ac$, $-b = A_0$ and $2a = B_0$, then $B_0 \mid n - A_0^2$, and the condition of the theorem is satisfied. In this case, all the A_i, B_i are integers. So the quadratic irrationals $(A_0 + \sqrt{n})/B_0$ for which A_i and B_i fail to be integers are solutions of quadratic equations with discriminant different from n. Even in this case, we can write the number in a different form so that the condition of the theorem is satisfied. (See Exercise 5.)

Example 11.4.2. Find the continued fraction expansion of $\sqrt{7}$.
    We tabulate the results of the first six steps

k	A_k	B_k	x_k	a_k
0	0	1	$\sqrt{7}$	2
1	2	3	$\frac{2+\sqrt{7}}{3}$	1
2	1	2	$\frac{1+\sqrt{7}}{2}$	1
3	1	3	$\frac{1+\sqrt{7}}{3}$	1
4	2	1	$2 + \sqrt{7}$	4
5	2	3	$\frac{2+\sqrt{7}}{3}$	1

Observe that when $k = 5$, we obtain the same terms as when $k = 1$. Since the computation of A_k and B_k depends only on the previous terms, the terms must repeat. Hence the continued fraction is

$$\sqrt{7} = [2, 1, 1, 1, 4, 1, 1, 1, 4, 1, 1, 1, 4, \ldots] = [2, \overline{1, 1, 1, 4}].$$

Example 11.4.3. Here is the result of the algorithm applied to $(2 + \sqrt{11})/3$. Using $A_0 = 2$ and $B_0 = 3$, we obtain

k	A_k	B_k	x_k	a_k
0	2	3	$\frac{2+\sqrt{11}}{3}$	1
1	1	10/3	$\frac{1+\sqrt{11}}{10/3}$	1
2	7/3	5/3	$\frac{7/3+\sqrt{11}}{5/3}$	3
3	8/3	7/3	$\frac{8/3+\sqrt{11}}{7/3}$	2
4	2	3	$\frac{2+\sqrt{11}}{3}$	1
5	1	10/3	$\frac{1+\sqrt{11}}{10/3}$	1

Continuing in this way, we see that $(2 + \sqrt{11})/3 = [1, 1, 3, 2, 1, 1, 3, 2, \dots]$. In this example, A_i and B_i are not integers because the condition $B_0 \mid n - A_0^2$ is not satisfied. This happens because $(2 + \sqrt{11})/3$ is not a root of quadratic polynomial of discriminant 11. (Verify?)

Exercise. Apply the algorithm of Theorem 11.4.1 to determine the continued fraction expansion of $\sqrt{11}$ and $(3 + \sqrt{5})/4$.

The computations performed so far indicate that the continued fraction of a quadratic irrational is eventually periodic. To prove the periodicity, we would like to show that some complete quotient repeats. Since we can assume that A_k and B_k are integers, it is enough to show that A_k and B_k are bounded; hence there are only finitely many choices for them, so they must repeat. When A_k and B_k repeat, the continued fraction begins to repeat from that point onwards.

The following lemmas are crucial to proving many properties of continued fractions of quadratic irrationals. If $x = a + b\sqrt{n}$, then we call $\bar{x} = a - b\sqrt{n}$ the **conjugate** of x. It is necessary to consider both the quadratic irrational and its conjugate to deduce properties of the continued fraction expansion. The key point in the proof of the periodicity is that we can find a complete quotient x_k such that $-1 < \overline{x_k} < 0$. This property leads to the desired estimates on the A's and B's. In the case of $\sqrt{n}$, we have $x_1 \overline{x_1} = -1/(n - a_0^2)$, hence $-1 < \overline{x_1} < 0$. This will prove that the continued fraction expansion of $\sqrt{n}$ repeats after the first term. If the property $-1 < \overline{x_1} < 0$ is satisfied for a single complete quotient, then it is satisfied for all subsequent complete quotients.

Lemma 11.4.4. *Suppose that $x > 1$ and $-1 < \bar{x} < 0$. If $x = a + 1/y$ with $a = \lfloor x \rfloor$, then $y > 1$ and $-1 < \bar{y} < 0$.*

PROOF. It is clear from the construction that $y > 1$. To prove the other inequality, we proceed as follows. After conjugating, $\overline{x} = a + 1/\overline{y}$ and $-1 < \overline{x} < 0$ imply that $-1 < a + 1/\overline{y} < 0$, so $-1 - a < 1/y < -a$ and we must have $\overline{y} < 0$. Also, since $a \geq 1$, it is true that $1/\overline{y} < -1$, which can be written as $-1 < \overline{y}$; therefore, $-1 < \overline{y} < 0$. ■

Lemma 11.4.5. *The complete quotients x_k of a quadratic irrational α satisfy $-1 < \overline{x_k} < 0$ and $x_k > 1$ for all k sufficiently large.*

PROOF. We first show that there exists a k such that $\overline{x_k} < 0$. Recall that for any k,

$$\alpha = \frac{x_k p_{k-1} + p_{k-2}}{x_k q_{k-1} + q_{k-2}}.$$

Taking conjugates of both sides, we obtain

$$\overline{\alpha} = \frac{\overline{x_k} p_{k-1} + p_{k-2}}{\overline{x_k} q_{k-1} + q_{k-2}}.$$

Solve for $\overline{x_k}$ to get

$$\overline{x_k} = -\frac{q_{k-2}}{q_{k-1}} \left(\frac{\overline{\alpha} - \frac{p_{k-2}}{q_{k-2}}}{\overline{\alpha} - \frac{p_{k-1}}{q_{k-1}}} \right).$$

As the convergents p_i/q_i approach α, the right-hand side of the equation approaches $-q_{k-2}/q_{k-1}$ for k large enough. This shows that all the $\overline{x_k}$ are eventually negative.

If $\overline{x_k} < 0$, then $-1 < x_{k+1} < 0$. This follows from the fact that $a_k \geq 1$ and the formula $\overline{x_{k+1}} = \dfrac{1}{\overline{x_k} - a_k}$. ■

Definition 11.4.6. We say that a continued fraction $[a_0, a_1, \ldots]$ is **eventually periodic** if there exist T and N such that $a_{i+T} = a_i$ for all $i \geq N$.

We can write an eventually periodic continued fraction in the form

$$[a_0, a_1, \ldots, a_{N-1}, \overline{b_1, \ldots, b_T}].$$

Remark. The periodic part and the period are not uniquely determined. We will usually take the smallest period and identify a period from its first occurrence. For example, $[1, 1, 2, 1, 2, 1, 2, 1, 2, \ldots]$ can be written as $[1, \overline{1, 2}]$, or $[1, 1, \overline{2, 1}]$, or $[1, 1, 2, \overline{1, 2}]$.

Theorem 11.4.7 (Lagrange). *The continued fraction expansion of a quadratic irrational is eventually periodic.*

PROOF. With the notation of Theorem 11.4.1, it is enough to show that some combination of A_k and B_k recurs, and hence the a_k's repeat from this point onwards. This is accomplished by showing that there are a finite number of possible combinations for A_k and B_k, while the continued fraction expansion must be infinite.

For k sufficiently large, we know that $x_k > 1$ and $-1 < \overline{x_k} < 0$, or equivalently,

$$\frac{A_k + \sqrt{n}}{B_k} > 1 \quad \text{and} \quad -1 < \frac{A_k - \sqrt{n}}{B_k} < 0.$$

By taking the difference, we see that $2\sqrt{n}/B_k > 0$, and hence $B_k > 0$.

Also $A_{k+1} = a_k B_k - A_k < x_k B_k - A_k = \sqrt{n}$; therefore, $A_{k+1} < \sqrt{n}$. Then $(A_k + \sqrt{n})/B_k > 1$ implies that $B_k < A_k + \sqrt{n} < 2\sqrt{n}$. From $-B_k < A_k - \sqrt{n}$, we obtain $A_k > -B_k + \sqrt{n} > -\sqrt{n}$. So we have obtained

$$-\sqrt{n} < A_k < \sqrt{n} \quad \text{and} \quad 0 < B_k < 2\sqrt{n}.$$

Hence there are only a finite number of possible complete quotients x_k and hence some value repeats. This proves that the continued fraction expansion of a quadratic irrational is eventually periodic. ∎

The converse of this assertion is also true. Every eventually periodic continued fraction is a quadratic irrational. The proof is left to the reader (Exercise 6). Here is an example that illustrates the technique of evaluating an eventually periodic continued fraction.

Example 11.4.8. Let $\alpha = [3, 2, 1, 2, 1, 2, 1, \dots]$. To compute α, separate the periodic part $x = [2, 1, 2, 1, \dots]$. The convergents of x are $2/1, 3/1, 8/3, \dots$. Using the relation $x = [2, 1, x]$, we can write

$$x = \frac{3x + 2}{x + 1}.$$

This is a quadratic equation in x, which has $x = 1 + \sqrt{3}$ as its only positive solution. Now $\alpha = [3, x]$, hence $\alpha = 3 + 1/x = \frac{5+\sqrt{3}}{2}$.

──────────── **Exercises for Section 11.4** ────────────

1. Show that $\left\lfloor \dfrac{A_k + \sqrt{n}}{B_k} \right\rfloor = \left\lfloor \dfrac{A_k + \lfloor \sqrt{n} \rfloor}{B_k} \right\rfloor$.

2. Determine the continued fraction expansion of $\sqrt{17}$, $\sqrt{19}$, and $\sqrt{21}$.

3. Determine the continued fraction expansions of $1 + \sqrt{5}$ and $(2 + \sqrt{7})/4$.

4. Evaluate the following continued fractions.

   (a) $[3, 2, 1, 3, 2, 1, \ldots]$.
   (b) $[1, 2, 2, 3, 2, 3, 2, 3, \ldots]$.

5. Given $(A_0 + \sqrt{n})/B_0$, show that there exist n', A_0', and B_0' such that $\frac{A_0 + \sqrt{n}}{B_0} = \frac{A_0' + \sqrt{n'}}{B_0'}$ and $B_0' \mid n' - A_0'^2$.

6. Prove that every eventually periodic continued fraction is a quadratic irrational.

7. Suppose α is a quadratic irrational. Show that there exists a positive constant C such that
$$\left| \alpha - \frac{p}{q} \right| > \frac{1}{Cq^2}.$$

8. Show that the second recurrence relation of Theorem 11.4.1 can be written as
$$B_k = B_{k-2} + a_{k-1}(A_{k-1} - A_k).$$

   This form does not require any divisions in computing the terms of the continued fraction expansion, a fact useful in the factoring algorithm discussed in Section 12.1.

9. Verify that the first complete quotient x_1 in the expansion of $\sqrt{n}$ satisfies $-1 < \overline{x_1} < 0$.

10. Write a computer program to implement the algorithm of Theorem 11.4.1 for the continued fraction expansion of $\sqrt{n}$. To detect the period, you may assume that the period occurs when B_k first equals 1 for $k \geq 1$. (This result is proved in Corollary 11.5.3.)

11. Use your program to determine the continued fractions of a few numbers of the form $\sqrt{n^2 - 1}$. What do you observe? Make a general conjecture and prove it. You may find the technique of Example 11.4.8 helpful in evaluating an eventually continued fraction.

12. Repeat Exercise 11 to numbers of the form $\sqrt{n^2 + 1}$, $\sqrt{n^2 \pm 2}$, ... Make conjectures regarding the form of the continued fractions and prove them.

13. Experiment with other forms of polynomials $P(n)$, where the expansion of $\sqrt{P(n)}$ has a nice pattern.

14. As in Exercises 11 and 12, many nice patterns can be discovered in the continued fraction of $\sqrt{n}$. Based on examples computed, what else do you notice regarding the shape of the periodic part or the period length of the continued fraction? Can you prove your conjectures?

15. What can you say about the period length of the expansion of $\sqrt{p}$ for p prime?

16. Make a table of continued fraction expansions of many quadratic irrationals α and their conjugates $\overline{\alpha}$. Do you see a relationship between the periodic parts of the continued fraction expansions of a quadratic irrational and its conjugate? Prove your conjectures.

11.5   Purely Periodic Continued Fractions

Our goal is to deduce further properties of the continued fractions of quadratic irrational numbers. We first study the purely periodic continued fractions because these occur in the continued fraction of all the quadratic irrationals. The crucial fact used in all the proofs in this section is the following lemma.

Lemma 11.5.1. *Suppose $x > 1$ is a quadratic irrational, satisfying $-1 < \overline{x} < 0$. If $x = a + 1/y$, $a = \lfloor x \rfloor$, then*

$$\left\lfloor \frac{-1}{\overline{y}} \right\rfloor = a.$$

PROOF. Since $0 < -\overline{x} < 1$, we have $a < a - \overline{x} < a+1$. Since $-1/\overline{y} = a - \overline{x}$, we obtain $a < -1/\overline{y} < a + 1$. Since a is an integer, the definition of the greatest integer function implies that $\lfloor -1/\overline{y} \rfloor = a$. ∎

A continued fraction x is **purely periodic** if $x = [\overline{a_0, a_1, \ldots, a_k}]$. The first term a_0 occurs again, so it must be positive, hence $x > 1$. These irrationals have the following simple characterization.

Theorem 11.5.2. *If a quadratic irrational x satisfies $x > 1$ and $-1 < \overline{x} < 0$, then its continued fraction expansion is purely periodic. Conversely, a purely periodic continued fraction converges to a quadratic irrational x satisfying $x > 1$, $-1 < \overline{x} < 0$.*

PROOF. Suppose $x > 1$ and $-1 < \overline{x} < 0$. Applying Lemma 11.4.4, we see that all the complete quotients x_k satisfy this property. Since the continued fraction is eventually periodic, there exists i and j, $i < j$ such that $x_i = x_j$. We can write

$$x_{i-1} = a_{i-1} + \frac{1}{x_i}$$

$$x_{j-1} = a_{j-1} + \frac{1}{x_j}.$$

Applying Lemma 11.5.1 to these equations and using $\frac{-1}{\overline{x_i}} = \frac{-1}{\overline{x_j}}$, we see that $a_{i-1} = a_{j-1}$, hence $x_{i-1} = x_{j-1}$. We can repeat this process to show that $x_{i-2} = x_{j-2}$ or $x_{i-k} = x_{j-k}$ for all k, $0 \leq k \leq i$. This is possible because all the complete quotients x_k satisfy the property $x_k > 1$, $-1 < \overline{x_k} < 0$. Hence there exists an m such that $x_0 = x_m$, that is, the continued fraction is purely periodic.

Conversely, if $x = [\overline{a_0, a_1, \ldots, a_k}]$, then

$$x = \frac{xp_k + p_{k-1}}{xq_k + q_{k-1}}.$$

We cross-multiply and simplify to see that x satisfies the quadratic equation $x^2 q_k + x(q_{k-1} - p_k) - p_{k-1} = 0$. Let $f(x) = x^2 q_k + x(q_{k-1} - p_k) - p_{k-1}$. Then $f(0) = -p_{k-1} < 0$ and

$$\begin{aligned} f(-1) &= q_k - q_{k-1} + p_k - p_{k-1} \\ &= (a_k - 1)q_{k-1} + q_{k-2} + (a_k - 1)p_{k-1} + p_{k-2} \\ &> 0. \end{aligned}$$

This implies that f has a root between -1 and 0. As x is the positive root of $f(x) = 0$ and $\overline{x}$ is the other root, we must have $-1 < \overline{x} < 0$. ∎

The technique of the proof can be adapted to the expansion of $\sqrt{n}$ to prove many of its properties. Our first property gives a simple criterion to detect when the period occurs, permitting easy computation of the continued fraction expansion.

Corollary 11.5.3. *The continued fraction expansion of $\sqrt{n}$ is of the form $\sqrt{n} = [a_0, \overline{a_1, a_2, \ldots, a_k}]$. Using the notation of Theorem 11.4.1, we find that the period length k is the smallest integer $k \geq 1$ such that $B_k = 1$.*

PROOF. If $x = \sqrt{n}$, then $x_1 = 1/(\sqrt{n} - a_0)$. Since $a_0 \geq 1$, we see that $\overline{x_1} = -1/(a_0 + \sqrt{n}) > -1$ and $\overline{x_1} < 0$. This implies that the continued fraction of x_1 is purely periodic; hence the expansion of $\sqrt{n}$ has the desired form.

If $x_{k+1} = x_1$, then $B_{k+1} = B_1$ and $A_{k+1} = A_1$. Using $x_{k+1} = x_1$ together with the equations $x_k = a_k + 1/x_{k+1}$ and $\sqrt{n} = a_0 + 1/x_1$ implies that $x_k = a_k - a_0 + \sqrt{n}$. Hence $A_k = a_k - a_0$ and $B_k = 1$. Conversely, if $B_k = 1$ for some k, then $x_k = A_k + \sqrt{n}$, hence $a_k = A_k + a_0$. This implies that $A_{k+1} = a_0$ and $B_{k+1} = n - a_0^2$, that is, $x_{k+1} = x_1$ and the expansion repeats. ∎

To detect the beginning of the period of the continued fraction of a quadratic irrational, we can use the theorem to check when the first complete quotient x_k lies between -1 and 0. The expansion repeats from this point onwards, and the period can be easily detected by comparing each complete quotient with x_k.

Here are a few more examples of the use of Lemma 11.5.1.

Example 11.5.4. Suppose $x = [\overline{a_0, a_1, \ldots, a_k}]$ is purely periodic; then the simple continued fraction of $\dfrac{-1}{x}$ is $[\overline{a_k, \ldots, a_0}]$.

We apply the lemma to the following equations.

$$x = x_0 = a_0 + \frac{1}{x_1} \qquad\qquad \overline{x}_0 = a_0 + \frac{1}{\overline{x}_1}$$

$$\vdots$$

$$x_k = a_k + \frac{1}{x_{k+1}} \qquad\qquad \overline{x}_k = a_k + \frac{1}{\overline{x}_{k+1}}$$

We start with the fact that $x_{k+1} = x$. The last equation implies that $\lfloor -1/\overline{x} \rfloor = a_k$ (by Lemma 11.5.1). Let $y = -1/\overline{x}$. The equation $y = a_k - \overline{x}_k$ implies that the first complete quotient y_1 of y is $-1/\overline{x}_k$. Again, using Lemma 11.5.1, we see that $\lfloor y_1 \rfloor = \lfloor -1/\overline{x}_k \rfloor = a_{k-1}$. The second complete quotient y_2 of y is $-1/\overline{x}_{k-1}$, and we have $y = [a_k, a_{k-1}, y_2]$. We continue to work backwards and see that the jth complete quotient of y, y_j, is $-1/\overline{x}_{k-j+1}$. From this we conclude that partial quotients in the expansion of $-1/\overline{x}$ are $a_k, a_{k-1}, \ldots, a_0$.

Example 11.5.5. If the expansion of $\sqrt{n}$ is $[a_0, \overline{a_1, a_2, \ldots, a_k}]$, then $a_k = 2a_0$.

Let x_i denote the ith complete quotient in the expansion of $\sqrt{n}$. Then $x_{k+1} = x_1$ and $x_k = a_k + 1/x_{k+1}$. Therefore, by using Lemma 11.5.1, we have that $x_k = a_k + 1/x_1$, implies

$$\left\lfloor \frac{-1}{\overline{x}_1} \right\rfloor = a_k.$$

On the other hand, using $\sqrt{n} = a_0 + 1/x_1$, we obtain $-1/\overline{x}_1 = \sqrt{n} + a_0$. This implies that $a_k = 2a_0$.

The technique used in these examples can be used to show more properties of the expansion of $\sqrt{n}$. These questions are explored in the exercises.

Exercises for Section 11.5
---

1. Show that the continued fraction expansion of $a_0 + \sqrt{n}$ with $a_0 = \lfloor \sqrt{n} \rfloor$ is purely periodic.

2. Determine the relation between the continued fractions of a quadratic irrational and its conjugate.

3. Suppose $\sqrt{n} = [a_0, \overline{a_1, \ldots, a_k}]$. Use the technique of Examples 11.5.4 and 11.5.5 to show that $a_1 = a_{k-1}$.

4. Using the fact that $x = a_0 + \sqrt{n}$ has a purely periodic expansion (Exercise 1), deduce that its period is symmetrical. Conclude that the expansion of $\sqrt{n}$ has the form $[a_0, \overline{a_1, a_2, \ldots, a_2, a_1, 2a_0}]$.

5.   (a) Suppose the p_k/q_k is the kth convergent of $\sqrt{n}$. Show that

$$p_k^2 - nq_k^2 = (-1)^{k+1} B_{k+1}.$$

  (b) Suppose the periodic part in the continued fraction expansion of $\sqrt{p}$ is of the form $[\overline{a_1, \ldots, a_n, a_n, \ldots, a_1, 2a_0}]$. Show that $p_{k-1}^2 \equiv -1$ (mod p), hence $p \equiv 1$ (mod 4). (The converse is also true, that is, if $p \equiv 1$ (mod 4) is prime, then the smallest period length of $\sqrt{p}$ is odd. This is proved in Section 14.5)

6. Suppose a prime $p \equiv 1$ (mod 4) is represented as a sum of two squares, $p = A^2 + B^2$, A and B positive integers. Show that $(A + \sqrt{p})/B$ and $(B + \sqrt{p})/A$ have purely periodic continued fractions. What can you say about the relationship between the continued fractions of these numbers and that of $\sqrt{p}$? Can you prove your conjectures?

11.6   Classical Continued Fraction Expansions

The problem of determining the continued fraction expansion of special numbers like e can be solved in many cases by determining the expansion of classical analytic functions. These expansions usually begin with the continued fraction expansion of the Hypergeometric function (Exercise 3). The hypergeometric function includes all the classical transcendental functions as special cases. A simpler form of the hypergeometric function is the conformal hypergeometric function. The properties of the conformal hypergeometric function are sufficient to yield the continued fraction expansion of e. Detailed analysis of the continued fraction expansions of classical functions can be found in the book of Perron.[3]

The conformal hypergeometric function $\psi(\gamma, x)$ is defined as

$$\psi(\gamma, x) = 1 + \frac{x}{\gamma \cdot 1} + \frac{x^2}{\gamma(\gamma + 1)2!} + \frac{x^3}{\gamma(\gamma + 1)(\gamma + 2)3!} + \cdots. \quad (1)$$

Replacing x by γx and letting $\gamma \to \infty$, we obtain the series for e^x.

Lemma 11.6.1. *The function $\psi(\gamma, x)$ satisfies*

$$\psi(\gamma, x) = \psi(\gamma + 1, x) + \frac{x}{\gamma(\gamma + 1)} \psi(\gamma + 2, x). \quad (2)$$

---

[3] O. Perron. *Die Lehre von den Kettenbruchen.* Stuttgart: Teubner, 1954.

PROOF. The proof is by comparing the coefficient of x^k on both sides for every $k \geq 0$. On the right-hand side, the coefficient is the sum of the coefficient of x^k term in $\psi(\gamma+1, x)$ and that of the x^{k-1} term in $\psi(\gamma+2, x)$. We obtain

$$\frac{x^k}{(\gamma+1)\cdots(\gamma+1+k-1)k!} + \frac{x^k}{\gamma(\gamma+1)\cdots(\gamma+2+k-2)(k-1)!},$$

$$= \frac{x^k}{(\gamma+1)\ldots(\gamma+k)(k-1)!}\left[\frac{1}{k}+\frac{1}{\gamma}\right]$$

$$= \frac{x^k}{(\gamma+1)\ldots(\gamma+k-1)(k)!},$$

which is the coefficient of x^k in $\psi(\gamma, x)$. ∎

This property is very useful to obtain the continued fraction expansion of the hypergeometric function. We start with the ratio $\gamma\dfrac{\psi(\gamma, x)}{\psi(\gamma+1, x)}$ and apply the identity of Equation (2).

$$\gamma\frac{\psi(\gamma, x)}{\psi(\gamma+1, x)} = \gamma + \cfrac{x}{(\gamma+1)\dfrac{\psi(\gamma+1, x)}{\psi(\gamma+2, x)}}$$

$$= \gamma + \cfrac{x}{(\gamma+1) + \cfrac{x}{(\gamma+2)\dfrac{\psi(\gamma+2, x)}{\psi(\gamma+3, x)}}}$$

$$= \gamma + \cfrac{x}{(\gamma+1) + \cfrac{x}{(\gamma+2) + \cfrac{x}{(\gamma+3) + \cdots}}}.$$

This is not a simple continued fraction. This technique needs to be modified to yield a simple continued fraction. A similar computation gives the continued fraction expansion (not necessarily simple) of the hypergeometric function and related classical transcendental functions.

The relation between the conformal hypergeometric function and the exponential function is obtained through the following identities.

Lemma 11.6.2.   (a) $\psi\left(\dfrac{1}{2}, \dfrac{x^2}{4}\right) = \dfrac{e^x + e^{-x}}{2}.$

(b) $x\psi\left(\dfrac{3}{2}, \dfrac{x^2}{4}\right) = \dfrac{e^x - e^{-x}}{2}$.

The assertions of the lemma can be verified using the Taylor series expansion of both sides, and we leave the details of the verification to the reader.

Theorem 11.6.3. *The continued fraction expansion of* $\dfrac{e^{2/y} + 1}{e^{2/y} - 1}$ *is*

$$\frac{e^{2/y} + 1}{e^{2/y} - 1} = [y, 3y, 5y, 7y, \dots], \tag{3}$$

and as a special case we obtain

$$\frac{e + 1}{e - 1} = [2, 6, 10, 14, 18, \dots]. \tag{4}$$

PROOF. To obtain a simple continued fraction, it is better to replace x by z^2 in equation (2). Further dividing by z yields

$$\frac{\gamma\psi(\gamma, z^2)}{z\psi(\gamma + 1, z^2)} = \frac{\gamma}{z} + \frac{1}{\dfrac{(\gamma + 1)}{z}\left[\dfrac{\psi(\gamma + 1, z^2)}{\psi(\gamma + 2, z^2)}\right]}.$$

The term in the fraction is of the same type as the ratio on the left-hand side, with γ replaced by $\gamma + 1$, and then we can repeat the same step using Lemma 11.6.1 with γ replaced by $\gamma + 1$. Repeated application of this yields

$$\frac{\gamma\psi(\gamma, z^2)}{z\psi(\gamma + 1, z^2)} = \frac{\gamma}{z} + \cfrac{1}{\dfrac{\gamma + 1}{z} + \cfrac{1}{\dfrac{\gamma + 2}{z} + \cdots}}.$$

The continued fraction expansion converges if the terms $\frac{\gamma + n}{z}$ remain positive. The substitutions given below can all be justified by writing a finite continued fraction, using the nth complete quotient and then extending it to the infinite fraction by a limit. The complete quotient is described by

$$\frac{\gamma\psi(\gamma, z^2)}{z\psi(\gamma + 1, z^2)} = \left[\frac{\gamma}{z}, \frac{\gamma + 1}{z}, \dots, \frac{\gamma + n}{z}, x_{n+1}\right], \tag{5}$$

where

$$x_{n+1} = \frac{\gamma + n + 1}{z}\left[\frac{\psi(\gamma + n + 1, z^2)}{\psi(\gamma + n + 2, z^2)}\right]. \tag{6}$$

Replace z by $z/2$ to get

$$\frac{2\gamma\psi(\gamma, z^2/4)}{z\psi(\gamma+1, z^2/4)} = \frac{2\gamma}{z} + \cfrac{1}{\cfrac{2(\gamma+1)}{z} + \cfrac{1}{\cfrac{2(\gamma+2)}{z} + \cdots}}.$$

Using Lemma 11.6.1, we can replace γ by $1/2$ and obtain

$$\frac{e^z + e^{-z}}{e^z - e^{-z}} = \frac{1}{z} + \cfrac{1}{\cfrac{3}{z} + \cfrac{1}{\cfrac{5}{z} + \cfrac{1}{\cfrac{7}{z} + \cdots}}}.$$

Next, let $\dfrac{1}{z} = y$ to obtain

$$\frac{e^{2/y} + 1}{e^{2/y} - 1} = y + \cfrac{1}{3y + \cfrac{1}{5y + \cfrac{1}{7y + \cdots}}}.$$

This is Lambert's continued fraction. We let $y = 2$ to get

$$\frac{e+1}{e-1} = 2 + \cfrac{1}{6 + \cfrac{1}{10 + \cfrac{1}{14 + \cdots}}},$$

which completes the proof of the theorem. ∎

We use this result to derive the continued fraction expansion of e.

Theorem 11.6.4. *The simple continued fraction of $e - 1$ is*

$$e - 1 = [1, 1, 2, 1, 1, 4, 1, 1, 6, \ldots]. \tag{7}$$

PROOF. The partial quotients, $a_0, a_1, \ldots$ of $[1, 1, 2, 1, 1, 4, \ldots]$ can be described by the equations

$$a_{3n} = 1$$
$$a_{3n+1} = 1$$
$$a_{3n-1} = 2n.$$

Let p_n/q_n be the convergents of $[1, 1, 2, 1, 1, 4, 1, 1, 6, \ldots]$. If r_n/s_n are the convergents of the expansion of $\dfrac{e+1}{e-1}$, then we will show that

$$\lim_{n \to \infty} p_n/q_n = \lim_{n \to \infty} \frac{\frac{r_n}{s_n} + 1}{\frac{r_n}{s_n} - 1}.$$

This will prove the theorem because $r_n/s_n \to \alpha = (e+1)/(e-1)$ and $e = (\alpha+1)/(\alpha-1)$. The convergents r_n/s_n satisfy the following recursion relations:

$$r_n = (4n+2)r_{n-1} + r_{n-2}$$
$$s_n = (4n+2)s_{n-1} + s_{n-2},$$

because the nth partial quotient of $(e+1)/(e-1)$ is $4n+2$. The numbers p_n satisfy the following relations:

$$p_{3n} = p_{3n-1} + p_{3n-2}, \tag{8}$$
$$p_{3n+1} = p_{3n} + p_{3n-1}, \tag{9}$$
$$p_{3n-1} = 2np_{3n-2} + p_{3n-3}, \tag{10}$$
$$p_{3n-2} = p_{3n-3} + p_{3n-4}, \tag{11}$$
$$p_{3n-3} = p_{3n-4} + p_{3n-5}. \tag{12}$$

There is a similar relation for the q_n's. We will show that

$$p_{3n+1} = r_n + s_n$$
$$q_{3n+1} = r_n - s_n.$$

The proof is by induction on n. It is clear for $n = 0$ because $p_1 = 2$, and $r_0 = 2$. Assume that the relation is true up to $p_{3n-2} = r_{n-1} + s_{n-1}$. To obtain the relation for p_{3n+1}, we add (8) and (9) and substitute twice (10) to get

$$p_{3n+1} = 2p_{3n-1} + p_{3n-2}$$
$$= 4np_{3n-2} + 2p_{3n-3} + p_{3n-2}.$$

We substitute (10) and (11) into this and simplify to get

$$p_{3n+1} = (4n+1)p_{3n-2} + 2p_{3n-4} + 2p_{3n-5}.$$

We can verify that $p_{3n-2} = 2p_{3n-4} + p_{3n-5}$, which yields, by the induction hypothesis,

$$p_{3n+1} = (4n+2)p_{3n-2} + p_{3n-5}$$
$$= (4n+2)(r_{n-1} + s_{n-1}) + (r_{n-2} + s_{n-2})$$
$$= (4n+2)r_{n-1} + r_{n-2} + (4n+2)s_{n-1} + s_{n-2}$$
$$= r_n + s_n.$$

This proves the relation for the p_{3n+1}. The assertion regarding the sequence q_{3n+1} is proved similarly. Since the convergents have a limit, any subsequence must converge to the same limit. Hence

$$\lim_{n\to\infty} \frac{p_n}{q_n} = \lim_{n\to\infty} \frac{p_{3n+1}}{q_{3n+1}} = \lim_{n\to\infty} \frac{r_n/s_n + 1}{r_n/s_n - 1} = \frac{\alpha+1}{\alpha-1} = e.$$

This completes the proof of the theorem. ∎

Corollary 11.6.5. *e is irrational.*

PROOF. This follows because we have shown that the simple continued fraction of e is infinite. ∎

The determination of the continued fraction of e shows that it is not the root of any quadratic equation with integer coefficients. Actually, e is not a solution of any polynomial equation with integer coefficients, but this fact is harder to establish.

——————————— **Exercises for Section 11.6** ———————————

1. Prove the identities of Lemma 11.6.2.

2. Show that $e^{2/q}$ is an irrational number.

3. The hypergeometric function is defined by the series

$$F(a, b; c; x) = 1 + \frac{ab}{c \cdot 1} z + \frac{a(a+1)b(b+1)}{c(c+1)1 \cdot 2} x^2 + \cdots .$$

   Show the following properties of the hypergeometric function.

   (a) $F(a,b;c;x) = F(a,b+1;c+1;x) - \frac{a(c-b)}{c(c+1)} x F(a+1,b+1;c+2;x)$.

   (b) $F(a,b+1;c+1;x) = F(a+1,b+1;c+2;x) - \frac{(b+1)(c-a+1)}{(c+1)(c+2)} x F(a+ 1,b+2;c+3;x)$.

4. Imitate the calculation given after Lemma 11.6.1 to produce a continued fraction expansion of the hypergeometric series. This is only a formal computation as the result will not be a simple continued fraction, and the results developed in this chapter do not apply to show the convergence. The convergence of some infinite continued fractions that are not simple is explored in the projects at the end of this chapter.

5. Using the definition of the hypergeometric series in Exercise 3, show that

$$\sin^{-1} x = xF(\frac{1}{2}, \frac{1}{2}; \frac{3}{2}; x^2)$$
$$\log(1 + x) = xF(1, 1, 2, -x)$$
$$(1 + z)^\alpha = F(-\alpha, 1; 1; -z).$$

You will need to use the Taylor series expansion of these functions.

---------------------------- **Projects for Chapter 11** ----------------------------

The first three projects are designed to explore some properties of continued fractions that are not simple. These fractions are important in approximating classical transcendental functions. The first project develops some of the basic properties of convergents of these fractions. This is a prerequisite for the next two projects. Projects 2 and 3 involve the study of special types of continued fractions, where it is possible to prove the convergence of the infinite continued fractions without using advanced techniques.

1. Denote the continued fraction

$$a_0 + \cfrac{b_1}{a_1 + \cfrac{b_2}{a_2 + \cfrac{b_3}{a_3 + \cfrac{\ddots}{+ \cfrac{b_{n-1}}{a_{n-1} + \cfrac{b_n}{a_n}}}}}}$$

by the expression $a_0 + \cfrac{b_1}{a_1+} \cfrac{b_2}{a_2+} \cfrac{b_3}{a_3+} \cdots \cfrac{b_n}{a_n}$. The kth convergent is defined as $\dfrac{p_k}{q_k} = a_0 + \cfrac{b_1}{a_1+} \cfrac{b_2}{a_2+} \cfrac{b_3}{a_3+} \cdots \cfrac{b_k}{a_k}$.

(a) Prove the following recurrence relations to compute p_k and q_k.

$$p_k = a_k p_{k-1} + b_k p_{k-2}$$
$$q_k = a_k q_{k-1} + b_k q_{k-2}$$

with initial conditions

$$p_{-1} = 1, \quad p_0 = a_0,$$
$$q_{-1} = 0, \quad q_0 = 1.$$

(b) Show that

$$p_k q_{k-1} - p_{k-1} q_k = (-1)^{k-1} b_1 b_2 \cdots b_k$$

$$p_k q_{k-2} - p_{k-2} q_k = (-1)^k a_k b_1 b_2 \cdots b_{k-1}$$

$$\left| \frac{p_k}{q_k} - \frac{p_{k-1}}{q_{k-1}} \right| = \frac{|b_1 b_2 \cdots b_k|}{|q_k q_{k-1}|} .$$

(c) Show that $a_0 + \dfrac{b_1}{a_1+} \dfrac{b_2}{a_2+} \dfrac{b_3}{a_3+} \cdots \dfrac{b_n}{a_n}$ is the same as the continued fraction $a_0 + \dfrac{b_1/a_1}{1+} \dfrac{b_2/a_1 a_2}{1+} \dfrac{b_3/a_2 a_3}{1+} \cdots \dfrac{b_n/a_{n-1} a_n}{1} .$

(d) We would like to define an infinite continued fraction as the limit of the convergents when the limit exists. Write a computer program to compute convergents using the formula of part (a). Compute the first 20 convergents in the following examples.

 (i) $a_i = 1$ and $b_i = -1$ for all i. Are all the convergents defined?

 (ii) $a_i = (-1)^i$ and $b_i = -1$. What do you conjecture to be the general formula for the convergents? Do the convergents have a limit? Do the odd or the even convergents have a limit?

 (iii) $a_i = 1$ and $b_i = 2$ for all i. Do the convergents have a limit? What property do the odd and even convergents satisfy?

 (iv) $a_i = 1$ and $b_i = (-1)^i 2$ for all i. Do the convergents have a limit? Do the odd and even convergents satisfy a monotonicity property?

 (v) Explore the convergence of the continued fraction for other values of a and b. Formulate conjectures regarding the convergence and test them extensively.

These examples show that the behavior of nonsimple continued fractions is more complicated. The general behavior is not well understood, but it is possible to show convergence in a few cases. The following exercises outline the proof of convergence in a special case. Assume that $b_i > 0$ and $a_i \geq 1$ for all i.

(e) Show that $p_k - p_{k-1} = (a_k - 1)p_{k-1} + b_k p_{k-2}$. Use this and the analogous property for q_k to show that p_k and q_k are increasing sequences.

(f) Show that the odd convergents form a decreasing sequence and that the even convergents form an increasing sequence. Also prove that every odd convergent is greater than all the following convergents and that every even convergent is less than all the following convergents.

(g) Show that the limit of the odd and even convergents exists. We say the continued fraction is convergent if the limits are equal.

(h) Show that

$$q_k > (a_k a_{k-1} + b_k) q_{k-2}$$

$$q_{k-1} > (a_{k-1} a_{k-2} + b_{k-1}) q_{k-3}$$

$$\vdots$$

$$q_3 > (a_3 a_2 + b_3) q_1$$

$$q_2 = (a_2 a_1 + b_2) q_0 .$$

If k is even conclude that

$$q_k q_{k-1} > q_0 q_1 (b_2 + a_1 a_2)(b_3 + a_2 a_3) \cdots (b_k + a_{k-1} a_k)$$

or

$$q_k q_{k-1} > q_0 q_1 b_2 b_3 \cdots b_k \left(1 + \frac{a_1 a_2}{b_2}\right) \cdots \left(1 + \frac{a_{k-1} a_k}{b_k}\right).$$

(i) Suppose z_n is a sequence of positive numbers. Prove that if $\sum z_n$ diverges, then $\prod (1 + z_n)$ also diverges. Use this fact to show that if $\sum \frac{a_k a_{k-1}}{b_k}$ **diverges**, then the continued fraction **converges**. (Hint: Use the formula for the difference of consecutive convergents proved in part (b) to show that the odd and even convergents have the same limit).

(j) Use this criterion for convergence to show that the periodic continued fraction $a + \frac{b}{a+} \frac{b}{a+} \cdots$, for $a \geq 1, b > 0$ converges to a root of the equation $x^2 - ax - b = 0$.

2. This project develops the continued fraction expansion of $\frac{\pi}{4}$. The expansion is not a simple continued fraction. You may assume the convergence result of project I, where it is shown that $a_0 + \dfrac{b_1}{a_1+} \dfrac{b_2}{a_2+} \dfrac{b_3}{a_3+} \cdots \dfrac{b_n}{a_n} \cdots$ converges when $a_i \geq 1$ and $b_i > 0$. The project uses the basic properties of Gauss's hypergeometric function, which you are asked to prove in the first exercise. (See Section 11.6 for the properties of the confluent hypergeometric function).

(a) The hypergeometric function is defined by the series

$$F(a, b; c; z) = 1 + \frac{ab}{c \cdot 1} z + \frac{a(a+1)b(b+1)}{c(c+1)1 \cdot 2} z^2 + \cdots .$$

Show the following properties of the hypergeometric function.

(i) $F(a, b; c; z) = F(a, b+1; c+1; z) - \frac{a(c-b)}{c(c+1)} z F(a+1, b+1; c+2; z)$

(ii) $F(a, b+1; c+1; z) = F(a+1, b+1; c+2; z) - \frac{(b+1)(c-a+1)}{(c+1)(c+2)} z F(a+1, b+2; c+3; z)$

(b) Prove that
$$\tan^{-1}(x) = x F(1/2, 1; 3/2; -x^2)$$

and apply the identities of the first exercise to the expansion of the quotient
$$\frac{F(a, b+1; c+1; x)}{F(a, b; c; x)}.$$

(c) Conclude that the following convergent continued fraction is $\tan^{-1}(x)$:

$$\frac{x}{1+} \frac{1^2 x^2}{3+} \frac{2^2 x^2}{5+} \frac{3^2 x^2}{7+} \frac{4^2 x^2}{9+} \cdots .$$

This continued fraction for $\tan^{-1}(x)$ converges for all real $x \neq 0$. Evaluate the continued fraction at $x = 1$ to obtain a formula for $\pi/4$.

(d) To illustrate the virtues of the continued fraction expansion, we compare the convergence with the Taylor series expansion at $z = 0$.

$$\tan^{-1} z = z - \frac{z^3}{3} + \frac{z^5}{5} - \frac{z^7}{7} + \cdots .$$

The series only converges for $|z| \leq 1$, $z \neq \pm i$.

   (i) To compute $\pi/4$ with an error $< 10^{-7}$, how many terms do we need in the Taylor series expansion?

   (ii) Verify that the 9th convergent of the continued fraction already approximates $\pi/4$ with an error less than 10^{-7}.

3. Here is another type of continued fraction that can be shown to be convergent. Let $b_i > 0$ and $C = \frac{b_1}{a_1-} \frac{b_2}{a_2-} \frac{b_3}{a_3-} \cdots$. Assume that $a_k \geq b_k + 1$ for all k.

   (a) Show the following properties of the convergents.

$$p_k - p_{k-1} \geq b_1 b_2 \cdots b_k$$
$$p_n \geq b_1 + b_1 b_2 + b_1 b_2 b_3 + \cdots + b_1 b_2 \cdots b_n$$
$$q_k - q_{k-1} \geq b_1 b_2 \cdots b_k$$
$$q_n \geq 1 + b_1 + b_1 b_2 + \cdots + b_1 b_2 \cdots b_n .$$

   (b) Show that $q_n - p_n \geq q_{n-1} - p_{n-1} \geq q_2 - p_2 \geq q_1 - p_1 \geq 1$. Conclude that $1 - p_n/q_n \geq 1/q_n$, and hence the continued fraction converges as the convergents form a bounded increasing sequence.

   (c) Show that if $a_k = b_k + 1$ for every k, then $C = 1 - 1/S$ with $S = 1 + b_1 + b_1 b_2 + \cdots$. Also prove that if strict inequality occurs at least once in $a_k \geq b_k + 1$, then $C < 1$.

   (d) Consider the periodic convergent fraction $\alpha = a - \frac{b}{a-} \frac{b}{a-} \cdots$ for $a \geq b + 1$. Show that α is a quadratic irrational.

   (e) Let $\psi(\gamma; x)$ be the confluent hypergeometric series defined in section 11.6. Prove the following formulas:

     (i) $\psi(1/2; -x^2/4) = \cos(x)$
     (ii) $\psi(3/2; -x^2/4) = \sin(x)/x$.

   Imitate the proof of Theorem 11.6.3 to derive a continued fraction expansion of $\tan(x)$.

4. **Decomposition of integral matrices of determinant ± 1.** We can use the formulas for p_k and q_k to prove an important result about the decomposition of integral matrices of determinant ± 1.

Recall that for $k \geq 2$, we can write

$$\begin{pmatrix} p_k & q_k \\ p_{k-1} & q_{k-1} \end{pmatrix} = \begin{pmatrix} a_k & 1 \\ 1 & 0 \end{pmatrix} \begin{pmatrix} p_{k-1} & q_{k-1} \\ p_{k-2} & q_{k-2} \end{pmatrix}. \tag{1}$$

Let $U = \begin{pmatrix} 1 & 1 \\ 0 & 1 \end{pmatrix}$, $S = \begin{pmatrix} 0 & 1 \\ 1 & 0 \end{pmatrix}$, and $S' = \begin{pmatrix} 0 & 1 \\ -1 & 0 \end{pmatrix}$.

(a) Show that $U^k = \begin{pmatrix} 1 & k \\ 0 & 1 \end{pmatrix}$ for any integer k and $S^{-1} = S$, $(S')^{-1} = -S'$.

(b) Show that if $p_k/q_k = [a_0, \ldots, a_k]$, then

$$\begin{pmatrix} p_k & q_k \\ p_{k-1} & q_{k-1} \end{pmatrix} = \begin{pmatrix} a_k & 1 \\ 1 & 0 \end{pmatrix} \begin{pmatrix} a_{k-1} & 1 \\ 1 & 0 \end{pmatrix} \cdots \begin{pmatrix} a_0 & 1 \\ 1 & 0 \end{pmatrix} \begin{pmatrix} 1 & 0 \\ 0 & 1 \end{pmatrix}$$

$$= \begin{pmatrix} a_k & 1 \\ 1 & 0 \end{pmatrix} \begin{pmatrix} a_{k-1} & 1 \\ 1 & 0 \end{pmatrix} \cdots \begin{pmatrix} a_0 & 1 \\ 1 & 0 \end{pmatrix}.$$

(c) Show that $\begin{pmatrix} k & 1 \\ 1 & 0 \end{pmatrix} = U^k S'$ and use this to show that $\begin{pmatrix} p_k & q_k \\ p_{k-1} & q_{k-1} \end{pmatrix}$ can be written as powers of U and S'. For example, $17/10 = [1, 1, 2, 3]$ implies that

$$\begin{pmatrix} 17 & 10 \\ 5 & 3 \end{pmatrix} = U^3 S' U^2 S' U S' U S'.$$

(d) Write any matrix $\begin{pmatrix} p & q \\ y & x \end{pmatrix}$ satisfying $q > 0$ and $0 < x < q$ in terms of U and S'. (See Example 11.2.5.)

(e) Use the classification of the solutions to a linear Diophantine equation to decompose any matrix $\begin{pmatrix} p & q \\ y & x \end{pmatrix}$ satisfying $px - qy = \pm 1$ into powers of U, S and S'.

5. This project investigates the relative occurrence of different integers as quotients in the Euclidean Algorithm. More precisely, we investigate the relative occurrence of an integer as a partial quotient in the simple continued fraction expansion of a number α, $0 < \alpha < 1$. We write $\alpha = [0, a_1, a_2, a_3, \ldots]$. If you did the project on the quotients in the Euclidean Algorithm in Chapter 2, then you can verify that the answers you obtained are correct.

(a) Given $k \geq 1$, determine the interval of real numbers α for which $a_1 = k$.

(b) Determine the interval of numbers α for which $a_1 = k$ and $a_2 = 1$. Similarly, determine the interval of numbers α for which $a_1 = k$ and $a_2 = m$.

(c) Suppose $I_1, I_2, \ldots, I_k, \ldots$ is a collection of disjoint intervals contained in $[0, 1]$. Then the probability that a number lies in one of the I_k is defined to be the sum of the lengths of the intervals. Show that the probability that $a_2 = 1$ is $2 \log(2) - 1$. Use the result of part (b) to estimate the probability that $a_2 = 2$.

(d) Determine the intervals on which $a_3 = 1$, and estimate the probability that $a_3 = 1$.

(e) To estimate the probability distribution of an integer occurring as some quotient, it is more convenient to adopt a continuous model. Write $\alpha = [0, a_1, a_2, \ldots, a_n, x_{n+1}]$, and define $f_n(x)$ to be probability that $x_{n+1} \geq 1/x$, for $0 < x \leq 1$.

(i) Show that $f_0(x) = x$.

(ii) Show that $x_{n+1} \geq x$ and $a_n = k$; then $k \leq x_n \leq k + x$, so

$$f_n(x) = \sum_{k \geq 1} \text{probability } (k \leq x_n \leq x_{n+1})$$

$$= \sum_{k \geq 1} f_{n-1}(1/k) - f_{n-1}(1/(k + x)).$$

Gauss conjectured that the probability distributions $f_n(x)$ approach the limit $f(x) = \log(1 + x)/\log(2)$. Hence the probability that a random quotient is 1 is $f(1) - f(1/2) = 1 - \log(3/2)/\log(2) = 1 - 0.584963$. (Does this agree with your computations?) Gauss's conjecture was proved by Kuz'min in 1928.

Chapter 12

Factoring Methods

There is probably no more absorbing pursuit in all of number theory than the resolution of a number into its prime factors.

— A. H. BEILER

12.1   Continued Fraction Factoring Method

The *continued fraction factoring method* is the first modern factoring method to make extensive use of computers. It was first proposed by D. H. Lehmer and R. E. Powers[1] in 1931. It was systematically explored and developed by M. A. Morrison and J. Brillhart,[2] who successfully factored the seventh Fermat number, $F_7 = 2^{128} + 1$, as

$$F_7 = 59649589127497217 \times 5704689200685129054721.$$

The main method in use during the years 1970–1985, it has since been supplanted by the quadratic sieve and, more recently, by the number field sieve.

The continued fraction method, the quadratic sieve, and the number field sieve use the following fact to determine a proper factor. They differ in how the data in the congruence is generated.

Lemma 12.1.1. *If $x^2 \equiv y^2 \pmod{n}$ and $x \not\equiv \pm y \pmod{n}$, then either $(x + y, n)$ or $(x - y, n)$ is a proper factor of n.*

---

[1]D. H. Lehmer and R. E. Powers. "On Factoring Large Numbers." *Bull. Amer. Math. Soc.,* 37(1931), 342–382.

[2]M. A. Morrison and J. Brillhart, "A Method of Factoring and the Factorization of F_7." *Mathematics of Computation.* 29(1975), 183–205.

PROOF. If p is a proper prime factor of n, then $p \mid (x^2 - y^2) = (x - y)(x + y)$. This implies that $p \mid (x - y)$ or $p \mid (x + y)$. But n doesn't divide $x - y$ nor $x + y$; hence $(x - y, n)$ or $(x + y, n)$ is divisible by p but not by n, and we obtain a proper factor of n. ∎

Recall the continued fraction expansion of $\sqrt{n}$ for n not a perfect square. The complete quotients x_k and the partial quotients a_k in the continued fraction expansion of $\sqrt{n}$ are

$$x_k = \frac{A_k + \sqrt{n}}{B_k}, \quad a_k = \lfloor x_k \rfloor,$$

where A_k and B_k are integers determined by

$$A_0 = 0, \ B_0 = 1, \ A_{k+1} = a_k B_k - A_k, \ B_{k+1} = \frac{n - A_{k+1}^2}{B_k}.$$

The utility of continued fractions in factoring is due to the following theorem.

Theorem 12.1.2. *Suppose that p_k / q_k is the kth convergent of $\sqrt{n}$. Then*

$$p_k^2 - nq_k^2 = (-1)^{k+1} B_{k+1}.$$

PROOF. Let x_k and a_k denote the complete and partial quotients of $\sqrt{n}$, respectively. Then $\sqrt{n} = [a_0, a_1, \ldots, a_k, x_{k+1}]$ for any k; therefore,

$$\sqrt{n} = \frac{x_{k+1} p_k + p_{k-1}}{x_{k+1} q_k + q_{k-1}}.$$

Since $x_{k+1} = (A_{k+1} + \sqrt{n})/B_{k+1}$, we have

$$\sqrt{n} = \frac{A_{k+1} p_k + \sqrt{n} p_k + p_{k-1} B_{k+1}}{A_{k+1} q_k + \sqrt{n} q_k + q_{k-1} B_{k+1}}.$$

Cross-multiply and write the $\sqrt{n}$ term on one side to get

$$nq_k - A_{k+1} p_k - p_{k-1} B_{k+1} = \sqrt{n}(p_k - A_{k+1} q_k - q_{k-1} B_{k+1}).$$

Both the left-hand side of the equation and the factor in parentheses on the right are integers; hence both must be 0, as $\sqrt{n}$ is irrational. We have

$$p_k - A_{k+1} q_k - q_{k-1} B_{k+1} = 0$$

and

$$nq_k - A_{k+1} p_k - p_{k-1} B_{k+1} = 0.$$

Multiplying the first equation by p_k and the second by q_k and equating them, we obtain

$$p_k^2 - nq_k^2 = B_{k+1}(p_k q_{k-1} - q_k p_{k-1}),$$

and using Proposition 11.2.4, we see that

$$p_k^2 - nq_k^2 = (-1)^{k+1} B_{k+1}. \qquad \blacksquare$$

This implies that $p_k^2 \equiv (-1)^{k+1} B_{k+1} \pmod{n}$. To apply Lemma 12.1.1, we need squares on both sides. The idea of the continued fraction method is to generate several pairs (p_k, B_{k+1}) and take suitable combinations to produce a square on the right and, hopefully, a factor of n. Recall that an integer is a perfect square if and only if the exponents in its prime factorization are all even. Thus, to find the products of B_k's that yield perfect squares, we obtain their prime factorization and combine them so that the exponents become even. The factorization of the B_k's is obtained by trial division.

First, we select a set of primes over which the B_k factor. If $p \mid B_k$, we have $p_k^2 \equiv nq_k^2 \pmod{p}$, so n is a quadratic residue modulo p. Select a set of primes p such that $\left(\dfrac{n}{p}\right) = 1$. This set of primes is called the *factor base*.

Example 12.1.3. The first 20 iterations with $n = 4141$ produce 13 values of B_k that can be factored over the factor base consisting of $2, 3, 5, 7, 11$. The values of B_k and their factorizations are:

$k+1$	p_k	B_{k+1}	B_{k+1} factored
2	129	77	$7^1 11^1$
3	193	20	$2^2 5^1$
6	814	36	$2^2 3^2$
8	3719	21	$3^1 7^1$
11	2266	84	$2^2 3^1 7^1$
12	3463	33	$3^1 11^1$
13	232	9	3^2
14	2570	5	5^1
15	2367	84	$2^2 3^1 7^1$
17	3959	4	2^2
18	3436	105	$3^1 5^1 7^1$
19	3254	21	$3^1 7^1$
20	3142	20	$2^2 5^1$

Notice that B_6 is a square, hence $p_5 \equiv (-1)^6 B_6 \pmod{n}$. Using the congruence $814^2 \equiv 6^2 \pmod{n}$, we compute $(814+6, 4141) = 41$, a proper factor of 4141. Row 12 with $k + 1 = 17$ also gives a square, but since 17 is odd, we get $3959^2 \equiv (-1)^{17} 2^2$. This cannot be used to obtain the prime factorization. Suppose we had not noticed that we had a square term. We could multiply row 5 ($k + 1 = 11$) and row 12 ($k + 1 = 19$) and obtain

$$p_{10}^2 p_{18}^2 \equiv (-1)^{11}(-1)^{19} B_{11} B_{19} \equiv 2^2 3^1 7^1 3^1 7^1 \equiv (2 \cdot 3 \cdot 7)^2 \pmod{n}.$$

We compute $(2266 \cdot 3254 - 42, n) = 41$, a proper factor of n.

The example shows the different possibilities for action once the prime factorization of B_k is obtained. We have to take into account the sign of the congruence given by the parity of $k + 1$. We may not be able to identify any perfect squares among the B_k's, so we need a systematic method to produce suitable combinations of the B_k's so that the exponents in the prime factorization are all even. Suppose we have the following factorization of the B_k's.

$$B_1 = p_1^{a_{11}} p_2^{a_{12}} \cdots p_r^{a_{1r}}$$

$$\vdots$$

$$B_s = p_1^{a_{s1}} p_2^{a_{s2}} \cdots p_k^{a_{sr}},$$

where $p_1, \ldots, p_r$ are the primes of the factor base with $p_1 = -1$. (We have to include $p_1 = -1$ to take into account the negative sign when $k + 1$ is odd.)

We have to find numbers $e_1, \ldots, e_s$ that are 0 or 1 such that

$$B_1^{e_1} \cdots B_s^{e_s}$$

is a perfect square. Writing this product using the prime factorization of the B_i's, we find that the numbers $e_1, \ldots, e_s$ are such that

$$e_i a_{1i} + e_2 a_{2i} + \cdots + a_{si} e_s$$

is even for all i. If we do the computations modulo 2, then we have to solve the equation $eA = 0$, where $e = (e_1, \ldots, e_s)$ and A is a matrix whose ijth entry is a_{ij}. This set of linear equations can be solved by Gaussian elimination modulo 2. If $s > r$, then we are guaranteed a nontrivial solution.

Example 12.1.4. The matrix of exponents modulo 2 in the previous example is given below.

$$
A =
\begin{array}{c}
\begin{array}{cccccc} -1 & 2 & 3 & 5 & 7 & 11 \end{array} \\
\left(
\begin{array}{cccccc}
0 & 0 & 0 & 0 & 1 & 1 \\
1 & 0 & 0 & 1 & 0 & 0 \\
0 & 0 & 0 & 0 & 0 & 0 \\
0 & 0 & 1 & 0 & 1 & 0 \\
1 & 0 & 1 & 0 & 1 & 0 \\
0 & 0 & 1 & 0 & 0 & 1 \\
1 & 0 & 0 & 0 & 0 & 0 \\
0 & 0 & 0 & 1 & 0 & 0 \\
1 & 0 & 1 & 0 & 1 & 0 \\
1 & 0 & 0 & 0 & 0 & 0 \\
0 & 0 & 1 & 1 & 1 & 0 \\
1 & 0 & 1 & 0 & 1 & 0 \\
0 & 0 & 0 & 1 & 0 & 0
\end{array}
\right)
\end{array}.
$$

The solutions to the equation $eA = 0$ are spanned by the following vectors.

$$
\begin{aligned}
v_1 &= (0,1,0,1,1,0,0,0,0,0,0,0,1) \\
v_2 &= (0,0,0,0,1,0,0,0,0,0,0,1,0) \\
v_3 &= (0,1,0,0,1,0,0,0,0,0,1,0,0) \\
v_4 &= (0,0,0,1,1,0,0,0,0,1,0,0,0) \\
v_5 &= (0,0,0,1,1,0,0,0,0,1,0,0,0) \\
v_6 &= (0,1,0,1,1,0,0,1,0,0,0,0,0) \\
v_7 &= (0,0,0,1,1,0,1,0,0,0,0,0,0) \\
v_8 &= (1,0,0,1,0,1,0,0,0,0,0,0,0) \\
v_9 &= (0,0,1,0,0,0,0,0,0,0,0,0,0)
\end{aligned}
$$

These solutions yield combinations that will produce congruences of the type

$$
p_{i_1}^2 p_{i_2}^2 \cdots p_{i_k}^2 \equiv B_{i_1+1} B_{i_2+1} \cdots B_{i_k+1} \pmod{n},
$$

where the expression on the right is a square. But it is possible (and happens often) that such a combination does not yield a proper factor. Consider the first element in the basis. It tells us that the product of the second, fourth, fifth, and last rows will give a perfect square. Computing the greatest common divisors, we see that

$$
\begin{aligned}
\gcd(193 \cdot 3719 \cdot 2266 \cdot 3142 - 20 \cdot 3 \cdot 2 \cdot 7, n) &= 4141 \\
\gcd(193 \cdot 3719 \cdot 2266 \cdot 3142 + 20 \cdot 3 \cdot 2 \cdot 7, n) &= 1.
\end{aligned}
$$

We don't obtain a factor with this combination of rows. Next, we consider the second element of this basis, which has 1's in the fifth and twelfth columns. This yields the congruence

$$2266^2 \cdot 3254^2 \equiv 84 \cdot 21 \pmod{n}.$$

We compute the greatest common divisors,

$$\gcd(2266 \cdot 3254 - 2 \cdot 3 \cdot 7, 4141) = 41$$
$$\gcd(2266 \cdot 3254 + 2 \cdot 3 \cdot 7, 4141) = 101$$

to get the factors of n.

Not all elements in the basis for the null space give combinations from which a factor can be deduced. We must try many vectors until a combination that gives a proper factor is found. If none of these yields a factor, then more data must be generated by enlarging the base over which the B_k's factor, or by computing more terms in the continued fraction expansion.

In many cases, it is necessary to compute the continued fraction expansion of $\sqrt{kn}$ for some multiple of n because the period length of the continued fraction expansion of $\sqrt{n}$ may be too short to yield enough data for the Gaussian elimination step.

Example 12.1.5. Consider the fifth Fermat number $F_5 = 2^{32} + 1$. We can easily verify that the number is composite using various compositeness tests. The period length of the continued fraction expansion of $\sqrt{F_5}$ is small, so we consider multiples of F_5. In this example, we take $n = 33F_5$. The factor base consists of the primes 2, 5, 7, 17, and the primes dividing 33, 3, and 11.

$k+1$	p_k	B_{k+1}	B_{k+1} factored
61	91872060437	20000	$2^5 5^4$
294	87626409706	24640	$2^6 11^1 5^1 7^1$
341	45970794893	1715	$5^1 7^3$
376	10149524055	31416	$2^3 3^1 11^1 7^1 17^1$
417	14215198189	161840	$2^4 5^1 7^1 17^2$
682	94951318839	8160	$2^5 3^1 5^1 17^1$
739	137816307510	159375	$3^1 5^5 17^1$
782	4294967773	226576	$2^4 7^2 17^2$
890	73655223917	635800	$2^3 11^1 5^2 17^2$
898	26849875367	286720	$2^3 5^1 7^1$
933	31250183037	199920	$2^4 3^1 5^1 7^2 17^1$
972	27270697390	40000	$2^6 5^4$

We notice that the last row yields a square, hence the congruence

$$27270697390^2 \equiv 40000 \pmod{33F_5},$$

and then we compute

$$(27270697390 - 2000, 33F_5) = 1923,$$

where $1923 = 3 \times 641$ is a proper factor of $33F_5$. Hence 641 must be a factor of F_5. Dividing F_5 by 641 yields

$$F_5 = 641 \times 6700417.$$

This example was special because we did not need the Gaussian elimination step.

————————————— **Exercises for Section 12.1** —————————————

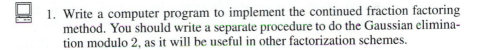

1. Write a computer program to implement the continued fraction factoring method. You should write a separate procedure to do the Gaussian elimination modulo 2, as it will be useful in other factorization schemes.

12.2   Quadratic Sieve

The quadratic sieve factoring method was invented by C. Pomerance in 1981,[3] extending earlier factorization methods of M. Kraitchik and J. Dixon. In order to find the factors of an integer n, this method (like the continued fraction method) tries to find a congruence $x^2 \equiv y^2 \pmod{n}$ such that $x \not\equiv \pm y \pmod{n}$. The difference between the quadratic sieve and the continued fraction method is the way in which the congruence is obtained. In the quadratic sieve, we compute the values of the polynomial $Q(x) = x^2 - n$ for several values of x, say $x_0, x_1, \ldots, x_k$. Suppose we find a subset $x_{i_1}, \ldots, x_{i_r}$ such that the product $Q(x_{i_1}) \cdots Q(x_{i_r})$ is a perfect square y^2. Since $Q(x_i) \equiv x_i^2 \pmod{n}$, we have a congruence

$$x_{i_1}^2 \cdots x_{i_r}^2 \equiv y^2 \pmod{n}.$$

If we are lucky, this congruence will yield a factor of n.

   We can find a subset of the x_i's such that a combination of the $Q(x_i)$'s gives a perfect square if we know the prime factorization of all the $Q(x_i)$'s.

———————————————

[3]C. Pomerance. "The Quadratic Sieve Factoring Algorithm." *Advances in Cryptology–Proceedings of EUROCRYPT 84 (LNCS 209)*, 1985, 169–182.

Factoring Machines

Numerous special purpose machines have been built to factor integers. The most successful of these devices are the sieve machines constructed by D. H. Lehmer. The figure displays the 16mm film based sieve machine constructed in 1934. Lehmer's sieve machines are part of the collection of The Computer Museum of Boston. A sieve machine is a device for rapidly obtaining a number that satisfies several congruences.

D. H. Lehmer's movie film sieve

Here is a simple example of factoring by sieves. The method is an extension of Fermat's factoring method of Section 2.4. Suppose we wish to factor 1817 by writing $x^2 - y^2 = 1817$. We have to take $x \geq \sqrt{1817} \approx 43$. We derive conditions on x so that $x^2 - 1817$ is likely to be a square. Now, $x^2 \equiv 1817 + y^2 \pmod{m}$ for any m. If $m = 3$, then $y^2 \equiv 0, 1 \pmod 3$; this implies that $x^2 \equiv 1817+0, 1817+1 \pmod 3$, that is, $x^2 \equiv 2, 0 \pmod 3$. Similarly, $y^2 \equiv 0, 1, 4 \pmod 5$ implies that $x^2 \equiv 2, 3, 1 \pmod 5$. These congruences imply that $x \equiv 0 \pmod 3$ and $x \equiv \pm 1 \pmod 5$. For each x greater than or equal to 43, we list whether it satisfies these congruences.

	43	44	45	46	47	48	49	50	51
$x \bmod 3$			✓			✓			✓
$x \bmod 5$	✓			✓			✓		✓

The first number satisfying the congruences is 51. We compute $51^2 - 1817 = 784 = 28^2$, from which we can compute the factorization of 1817. A sieve machine can be employed to solve these congruences. Each modulus m_i is represented by a ring with m_i slots. The positions with acceptable residues are tagged and the machines examine each position as the rings are advanced. All the rings are advanced one step in unison. If all the positions are tagged, then a solution is obtained. For additional information on sieve machines, consult R. F. Lukes, C. D. Patterson, and H. C. Williams. "Numerical Sieving Devices: Their History and Some Applications." *Nieuw Archief voor Wiskunde*, 13(1995), 113–139.

This is essentially the method used in the previous section. Recall that the appropriate combination is found using Gaussian elimination (modulo 2) on the exponents of the prime factorization. The main steps in the factoring method are the following.

1. Find values of x such that $Q(x)$ factors over a chosen set of small primes.

2. Determine a suitable subset $x_{i_1}, \ldots, x_{i_r}$ such that $Q(x_{i_1}) \cdots Q(x_{i_r}) = y^2$, a perfect square.

3. Check if $x_{i_1}^2 \cdots x_{i_r}^2 \equiv y^2 \pmod{n}$ gives a factorization of n.

The factoring methods based on the congruence $x^2 \equiv y^2 \pmod{n}$ differ in the way the first step is implemented. If $Q(x) = x^2 - n$, then $Q(x)$ should be small (compared to n) so that it is likely to factor completely over a set of small primes. For this, we choose x close to $\sqrt{n}$. Choose a bound B, and let x be in the range $\sqrt{n} - B \leq x \leq \sqrt{n} + B$. The interval $(\sqrt{n} - B, \sqrt{n} + B)$ is called the *sieving interval*.

The next step is to determine which of the $Q(x)$'s in the sieving interval factor. If $p \mid Q(x)$, then $x^2 \equiv n \pmod{p}$, so $\left(\dfrac{n}{p}\right) = 1$. We select a set of primes, called the *factor base*, satisfying $\left(\dfrac{n}{p}\right) = 1$ and $p < M$. The number M effects the number of primes in the factor base and can be varied depending on the size of n.

Suppose p is in a factor base for n; then we can solve the congruence $r^2 \equiv n \pmod{p}$ (using the procedure of Section 9.2) to find the two roots $r \equiv \pm a_p \pmod{p}$. If $p \mid Q(x)$, then $x^2 \equiv n \equiv a_p^2 \pmod{p}$, hence $x \equiv \pm a_p \pmod{p}$. This means that if $p \mid Q(x)$, then x lies in an arithmetic progression $kp \pm a_p$. This is the crucial step of the quadratic sieve. We select $Q(x)$'s that are likely to be factored over the factor base as those that lie in many of these arithmetic progressions. This is the sieving step of the quadratic sieve and also the step from which the method derives its speed. We are able to eliminate many $Q(x)$'s that do not factor over the factor base without obtaining a factorization of every $Q(x)$ by trial division. This is a significant improvement over the continued fraction method in which we had to factor every number by trial division. We give an example (albeit small) to illustrate this step.

Example 12.2.1. Let's take $n = 17819$. We choose a factor base of three primes, 2, 5, and 7 and let $B = 5$ so that there are 11 numbers in our sieving interval around $\sqrt{n} \approx 133$. Let x be in the interval $128 \leq x \leq 138$. Since $n \equiv 1 \pmod{2}$, $Q(x)$ is divisible by 2 for x odd. Since $n \equiv 4 \pmod{5}$, $Q(x)$ is divisible by 5 when $x \equiv 2, 3 \pmod{5}$. Also, $n \equiv 4 \pmod{7}$

implies that $Q(x)$ is divisible by 7 when $x \equiv 2, 5 \pmod 7$. The following table shows if 2, 5, or 7 is a factor of $Q(x)$.

x	2	5	7
128		✓	✓
129	✓		
130			
131	✓		✓
132		✓	
133	✓	✓	
134			
135	✓		✓
136			
137	✓	✓	
138		✓	✓

We can eliminate $x = 130, 134$, and 136, as $Q(x)$ for these values of x do not have any factors from the factor base. The values of $Q(x)$ for $x = 128$, 131, 133, 135, 137, and 138 are likely to factor over the factor base. Among these we see that only $Q(138)$ factors over the primes 5 and 7 as $Q(138) = 5^2 \cdot 7^2$. This term is a perfect square, yielding the congruence $138^2 \equiv 35^2 \pmod n$. From this congruence, we obtain $(138 - 35, 17819) = 103$ and $(138 + 35, 17819) = 173$ as the prime factors of n.

In the example, we were able to identify numbers that did not have any prime factor from the factor base. We want to be able to select only those $Q(x)$'s that are very likely to factor. To do this, we will have to divide $Q(x)$ by primes in the factor base. Since division takes time, with the observation of C. Pomerance to use logarithms, we subtract instead of dividing. For this we initialize a list S of numbers for $\sqrt{n} - B \leq x \leq \sqrt{n} + B$ with $S(x) = \log |Q(x)|$. For each prime p in the factor base, we go through the list and subtract $\log p$ whenever $p \mid Q(x)$ or, equivalently, p lies in the arithmetic progression $x \equiv \pm a_p \pmod p$. It is easy to do this operation because we start with the smallest value of x in the range lying in this arithmetic progression and then keep adding p. Let's consider the possible values of $S(x)$ after this is done for all the primes in the factor base.

If $Q(x)$ factors as a product of distinct primes in the factor base, then the value of $S(x)$ will be zero. This is because $Q(x) = p_1 \cdots p_r$; hence $S(x) = \log |Q(x)| = \sum_{i=1}^{r} \log(p_i)$, and subtracting $\log p$ whenever $Q(x)$ is divisible by p will make this zero. The other extreme case occurs when $Q(x)$ doesn't factor completely over the factor base; then it is divisible by a prime greater than the largest prime in the factor base. Let $p_{\max}$ be the largest prime in the factor base. Then $S(x) > \log(p_{\max})$. If $S(x) < \log(p_{\max})$, then we are assured that $Q(x)$ factors completely over the factor base, so we select those x for which $S(x) < \log(p_{\max})$.

Example 12.2.2. Let's take $n = 87463$ and a factor base of six primes consisting of 2, 3, 13, 17, 19, and 29. Take $B = 100$, so the sieve interval is $195 \leq x \leq 395$ (because $\sqrt{n} \approx 295$). We initialize the array S by $S(x) = \log|Q(x)|$. After the sieve step, we select the x such that $S(x) < \log(p_{\max}) = \log(29) = 3.3673$. This will give us terms that will factor completely over the factor base. The computation gives six values of x. These are $x = 265, 278, 296, 299, 307$, and 347. Here is the prime factorization of $Q(x)$'s.

x	$Q(x)$
265	$-1^1 2^1 3^1 13^2 17^1$
278	$-1^1 3^3 13^1 29^1$
296	$3^2 17^1$
299	$2^1 3^1 17^1 19^1$
307	$2^1 3^2 13^2 9^1$
347	$2^1 3^1 17^2 19^1$

By taking $S(x) < \log(p_{\max})$, we miss several values of $Q(x)$ that factor completely over the factor base. These are divisible by higher powers of the primes in the factor base. For example, $Q(316) = 3^6 17$ and $Q(242) = -3^2 13^2$ are missed. If we include values where $S(x) > \log(p_{\max})$, then we are likely to get many other terms that do not factor over the factor base.

If we want to obtain all the $Q(x)$'s that factor over the base, then there are two options. We can include x such that $S(x) > \log(p_{\max})$, or we can expand the sieve to include prime powers. Solve the congruences $x^2 \equiv n \pmod{p^s}$ for a few values of s. Suppose $a_{p,s}^2 \equiv n \pmod{p^s}$; then we can subtract $\log p$ from $S(x)$ whenever $x \equiv a_{p,s} \pmod{p^s}$. If $S(x)$ is close to 0, then we will know if $Q(x)$ factors over the factor base. In practice, it is efficient to sieve by the powers of a few small primes to increase the number of $Q(x)$'s which factor.

After we have factored the $Q(x)$'s, we must use the Gaussian elimination technique of the previous section to find a combination that yields a perfect square. In this step, we treat -1 as a prime, because we need an even number of negative integers to make a perfect square. The details of the Gaussian elimination are the same as those given in the previous section. If we have the factorizations

$$Q(x_1) = p_1^{a_{11}} p_2^{a_{12}} \cdots p_r^{a_{1r}}$$

$$\vdots$$

$$Q(x_k) = p_1^{a_{k1}} p_2^{a_{k2}} \cdots p_k^{a_{kr}},$$

where $p_1, \ldots, p_r$ are the primes of the factor base with $p_1 = -1$, we have to find numbers $e_1, \ldots, e_k$ that are 0 or 1 such that

$$Q(x_1)^{e_1} \cdots Q(x_k)^{e_k}$$

is a perfect square. If A is the $k \times r$ matrix with the ijth entry a_{ij} mod 2, then the exponent vector $(e_1, \ldots, e_k)$ satisfies $eA = 0$. Here the computations are performed modulo 2. The vector e can be found by Gaussian elimination.

Example 12.2.3. Using the data obtained for the factorization of $n = 87643$ in the previous example, we obtain the matrix

$$A = \begin{pmatrix} 1 & 1 & 1 & 0 & 1 & 0 & 0 \\ 1 & 0 & 1 & 1 & 0 & 0 & 1 \\ 0 & 0 & 0 & 0 & 1 & 0 & 0 \\ 0 & 1 & 1 & 0 & 1 & 1 & 0 \\ 0 & 1 & 0 & 1 & 0 & 0 & 1 \\ 0 & 1 & 1 & 0 & 0 & 1 & 0 \end{pmatrix}.$$

Here $x_1 = 265$, $x_2 = 278$, $x_3 = 296$, $x_4 = 299$, $x_5 = 307$, $x_6 = 347$. The row space of the matrix is spanned by the vectors $e_1 = (0, 0, 1, 1, 0, 1)$ and $e_2 = (1, 1, 1, 0, 1, 0)$. From e_1, we conclude that $Q(x_3)Q(x_4)Q(x_6)$ is a perfect square. Hence we compute

$$\gcd((296 \cdot 299 \cdot 347 - 2 \cdot 3^2 \cdot 17^2 \cdot 19), n) = 1.$$

The first combination fails to yield a factor. From e_2, we see that the product $Q(x_1)Q(x_2)Q(x_3)Q(x_5)$ is a perfect square. We compute

$$\gcd((265 \cdot 278 \cdot 296 \cdot 307 - 2 \cdot 3^4 \cdot 13^2 \cdot 17 \cdot 29), n) = 149,$$

which is a prime factor of n.

───────────────── **Exercises for Section 12.2** ─────────────────

1. Describe the quadratic sieve algorithm in algorithmic notation.

2. Write a computer program to implement the various steps of the quadratic sieve algorithm. You will need to write a procedure to compute square roots modulo p, and a procedure to perform the sieve. Programs to perform Gaussian elimination are available as part of many standard computing packages.

Chapter 13

Diophantine Approximations

13.1 Best Approximations

A fundamental problem in mathematics is to determine how closely we can approximate an irrational number by rational numbers. Every real number can be approximated by rational numbers so that the error is as small as we please. The problem is to control the size of the numerator and denominator in the rational number. The problem seems to have been first studied by Christian Huygens, the Dutch physicist, who used continued fractions to find rational approximations in the design of an automatic planetarium. In Huygens's planetarium, each of the six known planets were to be attached to a large circular gear driven by a smaller gear mounted on a common drive shaft. Approximating the year to be $365\frac{35}{144}$ days, Huygens calculated the ratios of the periods of each planet's orbit to the earth's year. For example, for Jupiter, he obtained $1247057/105190$. Since it was difficult to cut gear with more than 200 teeth, it was necessary to have an approximation p/q to this number with $0 \le p, q \le 200$. Huygens obtained these approximations by looking at the convergents in the continued fraction.

The basic approximation problem is the following.

> Given a real number α and a positive integer N, find the rational number with denominator q less than or equal to N that is nearest to α.

This question will be answered in the next section. First, we define a stronger notion of approximation, which is characterized in terms of the convergents of the continued fraction of α.

Definition 13.1.1. A rational number p/q is a **good approximation** to α if for every p'/q' with $1 \le q' < q$,

$$\left| \alpha - \frac{p}{q} \right| < \left| \alpha - \frac{p'}{q'} \right|.$$

Definition 13.1.2. A rational number p/q is a **best approximation** to α if for every p'/q', $1 \le q' < q$,

$$|q\alpha - p| < |q'\alpha - p'|.$$

Example 13.1.3. Let us consider $\pi \approx 3.1415927$. Since $10/3 - \pi \approx 0.1917 > \pi - 3 = 0.1415927$, there is no good approximation with denominator 3. It is clear that $13/4 = 3.25$ is a good approximation (because any rational numbers with denominator 2 and 3 are farther from π than $13/4$), and so is $16/5 = 3.2$. On the other hand,

$$5\pi \approx 15.707963$$

and

$$|5\pi - 16| > 0.1415927 = |\pi - 3|.$$

Hence $16/5$ is not a best approximation. Note that

$$6\pi \approx 18.8495$$

and

$$|6\pi - 19| > |\pi - 3|.$$

Hence there is no best approximation of the form $p/6$. Now

$$7\pi \approx 21.991149$$
$$|7\pi - 22| \approx 0.0088514$$

and so $22/7$ satisfies the definition of a best approximation. This well-known approximation has been in use since ancient times. The next best approximation to π is $333/106$, discovered in Europe in the sixteenth century. An even better best approximation is $355/113$, which was known in China in the fifth century. The diligent reader can verify that these are the first four convergents in the continued fraction expansion of $\pi = [3, 7, 15, 1, 292, 1, \dots]$. (The first convergent is $a_0 = 3$.)

The following lemma shows that the property of best approximation is stronger than the property of good approximation. We expect that it is easier to characterize the best approximations, and this is indeed the case. The numerator p of a best approximation is determined by the denominator q, unless the number $q\alpha$ happens to be halfway between integers. For irrational α, this cannot happen. For simplicity of arguments, we take α to be irrational.

Lemma 13.1.4. *If p/q is a best approximation to α, then it is a good approximation.*

PROOF. If p/q is a best approximation, then

$$|q\alpha - p| > |q'\alpha - p'|, \qquad 1 \le q' < q.$$

Dividing by q, we obtain

$$\left| \alpha - \frac{p}{q} \right| < \frac{q'}{q} \left| \alpha - \frac{p'}{q'} \right|.$$

Since $q'/q < 1$, we obtain

$$\left| \alpha - \frac{p}{q} \right| < \left| \alpha - \frac{p'}{q'} \right|.$$

This proves that p/q is a good approximation. ∎

Let $[a_0, a_1, a_2 \dots]$ be the simple continued fraction expansion of α and let $C_n = p_n/q_n$ be the convergents of α. In the proof of the following theorem, we use the inequality

$$\left| \alpha - \frac{p_n}{q_n} \right| < \frac{1}{q_n q_{n+1}}$$

and the estimate $|p/q - r/s| \ge 1/(qs)$ whenever $p/q \ne r/s$ for rational numbers p/q and r/s.

Theorem 13.1.5. *The best approximations to α are convergents of the continued fraction expansion of α, and, conversely, every convergent p_n/q_n, $n \ge 1$, of α is a best approximation.*

PROOF. Suppose p/q is a best approximation. If it is not a convergent, then we will show that it must lie between two convergents and then derive a contradiction. Clearly, p/q cannot be less than $a_0 = \lfloor \alpha \rfloor$ and $q > 1$. If p/q does not lie between two convergents, then $p/q > p_1/q_1$, hence $\alpha < p_1/q_1 < p/q$. Multiply by q and subtract p to obtain $q\alpha - p < qp_1/q_1 - p < 0$; thus,

$$|q\alpha - p| > \left| \frac{qp_1 - pq_1}{q_1} \right| \ge \frac{1}{q_1},$$

where we use the fact that $qp_1 - pq_1$ is a nonzero integer. Next, observe that, $\alpha - a_0 = 1/x_1 < 1/q_1$, and we have obtained

$$\alpha - a_0 < \frac{1}{q_1} < |q\alpha - p|,$$

contradicting the assumption that p/q is a best approximation.

We may assume that p/q lies between p_n/q_n and p_{n+2}/q_{n+2} that are on the same side of α as p/q.

$$
\begin{array}{ccccccc}
\dfrac{p_{n+1}}{q_{n+1}} & & \alpha & & \dfrac{p_{n+2}}{q_{n+2}} & \dfrac{p}{q} & \dfrac{p_n}{q_n}
\end{array}
$$

On one hand, we have

$$\frac{1}{qq_n} \leq \left| \frac{p}{q} - \frac{p_n}{q_n} \right| < \left| \frac{p_n}{q_n} - \frac{p_{n+1}}{q_{n+1}} \right| = \frac{1}{q_n q_{n+1}},$$

which implies that $qq_n > q_n q_{n+1}$, that is, $q > q_{n+1}$. (The first inequality follows because $pq_n - p_n q$ is a nonzero integer in the numerator. The second step is clear from the picture. The third inequality is derived similarly.)

On the other hand,

$$\frac{1}{qq_{n+2}} \leq \left| \frac{p}{q} - \frac{p_{n+2}}{q_{n+2}} \right| < \left| \frac{p}{q} - \alpha \right|.$$

This implies that $\dfrac{1}{q_{n+2}} < |q\alpha - p|$, but we know that

$$|q_{n+1}\alpha - p_{n+1}| < \frac{1}{q_{n+2}} < |q\alpha - p|.$$

Since p/q is a best approximation, the definition implies that $q < q_{n+1}$, but we showed that $q \geq q_{n+1}$. This contradicts our assumption that p/q is not a convergent.

To show the converse, suppose p_n/q_n, $n \geq 1$ is a convergent. We will show by induction that these are best approximations. To start the induction, suppose there exists p/q, with $1 \leq q < q_1$, such that $|q\alpha - p| < |q_1\alpha - p_1|$. Write $\alpha = a_0 + \dfrac{1}{x_1}$, where x_1 is the first complete quotient and $a_1 = \lfloor x_1 \rfloor$. This implies that $q_1 = a_1$ and $p_1 = a_0 a_1 + 1$. Then

$$\left| q\left(a_0 + \frac{1}{x_1}\right) - p \right| < \left| a_1\left(a_0 + \frac{1}{x_1}\right) - (a_0 a_1 + 1) \right|$$

simplifies to

$$\left| qa_0 - p + \frac{q}{x_1} \right| < \left| \frac{x_1 - a_1}{x_1} \right|.$$

We also have $|(x_1 - a_1)/x_1| < 1$ and $q/x_1 < 1$. Since $qa_0 - p$ is an integer, there are only two possibilities for $qa_0 - p$, either $qa_0 - p = 0$ or $qa_0 - p = -1$. If $qa_0 - p = 0$, then $q < |a_1 - x_1|$, that is, $q = 0$, and this is not possible. If $qa_0 - p = -1$, then $|q - x_1| < |a_1 - x_1|$, so q is closer to x_1 than a_1. This is only possible when $q = a_1 + 1$ and $x_1 > a_1 + 1/2$. This contradicts the assumption that $q < a_1$; hence we have shown that p_1/q_1 is a best approximation.

We assume now that $p_1/q_1, \ldots, p_k/q_k$ are best approximations; we must show that p_{k+1}/q_{k+1} is a best approximation. Let q be the smallest integer larger than q_k such that for a suitable p,

$$|q\alpha - p| < |q_k\alpha - p_k|. \tag{1}$$

This implies that p/q is a best approximation, since p_k/q_k has this property. Then the first part of the theorem implies that p/q is a convergent. Hence $q > q_k$ implies that $q \geq q_{k+1}$. If we show that

$$|q_{k+1}\alpha - p_{k+1}| < |q_k\alpha - p_k|, \tag{2}$$

the proof of the theorem will be complete, as q was chosen to be the smallest integer satisfying inequality (1), and then $q = q_{k+1}$. We prove (2) in the following lemma. ∎

Lemma 13.1.6. *For successive convergents p_n/q_n and p_{n+1}/q_{n+1} to α,*

$$|q_{n+1}\alpha - p_{n+1}| < |q_n\alpha - p_n|.$$

PROOF. If $\alpha = [a_0, a_1, \ldots a_{n+1}, x_{n+2}]$, then

$$\alpha = \frac{x_{n+2}p_{n+1} + p_n}{x_{n+2}q_{n+1} + q_n}.$$

We cross-multiply and solve for x_{n+2} to obtain

$$x_{n+2}(q_{n+1}\alpha - p_{n+1}) = p_n - \alpha q_n.$$

Now $x_{n+2} > 1$, as it is a complete quotient of the continued fraction; then

$$|q_{n+1}\alpha - p_{n+1}| < |q_n\alpha - p_n|.$$

This completes the proof of the lemma, and hence the theorem is proved. ∎

Corollary 13.1.7. *If a rational number p/q satisfies*

$$\left| \alpha - \frac{p}{q} \right| < \frac{1}{2q^2}$$

then p/q is a convergent of α.

PROOF. From the theorem, it suffices to show that p/q is a best approximation. Suppose that a rational number r/s satisfies,

$$|s\alpha - r| < |q\alpha - p| < \frac{1}{2q},$$

so we divide by s to get $|\alpha - r/s| < 1/(2qs)$. Also, $1/qs \le |r/s - p/q|$.
Add and subtract α and use the triangle inequality

$$\frac{1}{qs} \le \left|\frac{r}{s} - \alpha + \alpha - \frac{p}{q}\right|$$

$$\le \left|\alpha - \frac{r}{s}\right| + \left|\alpha - \frac{p}{q}\right|.$$

Then

$$\frac{1}{qs} \le \frac{1}{2qs} + \frac{1}{2q^2}$$

$$= \frac{q+s}{2q^2 s}.$$

Multiplying by $2q^2 s$, we obtain $q \le s$, that is, p/q is a best approximation. ∎

Example 13.1.8. The result of the corollary has applications to the solution
of Diophantine equations. Suppose we want to solve the equation $x^2 - 2y^2 = 1$ in the integers. By factoring the polynomial, we write $(x - \sqrt{2}y)(x + \sqrt{2}y) = 1$. We can assume that x and y are positive, so $x > y$, hence
$x + \sqrt{y} > 2y$. Therefore,

$$x - \sqrt{2}y = \frac{1}{x + \sqrt{2}y} < \frac{1}{2y},$$

and dividing by y, we get

$$\left|\frac{x}{y} - \sqrt{2}\right| < \frac{1}{2y^2}.$$

This inequality implies that x/y is a convergent of the continued fraction of
$\sqrt{2}$. We know that $\sqrt{2} = [1,2,2,2,2,\dots]$, so the convergents are 1, $3/2$,
$7/5$, $17/12$, ... and we see that $3^2 - 2\cdot 2^2 = 1$, $17^2 - 2\cdot 12^2 = 1$, and so on.
We leave it to the reader to prove that every odd convergent gives a solution
to the equation, and these give all the solutions except for the possibility of
changing the signs of x and y.

It is easy to find examples to see that not every convergent satisfies the inequality of the previous proposition. But we can show that of every two successive convergents at least one must satisfy it.

Proposition 13.1.9. *Of any two consecutive convergents of α, at least one must satisfy the inequality*

$$\left| \alpha - \frac{p}{q} \right| < \frac{1}{2q^2}.$$

PROOF. Consider p_n/q_n and p_{n+1}/q_{n+1}.

$$\begin{array}{ccc} \dfrac{p_{n+1}}{q_{n+1}} & \alpha & \dfrac{p_n}{q_n} \end{array}$$

Since they are on different sides of α, we have

$$\frac{1}{q_n q_{n+1}} = \left| \frac{p_n}{q_n} - \frac{p_{n+1}}{q_{n+1}} \right| = \left| \alpha - \frac{p_n}{q_n} \right| + \left| \alpha - \frac{p_{n+1}}{q_{n+1}} \right|.$$

If both of them do not satisfy the desired inequality, then we obtain

$$\frac{1}{q_n q_{n+1}} \geq \frac{1}{2q_n^2} + \frac{1}{2q_{n+1}^2}$$

$$\frac{1}{2q_n^2} + \frac{1}{2q_{n+1}^2} - \frac{1}{q_n q_{n+1}} \leq 0$$

$$\frac{1}{2} \left(\frac{1}{q_n} - \frac{1}{q_{n+1}} \right)^2 \leq 0,$$

which can only happen if

$$q_{n+1} = q_n.$$

Since $q_{n+1} = a_n q_n + q_{n-2}$ and $a_k \geq 1$ for $k \geq 1$, this can happen only if $n = 0$ and $a_1 = 1$, that is, $q_0 = q_1 = 1$. If $n = 0$ and $q_0 = q_1 = 1$, then $p_0/q_0 = a_0$ and $p_1/q_1 = a_0 + 1$, and we must have $|\alpha - a_0|^2 < 1/2$ or $|\alpha - (a_0 + 1)|^2 < 1/2$. This proves the proposition. ∎

Proposition 13.1.10. *A number α is irrational if and only if there exists infinitely many solutions to the inequality*

$$\left| \alpha - \frac{p}{q} \right| < \frac{1}{q^2}.$$

PROOF. If α is irrational, then every other convergent is a solution to the inequality; hence we have infinitely many solutions. Conversely, if α is a rational number a/b, then

$$\frac{1}{bq} \le \left| \frac{a}{b} - \frac{p}{q} \right| < \frac{1}{q^2}$$

implies that $q < b$. There are only finitely many possibilities for q and hence for p. Therefore, if the inequality has infinitely many solutions, α has to be an irrational number. ∎

The proposition is one way to prove the irrationality of a number.

———————————— **Exercises for Section 13.1** ————————————

1. Determine the best approximation p/q to $\sqrt{2}$ with denominator less than 10000. Can you give a good estimate for the error without computing it?

2. Show that $137/38$ is a convergent of $\sqrt{13}$ without computing the continued fraction expansion.

3. Make a table of values of $ne - m$ for $1 \le n \le 100$, where e is the base of the natural logarithms and m is the nearest integer to ne. Describe how to use this table to identify the convergents of the continued fraction of e.

4. Suppose that $\pi < p/q < 22/7$. Prove that $q > 106$.

5. Suppose that $|\pi - p/q| < |\pi - 22/7|$. Show that $q < 56$.

6. Verify that the following geometric interpretations of best approximations is justified.

   Imagine that there are pegs on the plane at each point (x, y), for x and y integers. (See Figure 13.1.1.) Given an irrational real number $\alpha > 0$, consider the line through the origin of slope α. This line does not pass through any point with integer coordinates. (Why?) Imagine that a piece of thin thread is tied to an infinitely remote point on the line, and we hold the other end in our hand. If the thread is pulled taut to the origin and then moved left, the thread will catch certain pegs above the line $y = \alpha x$. Prove that the pegs caught are the points (q, p), where p/q is an odd convergent, and if the thread is moved below the line, the pegs caught are the points (q, p) where p/q is an even convergent.

7. Show that if α and β have the same $n + 1$ initial terms $a_0, a_1, \ldots a_n$ in their continued fraction expansions, then any γ, $\alpha < \gamma < \beta$ has the same first $n + 1$ terms in its continued fraction expansion. Find the exact interval on the real line whose points have continued fraction expansions of the form $[2, 1, 2, 1, 1, 4, 1, 1, \ldots]$, where only the first 8 terms of the continued fraction are specified.

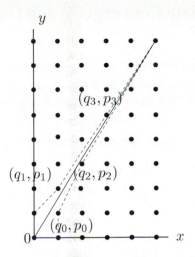

Figure 13.1.1 Line of slope α

8. Suppose that $p_1/q_1 < \alpha < p_2/q_2$ and $p_2q_1 - p_1q_2 = 1$. Show that either p_1/q_1 or p_2/q_2 is a convergent of α.

9. Show that every odd convergent p/q of $\sqrt{2}$ gives a solution to the equation $p^2 - 2q^2 = 1$.

10. Show that if $x^3 - 2y^3 = 1$, then x/y is a convergent of $\sqrt[3]{2}$.

11. We showed that if $|\alpha - p/q| < 1/(2q^2)$, then p/q is a convergent of α. Hurwitz showed that the value 2 in the denominator can be replaced by $\sqrt{5}$ and that this is the best possible result of this type. Prove Hurwitz's theorem by answering the following questions.

    (a) Suppose three consecutive convergents, C_{k-1}, C_k, and C_{k+1}, satisfy the inequality $\alpha - p/q \geq \frac{1}{\sqrt{5}q^2}$. Show that q_{k-1}/q_k, q_k/q_{k-1}, and q_{k+1}/q_k satisfy the inequality $x^2 - \sqrt{5}x + 1 \leq 0$.

    (b) Show that the three convergents must lie in the interval $(\frac{\sqrt{5}-1}{2}, \frac{\sqrt{5}+1}{2})$. This implies that $a_{k+1} < 1$, an impossibility. Hence one of every three convergents satisfies the inequality

$$\left| \alpha - \frac{p}{q} \right| < \frac{1}{\sqrt{5}q^2}.$$

12. (Suggested by Rachel Sendrovic) Show that if $1 + y^2 + z^2 = 3yz$ with $y < z$, then $y = f_{2k-1}$ and $z = f_{2k+1}$ for some k, where f_i is the ith Fibonacci number.

13. (I. Gessel, *Fibonacci Quarterly,* 1972, pp. 417–419.) Show that an integer n is a Fibonacci number if and only if $5n^2 + 4$ or $5n^2 - 4$ is a square.

13.2 Intermediate Convergents and Good Approximations

It follows from Lemma 13.1.4 that we can expect more good approximations than best approximations. These lie between the convergents of the continued fraction expansion, and they can be obtained from the convergents via the following construction.

Definition 13.2.1. If p_k/q_k are the convergents of α, then the intermediate convergents are

$$C_{k,a} = \frac{p_{k,a}}{q_{k,a}} = \frac{ap_{k-1} + p_{k-2}}{aq_{k-1} + q_{k-2}}$$

for $0 \le a \le a_k$.

Example 13.2.2.   1. For $\dfrac{1 + \sqrt{5}}{2} = [1, 1, 1, 1, 1 \ldots]$, since all the $a_k = 1$, the intermediate convergents are the same as the convergents.

2. For $\pi = [3, 7, 15, 1, \ldots]$, the first two convergents are $\dfrac{3}{1}, \dfrac{22}{7}$. The intermediate convergents corresponding to these convergents are $\dfrac{3a + 1}{a}$ and $\dfrac{22a' + 3}{7a' + 1}$, where $0 \le a \le 7$ and $0 \le a' \le 15$. Hence the intermediate convergents for π are

$$\frac{\mathbf{3}}{\mathbf{1}}, \frac{7}{2}, \frac{10}{3}, \frac{13}{4}, \frac{16}{5}, \frac{19}{6}, \frac{\mathbf{22}}{\mathbf{7}}, \frac{25}{8}, \frac{47}{15}, \ldots \ .$$

3. The largest intermediate convergent is $\dfrac{p_0 + p_{-1}}{q_0 + q_{-1}} = a_0 + 1$, and the smallest intermediate convergent is a_0. This can be verified using Lemma 13.2.4.

It is clear that $p_{k,0}/q_{k,0} = p_{k-2}/q_{k-2}$ and $p_{k,a_k}/q_{k,a_k} = p_k/q_k$. To understand the relationship between the intermediate convergents and the convergents, we need the following observation regarding fractions.

Lemma 13.2.3. *Suppose* $\dfrac{a}{b} < \dfrac{c}{d}$; *then for any* $r > 0$

$$\frac{a}{b} < \frac{ra + c}{rb + d} < \frac{c}{d},$$

and

$$\frac{a}{b} < \frac{a + rc}{b + rd} < \frac{c}{d}.$$

Further, if $r < s$, then

$$\frac{a + rc}{b + rd} < \frac{a + sc}{b + sd},$$

and

$$\frac{ra + c}{rb + d} > \frac{sa + c}{sb + d}.$$

PROOF. All the statements can be verified by cross multiplication. ∎

The lemma implies the following relations between the intermediate convergents.

Lemma 13.2.4. *For k even, $C_{k-2} < C_{k,1} < C_{k,2} < \ldots C_{k,a_k} = C_k$ and $q_{k-2} < q_{k-1,1} < \cdots < q_{k-1} < q_{k,1} < \cdots < q_k$.*

$$C_{k-2}\ C_{k,1}\ C_{k,2}\ \quad \cdots \quad\ C_{k,a_k}\ C_k$$

There is an analogous result of odd k that the reader is asked to state and prove in the exercises.

Lemma 13.2.5. $p_{k,a}q_{k,a+1} - p_{k,a+1}q_{k,a} = (-1)^{k+1}.$

PROOF. The proof is by using the definition of the intermediate convergents and the analogous property of the convergents. ∎

Lemma 13.2.5 implies that

$$\frac{p_{k,a}}{q_{k,a}} - \frac{p_{k,a+1}}{q_{k,a+1}} = (-1)^{k+1}\frac{1}{q_{k,a}q_{k,a+1}}.$$

Recall from the previous section that a good approximation to a real number α is a rational number p/q such that $|\alpha - p/q| < |\alpha - r/s|$, whenever $s < q$. We will show that all good approximations are intermediate convergents, but not every intermediate convergent is a good approximation.

Theorem 13.2.6. *If p/q is a good approximation to α, then it is an intermediate convergent of α.*

PROOF. Suppose p/q is a good approximation to α. Clearly, $p/q < p_0/q_0$ is not possible. If $\dfrac{p}{q} > \dfrac{p_0 + p_{-1}}{q_0 + q_{-1}} = a_0 + 1$, the largest intermediate convergent, then $a_0 + 1$ is a better approximation. Therefore, we can assume that p/q lies between two successive intermediate convergents and is not equal to either. Suppose p/q lies between $C_{k,a}$ and $C_{k,a+1}$ for k even. The proof will be similar for k odd.

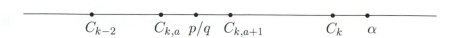

$$C_{k-2} \qquad C_{k,a} \ \ p/q \ \ C_{k,a+1} \qquad\qquad C_k \qquad \alpha$$

Lemma 13.2.5 implies that

$$\frac{1}{qq_{k,a}} \le \left| \frac{p}{q} - \frac{p_{k,a}}{q_{k,a}} \right| < \left| \frac{p_{k,a}}{q_{k,a}} - \frac{p_{k,a+1}}{a_{k,a+1}} \right| = \frac{1}{q_{k,a}q_{k,a+1}}.$$

This implies that $q > q_{k,a+1}$. This contradicts the assumption that p/q is a good approximation, because $C_{k,a+1}$ is closer to α and has a smaller denominator. The proof is similar for k odd. ∎

It is not true that every intermediate convergent is a good approximation. For π, $7/2$ and $10/3$ are intermediate convergents, but are farther away from π than 3 is. Only a subset of the intermediate convergents form good approximations, and these are determined in the following theorem.

Theorem 13.2.7. *An intermediate convergent $C_{k,a}$, $a \ne 0$ is a good approximation if and only if $2a > a_k$ or $2a = a_k$ and $[a_k, a_{k-1}, \ldots, a_1] > [a_k, a_{k+1}, \ldots]$.*

PROOF. Since $q_{k-2} < \cdots < q_{k-1} < q_{k,1} < q_{k,2} < \cdots < q_k$, it is clear that if $C_{k,a} = p_{k,a}/q_{k,a}$ is farther from α than $C_{k-1} = p_{k-1}/q_{k-1}$, then it is not a good approximation. Now we prove that if $C_{k,a}$ is closer to α than C_{k-1} is, then it is a good approximation.

Let $q_{k,a}$ be the smallest denominator such that $|\alpha - p_{k,a}/q_{k,a}| < |\alpha - p_{k-1}/q_{k-1}|$ If $C_{k,a}$ is not a good approximation, then there exists r_1/s_1 with $1 \le s_1 < q_{k,a}$ such that $|\alpha - p_{k,a}/q_{k,a}| < |\alpha - r_1/s_1|$. We must have $s_1 > q_{k-1}$ because for $q \le q_{k-1}$, $|\alpha - p/q| < |\alpha - p_{k-1}/q_{k-1}|$. Now, r_1/s_1 is not a good approximation because it is not an intermediate convergent (an intermediate convergent with denominator less than $q_{k,a}$ is farther from α than C_{k-1} is). Then there exists r_2/s_2 with $q_{k-1} < s_2 < s_1$ such that $|a - r_1/s_1| \ge |\alpha - r_2/s_2|$. Again r_2/s_2 is not a good approximation, and we can find r_3/s_3 with $q_{k-1}, s_3 < s_2$ that is closer to α. Proceeding in this manner, after at most $n = q_{k,a} - q_{k-1}$ steps, we must have $s_n < q_{k-1}$, a contradiction, so $p_{k,a}/q_{k,a}$ is a good approximation. The same argument can be used to prove that if $C_{k,a}$ is a good approximation, then $C_{k,a+1}$ is a good approximation. Then all the intermediate convergents that are closer to α than p_{k-1}/q_{k-1} is are good approximations and they verify

$$\left| \alpha - \frac{a p_{k-1} + p_{k-2}}{a q_{k-1} + q_{k-2}} \right| < \left| \alpha - \frac{p_{k-1}}{q_{k-1}} \right|,$$

which we write as

$$\left| \frac{a(\alpha q_{k-1} - p_{k-1}) + \alpha q_{k-2} - p_{k-2}}{a q_{k-1} + q_{k-2}} \right| < \left| \frac{\alpha q_{k-1} - p_{k-1}}{q_{k-1}} \right|. \tag{1}$$

Recall that $\alpha = [a_0, a_1, \ldots, a_{k-1}, x_k]$ implies that

$$\alpha = \frac{x_k p_{k-1} + p_{k-2}}{x_k q_{k-1} + q_{k-2}}.$$

Solving for x_k, we get

$$-x_k(\alpha q_{k-1} - p_{k-1}) = \alpha q_{k-2} - p_{k-2}.$$

We use this in Equation (1) to obtain

$$\left| \frac{(a - x_k)(\alpha q_{k-1} - p_{k-1})}{a q_{k-1} + q_{k-2}} \right| < \frac{\alpha q_{k-1} - p_{k-1}}{q_{k-1}},$$

which is equivalent to

$$|a - x_k| q_{k-1} < a q_{k-1} + q_{k-2}.$$

Since $a < a_k \leq x_k$, we can write

$$(x_k - a) q_{k-1} < a q_{k-1} + q_{k-2},$$

which is the same as

$$x_k q_{k-1} < 2a q_{k-1} + q_{k-2}. \tag{2}$$

If $2a > a_k + 1$, then

$$2a q_{k-1} + q_{k-2} > (a_k + 1) q_{k-1} + q_{k-2}$$
$$> (a_k + 1) q_{k-1}$$
$$> x_k q_{k-1}$$

satisfying (2).

For these a, the intermediate convergents $p_{k,a}/q_{k,a}$ are good approximations, as they are closer to α than p_{k-1}/q_{k-1} is.

If $2a \leq a_k - 1$, then

$$2aq_{k-1} + q_{k-2} \leq (a_k - 1)q_{k-1} + q_{k-2}$$
$$\leq a_k q_{k-1}$$
$$< x_k q_{k-1}.$$

For these a, $p_{k,a}/q_{k,a}$ is farther away from α than p_{k-1}/q_{k-1} is.

The only remaining case is $2a = a_k$. Equation (2) implies that

$$a_k q_{k-1} + q_{k-2} > x_k q_{k-1}$$
$$q_k > x_k q_{k-1}$$
$$\frac{q_k}{q_{k-1}} > x_k = [a_k, a_{k+1}, \ldots],$$

and since $\dfrac{q_k}{q_{k-1}} = [a_k, a_{k-1}, \ldots, a_1]$, we obtain the necessary condition in
the theorem. ∎

Example 13.2.8. We know that $\pi = [3, 7, 15, 1, 292, 1, 1, 1, \ldots]$, and its
convergents are

$$\frac{3}{1}, \frac{22}{7}, \frac{333}{106}, \frac{355}{113}, \ldots .$$

The intermediate convergents are

$$\frac{3a+1}{a}, \frac{22a'+3}{7a'+1}, \frac{333a''+22}{106a''+7}, \ldots ,$$

where $0 \leq a \leq 7 = a_1$, $0 \leq a' \leq 15 = a_2$, $0 \leq a'' \leq 1$. The theorem
implies that if $2a > 7$ or $a \geq 4$, then $\dfrac{3a+1}{a}$ is a good approximation.
Similarly, we need $a' \geq 8$ for good approximations. Using these values, we
get the following sequence of good approximations for π.

$$\frac{3}{1}, \frac{13}{4}, \frac{16}{5}, \frac{19}{6}, \frac{22}{7}, \frac{179}{57}, \frac{201}{64}, \frac{223}{71}, \ldots .$$

The approximation $223/71$ was known to Archimedes. The theorem shows
that no rational number p/q, with $7 < q < 57$, comes closer to π than $22/7$
does.

Example 13.2.9. The problem of constructing an accurate calendar arises because the tropical year (the time taken for the earth to complete one orbit) is not an integral number of days (which depends on the rotation of the earth about its axis). The Julian Calendar used $365\frac{1}{4}$ days as the approximation, which after sixteen centuries of use amounted to an error of about 10 days. The Gregorian Calendar was proposed to rectify this and uses the approximation $365\frac{97}{400}$. This amounts to adding 97 extra days every 400 years.

It was most recently measured that the tropical year is 31556925.9747 seconds, or in days,

$$\frac{315569259747}{864000000} = [365, 4, 7, 1, 3, 5, 1, 1, 3, 1, 7, 7, 1, 1, 1, 1, 2].$$

The problem of constructing the calendar is then reduced to finding a suitable approximation to the fractional part. The convergents of the fractional part are $1/4, 7/29, 8/33, 31/128, \ldots$. The approximation $365\frac{8}{33}$ was proposed by the Persian mathematician and poet, Omar Khayyam (1079 A.D.). Computing the intermediate convergents, one sees that the approximation used in the Gregorian calendar is not an intermediate convergent.

────────────────── **Exercises for Section 13.2** ──────────────────

1. Verify that the largest intermediate convergent is $\dfrac{p_0 + p_{-1}}{q_0 + q_{-1}} = a_0 + 1$, and the smallest intermediate convergent is a_0.

2. Compute the error in the Gregorian calendar. Design a calendar using the convergent 8/33. What is the accumulated error after 1000 years? After 1500 years?

3. Verify that the first seven intermediate convergents after 22/7 are farther away from π than 22/7 is, and the eighth 179/57 is closer than 22/7 is.

4. Use the expansion $\pi = [3, 7, 15, 1, 292, 1, 1, \ldots]$ to show that there are no good approximations to π between $\dfrac{333}{106}$ and $\dfrac{355}{113}$. What are the next three good approximations after $\dfrac{355}{113}$? What is the error in using these approximations?

5. Prove a property similar to Lemma 13.2.4 for the intermediate convergents that lie between the odd convergents. Using this, complete the proof of Theorem 13.2.6.

6. Determine the rational number p/q that comes closest to e under the condition that $q \leq 200$. What is the error?

7. Find the best rational approximation of $\sqrt{23}$ that differs from it by less than $1/100000$, and compare the size of the denominator with that of the rational approximation obtained by expressing $\sqrt{23}$ as a decimal fraction correct to 6 decimal places.

8. Determine the rational number p/q closest to $\sqrt{3}$ with $q \leq 200$.

13.3   Algebraic and Transcendental Numbers

We have seen that if α is an irrational number, then there are infinitely many solutions to the inequality $\left| \alpha - \frac{p}{q} \right| < \frac{1}{2q^2}$. This has applications to the solution of Diophantine equations. For example, we showed that the equation $x^2 - 2y^2 = 1$ has infinitely many solutions in the integers. For quadratic irrationals, it is also true that there is a constant $C > 0$ such that $\left| \alpha - \frac{p}{q} \right| > \frac{1}{Cq^2}$. This follows from the fact that the partial quotients of the continued fraction expansion are bounded.

It is an important problem to obtain similar inequalities for other irrational numbers. For example, to solve $x^3 - 2y^3 = 1$ in the integers, we factor $x^3 - 2y^3 = (x - \sqrt[3]{2}y)(x^2 + \sqrt[3]{2}xy + (\sqrt[3]{2}y)^2)$. We must have $|x| > |y|$, hence $x^2 + \sqrt[3]{2}xy + (\sqrt[3]{2}y)^2 > 3y^2$. This implies that

$$|x - \sqrt[3]{2}y| < \frac{1}{3y^2}$$

or

$$\left| \frac{x}{y} - \sqrt[3]{2} \right| < \frac{1}{3y^3}.$$

By a theorem of Thue, this inequality has a finite number of rational solutions; hence the equation $x^3 - 2y^3 = 1$ has a finite number of integer solutions. Thue proved his theorem in the context of algebraic numbers, which we now define.

Definition 13.3.1. A number α is said to be **algebraic** if it is a root of a polynomial equation with integer coefficients, $P(x) = 0$, where $P(x) = a_n x^n + a_{n-1} x^{n-1} + \cdots a_0$ and $a_i, 0 \leq i \leq n$ integers. If α is not algebraic, then it is said to be **transcendental**.

Example 13.3.2. The numbers $\sqrt{2}$ and i are both algebraic, as they are roots of $x^2 - 2 = 0$ and $x^2 + 1 = 0$, respectively.

A polynomial is said to be irreducible if it cannot be written as a product of polynomials of lower degree. If an algebraic number α is a root of an irreducible polynomial of degree n, then we say that α is of degree n. The degree of an algebraic number is well defined.

Thue proved that if α is an algebraic number of degree $n \geq 3$, then for any $c, \epsilon > 0$, the inequality

$$\left| \alpha - \frac{p}{q} \right| < \frac{c}{|q|^{\frac{n}{2}+1+\epsilon}} \tag{1}$$

holds only for a finite number of integers x, y. This implies that $\left| x/y - \sqrt[3]{2} \right| < 1/3y^3$ has only a finite number of solutions, so $x^3 - 2y^3 = 1$ has a finite number of solutions (unlike the equation $x^2 - 2y^2 = 1$.) The exponent of q in (1) was reduced by a number of mathematicians, culminating in the work of Roth, who proved that for any positive c and ϵ, the inequality

$$\left| \alpha - \frac{p}{q} \right| < \frac{c}{|q|^{2+\epsilon}}$$

has a finite number of solutions. For his work on this question, Roth won the Field's medal, the equivalent of the Nobel prize in Mathematics. The exponent 2 in Roth's theorem is the best possible because we know that there are infinitely many solutions to the inequality $|\alpha - p/q| < c/|q|^2$.

The proofs of the results of Thue and Roth are beyond the scope of this book. We will be content with proving Liouville's theorem, which has an interesting application to the construction of transcendental numbers. While it is simple to construct algebraic numbers, the definition does not give a way to construct transcendental numbers. It is not even clear that they exist. In general, it is difficult to determine if a given number, like e or π, is algebraic or transcendental. The fact that e is irrational was known to Euler, and only in 1873 was it proved that e is transcendental. This was a major breakthrough and soon led to the proof that π is transcendental. The transcendence of π settled the ancient problem of squaring the circle.[1]

The following result of Liouville was the first to show the existence of transcendental numbers.

Theorem 13.3.3 (Liouville). *If α is algebraic of degree $n > 1$, then there exists a constant C (depending on α) such that the inequality*

$$\left| \alpha - \frac{p}{q} \right| > \frac{C}{q^n} \tag{2}$$

holds for every rational number p/q with $q > 0$.

PROOF. Let $f(x) = a_n x^n + a_{n-1} x^{n-1} + \cdots + a_0$ be such that $f(\alpha) = 0$. The polynomial f has no rational roots because if p/q is a root of f, then we

---

[1]H. Dorrie. *100 Great Problems of Elementary Mathematics. Their History and Solution.* Translated by D. Antin. New York: Dover, 1965.

can write $f(x) = (x - p/q)g(x)$, where $g(x)$ is a polynomial of degree $n - 1$. This contradicts the assumption that the degree of α is n because $g(\alpha) = 0$.

For any rational p/q,

$$|q^n f(p/q)| = |a_n p^n + \ldots a_0 q^n| \geq 1, \tag{3}$$

and by the Mean Value Theorem,

$$f(p/q) = f(p/q) - f(\alpha) = (p/q - \alpha)f'(c) \tag{4}$$

for some c between α and p/q. Since c lies in a closed interval and f' is a polynomial, there exists C_1 such that $|f'| \leq C_1$. This implies that

$$\left| \alpha - \frac{p}{q} \right| = \frac{|q^n f(p/q)|}{q^n |f'(c)|} > \frac{1}{C_1 q^n}$$

and taking $C = 1/C_1$ proves the theorem. ∎

Proposition 13.3.4. *The number*

$$\alpha = 1 \pm \frac{1}{10^{1!}} \pm \frac{1}{10^{2!}} \pm \frac{1}{10^{3!}} \pm \frac{1}{10^{4!}} \pm \cdots$$

is transcendental for any choice of signs in the sum.

PROOF. For any choice of signs, the number is not rational because its decimal fraction expansion is not periodic. There are increasing numbers of 0's between successive digits, so the decimal expansion cannot be repeating.

Fix a choice of signs in α, and let p_k/q_k be the sum of the first $k + 1$ terms of the series. Then

$$\left| \alpha - \frac{p_k}{q_k} \right| = \left| \frac{1}{10^{(k+1)!}} \pm \frac{1}{10^{(k+2)!}} \pm \cdots \right| \leq \frac{1}{10^{(k+1)!}} + \frac{1}{10^{(k+2)!}} + \cdots$$

We will estimate the right-hand side by comparing it with the geometric series. For this, we see that $10^{(k+2)!} = (10^{(k+1)!})^{k+2} \geq (10^{(k+1)!})^2$ and $10^{(k+3)!} \geq 10^{(k+1)!})^3$ and so on. Using this we see that for any $i > 0$,

$$\frac{1}{10^{(k+i)!}} < \left(\frac{1}{10^{(k+1)!}} \right)^i$$

and hence

$$\left| \alpha - \frac{p_k}{q_k} \right| < \frac{1}{10^{(k+1)!}} + \left(\frac{1}{10^{(k+1)!}} \right)^2 + \left(\frac{1}{10^{(k+1)!}} \right)^3 + \cdots$$

$$< \frac{1}{10^{(k+1)!}} \sum_{i=0}^{\infty} \left(\frac{1}{10^{(k+1)!}} \right)^i$$

$$= \frac{1}{10^{(k+1)!} - 1}$$

$$< \frac{2}{10^{(k+1)!}}.$$

This implies that α cannot be algebraic, because if α was algebraic of degree n, then there is a constant C such that $|\alpha - p_k/q_k| > C/q_k^n$. This is not possible because we showed that $|\alpha - p_k/q_k| < 2/q_k^{k+1}$ and taking k to be much larger than n, we obtain a contradiction. ∎

The proposition shows that there are transcendental numbers. It is much more difficult to determine if a given number is algebraic or transcendental. The transcendence of e was established by Hermite and Lindemann showed that π is transcendental. The transcendence of many other numbers such as $\zeta(3)$ or Euler's constant is not known.

The above result can also be used to show that there are an uncountable number of transcendental numbers. A set S is said to be countable if there is a bijection between S and the set of natural numbers. If an infinite set is not countable, then we say that its cardinality is uncountable.

Proposition 13.3.5. *The set of transcendental numbers is uncountable.*

PROOF. The proof is by the classic diagonal argument of Cantor. Consider the set of numbers constructed in Proposition 13.3.4. If the set is countable, then we can index the numbers by the natural numbers. Let $\alpha_1, \alpha_2, \ldots$ be these numbers, where

$$\alpha_i = \sum_{n=0}^{\infty} a_{in} \frac{1}{10^{n!}}, \qquad a_{in} = \pm 1.$$

We construct another number α in the sequence by letting $\alpha = \sum_{n=0}^{\infty} b_n \frac{1}{10^{n!}}$ where $b_n = a_{nn}$. If the set is countable, then α occurs in the sequence. Suppose $\alpha = \alpha_k$; then $\alpha = \sum_{n=0}^{\infty} a_{kn} \frac{1}{10^{n!}}$. This implies that $b_k = a_{kk}$, but by choice $b_k = -a_{kk}$. This contradiction shows that the set of transcendental numbers cannot be countable. ∎

———————— **Exercises for Section 13.3** ————————

1. Determine the decimal expansion of the number α in Proposition 13.3.4 if (a) all the signs are taken to be negative, and (b) the signs are alternating.

2. Show that the set of rational numbers is countable.

3. Let x and y be real numbers in the interval $[0, 1]$ with continued fraction expansions $x = [0, a_1, a_2, a_3, \ldots]$ and $y = [0, b_1, b_2, b_3, \ldots]$. Let $z = [0, a_1, b_1, a_2, b_2, a_3, b_3, \ldots]$. Show that the function $(x, y) \mapsto z$ is a one-to-one correspondence of $[0, 1] \times [0, 1]$ onto $[0, 1]$, hence $[0, 1] \times [0, 1]$ has the same cardinality as $[0, 1]$.

4. Show that $[10, 10^{1!}, 10^{2!}, 10^{3!}, \ldots]$ is transcendental.

──────────────── **Projects for Chapter 13** ────────────────
---

1. **Farey Fractions:** This project uses some of the results about good approximations developed in Section 13.2. Alternately, it is possible to prove the results without resorting to this section. The proofs are based on the following facts. Suppose we wish to solve the Diophantine equation $px - qy = 1$ with $0 < x < q$. Let $p/q = [a_0, a_1, \ldots, a_n]$. The penultimate convergent p_{n-1}/q_{n-1} satisfies the condition $pq_{n-1} - qp_{n-1} = \pm 1$. To get the solution to the equation, we may have to take the penultimate convergent of $p/q = [a_0, a_1, a_2, \ldots, a_n - 1, 1]$. Show that these two are the only solutions of $px - qy = \pm 1$ with $0 < x < q$.

   The rational numbers h/k, $0 \le h \le k \le n$, $(h, k) = 1$ listed in increasing order are called the **Farey sequence of order** n. The sequence is denoted by $\mathcal{F}_n$. For example, the Farey sequence of order 4 is

   $$0 = \frac{0}{1}, \frac{1}{4}, \frac{1}{3}, \frac{1}{2}, \frac{2}{3}, \frac{3}{4}, \frac{1}{1} = 1.$$

   (a) Write down the Farey sequences of order 5 and 6. (Remember to place them in increasing order.)

   (b) The fraction p/q, $0 < p < q$ first appears in $\mathcal{F}_q$. Show that if $p/q = [a_0, a_1, \ldots, a_n]$ with $a_n > 1$, then its neighbors in $\mathcal{F}_q$ are C_{n-1} and $C_{n-1, a_n - 1}$, where C_{n-1} is the $(n-1)$st convergent of p/q and $C_{n-1, a_n - 1}$ is an intermediate convergent.

   (c) Show that the first term that appears between adjacent terms a/b and c/d is their mediant, $(a + c)/(b + d)$.

   (d) Write a program to generate the Farey sequence of any given order. Can you predict the neighbors of a/b in any $\mathcal{F}_n$ for $n > b$? Prove your conjecture.

   (e) Show that the number of elements in $\mathcal{F}_n$ is $1 + \phi(1) + \phi(2) + \cdots + \phi(n)$.

   (f) Show that if $|ps - qr| = 1$, then p/q and r/s are adjacent in $\mathcal{F}_n$ for $\max(q, s) \le n < q + s$, and they are separated by the single element $(p + r)/(q + s)$ in $\mathcal{F}_{q+s}$.

   (g) Suppose that, for each p/q such that $(p, q) = 1$, we construct a circle of radius $1/2q^2$ centered at $(p/q, 1/2q^2)$. These circles are known as Ford circles. Figure 13.3.1 gives an example of the Ford circles on the Farey series of order 4. Show that no two of these circles intersect at more than one point, and two such circles are tangent if and only if the corresponding fractions are neighbors in the Farey series of some order.

2. **Egyptian Fractions:** The *Papyrus of Ahmes* (also known as the *Rhind Papyrus*) is one of the most ancient mathematical documents. It describes the decomposition of a rational number as a finite sum of reciprocals of the integers. The ancient Egyptians represented rational numbers $p/q > 0$ as

   $$\frac{p}{q} = \frac{1}{x_1} + \frac{1}{x_2} + \cdots + \frac{1}{x_k},$$

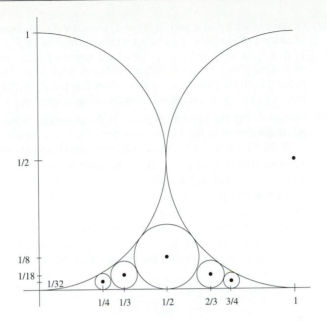

Figure 13.3.1 Ford Circles

where $x_1, \ldots, x_k$ are positive integers. This project is a study of the different methods of obtaining the Egyptian representation of a fraction p/q.

(a) Prove that every positive rational number has an Egyptian representation with distinct terms. (First, prove the theorem for rational numbers $p/q < 1$ by induction on the denominator.)

(b) Here is a method due to Fibonacci and Sylvester to find an Egyptian representation with distinct terms. The procedure returns a list of denominators in the Egyptian representation.

   (i) Let $q = a/b$.

   (ii) If $q = 0$ terminate the procedure; otherwise, select the smallest positive integer x such that $1/x < a/b$.

   (iii) Set $q = a/b - 1/x$, add x to the output list, and go to step 2.

Show that x in step 2 is $\lceil 1/q \rceil$. Prove that the algorithm always terminates.

(c) Write a program to implement the algorithm. Experiment with the representation of different rational numbers. The method tends to produce very large denominators; for example, $77/331$ has 8 terms with the largest denominator having 116 digits. There are other methods of representation that control the size of the denominators in the expansion.

(d) Among all rational numbers p/q with $0 \le p, q \le 500$, find the one whose Egyptian representation has the largest integer.

(e) Given p/q with $p < q$ and $(p, q) = 1$, consider $\mathcal{F}_q$, the Farey sequence of order q. Let $r/s < p/q$ be an adjacent fraction in $\mathcal{F}_q$; then $p/q = 1/qs + \frac{r}{s}$, $s < q$ and $r < p$. We have written p/q as the inverse of an integer plus a fraction with a smaller denominator. We repeat the process until the denominator is 1. Show that the algorithm terminates in $q - 1$ steps with each denominator being at most $q(q - 1)$.

(f) Implementing the algorithm given in the previous exercises requires generating parts of the Farey sequence. There is a better method due to M. N. Bleicher, [2] which uses the continued fraction expansion.

Suppose that $0 < p/q < 1$ has continued fraction $[0, a_1, a_2, \ldots a_n]$ with $a_n > 1$. Let p_k/q_k denote the convergents of p/q. If n is odd, then show that $p/q = p_n/q_n > p_{n-1}/q_{n-1}$ and $\frac{p}{q} = \frac{1}{qq_{n-1}} + \frac{p_{n-1}}{q_{n-1}}$. This relation is not valid if n is even, as p_{n-2}/q_{n-2} is the largest convergent less than p/q, and they are not adjacent terms in the Farey series. In this case we have to use the intermediate convergents to obtain the adjacent terms in the Farey series. Show that

$$\frac{p}{q} = \frac{p_{n-2}}{q_{n-2}} + \sum_{i=0}^{a_n-1} \frac{1}{(q_{n-2} + iq_{n-1})(q_{n-2} + (i+1)q_{n-1})}.$$

(g) Implement the algorithm of Bleicher to generate Egyptian fraction expansions. Compare with the method of Fibonacci–Sylvester described in part (b).

---

[2]M. N. Bleicher. "A New Algorithm for the Expansion of Egyptian Fractions." *Journal of Number Theory*, 4(1972), 342–382.

Chapter 14

Diophantine Equations

*I have found a very great number
of exceedingly beautiful theorems.*

— PIERRE DE FERMAT

14.1   Introduction

Diophantine problems, named after Diophantus of Alexandria (c. 250 A.D.), are concerned with the integral solutions of polynomial equations with integer coefficients. Diophantus was primarily interested in rational solutions and was content with finding one solution rather than all. The first mathematicians to study integral solutions of equations seem to be Aryabhata (700 A.D.) and Bhaskara (1100 A.D.). In addition to the linear equations, Bhaskara describes a procedure to obtain an infinite family of solutions to the equation $x^2 - dy^2 = 1$, where $d > 1$ is not a perfect square.

The modern era in the subject begins with Fermat. Fermat's interest was aroused by reading Bachet's translation of the surviving books of Diophantus. Fermat solved new problems, posed many challenges to other mathematicians, invented new methods, and in general was much more advanced than contemporary mathematicians. In Fermat's time, mathematicians were loath to publish their methods and challenged each other through problems. This partly accounts for the reason that so few of Fermat's methods are known.

In this section, we describe a few examples of solutions to Diophantine equations, including some discovered by Fermat. The rest of the chapter is concerned with some equations that can be solved by elementary techniques. Mathematicians have developed many methods to solve these problems, especially in the case of polynomial equations in low degree. The subject of

Diophantine equations is quite difficult. It is hard to tell if a given equation has solutions or not, and when it does, there may be no method to find all of them. It is difficult to tell which are easily solvable and which require advanced techniques.

In our study, when we ask to solve an equation, it is meant that we want all solutions to the equation. Such thorough analysis was first attempted by Fermat, and his contemporaries had trouble understanding this issue. Mathematicians were content to find one solution or a few examples. When Fermat asked for a solution to $x^2 - dy^2 = 1$, Wallis gave the trivial solution $x = 1$, $y = 0$. To his credit, Wallis later found infinitely many solutions to this equation. When Fermat asked for solutions to $x^3 + y^3 = a^3 + b^3$, Frenicle gave a few examples, including $9^3 + 10^3 = 1^3 + 12^3$. What Fermat desired in both cases was a general solution to the problem, not just a few examples.

The reader is familiar with Fermat's conjecture regarding the lack of non-trivial solutions to $x^n + y^n = z^n$ for $n > 2$. After centuries of attempts by countless mathematicians, the problem was finally solved by the work of Frey, Serre, Ribet, Taylor, and Wiles. The final solution involved the use of elliptic curves, another topic recognized by Fermat to be of fundamental importance. An elliptic curve is the locus of solutions to $y^2 = P(x)$, where $P(x)$ is a cubic polynomial. Fermat stated that the equation $y^2 = x^3 - 2$ has $y = 5$ and $x = 3$ as the only solution. Fermat published no proof of this assertion. A proof using representations of the binary quadratic form $x^2 + 2y^2$ is given in Exercise 18.6.4.

It may seem to the reader that Fermat only stated negative theorems. Here is an astonishing theorem of Fermat. In a letter to Brûlard de St. Martin, Fermat proposed the following problem:

> Find a Pythagorean triangle (that is, a right-angled triangle with sides a, b, and c measured in integers) such that the hypotenuse c is the square of an integer and the sum of the legs $a + b$ is also the square of an integer.

Fermat asserted that the problem has infinitely many primitive solutions of which the smallest is

$$a = 4565486027761$$
$$b = 1061652293520$$
$$c = 4687298610289.$$

A proof of this assertion is outlined in the projects at the end of this chapter.

Diophantine problems do not always have to be of polynomial type. There are many interesting, and very difficult, Diophantine problems which are not polynomial equations. The standard techniques of number theory do not apply to them, which partly accounts for their difficulty. For example, Catalan's conjecture asks if 8 and 9 are the only consecutive powers, or more precisely, if the equation $x^m - y^n = 1$ has only the solution $x = 3$, $m = 2$, $y = 2$, $n = 3$. Paulo Ribenboim's book *Catalan's conjecture* [1] is an excellent reference for this topic and a lot of other interesting Diophantine problems.

In another vein, Euler conjectured that it is impossible to have three fourth powers whose sum is a fourth power. For two hundred years, no one even knew of an example of four fourth powers adding up to a fourth power when R. Norrie found that

$$30^4 + 120^4 + 272^4 + 315^4 = 353^4.$$

Only recently, Noam Elkies disproved Euler's conjecture for fourth powers with the example

$$2682440^4 + 15365639^4 + 18796760^4 = 20615673^4.$$

It is reasons such as this that mathematicians insist on proofs of conjectures, however reasonable they may seem. Most conjectures can only be tested for the first few million integers, which are a tiny proportion of all the integers.

——————— **Exercises for Section 14.1** ———————

1. Write a computer program to search for triangular numbers that are squares. The nth triangular number is

$$T_n = 1 + 2 + \cdots + n = \frac{n(n+1)}{2}.$$

   Are there infinitely many such numbers? Do you see a pattern in the solutions? Can you guess a general form of the solutions?

2. Are there any triangular numbers that are perfect cubes?

3. Verify Fermat's solution to the Pythagorean triangle where the hypotenuse and the sum of the legs are squares.

4. On a visit to Ramanujan when he was sick, Hardy remarked that the number on the taxicab he rode, 1729, did not seem very interesting. Ramanujan replied that 1729 possessed a very interesting property. It is the smallest positive integer that can be expressed as a sum of two cubes in two different ways,

$$1729 = 1^3 + 12^3 = 9^3 + 10^3.$$

---
[1]P. Ribenboim. *Catalan's Conjecture*. Boston: Academic Press, 1994.

Find three other examples of numbers that can be expressed as sums of cubes in more than one way.

5. A tetrahedral number is of the form

$$T_n = \frac{n(n+1)(n+2)}{6}.$$

(These are the number of balls that can be stacked with triangular bases. You can derive this formula by considering sums of triangular numbers.)

It is known that there are three tetrahedral numbers that are perfect squares. Find them. Are there any tetrahedral numbers that are perfect cubes? Can you prove your conjecture?

6. E. Lucas asked for integers n such that $1^2 + 2^2 + \cdots + n^2$ is a square. It is known that all such n satisfy $|n| \leq 100$. Use this fact to find all solutions n, m to

$$1^2 + 2^2 + \cdots + n^2 = m^2.$$

7. Find a nontrivial solution ($x \neq y$) to the equation $x^y = y^x$. Can you find all the solutions? Prove your conjecture.

8. Find nontrivial integer solutions to the equation $x^3 + y^3 + z^3 = x + y + z$. Does the equation have infinitely many solutions?

9. Show that $x^y - y^x = 1$, $xy \neq 0$ has only one solution, $x = 3$, $y = 2$.

10. The equation $x^4 + y^4 = z^4 + t^4$ was solved by Euler who gave the following solution.

$$x = a^7 + a^5 b^2 - 2a^3 b^4 + 3a^2 b^5 + ab^6$$
$$y = a^6 b - 3a^5 b^2 - 2a^4 b^2 + a^2 b^5 + b^7$$
$$z = a^7 + a^5 b^2 - 2a^3 b^4 - 3a^2 b^5 + ab^6$$
$$t = a^6 b + 3a^5 b^2 - 2a^4 b^3 + a^2 b^5 + b^7.$$

(a) Verify that the above formulas give a solution to the equation.

(b) Write a computer program to find solutions to the equation. Determine the nontrivial solution such that $x^4 + y^4$ is smallest.

11. Euler solved the equation $A^3 + B^3 + C^3 = D^3$ by letting $A = (m-n)p+q^2$, $B = (m+n)p + q^2$, $C = p^2 - (m+n)q$, and $D = p^2 + (m-n)q$; and showing that $p = a^2 + 3b^2$, $q = c^2 + 3d^2$, $m = 3(bc \pm ad)$, and $\pm n = ac \mp 3bd$. Verify that the formulas for p, q, m, and n give solutions to the desired equation. Is the solution $3^3 + 4^3 + 5^3 = 6^3$ obtained by the formulas? How about the solution $1^3 + 6^3 + 8^3 = 9^3$?

14.2   Congruence Methods

If a Diophantine equation $f(x_1, x_2, \ldots, x_n) = 0$ has solutions in the integers, then it has solutions modulo m for every integer m. Hence, if the congruence $f(x_1, x_2, \ldots, x_n) \equiv 0 \pmod{m}$ is not solvable for an integer m, then $f(x_1, \ldots, x_n) = 0$ cannot have integer solutions. This is a useful technique to show that many equations do not have solutions.

Example 14.2.1. Show that $x^2 - 7y^2 = 3$ has no solutions in the integers.
   Working modulo 4, we have

$$x^2 - 7y^2 \equiv x^2 - 3y^2 \equiv x^2 + y^2 \quad (\text{mod } 4).$$

But $x^2 + y^2$ can only be 0, 1, or 2 modulo 4; hence $x^2 + y^2 \equiv 3 \pmod{4}$ has no solution. This implies that $x^2 - 7y^2 = 3$ has no solution in the integers.

Example 14.2.2. The equation $13x^2 + 5y^3 = 1892$ has no integral solutions.
   Modulo 13, the cubes are 0, 1, 5, 8, or 12. Then $5y^3 \equiv 0, 5, 12, 1, 8$ $(\text{mod } 13)$. But $1892 \equiv 7 \pmod{13}$; therefore, $13x^2 + 5y^3 = 1892$ has no integer solutions.

Example 14.2.3. $x^2 - dy^2 = -1$ has no solutions if $d \equiv 3 \pmod{4}$.
   Working modulo 4, we have that

$$x^2 - 3y^2 \equiv x^2 + y^2 \equiv -1 \quad (\text{mod } 4),$$

and we have already seen that this congruence has no solutions.

   The examples of the congruence method above give negative results. There is no general procedure to choose m, and there is a lot of unsuccessful work hidden behind the above examples. If we fail to find a modulus m for which the equation has no solutions, then it does not follow that the equation has integer solutions. We give an example to show that a Diophantine equation can have a solution modulo every m and still fail to have a solution in the integers. The example uses techniques of Section 3.4, Chapter 9, and a result about the Legendre symbol $\left(\dfrac{2}{p}\right)$ proved in Section 17.1.

Example 14.2.4. It is clear that the equation $x^2 + 34y^2 = 2$ has no integer solutions. We will show that the equation $x^2 + 34y^2 \equiv 2 \pmod{m}$ has a solution for every m.
   First, it suffices to show this for numbers m that are prime powers. If $p \equiv 1, 7 \pmod{8}$, then it is shown in Proposition 17.1.2 that $x^2 \equiv 2 \pmod{p}$ has a solution. Proposition 9.3.1 shows that $x^2 \equiv 2 \pmod{p^k}$ has a solution for every k. Hence for $p \equiv 1, 7 \pmod{8}$, by taking $y = 0$, the equation $x^2 + 34y^2 \equiv 2 \pmod{p^k}$ has a solution.

Suppose that $p \equiv 3, 5 \pmod{8}$. If $\left(\dfrac{34}{p}\right) = 1$, then $34^{-1} \bmod p^k$ is also a quadratic residue. Write $34^{-1} \equiv a^2 \pmod{p^k}$. We multiply y by a to see that $x^2 + 34y^2 \equiv 2 \pmod{p^k}$ has a solution if and only if $x^2 + y^2 \equiv 2 \pmod{p^k}$ has a solution. The latter has solutions $x = 1$, $y = 1$. If $\left(\dfrac{34}{p}\right) = -1$, then 17 is a quadratic residue modulo p^k (because 2 is a quadratic nonresidue for $p \equiv 3, 5 \pmod{8}$), and so is $17^{-1} \bmod p^k$. Hence $x^2 + 34y^2 \equiv 2 \pmod{p^k}$ has a solution if and only if $x^2 + 2y^2 \equiv 2 \pmod{p}$ has a solution. The latter equation has solutions $x = 0$, $y = 1$.

If $p = 2$, then it is easy to check that $x^2 + 34y^2 \equiv 2 \pmod{2^k}$ has solutions for $k = 1, 2, 3$. For $k \geq 3$, Proposition 9.3.3 shows that 17 is a quadratic residue modulo 2^k. In this case, multiplying y by the square root of 17^{-1} will reduce the equation to $x^2 + 2y^2 \equiv 2 \pmod{2^k}$, which clearly has solutions.

———————————— **Exercises for Section 14.2** ————————————

1. Show that $3x^2 - 7y^2 = 2$ has no solution in the integers.

2. Show that $7x^3 + 2 = y^3$ has no solution in the integers.

3. Show that $3x^2 + 2 = y^2$ has no solution in the integers.

4. Show that $x^2 - 5y^2 = 3z^2$ has no nontrivial solutions in the integers.

5. Show that $x^3 + y^3 = z^3$ has no solution when $3 \nmid xyz$.

6. Show that $x^3 + 2y^3 + 4z^3 = 9w^3$ has no nontrivial integer solutions.

7. Show that $5x^3 + 11y^3 + 13z^3 = 0$ has no nontrivial integral solutions.

8. Fill in all the details of Example 14.2.4. Show that the analysis still holds if 34 is replaced by $2q$, where q is any prime $q \equiv 1 \pmod{8}$.

   In many cases, the polynomial can be factored to obtain solutions to the Diophantine equation. The following problems involve this technique.

9. Solve $y^2 - x^4 = 9$ in the positive integers by factoring the left-hand side as $(y - x^2)(y + x^2)$.

10. Solve $x^2 + 12 = y^4$ in the integers.

11. Find all solutions to $x^3 - y^3 = 19$.

14.3 Pythagorean Triples

The problem of finding integer solutions to the equation $x^2 + y^2 = z^2$ is an ancient one. The first known description is in a Babylonian cuneiform from about 1500 B.C. A list of Pythagorean triples occurs there, including $(3, 4, 5)$ and $(4961, 6480, 8161)$. The list shows that they must have possessed a systematic method to solve the equation. The problem is treated by Euclid (300 B.C.) in the *Elements, Book X,* and again by Diophantus (250 A.D.).

Definition 14.3.1. A **Pythagorean triple** is a solution $\{x, y, z\}$ in the integers to the equation $x^2 + y^2 = z^2$. It is called **primitive** if $(x, y, z) = 1$.

It is clear that any Pythagorean triple is a multiple of a primitive triple; hence it suffices to determine these. In Project 1 of Chapter 2, a solution to the problem of determining the primitive Pythagorean triples is outlined. In the equation $x^2 + y^2 = z^2$ modulo 4, one of x or y must be even, say y. We can write $\left(\dfrac{y}{2}\right)^2 = \dfrac{z - x}{2}\dfrac{z + x}{2}$. Then one shows that if x and z are relatively prime and odd, then $(z - x)/2$ and $(z + x)/2$ are relatively prime. Since their product is a perfect square, unique factorization implies that each term is a perfect square. Solving these relations for x, y, and z leads to a characterization of Pythagorean triples.

We give a more geometric proof based on ideas of Diophantus to find rational solutions of equations. This technique will be very useful in the study of elliptic curves.

Theorem 14.3.2. *A Pythagorean triple $\{x, y, z\}$ with y even is primitive if and only if it is of the form*

$$x = r^2 - s^2, \quad y = 2rs, \quad z = r^2 + s^2,$$

where r and s are relatively prime positive integers.

PROOF. Dividing $x^2 + y^2 = z^2$ by z^2, we see that integer solutions to the Pythagorean equation are in correspondence with rational solutions to $u^2 + v^2 = 1$, with $u = x/z$ and $v = y/z$.

There are four obvious rational solutions to the equation $u^2 + v^2 = 1$. These are $(\pm 1, 0)$ and $(0, \pm 1)$. If (u_0, v_0) is a point on the circle with rational coordinates, then the slope of the line joining (u_0, v_0) to $(-1, 0)$ is rational. Conversely, if a line through $(-1, 0)$ with rational slope intersects the circle at another point (u_0, v_0), then u_0 and v_0 are rational.

Let t be a rational number. Consider the line with slope t through $(-1, 0)$ (see Figure 14.3.1). It has the equation $\dfrac{v - 0}{u + 1} = t$ or $v = t(u + 1)$. We substitute this in $u^2 + v^2 = 1$ to obtain $u^2 + t^2(u + 1)^2 = 1$ or $u^2(1 + t^2) + 2t^2 u + t^2 - 1 = 0$. We can use the quadratic formula to solve for u, or we observe that one root is -1 and the sum of the roots of the equation $au^2 + bu + c = 0$ is $-b/a$, hence

$$u - 1 = -\frac{2t^2}{1 + t^2},$$

Cuneiform Tablet of Pythagorean Triples

The construction of right triangles with integral sides was known in ancient Babylonia. The figure shows a cuneiform tablet dating from about 1900 B.C.

This tablet is in the Rare Book and Manuscript Library of Columbia University and is one of the oldest known mathematical documents.

If a and b are the legs of an integral right triangle and c the hypotenuse, then the table shows the values of $(c/a)^2$, b, and c in the first three columns. Babylonians used the sexadecimal (base 60) system

Plimpton Tablet 322

(remnants of which are still in use for measuring angles and time). The values in sexadecimal notation are shown in the following table. Numbers that are missing in the table are shown in parentheses.

$(c/a)^2$	b	c	
(1;59,0,)15	1,59	2,49	1
(1;56,56,)58,14,50,6,15	56,7	3,12,1	2
(1;55,7,)41,15,33,45	1,16,41	1,50,49	3
(1;)5(3,1)0,29,32,52,16	3,31,49	5,9,1	4
(1;)48,54,1,40	1,5	1,37	5
(1;)47,6,41,40	5,19	8,1	6
(1;)43,11,56,28,26,40	38,11	59,1	7
(1;)41,33,59,3,45	13,19	20,49	8
(1;)38,33,36,36	9,1	12,49	9
1;35,10,2,28,27,24,26,40	1,22,41	2,16,1	10
1;33,45	45	1,15	11
1;29,21,54,2,15	27,59	48,49	12
(1;)27,0,3,45	7,12,1	4,49	13
1;25,48,51,35,6,40	29,31	53,49	14
(1;)23,13,46,40	56	53	15

In the first row, $b = 1,59$, ($1 \cdot 60 + 59 = 119$ in the decimal system) and $c = 2,49$ ($2 \cdot 60 + 49 = 169$ in the decimal system). Then $(c/a)^2 = 1.9834028$, which is written as $1;59,0,15$ in the sexadecimal system. In the sexadecimal representation, the semicolon plays the role of the decimal point, so that $1; a_1 a_2 a_3 \cdots = 1 + a_1/60 + a_2/60^2 + \cdots$. The table is organized by decreasing values of $(c/a)^2$. Additional information about ancient Babylonian mathematics can be found in O. Neugebauer, *The Exact Sciences in Antiquity*, 2nd ed. New York: Dover Publications, 1969.

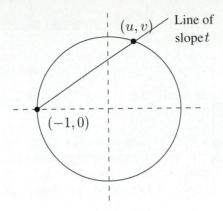

Figure 14.3.1 Rational points on the circle

or

$$u = \frac{1 - t^2}{1 + t^2}.$$

Let $t = s/r$ with $(r, s) = 1$. Then,

$$u = \frac{x}{z} = \frac{1 - s^2/r^2}{1 + s^2/r^2} = \frac{r^2 - s^2}{r^2 + s^2}.$$

Since $(x, z) = 1$, if $(r^2 - s^2, r^2 + s^2) = 1$, then

$$x = r^2 - s^2, \quad z = r^2 + s^2, \quad y = 2rs.$$

If $(r^2 - s^2, r^2 + s^2) \neq 1$, then we cannot take $x = r^2 - s^2, z = r^2 + s^2$. This is because $(r, s) = 1$ implies that $(r^2 - s^2, r^2 + s^2) = 1, 2$ (see Exercise 2). If $(r^2 - s^2, r^2 + s^2) = 2$, then

$$x = \frac{r^2 - s^2}{2}, \quad z = \frac{r^2 + s^2}{2}, \quad y = rs.$$

These equations can be rewritten in the form stated in the theorem. Observe that both r and s must be odd, so we can transform

$$z = \left(\frac{r + s}{2}\right)^2 + \left(\frac{r - s}{2}\right)^2$$

$$y = \left(\frac{r + s}{2}\right)^2 - \left(\frac{r - s}{2}\right)^2$$

$$x = 2\left(\frac{r + s}{2}\right)\left(\frac{r - s}{2}\right).$$

Letting $m = (r + s)/2$ and $n = (r - s)/2$, and switching x and y, we see that the solution is again of the form

$$x = m^2 - n^2, \quad y = 2mn, \quad z = m^2 + n^2.$$

Conversely, one can verify that for any $(m, n) = 1$, these formulas yield a Pythagorean triple. ∎

Many other Diophantine equations can be solved by this technique. For example, for $x^2 + 2y^2 = z^2$, we can employ either of the two techniques to show that all the solutions are of the form $x = m^2 - 2n^2$, $z = m^2 + 2n^2$, $y = 2mn$, for some m and n. Conversely, any triple of this form is a solution. (See Exercise 11.)

In the proof of the theorem, when $(r^2 - s^2, r^2 + s^2) = 2$, we were still able to transform the equation to get solutions of the same form as in the first case. This will not be possible in every case. There will usually be several families of solutions given by different formulas. Consider the following example.

Example 14.3.3. Find integral solutions to $x^2 + 5y^2 = z^2$.

Imitating the proof of the theorem, we write $u = x/z$ and $v = y/z$ and look for rational points on the ellipse $u^2 + 5v^2 = 1$. As before $(-1, 0)$ is a rational point on the curve. Consider a line L of slope t through $(-1, 0)$. If t is rational, the points of intersection of L with the ellipse are rational. Conversely, every rational point on the ellipse can be obtained this way. Letting $v = t(u + 1)$, we obtain

$$u^2(1 + 5t^2) + 10t^2u + 5t^2 - 1 = 0.$$

Solving for u, [either by using the quadratic formula, or by observing that the sum of the roots is $-(10t^2)/(1 + 5t^2)$], we obtain

$$u = \frac{1 - 5t^2}{1 + 5t^2}, \quad v = \frac{2t}{1 + 5t^2}.$$

Let $t = s/r$ with $(r, s) = 1$ so that

$$u = \frac{x}{z} = \frac{r^2 - 5s^2}{r^2 + 5s^2}.$$

Now, $(r^2 - 5s^2, r^2 + 5s^2)$ can be 1, 2, 5, or 10. If $(r^2 - 5s^2, r^2 + 5s^2) = 1$, we must have

$$x = r^2 - 5s^2, \quad z = r^2 + 5s^2, \quad y = 2rs.$$

If $(r^2 - 5s^2, r^2 + 5s^2) = 2$, then both r and s are odd, and we have

$$x = \frac{r^2 - 5s^2}{2}, \quad y = rs, \quad z = \frac{r^2 + 5s^2}{2}.$$

Unlike the corresponding step in the proof to the theorem, this solution cannot be written as $x = m^2 - 5n^2$, $y = 2mn$, $z = m^2 + 5n^2$. For example, take $r = 3$ and $s = 1$; then $z = (3^2 + 5 \cdot 1^2)/2 = 7$, but 7 cannot be written as $m^2 + 5n^2$. (This is clear by writing down a few values of the form $m^2 + 5n^2$.) In this case we have two families of solutions:

$$x = r^2 - 5s^2, \quad y = 2rs, \quad z = r^2 + 5s^2 \tag{1}$$

$$x = \frac{r^2 - 5s^2}{2}, \quad y = rs, \quad z = \frac{r^2 + 5s^2}{2}. \tag{2}$$

If $(r^2 - 5s^2, r^2 + 5s^2) = 5$, then r must be divisible by 5, and by writing $r = 5r'$ we see that the solution is of the first form

$$x = s^2 - 5r'^2, \quad y = 2r's, \quad z = s^2 + 5r'^2.$$

If $(r^2 - 5s^2, r^2 + 5s^2) = 10$, then r is divisible by 5 and r and s are both odd. Writing $r = 5r'$, we see that the solution is of the second form:

$$x = \frac{s^2 - 5r'^2}{2}, \quad y = r's, \quad z = \frac{s^2 + 5r'^2}{2}.$$

There is another way to write the second class of solutions to the equation $x^2 + 5y^2 = z^2$. Writing $m = (r + s)/2$ and $n = s$, we see that

$$\frac{r^2 + 5s^2}{2} = 2m^2 - 2mn + 3n^2,$$

$$\frac{r^2 - 5s^2}{2} = 2m^2 - 2mn - 2n^2,$$

$$rs = n(2m - n).$$

Hence the solutions to $x^2 + 5y^2 = z^2$ are in two families:

$$x = m^2 - 5n^2, \quad y = 2mn, \quad z = m^2 + 5n^2$$

$$x = 2m^2 - 2mn - 2n^2, \quad y = n(2m - n), \quad z = 2m^2 - 2mn + 3n^2.$$

Conversely, we can verify that these forms give a solution to the equation $x^2 + 5y^2 = z^2$.

Remark. There are many interesting problems related to Pythagorean triples. The problems include finding Pythagorean triangles with consecutive legs, or with specified area, or problems in which one is asked to determine the number of ways in which a given number can be a leg or a hypotenuse of a right triangle. We refer the reader to the book of Beiler for an extensive discussion of these problems.[2]

---

[2]A. H. Beiler. *Recreations in the Theory of Numbers*. New York: Dover Publications, 1964.

———————————————— **Exercises for Section 14.3** ————————————————

1. Find all Pythagorean triples with a side of length 20.

2. Show that if r and s are relatively prime, then $(r^2 - s^2, r^2 + s^2) = 1, 2$.

3. Let $\{x, y, z\}$ be a Pythagorean triple. Show that

   (a) At least one of x and y is divisible by 3.

   (b) At least one of x, y, and z is divisible by 5.

   (c) xyz is divisible by 60.

4. Prove that $\{3, 4, 5\}$ is the only Pythagorean triple whose terms are in arithmetic progression.

5. For every $n \geq 3$, show that there is a Pythagorean triple with n as one of its members.

6. Write a computer program to generate primitive Pythagorean triples $a = r^2 - s^2$, $b = 2rs$, and $c = r^2 + s^2$. For which values of r and s are two terms in the Pythagorean triple prime? Are there infinitely many triples with two primes?

7. Show that all right triangles with integer sides in which the hypotenuse exceeds a leg by one have sides that are $2n + 1$, $2n^2 + 2n$ and $2n^2 + 2n + 1$, where n is a positive integer.

8. Are there infinitely many Pythagorean triples in which the hypotenuse exceeds one of the legs by 2? What if the excess of the hypotenuse over one of the legs is 3?

9. Determine all primitive solutions to the equations

   (a) $x^2 + y^2 = 4z^2$.

   (b) $x^2 + 4y^2 = z^2$.

10. Verify that for odd integers m and n with $(m, n) = 1$,

$$x = 2m^2 - 2mn - 2n^2, \quad y = n(2m - n), \quad z = 2m^2 - 2mn + 3n^2,$$

   are solutions to the equation $x^2 + 5y^2 = z^2$.

11. Use the technique of Theorem 14.3.2 to determine all primitive solutions to the equation $x^2 + 2y^2 = z^2$.

12. Consider the equation $x^2 + 3y^2 = z^2$ and primitive solutions (x, y, z).

   (a) Show using congruences modulo 4 that x has to be odd.

(b) Show that if y is odd, then z is even. If $3 \mid z - x$, write $y^2 = \dfrac{z - x}{3}(z + x)$. Conclude that both $(z - x)/3$ and $z + x$ are perfect squares, and hence there exist r and s such that

$$x = \frac{r^2 - 3s^2}{2}, \quad y = rs, \quad z = \frac{r^2 + 3s^2}{2}.$$

(c) If y is even, show that there exist m and n such that

$$x = m^2 - 3n^2, \quad y = 2mn, \quad z = m^2 + 3n^2.$$

13. Using the technique of either Theorem 14.3.2 or the previous exercise, determine all primitive solutions to the equation $x^2 + py^2 = z^2$ for p an odd prime.

14. Show that all the primitive solutions (x, y, z) of $x^2 - 2y^2 = z^2$ are of the form
$$x = m^2 - 2n^2, \quad y = 2mn, \quad z = m^2 + 2n^2$$
for some m and n.

15. Show that if the equation $x^2 - dy^2 = 1$ has a nontrivial solution, then it has infinitely many solutions. (Here $(1, 0)$ is a trivial solution.)

16. Use the techniques of this section to solve $x^2 + t = u^2$, $x^2 - t = v^2$. Give examples of three squares in arithmetic progression.

14.4 Sums of Two Squares

Le principé de la démonstration de Fermat
est un des plus féconds dans toute la Théorie des Nombres,
et surtout dans celle des nombres entiers.[3]

— J. L. LAGRANGE

The classification of Pythagorean triples shows that the hypotenuse of a right triangle with integral sides is expressed as a multiple of sum of two squares. We would like to determine all integers that can be expressed as sums of two squares. This problem first appears in the work of Diophantus. Albert Girard (1595–1625) gave a complete description of such numbers in 1625. A few years later, Fermat gave a characterization of these numbers and asserted that he possessed a rigorous proof via the method of infinite descent. Euler was the first to publish a rigorous proof, which we describe in this section. It is believed that Fermat's proof was similar.

---

[3]Fermat's method of proof is one of the most fertile in the theory of numbers, and most of all in the integers.

Table 14.4.1 $x^2 + y^2$, $0 \leq x \leq y \leq 9$

0	1	4	9	16	25	36	49	64	81
	2	5	10	17	26	37	50	65	82
		8	13	20	29	40	53	68	85
			118	25	34	45	58	73	90
				32	41	52	65	80	97
					50	61	74	89	106
						72	85	100	117
							98	113	130
								128	145
									162

Table 14.4.1 gives a list of numbers expressible as sums of two squares. The pattern in the table is not obvious, but observe that every prime $p \leq 115$ such that $p \equiv 1 \pmod 4$ is expressible as a sum of two squares, and no prime congruent to 3 modulo 4 occurs in the table. The fundamental assertion of this section is that every prime $p \equiv 1 \pmod 4$ can be expressed as a sum of two squares. Once this is proved, it is not difficult to characterize all integers that are expressible as sums of two squares. This theorem has been proved before in Example 9.1.14 and also in Exercise 11.2.19. The proof given in this section illustrates Fermat's method of descent.

First, we take care of the primes congruent to 3 modulo 4.

Lemma 14.4.1. (a) *An integer n which is congruent to 3 modulo 4 cannot be expressed as a sum of two squares.*

(b) *If a prime p divides $x^2 + y^2$, and $p \equiv 3 \pmod 4$, then $p^2 \mid x^2 + y^2$.*

PROOF. (a) This is clear from the congruence $x^2 + y^2 \equiv n \pmod 4$, as the only squares modulo 4 are 0 and 1.

(b) If $p \mid x^2 + y^2$, then $x^2 + y^2 \equiv 0 \pmod p$, that is, $x^2 \equiv -y^2 \pmod p$. If $p \nmid x$, then -1 is a quadratic residue modulo p. By Proposition 9.1.11, this is not possible for $p \equiv 3 \pmod 4$. Hence $p \mid x$ and $p \mid y$; therefore, $p^2 \mid x^2 + y^2$. ∎

The techniques used in Euler's proof are related to the arithmetic of the Gaussian integers $\mathbb{Z}[i]$. The proof can be given without any reference to the Gaussian integers, but their use makes the identities occurring in the proof more transparent. A similar situation occurs with the proof of the four square theorem where it is convenient to utilize the arithmetic of integral Quaternions. (See the projects at the end of the chapter.) The three squares theorem is harder to prove because of the lack of similar identities.

Definition 14.4.2. The **Gaussian integers** $\mathbb{Z}[i]$ are the collection

$$\mathbb{Z}[i] = \{a + ib \ : \ a, b \in \mathbb{Z}, \ i^2 = -1\}.$$

Addition and multiplication are defined by:

$$(a + ib) + (x + iy) = (a + x) + i(b + y)$$
$$(a + ib)(x + iy) = (ax - by) + i(ay + bx).$$

Definition 14.4.3. We define the **conjugate** of $z = x + iy$ to be $\overline{z} = x - iy$. The **norm** of z is defined as $N(z) = z\overline{z} = x^2 + y^2$.

Exercise. Compute $(3 - 5i)(7 + 12i)$ and $N(1 + 12i)$.

Division in the Gaussian integers is performed by rationalizing the denominator.

$$\frac{x + iy}{a + ib} = \frac{(x + iy)(a - ib)}{(a + ib)(a - ib)} = \frac{(ax + by) + i(ay - bx)}{a^2 + b^2}.$$

The result is a Gaussian integer when $a^2 + b^2$ divides both $ax + by$ and $ay - bx$.

Exercise. Compute $\dfrac{6 - 4i}{1 + i}$.

Lemma 14.4.4. *The norm N satisfies $N(zw) = N(z)N(w)$ and $N\left(\dfrac{z}{w}\right) =$*
$\dfrac{N(z)}{N(w)}$.

PROOF. The assertion follows from the following. $N(zw) = zw\overline{zw} = zw\overline{z}\,\overline{w} = z\overline{z}w\overline{w}$. Here we use $\overline{zw} = \overline{z} \cdot \overline{w}$, a property that can be easily verified. The proof for $N(z/w)$ is similar. ∎

These basic properties show that the product of the sums of two squares is a sum of two squares.

Lemma 14.4.5. *If n and m are expressible as sums of two squares, then their product is also a sum of two squares.*

PROOF. Let $n = a^2 + b^2$ and $m = x^2 + y^2$; then

$$nm = (a^2 + b^2)(x^2 + y^2) = (ax \mp by)^2 + (ay \pm bx)^2.$$

This identity can be verified directly. This is also a direct consequence of the fact that $n = N(a + ib)$, $m = N(x + iy)$ and $nm = N((a + ib)(x + iy)) = N(ax - by + i(ay + bx))$. ∎

Similar identities are true for the number system $\mathbb{Z}[\sqrt{d}]$ or, equivalently, for representation by the form $x^2 - dy^2$. (See Exercise 12.)

We will use Fermat's method of descent to give another proof of the basic result that a prime p, $p \equiv 1 \pmod 4$, is a sum of two squares. For the method of descent, we need the following result expressing a quotient of sums of two squares as a sum of two squares when the denominator is a prime.

Proposition 14.4.6. *If a prime $q = a^2 + b^2$ divides $n = x^2 + y^2$, then n/q is expressible as a sum of two squares.*

PROOF. We use the identities derived for the norm,

$$
\frac{n}{q} = N\left(\frac{x + iy}{a + ib}\right) = N\left(\frac{ax + by + i(ay - bx)}{a^2 + b^2}\right)
$$

$$
= \frac{(ax + by)^2 + (ay - bx)^2}{q^2}. \tag{1}
$$

Written in a different way,

$$
\frac{n}{q} = N\left(\frac{x - iy}{a + ib}\right) = \frac{(ay + bx)^2 + (ax - by)^2}{q^2}. \tag{2}
$$

We will show that the terms on the right in (1) and (2) are integers. Since n/q is an integer, it is enough to show that $q \mid ay + bx$ or $q \mid ay - bx$.

Consider

$$
(ay - bx)(ay + bx) = a^2y^2 - b^2x^2
$$
$$
= (q - b^2)y^2 - b^2x^2
$$
$$
= qy^2 - b^2(x^2 + y^2)
$$
$$
= qy^2 - b^2 n.
$$

Since $q \mid n$, we see that $q \mid (ay - bx)(ay + bx)$ and since q is prime, we have $q \mid ay + bx$ or $q \mid ay - bx$. ∎

Theorem 14.4.7 (Fermat, Euler). *Every prime $p \equiv 1 \pmod 4$ is expressible as a sum of two squares.*

PROOF. The proof is in two steps. First, we show that if a prime divides a sum of two relatively prime squares, then it must be a sum of two squares. Then we use reciprocity to show that every $p \equiv 1 \pmod 4$ divides a sum of two relatively prime squares.

We will use Proposition 14.4.6 to show that if $p \mid x^2 + y^2$ with $(x, p) = 1$, then p must be a sum of two squares. To derive a contradiction, assume that p is not expressible as a sum of two squares. We can divide x and y by p to obtain remainders, whose absolute values are smaller than $p/2$. If $|x| > p/2$, write $x = pk + r$ with $0 < |r| < p/2$. Since $x^2 = p^2k^2 + 2pkr + r^2$ and $p \mid x^2 + y^2$, we see that $p \mid r^2 + y^2$. Similarly, we divide y by p, hence we can assume that $p \mid x^2 + y^2$, with $0 < |x|, |y| < p/2$. Then $x^2 + y^2 < p^2/4 + p^2/4 = p^2/2$. Here we use the fact that x is prime to p to see that $xy \neq 0$.

Since $p \mid x^2 + y^2$, we write $x^2 + y^2 = pd$ for some $d \leq p/2$. If every prime factor of d is expressible as a sum of two squares, then, by the previous proposition, we can divide pd by these prime factors and obtain a representation of p as a sum of two squares. Since we have assumed that this is not the case, there must be a prime factor q of d (q is smaller than p) that is not expressible as a sum of two squares. Repeating this argument, we obtain a strictly decreasing sequence of positive primes that are not sums of two squares, an impossibility.

So far, we have not used the hypothesis that $p \equiv 1 \pmod 4$. In the descent step, we only used the fact that p divides a sum of two relatively prime squares. Now, we use the properties of the Legendre symbol to show that if $p \equiv 1 \pmod 4$, then p divides a sum of two squares. If $p \equiv 1 \pmod 4$, then (using Euler's criterion) -1 is a quadratic residue modulo p, that is, $x^2 \equiv -1 \pmod p$ is solvable, that is, $p \mid x^2 + 1^2$.

As p divides a sum of two relatively prime squares, it must be a sum of two squares. This completes the proof. ∎

The following example illustrates the descent step.

Example 14.4.8. A solution to the congruence $x^2 \equiv -1 \pmod{89}$ is $x = 34$. Now, $89 \mid 34^2 + 1 = 13 \cdot 89$. Since $13 = 3^2 + 2^2$, we can divide $34^2 + 1$ by 13 to obtain a representation of 89 as a sum of two squares. Equation (1) gives $\dfrac{34 + i}{3 + 2i} = 8 - 5i$ and equation (2) gives $\dfrac{34 - i}{3 + 2i} = \dfrac{100 + 71i}{13}$. The first quotient has integral terms and we obtain $89 = 8^2 + 5^2$.

The proof of the theorem can be extended to representability of primes by the forms $x^2 + 2y^2$ and $x^2 + 3y^2$. For $x^2 + 5y^2$, the argument in the descent step falls short. If $p \mid x^2 + 5y^2$ and $|x| < p/2$, $|y| < p/2$, then $x^2 + 5y^2 < 3/2p^2$, so there is no guarantee that all prime divisors of $x^2 + 5y^2$ are less than p. For $x^2 + 5y^2$, it is easy to construct examples to show that the descent step fails; that is, a prime can divide a sum $x^2 + 5y^2$ without necessarily having to be of the form $x^2 + 5y^2$. This topic is explored further in Chapter 18.

Proposition 14.4.9. *The representation of a prime p, $p \equiv 1 \pmod 4$, as a sum of two squares is unique.*

PROOF. We count $p = a^2 + b^2$ and $b^2 + a^2$ as the same representation. We will show that representations as a sum of two squares are in bijection with

Fermat's Method of Descent

The method we employed to show that a prime of the form $4k+1$ is a sum of two squares is Fermat's method of descent. It is a powerful tool in the solution of Diophantine equations. The idea of the method is to start with a positive solution to an equation and construct a smaller positive solution. The procedure cannot continue indefinitely, so either we reach a smallest solution, or the equation has no solutions. For example, we employ this technique (in Section 14.6) to show that $x^4 + y^4 = z^4$ has no nontrivial solutions. The descent procedure gives a solution to $p = x^2 + y^2$ or to the equation $x^2 - dy^2 = 1$.

An interesting application of descent is given in Project 6 at the end of this chapter. In the project, we start with a positive solution to $2x^4 - y^4 = z^2$ and construct a smaller positive solution $2p^4 - q^4 = t^2$. The procedure is valid when $x > 1$, so by descent, we reach the smallest solution $x = 1$, $y = 1$. The descent steps can be reversed and all the solutions to $2x^4 - y^4 = z^2$ are obtained from this starting solution. In general, one can expect more than one starting solution. In 1921, L. J. Mordell proved (using descent) that the equation

$$ax^4 + bx^3y + cx^2y^2 + dxy^3 + ey^4 = fz^2$$

has a finite number of starting solutions, from which every solution can be obtained by reversing the descent step.

solutions to $x^2 \equiv -1 \pmod{p}$. Suppose $p = a^2 + b^2$; then $a^2 \equiv -b^2 \pmod{p}$, and $(ab^{-1})^2 \equiv -1 \pmod{p}$. If $x = ab^{-1} \bmod p$, then $bx \equiv a \pmod{p}$. Similarly, if $p = c^2 + d^2$ gives the same root x, then $dx \equiv c \pmod{p}$. Then we get $c(bx) - a(dx) \equiv ac - ca \pmod{p}$, that is, $(bc - ad)x \equiv 0 \pmod{p}$. Now all the terms a,b,c,d are less than $\sqrt{p}$, hence $|bc - ad| < p$; therefore, $bc - ad = 0$. This implies that $a = c$ and $b = d$ because $(a,b) = 1$. If $p = c^2 + d^2$ gives the root $-x$, then it is easy to check that $c = b$ and $d = a$. ∎

The previous results can be combined to yield the following.

Theorem 14.4.10. *A positive integer n is expressible as a sum of two squares if and only if every prime divisor of n of the form $4k + 3$ occurs to an even power in the factorization of n.*

PROOF. First, if n has the desired factorization, then each prime power in its factorization is a sum of two squares; therefore, n can be expressed as a sum of two squares. If n is a sum of two squares, then, by Lemma 14.4.1, every prime divisor $p \equiv 3 \pmod 4$ must appear to an even power in the prime factorization of n. ∎

Example 14.4.11. The numbers $715 = 5 \cdot 11 \cdot 13$ and $26264 = 8 \cdot 7^2 \cdot 67$ are not sums of two squares, while $2205 = 3^2 \cdot 5 \cdot 7^2$ is a sum of two squares.

We can use Proposition 14.4.9 to determine the number of representations of a number as a sum of two squares. This is outlined in the first project at the end of this chapter. To find a representation of n as $x^2 + y^2$, it is not practical to search for x and y sequentially, especially when n is large. The known methods depend on obtaining a prime factorization of n. We then represent the primes occurring as sums of squares and use Lemma 14.4.5 to obtain a representation of the product. The following algorithm expresses a prime $p \equiv 1 \pmod 4$ as a sum of two squares. Another algorithm, which also uses continued fractions, is given in Exercise 8.

Recall the continued fraction expansion of $\sqrt{p}$,

$$\sqrt{p} = [a_0, \overline{a_1, a_2, \ldots, a_2, a_1, 2a_0}].$$

It is shown in Proposition 14.5.11 that if $p \equiv 1 \pmod 4$, then the period length of $\sqrt{p}$ is odd, that is, it has an expansion of the form

$$\sqrt{p} = [a_0, \overline{a_1, \ldots, a_n, a_n, \ldots, a_1, 2a_0}].$$

Consider the continued fraction $\alpha = [\overline{a_n, \ldots, a_1, 2a_0, a_1, \ldots, a_n}]$. Because of the symmetry of the period, $-1/\overline{\alpha}$ has the same expansion, that is, $-1/\overline{\alpha} = \alpha$, or $\alpha \overline{\alpha} = -1$. Now, α is a complete quotient in the expansion of $\sqrt{p}$, ($\alpha = x_{n+1}$ in the notation of Section 11.3.). If we write $\alpha = (x + \sqrt{p})/y$, then

$$\alpha \overline{\alpha} = \left(\frac{x + \sqrt{p}}{y} \right) \left(\frac{x - \sqrt{p}}{y} \right) = \frac{x^2 - p}{y^2} = -1.$$

We can rewrite this as $x^2 + y^2 = p$. By using the notation of Theorem 11.4.1 for the continued fraction algorithm of $\sqrt{p}$,

$$\alpha = x_{n+1} = \frac{A_{n+1} + \sqrt{p}}{B_{n+1}}.$$

Hence

$$x = A_{n+1} \quad \text{and} \quad y = B_{n+1}.$$

The algorithm can store all the A_k and B_k in the computation of the continued fraction. Since we don't know the period in advance, we compute until $B_k = 1$ for the first time. Suppose $B_N = 1$, $N \geq 1$ is the smallest such integer; then the period length is $N = 2n+1$, and we return $A_{(N+1)/2}$ and $B_{(N+1)/2}$.

Algorithm 14.4.12. Represent $p \equiv 1 \pmod 4$ as a sum of two squares.

1. Let $A_0 = 0$, $B_0 = 1$, $k = 0$, and $a_0 = \lfloor \sqrt{p} \rfloor$.

2. Let $A_{k+1} = a_k B_k - A_k$, $B_{k+1} = (p - A_{k+1}^2)/B_k$, $a_{k+1} = \left\lfloor \frac{A_{k+1} + \sqrt{p}}{B_{k+1}} \right\rfloor$.

3. If $B_{k+1} = 1$, then $N = k + 1$ and $n = (N + 1)/2$; return A_n and B_n; otherwise, let $k = k + 1$, and go to step 2.

Example 14.4.13. Consider $p = 173$. The continued fraction expansion of $\sqrt{p}$ is $[13, \overline{6, 1, 1, 6, 26}]$. The complete quotients are the following: $x_1 = (13 + \sqrt{p})/4$, $x_2 = (11 + \sqrt{p})/13$ and $x_3 = (2 + \sqrt{p})/13$. The continued fraction of x_3 is purely periodic and symmetric, and this implies that $173 = 13^2 + 2^2$.

Exercise. Express $p = 29$ as a sum of two squares using Algorithm 14.4.12.

─────────────── **Exercises for Section 14.4** ───────────────

1. Determine which of the following are expressible as sums of two squares.

   (a) 282   (b) 379
   (c) 6997   (d) 7663

2. A solution to the congruence $x^2 \equiv -1 \pmod{509}$ is given by $x = 301$. The number 509 is prime. This implies that $509 \mid (301^2 + 1)$, and $301^2 + 1 = 2 \cdot 89 \cdot 509$. Use this information to work through the method of descent to find a representation of 509 as a sum of two squares.

 3. Express the following integers as sums of two squares.

   (a) 1229   (b) 5281
   (c) 19897   (d) 245813

4. (Fermat, 1630) Show that 21 cannot be expressed as the sum of the squares of two **rational** numbers.

5. Prove that a positive integer n is the hypotenuse of a Pythagorean triangle if and only if n is divisible by at least one prime of the form $4k + 1$.

6. Given z, show that the number of Pythagorean triples $\{x, y, z\}$ with $0 < x < y$ is $(S(z^2) - 4)/8$, where $S(z^2)$ is the (total) number of representations of z^2 as a sum of two squares. (In computing $S(n)$, count $n = x^2 + y^2$ and $n = y^2 + x^2$ as different representations unless $x = y$. Similarly, changing the sign of x or y counts as a different representation.)

7.   (a) Show that

$$(x^2 + 2y^2)(a^2 + 2b^2) = (ax \pm 2by)^2 + 2(bx \mp ay)^2.$$

(b) Imitate the proof of Theorem 14.4.7 to show that if $p \mid x^2 + 2y^2$ with $(x, y) = 1$, then p is of the form $x^2 + 2y^2$.

(c) We will prove in Section 17.1 that if $p \equiv 1, 3 \pmod 8$, then $p \mid x^2 + 2y^2$ for some x and y; hence p can be expressed in the form $x^2 + 2y^2$. Use this to characterize integers that can be written in the form $x^2 + 2y^2$.

8.   There is another method for representing a prime as a sum of two squares. First, we solve the equation $x^2 \equiv -1 \pmod p$ such that $0 < x < p/2$. Then p/x has the following symmetric continued fraction expansion

$$[a_0, a_1, \ldots, a_n, a_n, \ldots, a_0],$$

and $p = p_n^2 + p_{n-1}^2$, where p_k/q_k is the kth convergent of p/x. Justify this algorithm by showing that if p/x has a continued fraction expansion of this type, then $x^2 \equiv -1 \pmod p$. Use Exercise 11.2.19 to show that if $x^2 \equiv -1 \pmod p$, then p/x has the desired continued fraction expansion.

9.   Implement the algorithm of the previous exercise.

10. Implement Algorithm 14.4.12 to find a representation of p as the sum of two squares for $p \equiv 1 \pmod 4$.

11. Gauss constructed a representation of $p = x^2 + y^2$ for $p = 4k+1$ as follows. Take

$$x \equiv \frac{(2k)!}{2(k!)^2} \pmod p \quad \text{and} \quad y \equiv (2k)!x \pmod p,$$

where x and y are taken so that $|x|, |y| < p/2$. Verify Gauss's construction for $p = 29$ and $p = 41$.

12. Show that if n and m are represented by the form $x^2 - dy^2$, then their product is also represented by this form.

13. Suppose $n = a^2 + b^2$ with $(a, b) = 1$. Show that every divisor of n is a sum of two relatively prime squares.

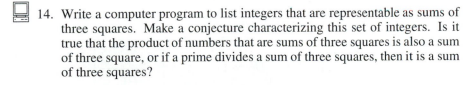

14. Write a computer program to list integers that are representable as sums of three squares. Make a conjecture characterizing this set of integers. Is it true that the product of numbers that are sums of three squares is also a sum of three square, or if a prime divides a sum of three squares, then it is a sum of three squares?

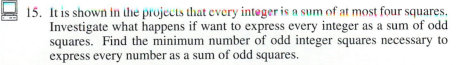

15. It is shown in the projects that every integer is a sum of at most four squares. Investigate what happens if want to express every integer as a sum of odd squares. Find the minimum number of odd integer squares necessary to express every number as a sum of odd squares.

14.5   The Brahmagupta–Bhaskara Equation

> *A person who can within a year solve the*
> *equation $x^2 - 92y^2 = 1$ is a mathematician.*
>
> — BRAHMAGUPTA

The equation $x^2 - dy^2 = 1$, for $d > 0$ squarefree, arises naturally in trying to approximate $\sqrt{d}$ by rational numbers. The first mathematicians to study this equation were Brahmagupta (c. 600 A.D.) and Bhaskara (c. 1100 A.D.). Bhaskara gave a procedure to compute infinitely many solutions of the equation, though he did not prove that all the solutions were obtained by his "cyclic method."

In the West, Fermat posed the problem as a challenge to the English mathematicians. After a false start, Brouncker and Wallis rose to the challenge and obtained infinitely many solutions by a method that is essentially the continued fraction expansion of $\sqrt{d}$. Among the challenge equations of Fermat were $x^2 - 61y^2 = 1$ and $x^2 - 109y^2 = 1$, whose smallest nontrivial solution is very large. This indicates that Fermat must have possessed a general method, though, as with many of his theorems, he did not publish a proof. Lagrange was the first to show that every solution of the equation is obtained from the continued fraction expansion of $\sqrt{d}$. Euler mistakenly attributed the solution of the equation to John Pell, and since then, it has also been known as Pell's equation.

The techniques of this section are based on the continued fraction expansion of $\sqrt{d}$ and the norm identity in the integers $\mathbb{Z}[\sqrt{d}]$. The algebra of $\mathbb{Z}[\sqrt{d}]$ is similar to that of the Gaussian integers introduced in Section 14.4.

First, we recall the relevant facts regarding the continued fraction expansion of $\sqrt{d}$. See Sections 11.4 and 11.5 for the proofs.

1. The continued fraction has the form $[a_0, \overline{a_1, a_2, \ldots, a_2, a_1, 2a_0}]$.

2. The expansion of $\sqrt{d}$ is computed by

$$A_0 = 0, \quad B_0 = 1, \quad a_k = \left\lfloor \frac{A_k + \sqrt{d}}{B_k} \right\rfloor$$

$$A_{k+1} = a_k B_k - A_k$$

$$B_{k+1} = \frac{d - A_{k+1}^2}{B_k}.$$

Bhaskara (1114–1185)

Bhaskara was born in 1114 A.D. in the Indian state of Mysore. He is the best-known Indian mathematician from the middle ages. His works include the *Lilavati*, the *Bija-Ganita*, and the *Siddhanta-Siromani*. The first two were textbooks of arithmetic, algebra, and geometry, and the latter, a treatise on astronomy. In number theory, his most important accomplishment was a general solution to the equation $x^2 - dy^2 = 1$ by a method called the Varga Prakrit ("the cyclic method"). This is similar to the method discovered by Brouncker five centuries later and shorter than the continued fraction method. Bhaskara gave a solution to $x^2 - 61y^2 = 1$ with $x = 1,766,319,049$ and $y = 226,153,980$ (not a trivial task in any century). Bhaskara refined the work of Brahmagupta and settled the question of division by zero by introducing the notion of infinity. In addition, Bhaskara had algebraic knowledge that was not matched in the West until the seventeenth century. For example, Bhaskara derived the formula

$$\sqrt{a \pm \sqrt{b}} = \sqrt{(a + \sqrt{a^2 - b})/2} \pm \sqrt{(a - \sqrt{a^2 - b})/2}.$$

Bhaskara also derived the addition formula for the sine function.

3. The convergents p_k/q_k of $\sqrt{d}$ satisfy

$$p_k^2 - dq_k^2 = (-1)^{k+1} B_{k+1}.$$

4. The shortest period length m of $\sqrt{d}$ is the smallest positive m such that $B_m = 1$. It is true that $B_k = 1$ if and only if k is a multiple of m.

Let $\mathbb{Z}[\sqrt{d}]$ be the numbers of the form $a + b\sqrt{d}$ for integers a and b. Addition and multiplication are defined by

$$(a + b\sqrt{d}) + (x + y\sqrt{d}) = (a + x) + (b + y)\sqrt{d}$$
$$(a + b\sqrt{d})(x + y\sqrt{d}) = (ax + dby) + (ay + bx)\sqrt{d}.$$

The conjugate of $\alpha = a + b\sqrt{d}$ is $\overline{\alpha} = a - b\sqrt{d}$, and the norm is $N(a + b\sqrt{d}) = (a + b\sqrt{d})(a - b\sqrt{d}) = a^2 - db^2$.

Example 14.5.1.   1. Let $z = 3 + 2\sqrt{2}$. Then $\overline{z} = 3 - 2\sqrt{2}$ and

$$N(z) = (3 + 2\sqrt{2})(3 - 2\sqrt{2}) = 3^2 - 2 \cdot 2^2 = 1.$$

2. We simplify $\dfrac{4 + 7\sqrt{2}}{3 + 2\sqrt{2}} = \dfrac{(4 + 7\sqrt{2})(3 - 2\sqrt{2})}{(3 + 2\sqrt{2})(3 - 2\sqrt{2})} = -16 + 13\sqrt{2}.$

Proposition 14.5.2 (Brahmagupta). $(a^2 - db^2)(x^2 - dy^2) = (ax \pm dby)^2 - d(ay \pm bx)^2$, *or, equivalently, if* $z, w \in \mathbb{Z}[\sqrt{d}]$, *then* $N(zw) = N(z)N(w)$.

PROOF. Because $N(z) = z\overline{z}$, then $N(z)N(w) = z\overline{z}w\overline{w} = zw\overline{z}\,\overline{w} = zw\overline{zw} = N(zw)$. If $z = a + b\sqrt{d}$, and $w = x + y\sqrt{d}$, then $N(z)N(w) = (a^2 - db^2)(x^2 - dy^2)$. On the other hand, $zw = (ax + dby) + (ay + bx)\sqrt{d}$ and $N(zw) = (ax + dby)^2 - d(ay + bx)^2$. If we use $z = a - b\sqrt{d}$ and $w = x - y\sqrt{d}$, then we obtain the result with the negative signs. ∎

The formula can also be verified directly without using the properties of the norm.

Exercise. Express $(2^2 - 2 \cdot 3^2)(5^2 - 2 \cdot 2^2)$ in the form $x^2 - 2y^2$.

With the preliminaries out of the way, we can solve the equation $x^2 - dy^2 = 1$. It is enough to consider x and y positive.

Proposition 14.5.3. *If* p *and* q *are positive integers satisfying* $p^2 - dq^2 = 1$, *then* p/q *is a convergent of* $\sqrt{d}$.

PROOF. If $p^2 - dq^2 = 1$, then $(p - q\sqrt{d})(p + q\sqrt{d}) = 1$. We can write this as $|p - q\sqrt{d}| = 1/|p + q\sqrt{d}|$, and dividing by q, we obtain

$$\left| \frac{p}{q} - \sqrt{d} \right| = \frac{1}{q|p + q\sqrt{d}|}.$$

Now, $p > q\sqrt{d}$, (as $p^2 > q^2 d$) and hence $p + q\sqrt{d} > 2q\sqrt{d}$; therefore,

$$\left| \frac{p}{q} - \sqrt{d} \right| = \frac{1}{q|p + q\sqrt{d}|} < \frac{1}{q2q\sqrt{d}} = \frac{1}{2q^2\sqrt{d}}.$$

Since $d > 1$, we have that $\left| \dfrac{p}{q} - \sqrt{d} \right| < \dfrac{1}{2q^2}$. Proposition 13.1.7 implies that p/q is a convergent of $\sqrt{d}$. ∎

Example 14.5.4. Since $3^2 - 2 \cdot 2^2 = 1$, the fraction $3/2$ must be a convergent of $\sqrt{2}$. It is the second convergent of $\sqrt{2} = [1, \overline{2}]$.

Theorem 14.5.5. *Let* p_k/q_k *be the* kth *convergent of* $\sqrt{d}$. *If the period length* m *of* $\sqrt{d}$ *is even, then the solutions to* $x^2 - dy^2 = 1$ *are* $x = p_{jm-1}$ *and* $y = q_{jm-1}$ *for any* $j \geq 0$. *If the period length* m *of* $\sqrt{d}$ *is odd, then the solutions are* $x = p_{jm-1}$ *and* $y = q_{jm-1}$ *for* j *even. In particular, if* d *is not a perfect square, the equation has infinitely many solutions.*

PROOF. Using Proposition 14.5.3, we see that every solution is a convergent. We know that p_k/q_k satisfies $p_k^2 - dq_k^2 = (-1)^{k+1}B_{k+1}$. If m is the period of the continued fraction, then $B_k = 1$ if and only if $m \mid k$. If this is the case, write $k = jm$ and obtain

$$p_{jm-1}^2 - dq_{jm-1}^2 = (-1)^j mB_{jm} = (-1)^{jm}.$$

If m is even, $(-1)^{jm} = 1$ and all the convergents p_{jm-1}/q_{jm-1} give solutions to the equation. If m is odd, $(-1)^{jm} = 1$ when j is even. ∎

Example 14.5.6. Consider the equation $x^2 - 7y^2 = 1$. The continued fraction of $\sqrt{7}$ is $\sqrt{7} = [2, \overline{1,1,1,4}]$. The period length is 4, and the convergents are:

k	0	1	2	3	4	5	6	7	8	9	10
p_k	2	3	5	8	37	45	82	127	590	717	1307
q_k	1	1	2	3	14	17	31	48	223	271	494

Since $p_3/q_3 = 8/3$, $8^2 - 7 \cdot 3^2 = 64 - 7 \cdot 9 = 1$ is the smallest nontrivial solution. The next solution is $p_7 = 127$, $q_7 = 48$, and we verify that $127^2 - 7 \cdot 48^2 = 1$.

Exercise. Which convergent gives the next solution to $x^2 - 7y^2 = 1$? Compute it.

Example 14.5.7. Consider $x^2 - 13y^2 = 1$. The continued fraction expansion of $\sqrt{13}$ is $\sqrt{13} = [3, \overline{1,1,1,1,6}]$. The period length is odd, so the solutions are $x = p_{5j-1}$ and $y = q_{5j-1}$ for j even. For j odd, (p_{5j-1}, q_{5j-1}) gives solutions to $x^2 - 13y^2 = -1$.

We compute the convergents p_k/q_k and $p_k^2 - 13q_k^2$.

k	p_k/q_k	$p_k^2 - 13q_k^2$
0	3/1	-4
1	4/1	3
2	7/2	-3
3	11/3	4
4	18/5	$\boxed{-1}$
5	119/33	4
6	137/38	-3
7	256/71	3
8	393/109	-4
9	649/180	$\boxed{1}$
10	4287/1189	-4
11	4936/1369	3
12	9223/2558	-3
13	14159/3927	4
14	23382/6485	$\boxed{-1}$
15	154451/42837	4

Exercise. Which convergent gives the next solution to the equation $x^2 - 13y^2 = 1$?

In practice, if the period is large, computing the convergents can take considerable time. Moreover, since we only want to calculate every mth convergent, it is wasteful to compute all the other convergents. It is more efficient to use Brahmagupta's identity to generate new solutions.

We denote a solution to $x^2 - dy^2$ by the pair (x, y). We can also write a solution as $z = x + y\sqrt{d}$ with x and y positive and $N(z) = 1$. If (a, b) and (u, v) are solutions to $x^2 - dy^2 = 1$, then

$$1 = (a^2 - db^2)(u^2 - dv^2) = (au \pm dbv)^2 - d(av \pm bu)^2;$$

that is, $(au \pm dbv, av \pm bu)$ are also solutions to $x^2 - dy^2 = 1$. This pair is obtained by taking the product in $\mathbb{Z}[\sqrt{d}]$; that is, if z and w in $\mathbb{Z}[\sqrt{d}]$ satisfy $N(z) = N(w) = 1$, then $N(zw) = 1$. If $z = a + b\sqrt{d}$ and $w = u + v\sqrt{d}$, then $zw = (au + dbv) + (av + bu)\sqrt{d}$.

In particular, if (a, b) is a solution and we write $(a + b\sqrt{d})^n = x_n + y_n\sqrt{d}$, then (x_n, y_n) is also a solution. Since powers of a solution are solutions, we could expect to write all solutions in terms of the smallest solution. We call the smallest nontrivial solution to $x^2 - dy^2 = 1$ the **fundamental solution**. If the period length m is even, this is (p_{m-1}, q_{m-1}), and if m is odd, it is (p_{2m-1}, q_{2m-1}).

Theorem 14.5.8. *If (a, b) is the fundamental solution to $x^2 - dy^2 = 1$, then all the solutions (x, y) with x and y positive are of the form (x_n, y_n), where*

$$x_n + y_n\sqrt{d} = (a + b\sqrt{d})^n \quad \text{for } n > 0.$$

PROOF. The proof is another illustration of Fermat's method of descent. The solutions to $x^2 - dy^2 = 1$ are a subset of the odd convergents; hence they form a decreasing sequence. The fundamental solution (a, b) is the largest among these solutions.

We have already seen that if $(a + b\sqrt{d})^n = x_n + y_n\sqrt{d}$, then (x_n, y_n) is also a solution.

Suppose that (p, q) is a solution, $p, q > 0$, but $z = p + q\sqrt{d}$ is not a power of $\epsilon = a + b\sqrt{d}$. Let p/q be the largest among all the convergents such that $p + q\sqrt{d}$ is not a power of ϵ. (It exists because there are only finitely many odd convergents greater than any odd convergent.) We have $z/\epsilon = (pa - dbq) + (qa - bp)\sqrt{d}$, and $(pa - dbq)^2 - d(qa - bp)^2 = 1$. We verify that $(pa - dbq)/(qa - bp) > p/q$, and, by the assumption on p/q, $(pa - dbq) + \sqrt{d}(qa - bp) = \epsilon^k$ for some k, but then $z/\epsilon = \epsilon^k$ or $z = \epsilon^{k+1}$.

Hence every solution can be obtained by taking powers of the fundamental solution ϵ. ∎

Example 14.5.9. The fundamental solution to $x^2 - 7y^2 = 1$ is given by $(8, 3)$. To compute other solutions, we compute the powers $z_n = (8 + 3\sqrt{7})^n$.

$$z_1 = 8 + 3\sqrt{7}$$
$$z_2 = 127 + 48\sqrt{7}$$
$$z_3 = 2024 + 765\sqrt{7}.$$

The reader can verify that these solutions correspond to the convergents p_3/q_3, p_7/q_7, and p_{11}/q_{11}.

Exercise. Determine $(8 + 3\sqrt{7})^4$. To which convergent of $\sqrt{7}$ does this solution correspond?

Remark. Our procedure defines multiplication on the set of solutions to $x^2 - dy^2 = 1$. The inverse of a solution $x + y\sqrt{d}$ can be defined by taking the reciprocal $(x + y\sqrt{d})^{-1} = x - y\sqrt{d}$. Readers familiar with abstract algebra will observe that this makes the set of solutions into a multiplicative group.

The algorithm for solving $x^2 - dy^2 = 1$ is then in two steps. First, we find the fundamental solution, and then we compute its higher powers to obtain all the solutions.

There are many other related equations that can be solved by the techniques of this section.

Proposition 14.5.10. *Let d be a squarefree integer. If the equation $x^2 - dy^2 = m$ has a nontrivial solution, then it has infinitely many solutions.*

PROOF. Suppose that $a^2 - db^2 = m$ and $x_0^2 - dy_0^2 = 1$; then

$$m = (a^2 - db^2)(x_0^2 - dy_0^2) = (ax_0 \pm dby_0)^2 - d(ay_0 \pm bx_0)^2,$$

so that $(ax_0 \pm dby_0, ay_0 \pm bx_0)$ is a solution to $x^2 - dy^2 = m$. Since there are infinitely many such (x_0, y_0), the equation $x^2 - dy^2 = m$ has infinitely many solutions. ■

Another equation of interest is $x^2 - dy^2 = -1$. Using the techniques of Proposition 14.5.3, we find that every solution (x, y) is obtained from a convergent x/y of $\sqrt{d}$. If the period length of the continued fraction of $\sqrt{d}$ is odd, then the equation $x^2 - dy^2 = -1$ has solutions. No general criterion is known to determine when the period length of $\sqrt{d}$ is odd. By congruences modulo 4, one can show that if $d \equiv 0, 3 \pmod 4$, then $x^2 - dy^2 = -1$ has no solution. If $d \equiv 1, 2 \pmod 4$, there is no general criterion to determine when the period length is odd. For example, $221 \equiv 1 \pmod 4$, but $\sqrt{221}$ has an even period; hence $x^2 - 221y^2 = -1$ has no solutions. If $p = 4k + 1$ is a prime, Lagrange showed that $x^2 - py^2 = -1$ has a solution. (Hence the period length of $\sqrt{p}$ is odd.)

Proposition 14.5.11 (Lagrange). *If p is a prime such that $p \equiv 1 \pmod{4}$, then $x^2 - py^2 = -1$ has a solution. Equivalently, the period length of the continued fraction of $\sqrt{p}$ is odd.*

PROOF. Let (a, b) be the fundamental solution to $x^2 - py^2 = 1$; then a must be odd and b even. (Look at $a^2 - pb^2 \equiv 1 \pmod{4}$.) Write $pb^2 = a^2 - 1 = (a - 1)(a + 1)$, or

$$p \left(\frac{b}{2} \right)^2 = \frac{a - 1}{2} \frac{a + 1}{2}.$$

Suppose $p \mid (a - 1)/2$; then $(a - 1)/2p$ and $(a + 1)/2$ are relatively prime; hence they are perfect squares, $(a - 1)/2p = u^2$ and $(a + 1)/2 = v^2$, and they satisfy $u^2 - pv^2 = 1$. The solution (u, v) is smaller than (a, b), a contradiction. Therefore, we must have $p \mid (a + 1)/2$ and that $(a + 1)/2p$ and $(a - 1)/2$ are perfect squares: $(a + 1)/2p = r^2$ and $(a - 1)/2 = s^2$, or $a + 1 = 2pr^2$ and $a - 1 = 2s^2$. Since $a - 1 + 2 = a + 1$, we write $2s^2 + 2 = 2pr^2$ or $s^2 - pr^2 = -1$, that is, the equation $x^2 - py^2 = -1$ has a nontrivial solution. ∎

Since the period length of $\sqrt{p}$ must be odd, the solutions to $x^2 - py^2 = -1$ are the convergents (p_{jm-1}, q_{jm-1}) for j odd.

Example 14.5.12. Find the smallest nontrivial solution to $x^2 - 29y^2 = -1$.

Since 29 is a prime congruent to 1 modulo 4, we know that the equation has solutions. We compute $\sqrt{29} = [5, \overline{2, 1, 1, 2, 10}]$, and the convergents are

k	0	1	2	3	4	5	6
p_k/q_k	$5/1$	$11/2$	$16/3$	$27/5$	$70/13$	$727/135$	$1524/283$
$p_k^2 - 29q_k^2$	-4	5	-5	4	-1	4	-5

The smallest nontrivial solution is $(70, 13)$.

──────────────────── **Exercises for Section 14.5** ────────────────────

1. Write a computer program to find the fundamental solution to $x^2 - dy^2 = 1$ using the continued fraction expansion of $\sqrt{d}$.

2. Write a program to multiply numbers in $\mathbb{Z}[\sqrt{d}]$ and to compute powers of $a + b\sqrt{d}$. You can represent a number $a + b\sqrt{d}$ in $\mathbb{Z}[\sqrt{d}]$ by a pair (a, b).

3. Find the fundamental solutions to the following equations:

    (a) $x^2 - 8y^2 = 1$.

(b) $x^2 - 47y^2 = 1$.

(c) $x^2 - 61y^2 = 1$.

4. Find the first five nontrivial solutions to the equations in the previous exercise.

5. Verify that among all d, $2 \le d \le 100$, the fundamental solution to $x^2 - 61y^2 = 1$ is the largest. [One measure of the size of the fundamental solution (a, b) is $\log |a + b\sqrt{d}|$.]

6. The nth triangular number T_n is $1 + 2 + \cdots + n = n(n+1)/2$. (This is the number of lattice points in a triangle with side n.) A square number is $S_m = m^2$. Determine all the triangular numbers that are square.

7. Show that if $x^2 - dy^2 = -1$, then x/y is a convergent of $\sqrt{d}$.

8. Find examples of $d \equiv 1, 2 \pmod 4$ such that $x^2 - dy^2 = -1$ has no solutions.

9. Find the smallest positive solutions of $x^2 - 47y^2 = -1$ and $x^2 - 37y^2 = -1$.

10. Show that if $x^2 - dy^2 = -1$ has a solution, then d is representable as a sum of two squares.

11. Find a nontrivial solution to $x^2 - 2y^2 = 7$. Use Proposition 14.5.10 to find five more solutions.

12. Suppose (a, b) is a solution to $x^2 - dy^2 = N$, and (u, v) a solution to $x^2 - dy^2 = M$. Show that $(au + dbv, av + bu)$ is a solution to $x^2 - dy^2 = MN$.

13. If (a, b) is the least positive solution of $x^2 - dy^2 = -1$, show that $(2a^2 + 1, 2ab)$ is the fundamental solution to $x^2 - dy^2 = 1$.

14. Show that there are infinitely many positive integers a such that both $a + 1$ and $3a + 1$ are perfect squares.

15. A Pythagorean triple $\{a, b, c\}$ has consecutive legs if a and b are consecutive integers. Show that there are infinitely many primitive Pythagorean triples with consecutive legs. Find the first 10 such triples.

16. In his biography of Ramanujan, Kanigel[4] describes an incident where Ramanujan casually solved the following puzzle.

> The problem is to find an address, a house number n on a street with numbers 1 through N, such that the house numbers on one side of the address add up to the same number as all the numbers on the other side of the address. Ramanujan answered by dictating to his friend Mahalonobis a continued fraction and gave the explanation: "Immediately I heard the problem, it was clear that the solution should obviously be a continued fraction;

---

[4] R. Kanigel. *The Man Who Knew Infinity: A Life of the Genius Ramanujan.* New York: Scribner's, 1991.

I then thought, Which continued fraction? and the answer came to my mind."

If you lack such divine inspiration, solve the puzzle by answering the following questions.

(a) The condition states that

$$1 + 2 + \cdots + n - 1 = n + 1 + n + 2 + \cdots + N.$$

Show that this is equivalent to $n^2 = (N^2 + N)/2$.

(b) Complete the square with respect to N and show that $(2N+1)^2 - 1 = 8n^2$. Write $X = 2N+1$ and $Y = n$ so that $X^2 - 8Y^2 = 1$. Show that there are infinitely many solutions to this equation, hence infinitely many solutions to the puzzle. Compute the first seven solutions to the puzzle.

(c) Explain Ramanujan's statement, "the solution should obviously be a continued fraction."

14.6   Fermat's Last Theorem

> *It is impossible ... for any number which is a*
> *power greater than the second to be written as*
> *a sum of two like powers.*
> *I have a truly marvelous demonstration of this proposition*
> *which this margin is too narrow to contain.*
>
> — PIERRE DE FERMAT.

The most celebrated of all Diophantine problems is Fermat's Last Theorem (FLT), which asserts that the equation $x^n + y^n = z^n$ has no nontrivial solutions in the integers when n is larger than 2. The assertion was first made by Fermat in the late 1630s in the margin of Bachet's translation of Diophantus's *Arithmetic*. Fermat offered no proof of it except in the case $n = 4$. It is very likely that he was mistaken about having possessed a proof, as the solution had eluded some of the greatest mathematicians for centuries. Fermat himself may have realized his error, as, unlike his other theorems, he never mentioned this (FLT) to any of his numerous correspondents. The problem was finally settled by Wiles in 1994 building on the work of Frey, Serre, Ribet, and Taylor and Wiles. It is interesting to note that the proof of FLT involves the theory of elliptic curves, another of Fermat's inventions.

We prove the simplest case of FLT, for $n = 4$. The proof is by the technique of infinite descent. We used this technique in Section 14.4 to show that every prime of the form $4k + 1$ is represented as a sum of two squares.

The following is a modification of Fermat's own published proof in which he showed that there is no Pythagorean triangle whose area is a square. (See Exercise 4.) The proof makes repeated use of the classification of Pythagorean triples and the fact that if $ab = m^2$ with $(a, b) = 1$, then a and b are perfect squares.

Theorem 14.6.1. *The equation $x^4 + y^4 = z^4$, $xyz \neq 0$, has no solutions in the integers.*

PROOF. We will show that the equation $x^4 + y^4 = z^2$ has no nontrivial solutions. This will imply that $x^4 + y^4 = z^4$ has no nontrivial solutions. The proof is by Fermat's method of descent. Assuming that x, y, and z are positive integers satisfying $x^4 + y^4 = z^2$, we will construct another solution (u, v, w) verifying $u^4 + v^4 = w^2$ with $0 < w < z$. Since this process cannot continue indefinitely, we reach a contradiction.

Suppose x, y, and z with $xyz \neq 0$ satisfy

$$x^4 + y^4 = z^2.$$

Then (x^2, y^2, z) is a Pythagorean triple. We can assume that x, y and z are relatively prime. Using Theorem 14.3.2, we note that there exist m and n relatively prime such that

$$x^2 = m^2 - n^2 \tag{1}$$

$$y^2 = 2mn \tag{2}$$

$$z = m^2 + n^2. \tag{3}$$

In (1), (x, n, m) is a Pythagorean triple, $x^2 + n^2 = m^2$. As x is odd, n must be even. By the classification of Pythagorean triples, there exist r and s relatively prime such that

$$x = r^2 - s^2 \tag{4}$$

$$n = 2rs \tag{5}$$

$$m = r^2 + s^2. \tag{6}$$

As n is even, m must be odd, hence $2n$ and m are relatively prime. From Equation (2), $2n$ and m must be perfect squares, that is, $m = w^2$ and $2n = 4t^2$, or $n = 2t^2$. Equation (5) implies that $rs = t^2$, so there exist u and v relatively prime such that $r = u^2$ and $s = v^2$. Equation (6) implies

$$w^2 = u^4 + v^4.$$

But $0 < w < w^2 = m < m^2 + n^2 = z$, hence we have constructed a smaller solution. This process cannot be repeated indefinitely, because a decreasing sequence of positive integers must terminate. Hence the equation $x^4 + y^4 = z^2$ has no nontrivial solution in the integers, and so $x^4 + y^4 = z^4$ has no nontrivial solutions. ∎

The theorem reduces the proof of FLT to odd prime exponents. The case $n = 3$ was settled by Euler using the technique of infinite descent. For $n = 5$, Sophie Germain showed that if $x^5 + y^5 = z^5$, then one of the numbers x, y, or z must be divisible by 5. In fact, she proved a more general theorem.

Theorem 14.6.2 (Germain). *If p is a prime such that $q = 2p + 1$ is prime, then $x^p + y^p = z^p$ has no solutions when $p \nmid xyz$.*

PROOF. We prove the theorem for $p = 5$. The reader should have no trouble in extending the proof to the general case.

We can replace z by $-z$ and obtain a symmetric equation $x^5 + y^5 + z^5 = 0$. We can assume that x, y, and z are relatively prime. Write $-x^5 = y^5 + z^5$ and factor the right-hand side.

$$-x^5 = (y + z)(y^4 - y^3 z + y^2 z^2 - yz^3 + z^4) \tag{7}$$

We claim that $y + z$ and $y^4 - y^3 z + y^2 z^2 - yz^3 + z^4$ are relatively prime. Suppose r is a common prime factor, $r \mid y + z$, that is, $y \equiv -z \pmod{r}$. Then $y^4 - y^3 z + y^2 z^2 - yz^3 + z^4 \equiv 5z^4 \pmod{r}$. Hence $r \mid 5z^4$, so $r \mid z$ or $r = 5$, which is not possible, as 5 does not divide x (by assumption). If $r \mid z$, then $r \mid y$, and, again, this is not possible.

Unique factorization in the integers implies that the two terms on the right of (7) are fifth powers. Let $y + z = A^5$ and $y^4 - y^3 z + y^2 z^2 - yz^3 + z^4 = T^5$. We can repeat this calculation by starting with y^5 or z^5 in (7) and obtain $x + z$ and $x + y$ as perfect powers.

$$x + z = B^5 \tag{8}$$

$$x + y = C^5 \tag{9}$$

$$y + z = A^5. \tag{10}$$

Fermat's Theorem implies that $a^5 \equiv \pm 1 \pmod{11}$ for $(a, 11) = 1$. Hence $x^5 + y^5 + z^5 \equiv 0 \pmod{11}$ is only possible if one of x, y, or z is divisible by 11. (Why?) Suppose $11 \mid x$. We subtract (10) from the sum of (8) and (9) to obtain

$$B^5 + C^5 - A^5 = 2x.$$

For the same reason, the congruence $B^5 + C^5 - A^5 \equiv 0 \pmod{11}$ is not possible unless $11 \mid ABC$. This implies that $11 \mid A$ because $11 \mid x$, and $11 \mid B$ implies $11 \mid z$, contradicting $(x, z) = 1$. (Show that $(x, y, z) = 1$, and $x^5 + y^5 + z^5 = 0$ implies that $(x, z) = 1$.) Similarly, $11 \mid x$, and $11 \mid C$ implies $11 \mid y$. But if $11 \mid A$, then $11 \nmid T$; hence $\pm 1 \equiv T^5 \equiv 5z^4 \pmod{11}$. But $z \equiv B^5 \pmod{11}$, so $z \equiv 1 \pmod{11}$. This implies that $\pm 1 \equiv 5 \pmod{11}$, a contradiction. ∎

Sophie Germain (1776–1831)

Germain

Sophie Germain was born in Paris into a prosperous family. She became interested in mathematics after reading about Archimedes; stuck down at the siege of Syracuse as he meditated over his diagrams. Initially, her parents were opposed to her mathematical studies, but she persisted and eventually her parents supported her. She taught herself mathematics by reading the works of great mathematicians, as formal university education was not open to women.

Sophie Germain maintained an extensive correspondence (under the penname "M. LeBlanc") with leading mathematicians including Legendre and Gauss. She earned the respect of both men, who continued to praise her work after learning her true identity. Apart from her contribution to Fermat's last problem, her most important work involved the mathematical theory of elasticity, for which she won a prize from the Paris Academy of Sciences.

A prime p such that $2p+1$ is prime is known as a Germain prime. It is not known if there are infinitely many such primes. Germain's theorem leaves the solution of Fermat's conjecture to the case when $5 \mid xyz$. Surprisingly, this case turns out to be much more difficult. The book of Edwards is an excellent source for the history and techniques regarding Fermat's conjecture.[5]

The technique of infinite descent was used by Dirichlet and Legendre to solve the problem for $n = 5$, and by Lamé for the case $n = 7$.

The most important contribution to FLT in the nineteenth century was due to Kummer. As a byproduct of his ideal theory, Kummer was able to give conditions on a prime so that FLT was satisfied. These were verified for a large number of cases but not enough to settle the problem in general. By 1980, FLT had been verified for $n \leq 125000$.

In the mid 1980s Gerhard Frey proposed a completely new method for the solution of FLT. From a solution $a^p + b^p = c^p$, Frey constructed the curve $y^2 = x(x - a^p)(x - c^p)$. Frey conjectured that it should be possible to show that a curve of this type cannot exist using a major unsolved problem in number theory that relates elliptic curves to modular forms. Ken Ribet

---

[5]H. M. Edwards. *Fermat's Last Theorem.* New York: Springer–Verlag, 1977.

showed that if a conjecture made by Shimura and Taniyama were true, then FLT would follow. Finally, Andrew Wiles proved the Shimura–Taniyama conjecture in 1994, completing the proof of FLT. We will study elliptic curves in the last chapter, but a description of the Shimura–Taniyama conjecture is beyond the scope of this book.

—————————— **Exercises for Section 14.6** ——————————

1. Show that for any even n, $n > 1$, $x^{2n} + y^{2n} = z^{2n}$ has no nontrivial solutions.

2. Show that the equation $x^4 - y^4 = z^2$ has no nontrivial solutions.

3. Prove that any Pythagorean triple has at most one perfect square.

4. Apply the method of infinite descent to show that the equation $x^4 - 4y^4 = z^2$ has no nontrivial solutions. Use this fact to show that the area of a Pythagorean triangle can never be a perfect square.

5. Using congruences modulo 25, show that if $x^5 + y^5 = z^5$, then one of x, y, or z is divisible by 5. Does a similar calculation modulo 49 prove that $x^7 + y^7 = z^7$ has no solutions when $7 \nmid xyz$?

6. Modify the proof given in the text to prove Sophie Germain's theorem for case when p is a prime such that $2p + 1$ is prime.

7. Imitate the proof of Theorem 14.6.2 to show that $x^7 + y^7 = z^7$ has no solution in the integers when $7 \nmid xyz$. (Hint: Use $4 \cdot 7 + 1 = 29$ as the auxiliary prime.)

8. Use the technique of infinite descent to show that the equation $x^4 + 2y^4 = z^2$ has no nontrivial solutions.

—————————— **Projects for Chapter 14** ——————————

1. **The Number of Representations as a Sum of Two Squares:** We showed that every prime $p \equiv 1 \pmod 4$ has a unique representation as a sum of two squares. This project will investigate the number of solutions to $x^2 + y^2 = n$.

   (a) Write a computer program to compute the solutions (x, y) to $x^2 + y^2 = n$ with $x > 0$ and $y \geq 0$. (We count $3^2 + 4^2$ and $4^2 + 3^2$ as different solutions, but $5^2 + 0^2$ is the same as $0^2 + 5^2$.)

   (b) Let $R(n)$ be the number of solutions to $x^2 + y^2 = n$ with $x > 0$ and $y \geq 0$. Determine $R(2^k)$ and $R(p^k)$ for p prime.

   (c) What can you say about $R(p^r q^s)$, where p and q are distinct primes? Prove your conjecture.

(d) Give a formula for $R(n)$ in terms of the prime divisors of n.

(e) Show that the total number of right triangles with hypotenuse z is

$$(R(z^2) - 1)/2.$$

(f) In a letter to Mersenne, Fermat asked for a number n that is the hypotenuse of precisely 367 right triangles. Find n.

2. **Sums of Four Squares:** This project proves Lagrange's theorem that every positive integer is a sum of four squares. The proof uses an identity that is derived using the arithmetic of the Quaternions. (Recall that we used the Gaussian integers to derive the necessary identities for the results on sums of two squares.) The integral quaternions, $\mathbb{H}$, are defined as

$$\mathbb{H} = \{a + bi + cj + dk : \quad a, b, c, d \in \mathbb{Z}\}.$$

Addition of quaternions is defined term-wise by

$$(a+bi+cj+dk)+(x+yi+zj+uk) = a+x+(b+y)i+(c+z)j+(d+u)k,$$

and multiplication is defined to satisfy the distributive laws, with i, j, k satisfying the following properties.

1. $i^2 = -1, j^2 = -1, k^2 = -1$.
2. $i \cdot j = -j \cdot i = k$.
3. $j \cdot k = -k \cdot j = i$.
4. $k \cdot i = -i \cdot k = j$.

If $w = a+bi+cj+dk$, then its conjugate is defined as $\overline{w} = a-bi-cj-dk$. The norm of w is $N(w) = w \cdot \overline{w}$.

(a) Compute the norm of w. Show that $\overline{\alpha \cdot \beta} = \overline{\beta} \cdot \overline{\alpha}$. Use this to show that the product of two numbers that are sums of four squares is also a sum of four squares.

(b) This step will prove that if a prime p divides a sum of four squares, then it is a sum of four squares.

   (i) Let $p \mid x^2 + y^2 + z^2 + u^2 = m$. Explain why m can be taken to be less than p^2. Write $m = np$ with $n < p$.

   (ii) Let $a = x \bmod n$, $b = y \bmod n$, $c = z \bmod n$, and $d = u \bmod n$. Let $w = x+yi+zj+uk$ and $v = a+bi+cj+dk$. Show that $N(v) = nk$ with $k < n$, and use the identity of part (a) to show that $np \cdot nk$ is a sum of four squares, each of which is divisible by n^2. This will express pk with $k < n$ as a sum of four squares. Explain how the method of descent can be used to conclude that $k = 1$, and hence p is a sum of four squares.

(c) To finish the proof, show that p divides a sum of four squares by showing that the congruence $x^2 + y^2 + 1 \equiv 0 \pmod{p}$ has a solution.

3. **Sums of k Squares:** Let $S_k(n)$ be the number of representations of n as a sum of k squares. Here $S_k(n)$ counts k-tuples of integers $(x_1, x_2, \ldots, x_k)$ such that $x_1^2 + \cdots + x_k^2 = n$.

   (a) Show that $S_k(2) = 4\binom{k}{2}$.

   (b) Show that $S_k(3) = 8\binom{k}{3}$.

   (c) Determine $S_k(6)$.

   (d) In the first project above, it is shown that $R(n) = S_2(n)/4$ is multiplicative. Show that $R_k(n) = S_k(n)/2k$ can be multiplicative only when $k = 1, 2, 4$, or 8. (It is possible to show that R_4 and R_8 are multiplicative. For four squares, this is due to the presence of the Quaternions, and for eight squares, the Cayley numbers provide similar identities to prove the multiplicativity.)

   (e) The function $S_4(n)$ is related to the sum of divisors of n. Can you find this relation?

4. **The Markoff Equation:** The solutions to the equation $x^2 + y^2 + z^2 = 3xyz$ have many beautiful patterns among them. Write a computer program to find integer solutions to this equation. Explain why it is sufficient to find only the positive solutions. Your program should print solutions $0 < x \le y \le z$ because all other solutions can be obtained from these.

   (a) Determine all the solutions in which at least two terms are equal.

   (b) Are there any solutions with $y = z$ satisfying $x < y$?

   (c) Use your program to find all solutions with $x = 1$ for $z \le 4000$. Do you see a pattern in the solutions? Make a conjecture and prove it.

   (d) Do you see a pattern in the solutions for $x = 2$ and $x = 5$?

   (e) Compute more solutions and investigate them for additional patterns and observation. There are rules by which all the solutions can be generated. Can you find these rules?

   (f) A generalization of the Markoff equation is the Hurwitz equation

   $$x_1^2 + \cdots + x_n^2 = a x_1 \ldots x_n.$$

   Investigate the solutions to this equation for different values of a and n.

5. **Liouville's Theorem:** Liouville proved the analog of Fermat's Last Theorem for polynomials. The proof is surprisingly simple and is reproduced in the following questions. The goal is to show that $x(t)^n + y(t)^n = z(t)^n$ has no solution for $n > 2$, where x, y, z are polynomials in one variable, t.

   (a) Differentiate with respect to t and eliminate x and x' to show that

   $$y^{n-1}(yx' - y'x) = z^{n-1}(zx' - xz').$$

(b) Let $\deg(x)$ denote the degree of the polynomial. Show that we can take x, y, and z such that they have no common factors. Conclude that y^{n-1} divides $zx' - xz'$; hence

$$(n-1)\deg(y) \leq \deg(z) + \deg(x) - 1.$$

(You have to show that if x and y have no common factors, then $xy' - yx'$ is a nonzero polynomial.)

(c) Obtain similar inequalities for $\deg(x)$ and $\deg(z)$ and show that

$$(n-1)[\deg(x)+\deg(y)+\deg(z)] \leq 2[\deg(x)+\deg(y)+\deg(z)]-3;$$

hence $n > 2$.

6. **An Equation in which Descent becomes Ascent:** In a letter to Mersenne, Fermat asked for a right triangle with integer sides such that the hypotenuse and the sum of the legs are squares. If $\{a, b, c\}$ is the Pythagorean triple with hypotenuse c, then the condition is

$$a^2 + b^2 = c^2, \quad c = x^2, \quad a + b = y^2.$$

1. Show that $2x^4 - y^4 = z^2$ and $(a - b)^2 = z^2$. Explain how to find $\{a, b, c\}$ from the solutions to $2x^4 - y^4 = z^2$. (In the problem, we need positive values of a, b and c.)

2. Assume that a is odd. Use the classification of Pythagorean triples to conclude that there exist m and n such that

$$a = m^2 - n^2, \quad b = 2mn, \quad c = m^2 + n^2.$$

Show that $x^2 = m^2 + n^2$, hence there exist r and s such that

$$m = r^2 - s^2, \quad n = 2rs, \quad x = r^2 + s^2.$$

3. Use the classification of solutions to the equation $X^2 + 2Y^2 = Z^2$ to conclude that there exist u and v such that

$$y = u^2 - 2v^2, \quad n = 2uv, \quad m + n = u^2 + 2v^2.$$

This implies that $rs = uv$. Let $r/u = v/s = p/q$ in lowest form with $r = kp$, $u = kq$, $v = dp$ and $s = dq$ for some k and d. Equating the two expressions for m gives a quadratic equation in d/k which has rational solutions. Conclude that $(2p^4 - q^4)$ must be a perfect square. This is the descent step. Show that if $x \neq 1$, then $0 < p < x$.

4. Now, convert the descent into an ascent by starting with the smallest solution $(1, 1, 1)$ to $2x^4 - y^4 = z^2$. Take $p = 1$, $q = 1$ and determine k and d. Work through all the equations to find x, y, and z and then $\{a, b, c\}$. This procedure can be applied again to the new values of (x, y, z) and we see that an infinite set of solutions can be obtained. Determine the first 5 solutions to $2x^4 - y^4 = z^2$ and the first two solutions to Fermat's problem. Do we obtain all the solutions by this method?

Chapter 15

Arithmetical Functions and Dirichlet Series

15.1 Arithmetical Functions

In this chapter, we explore the relation between number theoretic functions and analytic functions. The generating functions of arithmetic functions, the Dirichlet series, have many nice properties from which additional information can be derived about functions of interest. The analytic theory is an important part of number theory. We sketch a proof of Dirichlet's Theorem on the infinitude of primes in arithmetic progressions. Since Dirichlet series and Euler products are necessary for this theorem, we have developed them first.

Definition 15.1.1. A real- or complex-valued function on the natural numbers is called an **arithmetical function.**

The most interesting arithmetical functions are the number-theoretic functions, some of which have already been encountered in previous chapters.

Example 15.1.2. 1. The number of positive divisors of n is an arithmetical function denoted by $\nu(n)$. Other examples of arithmetical functions are: the sum of the positive divisors of n, $\sigma(n)$, and the number of invertible elements modulo n, $\phi(n)$.

2. The number of ways to write n as a sum of two squares $x^2 + y^2$ with $x > 0, y \geq 0$ is an arithmetical function denoted by $R_2(n)$.

The interesting arithmetical functions are related to the divisibility properties of the integers, and hence, to the set of divisors of an integer. We will use the notation $\sum_{d|n}$ to mean the sum over all the positive divisors of

n, including 1 and n. Using this notation, we can write $\nu(n) = \sum_{d|n} 1$ and

$$\sigma(n) = \sum_{d|n} d.$$

The divisibility properties of the integers are intimately related to the prime factorization. Hence, many interesting arithmetical functions can be understood in terms of their values on primes and prime powers. We have seen this in the case of $\nu(n)$ (Proposition 2.3.2) and $\phi(n)$ (Corollary 4.2.4). The basic property that assures good behavior with respect to the prime factorization is the multiplicativity.

Definition 15.1.3. An arithmetical function f is said to be **multiplicative** if $f(mn) = f(m)f(n)$ for all $(m, n) = 1$. If $f(mn) = f(m)f(n)$ for all m and n, then we say that f is **completely multiplicative.**

We will prove below that $\nu(n)$ and $\sigma(n)$ are multiplicative. We showed that $\phi(n)$ is multiplicative. The multiplicativity of $R_2(n)$ was studied in the first project of Chapter 14. Here are some more examples of multiplicative functions.

Example 15.1.4.   1. The functions $f(n) = n$ and $g(n) = 1$ are multiplicative.

2. If f is a multiplicative function, then f^2, f^3, and higher powers of f are multiplicative functions.

3. The product $f(n)g(n)$ of two multiplicative functions f and g is multiplicative, and so is the quotient $f(n)/g(n)$, if $g(n) \neq 0$ for every n.

4. A multiplicative function is not necessarily completely multiplicative. For example, $\phi(n)$ is not completely multiplicative.

5. Suppose $\chi(n) = 1$ if $n \equiv 1 \pmod 4$; $\chi(n) = -1$ if $n \equiv -1 \pmod 4$ and $\chi(n) = 0$ for $n \equiv 0 \pmod 2$. Then χ is also a multiplicative function. It is easy to verify that $\chi(n)$ is completely multiplicative.

Exercise. Show that if f is multiplicative, then so is $f(n)/n^s$ for all s.

Exercise. Verify that $\chi(n)$ is a multiplicative function.

Now we turn our attention to proving the properties of the divisor functions. We start with $\nu(n)$, the number of divisors of n. The value of ν can be easily determined from the prime factorization of n. For a prime number p, $\nu(p) = 2$. For a prime power, $n = p^e$, it is clear that the only positive divisors are $1, 1, p, \ldots, p^e$, a total of $e + 1$. Hence $\nu(p^e) = e + 1$. We can determine ν for other integers using the following lemma.

Lemma 15.1.5. *If $n = p_1^{e_1} p_2^{e_2} \cdots p_k^{e_k}$ is the prime factorization of n, then*

$$\nu(n) = (e_1 + 1)(e_2 + 1) \cdots (e_k + 1).$$

PROOF. Any divisor d of n has factorization $d = p_1^{a_1} p_2^{a_2} \cdots p_k^{a_k}$, where $0 \le a_i \le e_i$, for $1 \le i \le k$. There are $(e_i + 1)$ ways to choose each a_i and each choice gives a rise to a different divisor, so the total number of divisors is the product of the $(e_1 + 1)$. ∎

Corollary 15.1.6. *If n and m are relatively prime, $\nu(nm) = \nu(n)\nu(m)$.*

PROOF. If $n = p_1^{e_1} p_2^{e_2} \cdots p_k^{e_k}$ and $m = q_1^{f_1} \ldots q_r^{f_r}$, then $(m, n) = 1$ implies that none of the p's occur among the q's, and vice versa. Then $\nu(mn) = (e_1 + 1) \cdots (e_k + 1)(f_1 + 1) \cdots (f_r + 1) = \nu(n)\nu(m)$. ∎

Example 15.1.7. 1. To determine $\nu(72)$, we factor $72 = 2^3 3^2$. Lemma 15.1.5 implies that $\nu(72) = 4 \cdot 3 = 12$. We can verify this by enumerating the divisors, which are $\{1, 2, 3, 4, 6, 8, 9, 12, 24, 36, 72\}$.

2. Suppose that $\nu(n) = 6$. Since $6 = 1 \cdot 6$ or $2 \cdot 3$, using Lemma 15.1.5 above, we find that n is either p^5 or pq^2 for some primes p and q. The smallest choices for p and q are 2 and 3. Therefore, the smallest value of n for which $\nu(n) = 6$ must be $n = 2^2 \cdot 3 = 12$.

3. We note that $\nu(4) = 3$, $\nu(9) = 3$, $\nu(25) = 3$, $\nu(36) = 9$ are all odd numbers. Hence we may be led to expect that $\nu(n)$ is odd when n is a perfect square. Do you see why?

Exercise. When is $\nu(n) = 3$?

Now we study the function $\sigma(n)$, the sum of the positive divisors of n. This function also depends on the prime factorization of n, so we expect it to be multiplicative. For example, $\sigma(6) = 1 + 2 + 3 + 6 = 12$, $\sigma(7) = 1 + 7 = 8$, and $\sigma(42) = 1 + 2 + 3 + 4 + 6 + 7 + 14 + 21 + 42 = 96 = 8 \cdot 12$. For a prime number p, $\sigma(p) = 1 + p$ and

$$\sigma(p^e) = 1 + p + p^2 + \cdots + p^e = \frac{p^{e+1} - 1}{p - 1}. \tag{1}$$

Although it is possible to derive the formula for $\sigma(n)$ directly, we prefer to show the multiplicativity of σ and then deduce the formula for $\sigma(n)$. Towards this goal, we prove the following proposition, which is useful to prove the multiplicativity of many functions.

Proposition 15.1.8. *If f is a multiplicative function, then $F(n) = \sum_{d \mid n} f(d)$*

is also a multiplicative function.

PROOF. A divisor d of nm can be written uniquely as $d_1 d_2$ where $d_1 \mid n$ and $d_2 \mid m$ and $(d_1, d_2) = 1$. Conversely, if $d_1 \mid n$ and $d_2 \mid m$, then $d_1 d_2 \mid mn$ as $(m, n) = 1$. Using this decomposition, we obtain

$$F(nm) = \sum_{d \mid nm} f(d)$$

$$= \sum_{d_1 \mid n,\, d_2 \mid m} f(d_1 d_2).$$

Using the multiplicativity of f,

$$F(nm) = \sum_{d_1 \mid n,\, d_2 \mid m} f(d_1) f(d_2)$$

$$= \sum_{d_1 \mid n} f(d_1) \sum_{d_2 \mid m} f(d_2)$$

$$= F(n) F(m),$$

which proves the proposition. ∎

Corollary 15.1.9. *The sum of the positive divisors, σ, is a multiplicative function.*

PROOF. This follows by applying the previous proposition with $f(n) = n$ and observing that $\sigma(n) = \sum_{d \mid n} f(d)$. ∎

Corollary 15.1.10. *If $n = p_1^{e_1} p_2^{e_2} \cdots p_k^{e_k}$ is the prime factorization of n, then*

$$\sigma(n) = \frac{p_1^{e_1+1} - 1}{p_1 - 1} \cdots \frac{p_k^{e_k+1} - 1}{p_k - 1}.$$

PROOF. This follows from the multiplicativity of σ and Equation (1). ∎

We note that $\nu(n) = \sum_{d \mid n} 1$, and Proposition 15.1.8 implies that ν is multiplicative. This gives another proof of Corollary 15.1.6.

Example 15.1.11. The proposition implies that $\sum_{d \mid n} \chi(d)$ is multiplicative, where $\chi(n) = 1$ if $n \equiv 1 \pmod 4$; $\chi(-1) = -1$ if $n \equiv -1 \pmod 4$; and 0 otherwise. It turns out that $R_2(n) = \sum_{d \mid n} \chi(d)$. This can be verified by observing that since both sides are multiplicative, it is sufficient to evaluate both sides for prime powers. If $p \equiv 1 \pmod 4$, then $R_2(p^k) = k + 1$ and so is $\sum_{d \mid p^k} \chi(d)$. If $p \equiv 3 \pmod 4$, then $R_2(p^k) = 1$ when k is even and 0 when k is odd. The same is true for $\sum_{d \mid p^k} \chi(d)$. If $p \equiv 1 \pmod 4$, then $\sum_{d|p^k} \chi(d) = k + 1$, because p^k has $k + 1$ divisors, $1, p, p^2, \ldots, p^k$. Similarly, if $p \equiv 3 \pmod 4$, then $\sum_{d|p^k} \chi(d) = 1$ when k is even and 0 when k is odd, because in this case $\chi(p^i) = (-1)^i$. Finally, $R_2(2^k) = 1$ and, because $\chi(1) = 1$ and $\chi(2^i) = 0$ for $i > 1$, the result follows for $\sum_{d|2^k} \chi(d)$. This gives another formula for computing the number of representations of n as a sum of two squares.

Example 15.1.12. 1. **Perfect numbers:** The numbers 6 and 28 were considered special by ancient Greek geometers because these numbers are equal to the sum of their divisors (excluding the number itself in the sum). If $\sigma(n) = 2n$, then n is perfect. Perfect numbers that are even (known as even perfect numbers) were characterized in terms of Mersenne primes by Euclid and Euler (see the projects for Chapter 2). There are 34 perfect numbers known corresponding to the 34 known Mersenne primes. It is conjectured that there are infinitely many even perfect numbers. No odd perfect numbers are known, and it is conjectured that there are none. It is possible to prove that if n is an odd perfect number, then it is has at least 10 prime factors. Many other results are known about odd perfect numbers. See Exercises 10 and 11.

2. **Amicable Numbers:** Two distinct natural numbers m and n are said to form an **amicable pair** if $\sigma(m) = \sigma(n) = m + n$. This means that m is the sum of all the divisors less than n of n, and vice versa. The numbers 220 and 284 form an amicable pair. It is not known if there are infinitely many amicable numbers. Amicable numbers have a long history, with the earliest references occurring in the Bible.[1]

_____ **Exercises for Section 15.1** _____

1. Find the smallest positive solution for each of the equations $\nu(n) = 10$, $\nu(n) = 50$, and $\nu(n) = 100$.

2. Find all n such that $\sigma(n) = 31$ and $\sigma(n) = 48$.

3. (a) Characterize the integers n such that $\nu(n)$ is odd.
   (b) Characterize the integers n such that $\sigma(n)$ is odd.

4. Let f be a multiplicative arithmetical function such that $f(r) \neq 0$ for some r. Show that $f(1) = 1$.

5. Show that if $m \mid n$, then $\phi(mn) = m\phi(n)$.

6. Let $s(n) = \sigma(n) - n$. Show that $s(n) = 5$ has no solutions. Can you find other integers r such that $s(n) = r$ has no solutions? (Such numbers are called **untouchable**.)

[1] L. E. Dickson. *History of the Theory of Numbers.* Vol. 1. New York: Chelsea, 1952.

7. Find a number n such that $\sigma(n)/n = 13/5$.

8. The function $s(n) = \sigma(n) - n$ can be iterated to produce an **aliquot se-quence** $\{s^k(n)\}$, $k = 0, 1, \ldots$. Here $s^2(n) = s(s(n))$, and so on. A number n is said to be **sociable** if its aliquot sequence is cyclic. Write a program to detect if a number is sociable. Verify that 14316 is sociable.

9. The kth power divisor function $\sigma_k(n)$ is defined by $\sigma_k(n) = \sum_{d\mid n} d^k$. Show that σ_k is a multiplicative function. Determine a formula for $\sigma_k(n)$ in terms of the prime factorization of n.

10. Suppose $n = p^a q^b$, where p and q are distinct odd primes. Show that $\sigma(n) < \dfrac{n}{(p-1)(q-1)}$. Estimate the maximum value of $\dfrac{pq}{(p-1)(q-1)}$ to conclude that n cannot be perfect. Hence any odd perfect number must have at least three distinct divisors.

11. Suppose n is an odd perfect number. Show that n is of the form $p^e m^2$, where p is prime and $p \equiv e \equiv 1 \pmod 4$.

12. Determine all n such that $\phi(n) + \sigma(n) = 2n$.

13. Let $k > 1$ and $a = 3 \cdot 2^k - 1$, $b = 3 \cdot 2^{k-1} - 1$, and $c = 3 \cdot 2^{2k-1} - 1$. Suppose a, b, and c are prime. Let $m = 2^k ab$ and $n = 2^k c$. Show that m, n form an amicable pair. Verify that $k = 2$ gives 220 and 284. Find all values of $k \le 100$ that will give amicable pairs by this method. These formulas are due to Thabit ibn Qurra (826–901 A.D.), though he was aware of only one example, 220, 284.

14. If three numbers a, b, and c satisfy $\sigma(a) = b$, $\sigma(b) = c$, and $\sigma(c) = a$, then they are said to form an **amicable triple**. Write a program to find amicable triples.

15. Write a computer program to find integers n such that $\sum_{d\mid n} d$ is a perfect square. (An example of such a number is 22, which satisfies $1 + 2 + 11 + 22 = 36$.)

16. Use a computer to find integers n such that $\sum_{d\mid n, d\neq n} d$ is a perfect cube.

17. Show that if $\sigma(n)$ is a power of 2, then n is a product of distinct Mersenne primes.

18. Show that $f(n) = 2^{\omega(n)}$ is a multiplicative function, where $\omega(n)$ is the number of distinct prime factors of n and $f(1) = 1$.

19. The Möbius function μ is defined by:

$$\mu(1) = 1$$

$$\mu(n) = \begin{cases} 0 & \text{if } d^2 \mid n \text{ for some } d > 1. \\ (-1)^r & \text{if } n \text{ is the product of } r \text{ distinct primes.} \end{cases}$$

(a) Show that μ is multiplicative.

(b) Show that $\sum_{d|n} \mu(d) = \begin{cases} 1 & \text{if } n = 1 \\ 0 & \text{if } n > 1. \end{cases}$

20. Show that

$$\sum_{d|n} \nu^3(d) = \left(\sum_{d|n} \nu(d) \right)^3.$$

21. Show that $\nu(n) \le 2\sqrt{n}$.

22. Determine $\sum_{d|n} \phi(d)$ and $\sum_{d|n} \nu(d)$.

23. Show that $\sum_{d|n} \dfrac{1}{d} = \dfrac{\sigma(n)}{n}$.

24. A number n is said to be **weird** if $\sigma(n) > 2n$ and no subset of divisors of n (excluding n) adds to n. Verify that 70 is a weird number. (It is the smallest weird number.) Show that $70p$ is weird if $p > \sigma(70) = 144$, and p is prime. (This shows that there are infinitely many weird numbers.) Write a computer program to verify if a number is weird. What is the smallest weird number larger than 70?

15.2 Dirichlet Series

To obtain a better understanding of the arithmetical functions including results about their distribution and asymptotic behavior, it is convenient to define their generating functions. These generating functions are the Dirichlet series, first used by Dirichlet in his proof of the infinitude of primes in arithmetic progression. Some examples of these series occur in the work of Euler, who used the divergence of the sum $\sum_{n=1}^{\infty} 1/n$ to prove the infinitude of primes. Euler also used similar techniques to prove the infinitude of primes of the form $4n + 1$ and $4n + 3$.

Definition 15.2.1. If f is an arithmetical function, then the **Dirichlet series** associated to f is

$$F(s) = \sum_{n=1}^{\infty} \frac{f(n)}{n^s},$$

where s is a complex number.

Although we will be concerned primarily with formal aspects of Dirichlet series, it is important to know where the series converge and define differentiable functions of the variable s. For a complex number s, we define n^s in the following way. Write $s = \sigma + it$ with σ and t real numbers. Here σ is the real part, denoted by $\mathrm{Re}(s)$, and t the imaginary part, denoted by $\mathrm{Im}(s)$. Using the properties of the exponential function, we have

$$n^s = e^{s\log n} = e^{\sigma\log n}e^{it\log n} = n^\sigma e^{it\log n},$$

where $e^{ix} = \cos x + i \sin x$ for a real number x. Using the basic trigonometric identities, we find that $\left|e^{ix}\right| = \cos^2 x + \sin^2 x = 1$ and $|n^s| = n^\sigma$.

Example 15.2.2. We evaluate $2^{1+(\pi/4)i}$ by

$$2^1[\cos(\pi\log(2)/4) + i\sin(\pi\log(2)/4)].$$

Definition 15.2.3. We say that $f = O(g)$ if there is a constant C such that $|f(n)| < Cg(n)$ for all n sufficiently large.

The O-notation is useful when we want to describe the principal term of an expression. (See Appendix C.)

Example 15.2.4.   1. We can write $\sin x = O(1)$ as $|\sin x| \le 1$.

2. Since $\phi(n) < n$, we write $\phi(n) = O(n)$.

Proposition 15.2.5. *If $f = O(n^r)$ for some real number r, then $F(s)$ converges in the half-plane $\mathrm{Re}(s) > r + 1$.*

PROOF. We compute

$$
\begin{aligned}
|F(s)| &= \left|\sum_{n=1}^{\infty} \frac{f(n)}{n^s}\right| \\
&\le \sum_{n=1}^{\infty} \frac{|f(n)|}{|n^s|} \\
&\le \sum_{n=1}^{\infty} \frac{n^r}{n^{\mathrm{Re}(s)}} \\
&= \sum_{n=1}^{\infty} \frac{1}{n^{\mathrm{Re}(s)-r}}.
\end{aligned}
$$

Since $\sum_{n=1}^{\infty} 1/n^t$ converges if $t > 1$ (by the integral test), $F(s)$ converges if $\mathrm{Re}(s) - r > 1$ or $\mathrm{Re}(s) > r + 1$.   ∎

It is possible for the series to converge in a larger region than one given by the proposition. For example, we can show that the series defined by $\sum_{n=1}^{\infty}(-1)^n/n^s$ converges for $\mathrm{Re}(s) > 0$. It is not enough to know the convergence of the series. For applications, especially to Dirichlet's Theorem, we must determine if the series converges uniformly.

Definition 15.2.6. A series $\sum_{n=1}^{\infty} a_n$ **converges uniformly** on a closed interval I if for every $\epsilon > 0$ there exists N_0 such that $\left| \sum_{n=k}^{l} a_n \right| < \epsilon$ for all $k, l \geq N_0$ and N_0 is independent of s in I.

Lemma 15.2.7. *Suppose $\sum_{n=1}^{\infty} a_n/n^s$ converges uniformly on a closed interval I; then it converges to a continuous function on I.*

The proof can be found in any standard complex analysis text.[2]

Lemma 15.2.8. *Suppose $\sum a_n/n^s$ is such that $\left| \sum_{n=k}^{l} a_n \right| \leq C$ for some C and all k, l; then the Dirichlet series is uniformly convergent for $\sigma > a$ for any $a > 0$.*

PROOF. Let $A_k = \sum_1^k a_n$ so that $a_n = A_n - A_{n-1}$; then we can write

$$\left| \sum_{n=k}^{l} \frac{a_n}{n^s} \right| = \left| \sum_{n=k}^{l} \frac{A_n - A_{n-1}}{n^s} \right|$$

$$= \left| \frac{A_k - A_{k-1}}{k^s} + \frac{A_{k+1} - A_k}{(k+1)^s} + \cdots + \frac{A_l - A_{l-1}}{k^s} \right|$$

$$= \left| \sum_{n=k}^{l} A_n \left(\frac{1}{n^s} - \frac{1}{(n+1)^s} \right) + \frac{A_l}{l^s} - \frac{A_{k-1}}{k^s} \right|.$$

Note that $|A_n| \leq C$, so we can write

$$\leq C \left| \sum_{n=k}^{l} \left(\frac{1}{n^s} - \frac{1}{(n+1)^s} \right) \right| + \frac{2C}{k^s}$$

$$\leq C \left| \frac{1}{k^s} - \frac{1}{(k+1)^s} + \frac{1}{(k+1)^s} - \frac{1}{(k+2)^s} + \cdots - \frac{1}{(l+1)^s} \right|$$

$$+ \frac{2C}{k^s}$$

$$\leq 3C \left| \frac{1}{k^s} \right| = 3C \left| k^{-s} \right|$$

$$\leq 3C k^{-\sigma} \quad \text{for } s > \sigma.$$

Given $\epsilon > 0$, we choose N so that $3C k^{-\sigma} < \epsilon$ for $k \geq N$, which proves the uniform convergence of the series. ■

---

[2]J. B. Conway. *Functions of One Complex Variable.* New York: Springer–Verlag, 1980.

Corollary 15.2.9. *If the series $\sum_{n=1}^{\infty} a_n/n^s$ is such that the partial sums $\left|\sum_{n=1}^{N} a_n\right|$ are bounded for all N, then it converges uniformly for* $\mathrm{Re}(s) > 0$.

Example 15.2.10.   1. If $f(n) = n$, then the series $\displaystyle\sum_{n=1}^{\infty} \frac{n}{n^s} = \sum_{n=1}^{\infty} \frac{1}{n^{s-1}} = \zeta(s-1)$ converges for $\mathrm{Re}(s) > 2$.

2. The series $\displaystyle\sum_{n=1}^{\infty} \frac{\nu(n)}{n^s}$ and $\displaystyle\sum_{n=1}^{\infty} \frac{\phi(n)}{n^s}$ converge absolutely for $\mathrm{Re}(s) > 2$.
    This is because $\nu(n) < n$ and $\phi(n) < n$.

3. The series $\displaystyle\sum_{n=1}^{\infty} \frac{(-1)^n}{n^s}$ converges uniformly for $\mathrm{Re}(s) > 0$ as the partial sums of the coefficients are bounded.

One reason the Dirichlet series are useful in number theory is due to the following property. We consider the product $F(s)G(s)$ of two Dirichlet series, where $F(s)$ is associated to $f(n)$ and $G(s)$ to $g(n)$. The goal is to express $F(s)G(s)$ as a Dirichlet series $\sum_{n=1}^{\infty} a_n/n^s$. To see how this is accomplished, consider the coefficient of $1/12^s$ in $F(s)G(s)$. To obtain $1/12^s$ as a product of terms from $F(s)$ and $G(s)$, we only take terms that are divisors of 12. We have the following terms

$$\frac{f(1)g(12)}{12^s} + \frac{f(2)g(6)}{2^s 6^s} + \frac{f(3)g(4)}{3^s 4^s} + \frac{f(4)g(3)}{4^s 3^s} + \frac{f(6)g(2)}{6^s 2^s} + \frac{f(12)g(1)}{12^s},$$

and it is clear that these are the only terms that can give a denominator of 12^s in $F(s)G(s)$.

Proposition 15.2.11. *Suppose f and g are arithmetical functions, and $F(s)$ and $G(s)$, the associated Dirichlet series. Then*

$$F(s)G(s) = \sum_{n=1}^{\infty} \sum_{d|n} \frac{f(d)g(n/d)}{n^s}.$$

The equation is valid in the half-plane of absolute convergence of the two series $F(s)$ and $G(s)$.

PROOF. The proof is by writing down the product and rearranging the terms. The rearrangement is justified by the absolute convergence of the two series.

$$\begin{aligned}
F(s)G(s) &= \sum_{d=1}^{\infty} \frac{f(d)}{d^s} \sum_{m=1}^{\infty} \frac{d(m)}{m^s} \\
&= \sum_{d,m} \frac{f(d)g(m)}{d^s m^s}.
\end{aligned} \tag{1}$$

We group terms with the same denominator by letting $dm = n$, to get

$$F(s)G(s) = \sum_{n=1}^{\infty} \sum_{dm=n} \frac{f(d)g(m)}{n^s}$$

$$= \sum_{n=1}^{\infty} \sum_{d|n} \frac{f(d)g(n/d)}{n^s}. \qquad \blacksquare$$

Definition 15.2.12. If f and g are arithmetic functions, we define their **convolution product** $f * g$ by

$$(f * g)(n) = \sum_{d|n} f(n)g(n/d).$$

The proposition above states that the Dirichlet series of a convolution product is the product of the Dirichlet series.

Example 15.2.13.   1. Let $I(n) = 1$ for all n; then $I * I = \nu(n)$.

2. Let $J(n) = n$ for all n; then $I * J = \sigma(n)$.

The convolution allows us to compute the Dirichlet series of many arithmetical functions.

Example 15.2.14.   1. Since $\nu(n) = \sum_{d|n} 1$, the series $\sum_{n=1}^{\infty} \frac{\nu(n)}{n^s}$ can be seen to be equal to $\zeta(s)^2$ by taking f and g to be 1 for all positive integers n.

2. Since $\sigma(n) = \sum_{d|n} d$, we take $f(n) = n$ and $g(n) = 1$ in the above proposition. The series for f is $F(s) = \zeta(s-1)$ and that of g, $\zeta(s)$. Hence

$$\sum_{n=1}^{\infty} \frac{\sigma(n)}{n^s} = \zeta(s-1)\zeta(s).$$

───────────────── **Exercises for Section 15.2** ─────────────────

1. Show that $O(O(g(r))) = O(g(r))$ and $(O(g(r)))^2 = O(g^2(r))$.

2. Show that $\sigma(n) = O(n^2)$. What can you conclude about the region of convergence of $\sum \sigma(n)/n^s$?

3. Determine the Dirichlet series of $\sigma_k(n)$ in terms of $\zeta(s)$. Where does this series converge?

4. Suppose the arithmetical functions f and g have Dirichlet series $F(s)$ and $G(s)$. If $F(s) = \zeta(s)G(s)$, what is the relation between f and g?

5. If f and g are multiplicative functions, prove that their convolution product $f * g$ is also a multiplicative function.

6. Show that $f * g = g * f$.

7. Liouville's function $\lambda(n)$ is defined by

$$\sum_{d|n} \lambda(d) = \begin{cases} 1 & \text{if } n \text{ is a square;} \\ 0 & \text{otherwise.} \end{cases}$$

    (a) What is $\lambda(p^r)$?

    (b) Is λ a multiplicative function?

    (c) Let $\Lambda(s)$ be the Dirichlet series associated to $\lambda(n)$. Show that

$$\zeta(s)\Lambda(s) = \zeta(2s).$$

8. Let $F(s)$ be the Dirichlet series associated to the function $f(n)$. Show that $F'(s)$ (the derivative as a function of s) is a Dirichlet series of the function $f(n) \log n$; thus, we may define the derivative of f to be $f'(n) = f(n) \log n$. Prove that

$$(f + g)' = f' + g'$$

and

$$(f * g)' = f' * g + f * g'.$$

9. Use the fact that $\sum_{d|n} \phi(d) = n$ to determine the Dirichlet series of Euler's ϕ-function.

15.3   Euler Products

Euler was the first to study the series $\zeta(s) = \displaystyle\sum_{n=1}^{\infty} \frac{1}{n^s}$ and its associated infinite

product expansion $\prod_p \left(1 - \frac{1}{p^s}\right)^{-1}$. Since $\zeta(s) \to \infty$ as $s \to 1^+$, Euler

was able to deduce the infinitude of primes by proving that the series $\displaystyle\sum_p \frac{1}{p}$

(obtained by taking the logarithm of the infinite product) diverges, where the sum is over all the primes.

To see how one obtains this product, write the series for $\zeta(s)$ in two parts, a sum over the even numbers and another over the odd numbers. This rearrangement of the terms of the series is justified in the region of absolute convergence.

$$\zeta(s) = 1 + \frac{1}{2^s} + \frac{1}{3^s} + \frac{1}{4^s} + \cdots$$

$$= 1 + \frac{1}{3^s} + \frac{1}{5^s} + \frac{1}{7^s} + \cdots + \frac{1}{2^s} + \frac{1}{4^s} + \frac{1}{6^s} + \cdots .$$

By factoring 2 from the second sum, we get

$$\zeta(s) = 1 + \frac{1}{3^s} + \frac{1}{5^s} + \cdots + \frac{1}{2^s}\left(1 + \frac{1}{2^s} + \frac{1}{3^s} + \cdots\right)$$

$$= 1 + \frac{1}{3^s} + \frac{1}{5^s} + \cdots + \frac{1}{2^s}\zeta(s).$$

∎

Moving the term involving $\zeta(s)$ to the left yields

$$\zeta(s)\left(1 - \frac{1}{2^s}\right) = 1 + \frac{1}{3^s} + \frac{1}{5^s} + \cdots .$$

Now we can group the multiples of 3 on the right and obtain

$$\zeta(s)\left(1 - \frac{1}{2^s}\right) = \frac{1}{3^s}\zeta(s)\left(1 - \frac{1}{2^s}\right) + 1 + \frac{1}{5^s} + \frac{1}{7^s} + \frac{1}{11^s} + \cdots . \tag{1}$$

If we move the term on the right involving zeta to the left, we obtain

$$\zeta(s)\left(1 - \frac{1}{2^s}\right)\left(1 - \frac{1}{3^s}\right) = 1 + \frac{1}{5^s} + \frac{1}{7^s} + \cdots ,$$

where the sum on the right does not involve any multiple of 2 and 3. Next we can factor out the multiples of 5 on the right, and continuing this way, obtain

$$\zeta(s)\prod_p\left(1 - \frac{1}{p^s}\right) = 1;$$

dividing by the product expression, we obtain

$$\zeta(s) = \prod_p\left(1 - \frac{1}{p^s}\right)^{-1}.$$

Exercise. Verify Equation (1).

Exercise. Where do we use the unique factorization into primes in the computation above?

We have to make sense of an infinite product in the computation above. Just as with infinite sums, we define an infinite product as a limit of finite products.

Definition 15.3.1. Suppose $\{a_n\}$ is a sequence of numbers, and let $P_N = \prod_{n=1}^{N} a_n$. If the sequence $\{P_N\}$ converges to a limit P, then we say that $\prod_{n=1}^{\infty} a_n$ converges and is equal to P.

The following proposition shows that the convergence of an infinite product can be determined by the convergence of the series obtained by taking the logarithm.

Proposition 15.3.2. *Let* $\operatorname{Re} a_n > 0$ *for all* $n > 0$. *Then* $\prod_{n=1}^{\infty}$ *converges if and only if the series* $\sum_{n=1}^{\infty} \log a_n$ *converges.*

The proof can be found in any standard complex analysis text.

Definition 15.3.3. If $\operatorname{Re} a_n > 0$ for all $n > 0$, then the infinite product $\prod_{n=1}^{\infty} a_n$ is said to **converge absolutely** if the series $\sum_{n=1}^{\infty} a_n$ converges absolutely.

We will be interested in infinite products indexed by the set of prime numbers. These products are called **Euler products**. The utility of Euler products in number theory is due to the following result.

Theorem 15.3.4. *If f is a multiplicative function and its Dirichlet series converges absolutely for* $\operatorname{Re}(s) > a$, *then*

$$\sum_{n=1}^{\infty} \frac{f(n)}{n^s} = \prod_{p} \left(1 + \frac{f(p)}{p^s} + \frac{f(p^2)}{p^s} + \cdots \right)$$

and the product is also absolutely convergent for $\operatorname{Re}(s) > a$.

PROOF. The proof is similar to the Euler product expansion for $\zeta(s)$ and makes use of the unique factorization into primes for integers and the multiplicativity of f. A formal proof can be given that doesn't require any convergence considerations. We need to know the domain of validity of the Euler product expansion for applications, so it is necessary to pay attention to the convergence. We write $g(n) = f(n)/n^s$. Then g is also a multiplicative function. If the Dirichlet series for f is absolutely convergent, then so is the sum $1 + f(p)/p^s + f(p^2)/p^{2s} + \cdots = 1 + g(p) + g(p^2) + \cdots$, as it is a subsum of an absolutely convergent series. Then the product of these over a finite number of primes is also absolutely convergent;

$$P(x) = \prod_{p \leq x} (1 + g(p) + g(p^2) + \cdots) = \sum_{n \in A} g(n),$$

where A consists of all numbers having all prime factors less than or equal to x. We use the fact that the left-hand side consists of factors of the form $g(p_1^{e_1}) \cdots g(p_k^{e_k}) = g(p_1^{e_1} \cdots p_k^{e_k})$ by the multiplicativity of g. Then

$$\sum_{n=1}^{\infty} g(n) - P(x) = \sum_{n \in B} g(n),$$

where B consists of all numbers with a prime factor larger than x. The series on the left-hand side converges absolutely as it is the Dirichlet series of f. Then, the sum on the right-hand side approaches 0 as $x \to \infty$. This implies that $P(x) \to \sum_{n=1}^{\infty} g(n)$, which proves the Euler product expansion. The absolute convergence of the product then follows from that of the series. ∎

The Euler product representation can be used to determine the Dirichlet series associated to multiplicative functions. It is also true that if the Dirichlet series for an arithmetical function f has an Euler product expansion, then f is multiplicative. This can be proved by using unique factorization into primes, and deriving a formula for $f(n)$ in terms of the prime factors of n. We leave the details of the proof to the reader.

Example 15.3.5. Consider the completely multiplicative function defined by $\chi(p) = 1$ if $p \equiv 1 \pmod{4}$ and $\chi(p) = -1$ if $p \equiv -1 \pmod{4}$. We define $\chi(2)$ to be 0. Then the Euler product of its Dirichlet series can be obtained by

$$\sum_{n=1}^{\infty} \frac{\chi(n)}{n^s} = 1 - \frac{1}{3^s} + \frac{1}{5^s} - \frac{1}{7^s} + \cdots$$

$$= \prod_{p}(1 + \chi(p)p^{-s} + \chi(p^2)p^{-2s} + \cdots)$$

$$= \prod_{p}(1 - \chi(p)p^{-s})^{-1}$$

$$= \prod_{p \equiv 1 (\mathrm{mod}\, 4)} (1 - p^{-s})^{-1} \prod_{p \equiv 3 (\mathrm{mod}\, 4)} (1 + p^{-s})^{-1}.$$

This series was also first considered by Euler to prove the infinitude of primes of the form $4n + 1$ and $4n - 1$. Dirichlet's proof uses generalizations of these series for primes in a general progression $a + kb$.

Example 15.3.6. Since $\nu(n)$ is multiplicative, we obtain

$$\sum_{n=1}^{\infty} \frac{\nu(n)}{n^s} = \prod_{p}\left(1 + \frac{\nu(p)}{p^s} + \frac{\nu(p^2)}{p^{2s}} + \cdots + \frac{\nu(p^k)}{p^{ks}} + \cdots\right).$$

Using $\nu(p^k) = k + 1$, we have

$$\sum_{n=1}^{\infty} \frac{\nu(n)}{n^s} = \prod_{p}\left(1 + \frac{2}{p^s} + \frac{3}{p^{2s}} + \cdots + \frac{k+1}{p^{ks}} + \cdots\right)$$

$$= \prod_{p}\left(1 + 2p^{-s} + 3(p^{-s})^2 + \cdots + (k+1)(p^{-s})^k + \cdots\right)$$

$$= \prod_{p}\left(\frac{1}{(1 - p^{-s})^2}\right)$$

$$= \left(\prod_{p}\frac{1}{(1 - p^{-s})}\right)^2$$

$$= \zeta(s)^2.$$

We use the fact that the sum of the series $1 + 2x + 3x^2 + 4x^3 + \cdots$ is $1/(1-x)^2$, which is proved by differentiating the geometric series, $1 + x + x^2 + x^3 + \cdots = 1/(1-x)$.

We can determine the Dirichlet series for the Euler ϕ-function using the Euler product expansion.

Proposition 15.3.7. *The Dirichlet series for the ϕ-function is*

$$\sum_{n=1}^{\infty} \frac{\phi(n)}{n^s} = \frac{\zeta(s-1)}{\zeta(s)}.$$

PROOF. This can be proved using the above theorem and the value $\phi(p^k) = p^k - p^{k-1}$.

$$\sum_{n=1}^{\infty} \frac{\phi(n)}{n^s} = \prod_{p} \left(1 + \frac{\phi(p)}{p^s} + \frac{\phi(p^2)}{p^{2s}} + \cdots \right)$$

$$= \prod_{p} \left(1 + \frac{p-1}{p^s} + \frac{p^2 - p}{p^{2s}} + \cdots + \frac{p^k - p^{k-1}}{p^{ks}} + \cdots \right)$$

$$= \prod_{p} \left(\sum_{k=0}^{\infty} \frac{p^k}{p^{ks}} - \sum_{k=1}^{\infty} \frac{p^{k-1}}{p^{ks}} \right).$$

We again utilize the formula for the sum of a geometric series to get

$$\sum_{n=1}^{\infty} \frac{\phi(n)}{n^s} = \prod_{p} \left(\frac{1}{1 - p^{-s+1}} - \frac{1}{p} \frac{p^{-s+1}}{(1 - p^{-s+1})} \right)$$

$$= \prod_{p} \frac{1 - p^{-s}}{1 - p^{-s+1}}$$

$$= \prod_{p} (1 - p^{-s}) \prod_{p} \frac{1}{1 - p^{-(s-1)}}$$

$$= \frac{\zeta(s-1)}{\zeta(s)}. \qquad \blacksquare$$

Corollary 15.3.8. *The Euler ϕ-function satisfies*

$$\sum_{d|n} \phi(d) = n.$$

PROOF. This follows from the Dirichlet series of $\phi(n)$ in Proposition 15.3.7. Since $\sum_{n=1}^{\infty} \frac{\phi(n)}{n^s} = \zeta(s-1)/\zeta(s)$, we obtain $\zeta(s)\sum_{n=1}^{\infty} \frac{\phi(n)}{n^s} = \zeta(s-1) = \sum_{n=1}^{\infty} \frac{n}{n^s}$. We apply the convolution product identity (Proposition 15.2.11) to conclude that $n = \sum_{d|n} \phi(d)$. $\qquad \blacksquare$

Recall that we proved this corollary in Chapter 7 as a result of determining elements of various orders. The corollary can also be proved directly by observing that $\sum_{d|n} \phi(d)$ is a multiplicative function, hence it is enough to determine it for prime powers.

Example 15.3.9. To illustrate Corollary 15.3.8, take $n = 24$. Then the set of divisors is $\{1, 2, 3, 4, 6, 8, 12, 24\}$ and the associated ϕ values are $\{1, 1, 2, 2, 2, 4, 4, 8\}$. Then $1 + 1 + 2 + 2 + 2 + 4 + 4 + 8 = 24$ as expected.

————————————— **Exercises for Section 15.3** —————————————

1. Suppose f is an arithmetical function such that

$$\sum_{n=1}^{\infty} \frac{f(n)}{n^s} = \prod_p \left(1 + \frac{f(p)}{p^s} + \frac{f(p^2)}{p^{2s}} + \cdots\right).$$

Show that f is multiplicative.

2. Prove that f is a completely multiplicative function if and only if its Dirichlet series has the Euler product expansion

$$\prod_p \left(1 - f(p)p^{-s}\right)^{-1}.$$

3. Suppose an arithmetic function f has Dirichlet series $1/\zeta(s)$. Determine f by considering the Euler product of $\zeta(s)$.

4. Find the Euler product expansion of $\sum_{n=1}^{\infty} \frac{\sigma(n)}{n^s}$ using the multiplicativity of σ and the formula $\sigma(p^k) = \dfrac{p^{k+1} - 1}{p - 1}$.

5. Show that

$$\sum_{n=1}^{\infty} \frac{\nu(n^2)}{n^s} = \frac{\zeta^3(s)}{\zeta(2s)}$$

and

$$\sum_{n=1}^{\infty} \frac{2^{\omega(n)}}{n^s} = \frac{\zeta^2(s)}{\zeta(2s)}.$$

6. Determine the Dirichlet series of $\sum_{d|n} \chi(d)$.

7. By computing the Euler products of both sides, show that

$$\sum_{d|n} \nu^3(d) = \sum_{d|n} (\nu(d))^3.$$

8. Show that

$$\sum_{n=1}^{\infty} \frac{\nu^2(n)}{n^s} = \frac{\zeta^4(s)}{\zeta(2s)}.$$

9. Suppose $\sum \dfrac{f(n)}{n^s} = \dfrac{\zeta(s)}{\zeta(2s)}$. Determine the Euler product expansion of the right-hand side and use this to describe f. Similarly, describe the arithmetical function arising in $\dfrac{\zeta(s)}{\zeta(ks)}$ for $k > 2$.

10. Show that

$$\sum_{d|n} \nu(d)\phi\left(\frac{n}{d}\right) = \sigma(n).$$

15.4   Möbius Inversion Formula

If $f(n)$ and $g(n)$ are multiplicative functions, then we know that $h = f * g$ is also a multiplicative function. In the case $g(n) = 1$, we have $h(n) = \sum_{d|n} f(d)$ is multiplicative. Our goal is to describe a formula to compute f from h.

Let $F(s)$ and $H(s)$ denote the Dirichlet series associated to f and h. Using Proposition 15.2.11, we have $H(s) = \zeta(s)F(s)$. Thus, to determine $f(n)$, we first find $F(s)$ by dividing by $\zeta(s)$ and obtain

$$F(s) = \frac{1}{\zeta(s)} H(s).$$

Hence our problem is reduced to write $\dfrac{1}{\zeta(s)}$ as a Dirichlet series $\displaystyle\sum_{n=1}^{\infty} \frac{a_n}{n^s}$. Then we can use Proposition 15.2.11 to obtain $f(n)$ as a convolution of the a_n and the $h(n)$. The reciprocal of the ζ-function is determined in the following proposition.

Proposition 15.4.1. *The Dirichlet series expansion of the reciprocal of the ζ-function is*

$$\frac{1}{\zeta(s)} = \sum_{n=1}^{\infty} \frac{\mu(n)}{n^s},$$

where $\mu(n)$ is a multiplicative function determined by the property

$$\sum_{d|n} \mu(d) = \begin{cases} 1 & \text{if } n = 1; \\ 0 & \text{otherwise.} \end{cases} \tag{1}$$

PROOF. The absolute convergence of the series and Euler product expansions of $\zeta(s)$ imply that for $\text{Re}(s) > 1$,

$$\frac{1}{\zeta(s)} = \prod_p (1 - p^{-s}) = \sum_{n=1}^{\infty} \frac{\mu(n)}{n^s},$$

which implies that the function $\mu(n)$ is multiplicative as its Dirichlet series has an Euler product expansion. The series is defined by the Euler product, so it is absolutely convergent for $\text{Re}(s) > 1$. Then using Proposition 15.2.11 for the formula $\zeta(s) \sum_{n=1}^{\infty} \frac{\mu(n)}{n^s} = 1$, we obtain the stated property for $\mu(n)$. ∎

The function $\mu(n)$ is known as the Möbius function. Equation (1) allows us to determine the value of the μ function.

Proposition 15.4.2. *We have $\mu(1) = 1$, and $\mu(n) = (-1)^k$ if n is the product of k distinct primes; otherwise, $\mu(n) = 0$.*

PROOF. Since $\sum \frac{\mu(n)}{n^s} = \frac{1}{\zeta(s)} = \prod_p (1 - p^{-s})$, the function μ is multiplicative. We have $\mu(1) = 1$, $\mu(p) = -1$, and $\mu(p^k) = 0$ for $k > 1$. The value of μ stated in the proposition follows from the multiplicativity. ∎

The Möbius function is nonzero only on squarefree integers.

Example 15.4.3.   1. We have $\mu(24) = \mu(8)\mu(3) = 0$, since $8 = 2^3$. Also, $\mu(30) = \mu(2 \cdot 3 \cdot 5) = (-1)^3 = -1$.

2. To illustrate Equation (1), consider $\sum_{d|12} \mu(d) = \mu(1) + \mu(2) + \mu(3) + \mu(4) + \mu(6) + \mu(12) = 1 - 1 - 1 + 0 + 1 + 0 = 0$.

Möbius Inversion Formula. *Suppose $g(n) = \sum_{d|n} f(d)$ is multiplicative; then*

$$f(n) = \sum_{d|n} \mu(d)g(n/d) = \sum_{d|n} g(d)\mu(n/d)$$

and the function f is multiplicative.

PROOF. This follows from the formula

$$\left(\sum_{n=1}^{\infty}\frac{f(n)}{n^s}\right)\zeta(s) = \sum_{n=1}^{\infty}\frac{g(n)}{n^s}$$

$$\sum_{n=1}^{\infty}\frac{f(n)}{n^s} = \sum_{n=1}^{\infty}\frac{\mu(n)}{n^s}\sum_{n=1}^{\infty}\frac{g(n)}{n^s}.$$

Applying Proposition 15.2.11, we obtain

$$f(n) = \sum_{d|n}\mu(d)g(n/d). \qquad \blacksquare$$

We apply the inversion formula to the functions σ, ν, and ϕ to obtain the following corollaries.

Corollary 15.4.4. *The following formulas hold:*

(a) $\displaystyle\sum_{d|n}\sigma(d)\mu(n/d) = \sum_{d|n}\mu(d)\sigma(n/d) = n.$

(b) $\displaystyle\sum_{d|n}\nu(d)\mu(n/d) = 1.$

PROOF. (a) This is Möbius inversion formula applied to the relation $\sigma(d) = \displaystyle\sum_{d|n}d$. The proof of (b) is similar. $\blacksquare$

Example 15.4.5. Let $n = 24$; its divisors are $\{1,2,3,4,6,8,12,24\}$, and the formula reads

$$\sum_{d|24}\mu(d)\nu(n/d) = \mu(1)\nu(24) + \mu(2)\nu(12) + \mu(3)\nu(8) + \mu(4)\nu(6)$$

$$+ \mu(6)\nu(4) + \mu(8)\nu(3) + \mu(12)\nu(2) + \mu(24)\nu(1)$$
$$= 1\cdot 8 - 1\cdot 6 - 1\cdot 4 + 1\cdot 3$$
$$= 1.$$

Corollary 15.4.6. *The Riemann ζ-function satisfies*

$$\frac{\zeta(s)}{\zeta(2s)} = \sum_{n=1}^{\infty}\frac{|\mu(n)|}{n^s}.$$

PROOF. This follows from the Euler product expansion of the left-hand side and then comparing it with the right-hand side.

$$\frac{\zeta(s)}{\zeta(2s)} = \frac{\prod_p (1 - p^{-s})^{-1}}{\prod_p 1 - p^{-2s})^{-1}}$$

$$= \prod_p \frac{(1 - p^{-s})^{-1}}{(1 - p^{-s})^{-1}(1 + p^{-s})^{-1}}$$

$$= \prod_p (1 + p^{-s})^{-1}$$

$$= \prod_p (1 + |\mu(p)|p^{-s})^{-1},$$

where we use the fact that $|\mu(n)|$ is also a multiplicative function. ∎

——————————— **Exercises for Section 15.4** ———————————

1. Suppose $F(n) = \sum\limits_{d|n} f(d)$. Find an expression for $f(72)$ in terms of the values of F.

2. Show that

$$\phi(n) = \sum_{d|n} d\mu(n/d) = \sum_{d|n} \mu(d)n/d = n \sum_{d|n} \mu(d)/d.$$

3. Determine $\sum\limits_{d|n} \mu(d)\sigma(n/d)$ and $\sum\limits_{d|n} \mu(n/d)\sigma_k(d)$.

4. Show that

$$\sum_{d|n} \mu(d)f(d) = \prod_{k=1}^{r} (1 - f(p_i)),$$

where n has the prime factorization $n = p_1^{e_1} \cdots p_r^{e_r}$. Use this to compute

(a) $\sum\limits_{d|n} |\mu(d)|$ (b) $\sum\limits_{d|n} \mu(d)\nu(d)$ (c) $\sum\limits_{d|n} \mu(d)\sigma(d)$

5. Let $\Lambda(n) = \log p$ if n is a power of a prime p, 0 otherwise. Prove that

$$\log n = \sum_{d|n} \Lambda(d)$$

and

$$\Lambda(n) = -\sum_{d|n} \mu(d) \log(d).$$

6. A series of the form $\sum_{n=1}^{\infty} f(n)x^n/(1-x^n)$ is called the **Lambert series** of $f(n)$. Show that

$$\sum_{n=1}^{\infty} f(n)\frac{x^n}{1-x^n} = \sum_{n=1}^{\infty} g(n)x^n,$$

where $g(n) = \sum_{d|n} f(d)$. Apply this result to determine the Lambert series of $\mu(n)$, n^k, and Liouville's function $\lambda(n)$.

7. Show that

$$\frac{1}{10} = \frac{1}{9} - \frac{1}{99} - \frac{1}{999} - \frac{1}{9999} - \frac{1}{99999} - \cdots .$$

8. Suppose f is a function of a real variable. Define $F(x) = \sum_{n \le x} f(x/n)$, where the sum is over positive integers less than or equal to x. Prove the general Möbius inversion formula for this sum.

$$f(x) = \sum_{n \le x} \mu(n) F(x/n).$$

Answer the following questions.

(a) If $f(x) = 1$ for all x, what is F? What identity does the inversion formula imply?

(b) If $f(x) = x$ for all x, determine F. What identity does the inversion formula imply?

9. Suppose $S(x)$ is the number of positive squarefree integers less than or equal to x. Determine $\sum_{n^2 \le x} S(x/n^2)$. Can you apply an inversion formula to this sum to give a formula for $S(x)$?

10. Show that the number of positive integers less than or equal to x that are perfect powers is

$$\sum_{2^n \le x} \mu(n)(x^{1/n} - 1).$$

15.5 Dirichlet's Theorem

In this section, we will outline a proof of Dirichlet's Theorem that there are infinitely many primes in an arithmetic progressions $5n + a$, $a = 1, 2, 3, 4$. The general proof is similar in spirit.[3] We first discuss Euler's proof of the infinitude of primes.

Theorem 15.5.1. *There are infinitely many primes.*

PROOF. (Euler) We will use the fact that $\lim_{s \to 1+} \zeta(s) = \infty$. Write $\zeta(s)$ as a product

$$\zeta(s) = \prod_p \frac{1}{1 - p^{-s}}$$

$$= \prod_p \left(1 + \frac{1}{p^s} + \frac{1}{p^{2s}} + \cdots\right).$$

Taking logarithms on both sides

$$\log(\zeta(s)) = -\sum_p \log(1 - p^{-s}). \tag{1}$$

We use the Taylor series $-\log(1 - x) = x + \frac{x^2}{2} + \frac{x^3}{3} + \frac{x^4}{4} + \cdots$ in each term of the right-hand side of (1)

$$\log(\zeta(s)) = \sum_p \left((p^{-s}) + \frac{(p^{-s})^2}{2} + \frac{(p^{-s})^3}{3} + \cdots\right).$$

Break the sum into two parts

$$= \sum_p p^{-s} + \sum_p \sum_{m>1} \frac{p^{-ms}}{m}.$$

As s approaches 1 from the right, the second sum converges, since

$$\sum_p \sum_{m=2}^{\infty} \frac{1}{m} p^{-m} < \sum_p \sum_{m=2}^{\infty} p^{-m} = \sum_p \frac{1}{p(p-1)}$$

$$< \sum_{n=1}^{\infty} \frac{1}{n(n-1)} < 1.$$

---

[3]J. P. Serre. *A Course in Arithmetic.* New York: Springer–Verlag, 1973.

$\zeta(s)$ approaches ∞ as s approaches 1 from the right (see Lemma 15.5.2 below), so

$$\lim_{s \to 1+} \sum_p p^{-s} = \infty.$$

This implies the infinitude of primes. ∎

The crux of the proof is to write $\zeta(s)$ as two sums: the first, a sum over the primes, and the second, an absolutely convergent sum. We use that $\lim_{s \to 1+} \zeta(s)$ is infinite and the second series converges, so the sum over the primes must diverge as $s \to 1^+$. Dirichlet used a similar idea, but here one needs to consider many more functions to be able to isolate the primes in a given arithmetic progression.

To justify the limit, that is, that $\lim_{s \to 1+} \zeta(s) = +\infty$, we must prove the following result.

Lemma 15.5.2. *The Riemann ζ-function satisfies*

$$\zeta(s) = \frac{1}{s-1} + \zeta_0(s),$$

where $\zeta_0(s)$ is continuous and differentiable for $s > 0$.

PROOF. Note that

$$\frac{1}{s-1} = \int_1^\infty t^{-s} dt = \sum_{n=1}^\infty \int_n^{n+1} t^{-s}\, dt.$$

Hence we can write

$$\zeta(s) = \frac{1}{s-1} + \sum_{n=1}^\infty \left(\frac{1}{n^s} - \int_n^{n+1} t^{-s}\, dt \right)$$

$$= \frac{1}{s-1} + \sum_{n=1}^\infty \int_n^{n+1} (n^{-s} - t^{-s})\, dt,$$

and each term inside the sum is defined and continuous for $\mathrm{Re}(s) > 0$, and hence the same applies to the sum because of the absolute convergence of the series. ∎

This lemma shows that the limit operations in Euler's proof of the infinitude of primes are justified.

Definition 15.5.3. A Dirichlet character χ modulo m is a complex-valued function such that

$$\chi(ab) = \chi(a)\chi(b)$$
$$\chi(n) = 0 \quad \text{if } (n, m) > 1$$
$$\chi(1) = 1$$
$$\chi(a) = \chi(b) \quad \text{if } a \equiv b \pmod{m}.$$

Example 15.5.4. 1. Modulo 2, all odd numbers are congruent to 1, so there is only one Dirichlet character satisfying

$$\chi(n) = \begin{cases} 1 & \text{if } n \text{ is odd} \\ 0 & \text{if } n \text{ is even.} \end{cases}$$

2. Modulo 3, as $4 \equiv 1 \pmod{3}$, the multiplicativity implies that

$$\chi(2 \cdot 2) = \chi^2(2) = \chi(4)\chi(1) = 1,$$

so $\chi(2)$ must be ± 1. We have two possible characters, χ_0 and χ_1:

$$\chi_0(0) = 0, \qquad \chi_0(1) = 1, \qquad \chi_0(2) = 1$$
$$\chi_1(0) = 0, \qquad \chi_1(1) = 1, \qquad \chi_1(2) = -1.$$

Note that both characters are constant on the arithmetic progressions $3k + r$, $r = 0, 1, 2$.

3. Since 5 is a prime, by Fermat's Theorem, $a^4 \equiv 1 \pmod{5}$ for all $(a, 5) = 1$. Hence $\chi(a^4) = \chi(a)^4 = 1$, that is, $\chi(a)$ is a fourth root of unity. Then the values of χ can only be 1, -1, i, -1, and we can determine all the characters modulo 5 by assigning values to $\chi(2)$ as 2 is a primitive root modulo 5. In this way we obtain the following four characters. For each of the characters x_i, $i = 0, \dots, 3$, we list the values of $x_i(n \bmod 5)$.

n	1	2	3	4	5
χ_0	1	1	1	1	0
χ_1	1	-1	-1	1	0
χ_2	1	i	$-i$	-1	0
χ_3	1	$-i$	i	-1	0

These functions can be extended to all the integers by defining $\chi(n) = \chi(n \bmod 5)$.

Gustav Peter Lejeune Dirichlet (1805–1859)

Dirichlet

Dirichlet was born in 1805, near the city of Cologne. After his early education, he went to college in Cologne, where one of his teachers was the physicist Georg Ohm. For graduate work, Dirichlet went to Paris, the leading mathematical center in the world. There he was influenced by the French mathematicians Fourier, Legendre, Laplace, and Lagrange. After spending many years as a professor in Berlin, Dirichlet succeeded Gauss in Göttingen upon the latter's death.

Dirichlet made fundamental contributions to the theory of Fourier series, being the first to rigorously establish their convergence. He also studied the solutions of Laplace's equation $\partial^2 f / \partial x^2 + \partial^2 f / \partial y^2 + \partial^2 f / \partial z^2 = 0$, in regions where the boundary values of f is specified. The Dirichlet problem, as this is now known, is of great importance in physics. Dirichlet's first love was number theory, kindled by Gauss's great classic, the *Disquisitiones Arithmeticae*. Dirichlet settled the cases $n = 5$ (with Legendre) and $n = 14$ of Fermat's Last Theorem. His most important contribution is the introduction of analytical functions (the Dirichlet series) to prove the infinitude of primes in an arithmetic progression and to derive an explicit formula for the class number of binary quadratic forms.

All the χ's that we have obtained are completely multiplicative functions.

Exercise. Determine the Dirichlet characters modulo 7.

Lemma 15.5.5. *Modulo m, $\chi(n)$ is either 0 or a root of unity.*

PROOF. This follows from Euler's Theorem, since $(a, m) = 1$ implies that $a^{\phi(m)} \equiv 1 \pmod{m}$; therefore, $\chi(a)^{\phi(m)} = 1$, so $\chi(a)$ is a $\phi(m)$th root of unity. ∎

For each Dirichlet character, we define the Dirichlet L-function

$$L(s, \chi) = \sum_{n=1}^{\infty} \frac{\chi(n)}{n^s}.$$

Lemma 15.5.6.   (a) *Let $\chi_0(n) = 1$ for all $(n, m) = 1$ and 0 otherwise. Then $L(s, \chi_0)$ is absolutely convergent in the set $\mathrm{Re}(s) > 1$ and*

$$\lim_{s \to 1+} L(s, \chi_0) = \infty.$$

(b) *If χ is a Dirichlet character modulo p, $\chi \neq \chi_0$, then $L(s, \chi)$ is uniformly convergent for $\mathrm{Re}(s) > 0$.*

(c) $L(s, \chi) = \prod_p \left(1 - \chi(p)p^{-s}\right)^{-1}.$

PROOF.   (a) Let $L_0 = L(s, \chi_0) = \displaystyle\sum_{n, (n,m)=1} \frac{1}{n^s}$. This series is convergent

for $\mathrm{Re}(s) > 1$, as it is a subsum of the series for $\zeta(s)$. The complete multiplicativity of χ_0 implies a convergent Euler product

$$L(s, \chi_0) = \prod_p \left(1 + \frac{\chi(p)}{p^s} + \frac{\chi^2(p)}{p^{2s}} + \cdots\right).$$

If $p \mid m$, then $\chi(p) = 0$, so

$$L(s, \chi_0) = \prod_{p \nmid m} \left(1 - p^{-s}\right)^{-1} = \zeta(s) \prod_{p \mid m} \left(1 - \frac{1}{p^s}\right).$$

Note that on the right-hand side there are a finite number of terms $1 - 1/p^s$ that are defined at $s = 1$. Then Lemma 15.5.2 implies that $L(s, \chi_0)$ approaches ∞ as s approaches 1 from the right.

(b) This follows from the criterion for convergence given in Proposition 15.2.8. There exists a primitive root a modulo p; hence a complete residue system modulo p is given by $0, 1, a, a^2, \ldots, a^{p-1}$. But $\chi^{p-1}(a) = 1$, and $\chi(a) \neq 1$; therefore, $1 + \chi(a) + \chi(a^2) + \cdots + \chi(a^{p-1}) = 0$. Then, $\sum_{n=1}^{N} \chi(n)$ is bounded by p for any N, and the convergence follows.

(c) The Euler product expansion follows from the convergence and complete multiplicativity of χ.   ∎

Now we prove a special case of Dirichlet's theorem. The general case is similar, but requires more care in handling all the Dirichlet characters.

Theorem 15.5.7. *There are infinitely many primes of the form $5n+1$, $5n+2$, $5n + 3$, and $5n + 4$.*

PROOF. As in Euler's proof, we want a sum $\sum 1/p^s$, where the primes are in a progression $5n + a$. To achieve such a sum we use the logarithms of the L-functions of the Dirichlet characters modulo 5, and take suitable linear combinations of them.

The corresponding Euler product expansions are

$$L_0(s) = (1 - 5^{-s})\zeta(s) = \prod_{p \neq 5}(1 - p^{-s})^{-1}$$

$$L_1(s) = L(s, \chi_1) = \prod_{\substack{p \equiv 1(\bmod 5) \\ p \equiv 4(\bmod 5)}}(1 - p^{-s})^{-1} \prod_{\substack{p \equiv 2(\bmod 5) \\ p \equiv 3(\bmod 5)}}(1 + p^{-s})^{-1}$$

$$L_2(s) = L(s, \chi_2) = \prod_{p \equiv 1(\bmod 5)}(1 - p^{-s})^{-1} \prod_{p \equiv 2(\bmod 5)}(1 - ip^{-s})^{-1}$$
$$\prod_{p \equiv 3(\bmod 5)}(1 + ip^{-s})^{-1} \prod_{p \equiv 4(\bmod 5)}(1 + p^{-s})^{-1}$$

$$L_3(s) = L(s, \chi_3) = \prod_{p \equiv 1(\bmod 5)}(1 - p^{-s})^{-1} \prod_{p \equiv 2(\bmod 5)}(1 + ip^{-s})^{-1}$$
$$\prod_{p \equiv 3(\bmod 5)}(1 - ip^{-s})^{-1} \prod_{p \equiv 4(\bmod 5)}(1 + p^{-s})^{-1}.$$

We write C.S. for a convergent series whose actual form is not important for our proof. The logarithms of the L-functions are

$$\log L_0(s) = \sum_p \frac{1}{p^s} + \text{C.S.}$$

$$\log L_1(s) = \sum_{p \equiv 1,4(\bmod 5)} \frac{1}{p^s} + \sum_{p \equiv 2,3(\bmod 5)} \frac{-1}{p^{-s}} + \text{C.S.}$$

$$\log L_2(s) = \sum_{p \equiv 1(\bmod 5)} \frac{1}{p^s} + \sum_{p \equiv 2(\bmod 5)} \frac{1}{ip^s} + \sum_{p \equiv 3(\bmod 5)} \frac{-1}{ip^s}$$
$$+ \sum_{p \equiv 4(\bmod 5)} \frac{-1}{p^{-s}} + \text{C.S.}$$

$$\log L_3(s) = \sum_{p \equiv 1(\bmod 5)} \frac{1}{p^s} + \sum_{p \equiv 2(\bmod 5)} \frac{-1}{ip^{-s}} + \sum_{p \equiv 3(\bmod 5)} \frac{1}{ip^{-s}}$$
$$+ \sum_{p \equiv 4(\bmod 5)} \frac{-1}{p^{-s}} + \text{C.S.}$$

We are going to manipulate these series to obtain sums of $1/p^s$ over primes congruent to $1, 2, 3, 4 \bmod 5$, plus some term that is convergent at $s = 1$. (We assume that we are working in the region of absolute convergence $\mathrm{Re}(s) > 1$, where these manipulations and rearrangement of terms are justified.)

$$\log L_0 + \log L_1 + \log L_2 + \log L_3 = 4 \sum_{p \equiv 1 (\mathrm{mod}\, 5)} \frac{1}{p^s} + \text{C.S.} \qquad (1)$$

$$\log L_0 + \log L_1 - \log L_2 - \log L_3 = 4 \sum_{p \equiv 4 (\mathrm{mod}\, 5)} \frac{1}{p^s} + \text{C.S.} \qquad (2)$$

To obtain the other two cases, consider

$$\log L_2 - \log L_3 = \sum_{p \equiv 2 (\mathrm{mod}\, 5)} \frac{1}{ip^s} + \sum_{p \equiv 3 (\mathrm{mod}\, 5)} \frac{-1}{ip^s} + \text{C.S.}$$

$$\log L_0 - \log L_1 = \sum_{p \equiv 2 (\mathrm{mod}\, 5)} \frac{1}{p^s} + \sum_{p \equiv 3 (\mathrm{mod}\, 5)} \frac{1}{p^s} + \text{C.S.}$$

Then we have

$$\log L_0 - \log L_1 + i(\log L_2 - \log L_3) = 2 \sum_{p \equiv 2 (\mathrm{mod}\, 5)} \frac{1}{p^s} + \text{C.S.} \qquad (3)$$

$$\log L_0 - \log L_1 - i(\log L_2 - \log L_3) = 2 \sum_{p \equiv 3 (\mathrm{mod}\, 5)} \frac{1}{p^s} + \text{C.S.} \qquad (4)$$

The infinitude of primes of each type above would follow if we can show that the expressions on the left-hand side approach ∞ as $s \to 1^+$, since the right-hand side involves the desired term and a convergent sum. The left-hand side can be written as a logarithm of products involving L_1, L_2, L_3, and L_4. For example, in Equation (2), we can write the left-hand side as $\log[(L_0 e^i L_2)/(L_1 L_3)]$. Since $\zeta(s) \to \infty$ as $s \to 1^+$, $L_0 \to \infty$ as $s \to 1^+$. If we show that L_1, L_2, and L_3 are not zero as $s \to 1^+$; then each term on the left will approach ∞. To show that L_1, L_2, and L_3 are nonzero as $s \to 1^+$, consider the product $L_0 L_1 L_2 L_3$. It is not difficult to verify (using the Euler product expansions) that

$$L_0 L_1 L_2 L_3 = \prod_{p \equiv 1 (\mathrm{mod}\, 5)} (1 - p^{-s})^4 \prod_{p \equiv 2,3 (\mathrm{mod}\, 5)} (1 - p^{-4s})^{-1}$$

$$\times \prod_{p \equiv 4 (\mathrm{mod}\, 5)} (1 - p^{-2s})^2. \qquad (5)$$

It can be seen by expanding the Euler product that this is a Dirichlet series with positive coefficients. Using Lemma 15.5.2, we write

$$\zeta(s) = \frac{1}{s-1} + \zeta_0(s),$$

where $\zeta_0(s)$ is defined at $s = 1$.

Suppose $L_2(1)$ is zero. Then we can write $L_2(s) = (s-1)f(s)$ for some differentiable function f.[4] We can write $L_0(s) = \zeta(s)(1 - 1/5^s)$. If we denote $h(s) = 1 - 1/5^s$, then the left-hand side is

$$L_0 L_1 L_2 L_3 = h(s)(f(s)L_1 L_3 + \zeta_0(s)L_1 L_2 L_3),$$

which is defined for all s such that $\mathrm{Re}(s) > 0$. We derive a contradiction by showing that $L_0 L_1 L_2 L_3$ cannot be defined for all s satisfying $\mathrm{Re}(s) > 0$.

Take s to be a positive real number. Then $p^{4s} > p^s$. Hence $p^{-s} > p^{-4s}$. This implies that $1 - p^{-s} < 1 - p^{-4s}$, that is, $(1 - p^{-s})^{-1} > (1 - p^{-4s})^{-1}$. Similarly, $(1 - p^{-2s})^{-1} > (1 - p^{-4s})^{-1}$. All the terms in the Euler product of Equation (6) are greater than 1, so the following inequalities are justified.

$$L_0 L_1 L_2 L_3 = \prod_{p\equiv 1(\mathrm{mod}\,5)} (1 - p^{-s})^{-4} \prod_{p\equiv 2,3(\mathrm{mod}\,5)} (1 - p^{-4s})^{-1}$$

$$\times \prod_{p\equiv 4(\mathrm{mod}\,5)} (1 - p^{-2s})^{-2}$$

$$\geq \prod_{p\equiv 1(\mathrm{mod}\,5)} (1 - p^{-4s})^{-4} \prod_{p\equiv 2,3(\mathrm{mod}\,5)} (1 - p^{-4s})^{-1}$$

$$\times \prod_{p\equiv 4(\mathrm{mod}\,5)} (1 - p^{-4s})^{-2}$$

$$\geq \prod_{p\equiv 1(\mathrm{mod}\,5)} (1 - p^{-4s})^{-1} \prod_{p\equiv 2,3(\mathrm{mod}\,5)} (1 - p^{-4s})^{-1}$$

$$\times \prod_{p\equiv 4(\mathrm{mod}\,5)} (1 - p^{-4s})^{-1}$$

$$\geq \zeta(4s).$$

We know $\zeta(4s)$ diverges at $s = 1/4$, hence $L_0 L_1 L_2 L_3$ cannot be convergent for $\mathrm{Re}(s) > 0$, contradicting the fact that $L_0 L_1 L_2 L_3$ was convergent for all $s > 0$. This contradicts the assumption that L_2 was zero at 1. A similar contradiction can be derived if L_1 or L_3 are zero at $s = 1$. Hence all these are

---

[4] See any textbook on complex analysis for a proof of this fact.

nonzero. This shows that the left-hand sides of Equations (1)–(4) approach infinity as $s \to 1$. Hence the sum of the reciprocal of the primes $\sum 1/p$, over each of the arithmetic progressions, diverges. Hence there are infinitely many primes in each progression $5n + 1, 5n + 2, 5n + 3$, and $5n + 4$. ∎

—————————— **Exercises for Section 15.5** ——————————

1. Verify the following identity appearing in Theorem 15.5.7:

$$L_0 L_1 L_2 L_3 = \prod_{p \equiv 1 (\mathrm{mod}\, 5)} \left(1 - p^{-s}\right)^{-4} \prod_{p \equiv 2,3 (\mathrm{mod}\, 5)} \left(1 - p^{-4s}\right)^{-1}$$
$$\times \prod_{p \equiv 4 (\mathrm{mod}\, 5)} \left(1 - p^{-2s}\right)^{-2}$$

2. What are the Dirichlet characters modulo 4? Imitate the proof of Theorem 15.5.7 to show the infinitude of primes in the arithmetic progressions $4n+1$ and $4n + 3$.

3. (a) Prove that there are four characters modulo 8. Show that the values of the character for odd values of n mod 8 are as shown below.

n	1	3	5	7
χ_0	1	1	1	1
χ_1	1	−1	1	−1
χ_2	1	1	−1	−1
χ_3	1	−1	−1	1

   (b) Show that $L(s, \chi_0) = (1 - 2^{-s}) \zeta(s)$.
   (c) Determine the Euler product expansion of $L(s, \chi_i)$ for $i = 1, 2, 3$.
   (d) Let $L_i = L(s, \chi_i), i = 0, 1, 2, 3$. Determine $L_0 L_1 L_2 L_3$.
   (e) Obtain identities involving the L_i's to separate the following series: $\sum_{p \equiv a (\mathrm{mod}\, 8)} 1/p^s, a = 1, 3, 5, 7$. For example,

   $$\frac{L_2 L_3}{L_0 L_1} = \prod_{p \equiv 5 (\mathrm{mod}\, 8)} \frac{(1 - p^{-s})^4}{(1 - p^{-2s})^2}.$$

   (f) Show the infinitude of primes in each of the arithmetic progressions $8n + 1, 8n + 3, 8n + 5$, and $8n + 7$.

Chapter 16

Distribution of Primes

16.1   Counting Primes

The problem of understanding the distribution of the prime numbers has been an important one since ancient times. Euclid showed that there are infinitely many primes. While numerous tables of primes were constructed, nothing else was proved about the primes until the middle of the nineteenth century, pointing to the great difficulty of the subject. Knowledge of the distribution of primes is of fundamental importance in number theory. The majority of results in the theory are accessible only by the use of complex function theory.[1] There are some results that do not require complex analysis, and we study a few such theorems in this chapter. We will obtain some estimates on the prime number function and prove Bertrand's postulate that there is a prime number between a natural number n and $2n$.

We denote by $\pi(x)$ the number of primes less than or equal to x. There are large gaps in the sequence of primes, and it is also expected that there are infinitely many twin primes. This illustrates that $\pi(x)$ is not a simple function to analyze. There are two problems associated with $\pi(x)$. One is to compute $\pi(x)$ for a given value of x (say 10^6 or 10^8) without listing all the primes. The second is to describe the laws governing $\pi(x)$. The first problem is easier to solve, though the actual computation can be time consuming. The second problem is more difficult, and we discuss some results giving bounds on $\pi(x)$ for large x.

The first person to give a formula for $\pi(x)$ without listing all the primes was Legendre. His formula is based on the Principle of Inclusion–Exclusion.

Principle of Inclusion–Exclusion.   *Let $S_1, \ldots, S_k$ be subsets of a set S. The*

---

[1]See T. M. Apostol. *Introduction to Analytic Number Theory.* New York: Springer–Verlag, 1975.

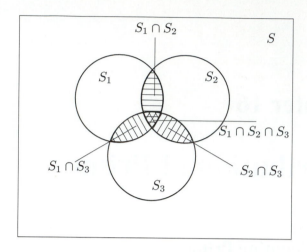

Figure 16.1.1 Inclusion–Exclusion with three subsets

number of elements of S that are not in any of the S_i is

$$|S| - \sum_{1 \le i \le k} |S_i| + \sum_{1 \le i < j \le k} |S_i \cap S_j|$$

$$- \sum_{1 \le i < j < l \le k} |S_i \cap S_j \cap S_l| + \cdots + (-1)^k |S_1 \cap \cdots \cap S_k|.$$

PROOF. Suppose $x \in S$ appears in exactly t sets, $S_1, \ldots, S_t$ for $t \ge 1$. Then x appears once in $|S|$, t times in $\sum_{1 \le i \le k} |S_i|$, $\binom{t}{2}$ times in $\sum_{1 \le i < j \le k} |S_i \cap S_j|$, and so on. It doesn't appear in any intersections of more than t subsets. Then the contribution of x to the sum is

$$1 - \binom{t}{1} + \binom{t}{2} + \cdots + (-1)^t \binom{t}{t},$$

which is $(1 - 1)^t = 0$ by the Binomial Theorem. Therefore, only the x's not in any of the S_i make a nonzero contribution to the sum, and these are counted once in the first term. ∎

Corollary 16.1.1 (Legendre's Formula). *Let $p_1, \ldots, p_k$ be the primes less than or equal to $\sqrt{x}$; then*

$$\pi(x) - \pi(\sqrt{x}) + 1 = x - \sum_{p_i \le \sqrt{x}} \left\lfloor \frac{x}{p_i} \right\rfloor + \sum_{p_i < p_j \le \sqrt{x}} \left\lfloor \frac{x}{p_i p_j} \right\rfloor - \cdots .$$

PROOF. Let S_i be the set of multiples of p_i (including p_i) less than or equal to x, and let S be the set integers from 1 to x. The primes between $\sqrt{x}$ and x are precisely the elements of S that are not in any of the S_i, as there cannot be a composite number less than x whose prime factors are all greater than $\sqrt{x}$. By the Principle of Inclusion–Exclusion, we have

$$\pi(x) - \pi(\sqrt{x}) + 1 =$$
$$|S| - \sum_{1 \le i \le k} |S_i| + \sum_{1 \le i < j \le k} |S_i \cap S_j| + \cdots + (-1)^k |S_1 \cap \cdots \cap S_k| \quad (1)$$

The term 1 on the left-hand side appears because 1 is not a prime and is not in any of the S_i. To compute $|S_{i_1} \cap \cdots \cap S_{i_r}|$, recall that the number of multiples of d less than or equal to x is $\lfloor x/d \rfloor$. Hence $|S_{i_1} \cap \cdots \cap S_{i_r}| = \lfloor x/p_{i_1} \cdots p_{i_r} \rfloor$. Substituting in (1), we obtain the desired result. ∎

Example 16.1.2. Let's compute $\pi(100)$ using Legendre's formula. There are four primes (2, 3, 5, and 7) up to $\sqrt{100} = 10$. Using Legendre's formula,

$$\pi(100) - \pi(10) + 1 = 100 - \left\lfloor \frac{100}{2} \right\rfloor - \left\lfloor \frac{100}{3} \right\rfloor - \left\lfloor \frac{100}{5} \right\rfloor - \left\lfloor \frac{100}{7} \right\rfloor$$
$$+ \left\lfloor \frac{100}{6} \right\rfloor + \left\lfloor \frac{100}{10} \right\rfloor + \left\lfloor \frac{100}{14} \right\rfloor$$
$$+ \left\lfloor \frac{100}{15} \right\rfloor + \left\lfloor \frac{100}{21} \right\rfloor + \left\lfloor \frac{100}{35} \right\rfloor$$
$$- \left\lfloor \frac{100}{30} \right\rfloor - \left\lfloor \frac{100}{42} \right\rfloor - \left\lfloor \frac{100}{70} \right\rfloor - \left\lfloor \frac{100}{105} \right\rfloor$$
$$+ \left\lfloor \frac{100}{210} \right\rfloor.$$

Hence

$$\pi(100) - 3 = 100 - 50 - 33 - 20 - 14$$
$$+ 16 + 10 + 7 + 6 + 4 + 2$$
$$- 3 - 2 - 1 - 0$$
$$+ 0,$$

which implies that $\pi(100) = 25$.

Exercise. Write down the terms that occur in Legendre's formula to calculate $\pi(200)$.

Legendre's formula is very time consuming to carry out in practice. It can be implemented on a computer to calculate $\pi(x)$ for x up to 10^{10}. The formula requires a table of primes that can be computed using the Sieve of Eratosthenes. There are many improvements of the formula due to Meissel and Lehmer, which are more complicated to describe but are faster in practice. An excellent treatment of these can be found in Riesel's book.[2]

——————————— **Exercises for Section 16.1** ———————————

 1. Write a computer program to implement Legendre's formula. Use it to compute $\pi(10^7)$, $\pi(10^8)$, and $\pi(10^9)$.

2. Let $p_1, \ldots, p_k$ be the distinct prime factors of n. Use the Principle of Inclusion–Exclusion to derive a formula for $\phi(n)$.

3. Use the Principle of Inclusion–Exclusion to give a formula for $S(x)$, the number of squarefree integers less than or equal to x.

4. A permutation of n letters $1, 2, \ldots, n$ is called a **derangement** if it does not fix any letter; that is, if f is the permutation viewed as a one-to-one function from n letters to n letters, then there is no i such that $f(i) = i$. Determine the number of permutations fixing a single letter, two letters, three letters, and so on. Use this and the Principle of Inclusion–Exclusion to show that the number of derangements of n letters is $\sum\limits_{i=1}^{n} (-1)^i \dfrac{n!}{i!}$.

16.2 Chebyshev's Functions

Chebyshev was the first mathematician to prove something nontrivial about the number of primes. He showed that there exists constants c and C, explicitly determined, such that $c(x/\log x) < \pi(x) < C(x/\log x)$. Moreover, Chebyshev showed that if $disp\dfrac{\pi(x)}{x/\log x}$ has a limit, then it must be 1. To prove his theorem, Chebyshev defined two functions that are easier to analyze than $\pi(x)$. The statements of the prime number theorem or any estimates regarding $\pi(x)$ can be given for these functions, and then the corresponding formula is derived for $\pi(x)$. We use the convention that $\sum_{p \leq x}$ is the sum over the prime numbers less than or equal to x.

Definition 16.2.1. The functions $\Theta(x)$ and $\psi(x)$ are defined as

$$\Theta(x) = \sum_{p \leq x} \log p$$

$$\psi(x) = \sum_{p^m \leq x} \log p$$

—————————————
[2]H. Riesel. *Prime Numbers and Computer Methods of Factorization.* Boston: Birkhauser, 1985.

where the second sum is over all primes and positive integers m such that $p^m \leq x$.

Example 16.2.2.   1. $\Theta(12) = \log 2 + \log 3 + \log 5 + \log 7 + \log 11$.

2. Since 2, 2^2, 2^3, 3, 3^2, 5, 7, and 11 are the prime powers less than 12, $\psi(12) = 3 \log 2 + 2 \log 3 + \log 5 + \log 7 + \log 11$.

3. $\psi(30) = 4 \log 2 + 3 \log 3 + 2 \log 5 + \log 7 + \log 11 + \log 13 + \log 17 + \log 19 + \log 23 + \log 29 = 28.4765$.

The functions Θ and ψ are step functions with jumps at the primes.

Lemma 16.2.3. *The functions $\psi(x)$ and $\Theta(x)$ satisfy*

$$\psi(x) = \sum_{p \leq x} \left\lfloor \frac{\log x}{\log p} \right\rfloor \log p = \sum_{p} \left\lfloor \frac{\log x}{\log p} \right\rfloor \log p, \tag{1}$$

$$\psi(x) = \Theta(x) + \Theta(x^{1/2}) + \Theta(x^{1/3}) + \Theta(x^{1/4}) + \cdots . \tag{2}$$

PROOF. If p^m is less than x, then $m < \log x / \log p$. The largest power of p, p^r, which is less than or equal to x, has the exponent $r = \lfloor \log x / \log p \rfloor$. This proves (1) for ψ, as we count the $\log p$ factor r times in the sum. If $p > x$, then $\lfloor \log x / \log p \rfloor = 0$, so we can drop the restriction on p in the sum.

For the proof of (2), we observe that if $\log p$ is included in the sum for $\Theta(x^{1/k})$, then $p \leq x^{1/k}$, that is, $p^k \leq x$. Hence the largest value of k for which $\log p$ occurs in the expression $\Theta(x^{1/k})$ is $k = \lfloor \log x / \log p \rfloor$. This, together with (1), proves the second formula for $\psi(x)$. ∎

The relation between the functions Θ, ψ and $\pi(x)$ is given by the following lemma. Using these, we find that the Prime Number Theorem is equivalent to showing that $\psi(x) \sim x$, where the symbol $f(x) \sim g(x)$ means that the ratio of the functions approaches 1 as $x \to \infty$.

Lemma 16.2.4. *The functions Θ and ψ satisfy*

$$\Theta(x) \leq \psi(x) \leq \pi(x) \log(x).$$

PROOF. $\Theta(x) \leq \psi(x)$ is clear. Using lemma 16.2.3 we get

$$\psi(x) = \sum_{p \leq x} \left\lfloor \frac{\log x}{\log p} \right\rfloor \log p \leq \sum_{p \leq} \frac{\log x}{\log p} \log p = \pi(x) \log x.$$

∎

Chebyshev proved his theorem for ψ. Once inequalities are obtained for ψ, then $\Theta(x)$ can be estimated using the inequalities in Lemma 16.2.5. From Θ, a lower bound for $\pi(x)$ follows easily; to get an upper bound, we will use the Abel summation method.

Lemma 16.2.5. $\Theta(x)$ *is bounded above and below by*

$$\psi(x) - 2\psi(\sqrt{x}) \leq \Theta(x) \leq \psi(x) - \psi(\sqrt{x}).$$

PROOF. This follows from lemma 16.2.3 by writing

$$\begin{aligned}
\psi(x) - 2\psi(\sqrt{x}) &= \Theta(x) + \Theta(x^{1/2}) + \Theta(x^{1/3} + \cdots \\
&\quad - 2\Theta(x^{1/2} - 2\Theta(x^{1/4})_2\Theta(x^{1/6}) - \cdots \\
&= \Theta(x) - \Theta(x^{1/2}) + \Theta(x^{1/3}) - \Theta(x^{1/4}) - \cdots .
\end{aligned}$$

The series is alternating with decreasing terms, so the lower bound on Θ follows. The upper bound on $\Theta(x)$ can be obtained by looking at $\psi(x) - \psi(\sqrt{x})$ and using Lemma 16.2.3. ∎

Lemma 16.2.6. *Let* $T(x) = \log(\lfloor x \rfloor!)$, *the logarithm of the factorial of the greatest integer less than or equal to* x. *Then,*

$$T(x) = \psi(x) + \psi(x/2) + \psi(x/3) + \psi(x/4) + \cdots$$

PROOF. The function $T(x)$ is easy to compute, as we know the prime factorization of $\lfloor x \rfloor!$. Suppose $p \leq x$; then the exponent e of p in the prime factorization is given by

$$e = \left\lfloor \frac{\lfloor x \rfloor}{p} \right\rfloor + \left\lfloor \frac{\lfloor x \rfloor}{p^2} \right\rfloor + \left\lfloor \frac{\lfloor x \rfloor}{p^3} \right\rfloor + \cdots . \tag{3}$$

Using $\lfloor \lfloor x \rfloor / n \rfloor = \lfloor x/n \rfloor$, we can write this as

$$e = \left\lfloor \frac{x}{p} \right\rfloor + \left\lfloor \frac{x}{p^2} \right\rfloor + \left\lfloor \frac{x}{p^3} \right\rfloor + \cdots .$$

This series is finite and the last nonzero term is $\lfloor x/p^r \rfloor$, where r is the integral part of $\log x / \log p$. Therefore,

$$T(x) = \sum_{p \leq x} \left(\left\lfloor \frac{x}{p} \right\rfloor + \left\lfloor \frac{x}{p^2} \right\rfloor + \cdots \right) \log p.$$

Using Lemma 16.2.3, we can write the left-hand side as

$$\Theta(x) + \Theta(x^{1/2}) + \Theta(x^{1/3}) + \Theta(x^{1/4}) + \cdots$$
$$\Theta(x/2) + \Theta((x/2)^{1/2}) + \Theta((x/2)^{1/3}) + \cdots$$
$$\Theta(x/3) + \Theta((x/3)^{1/2}) + \Theta((x/3)^{1/3}) + \cdots$$
$$\cdots$$

We count the number of times $\log p$ occurs in this expression. In the first row $\log p$ occurs in the first r terms, from $\Theta(x)$ to $\Theta(x^{1/r})$. Hence $\log p$ cannot occur in more columns than the first r. In the first column, $\log p$ occurs in $\Theta(x/k)$ if $p \leq x/k$, that is, $k \leq x/p$. Hence the largest value of k is $\lfloor \lfloor x \rfloor / p \rfloor$, the first term in Equation (3). In the jth column, $\log p$ occurs in $\Theta((x/k)^{1/j})$ when $p < (x/k)^{1/j}$, that is, $p^j < x/k$ or $k < x/p^j$. The largest such value is $\lfloor x/p^j \rfloor$. This is the jth term in (3). This proves the Lemma. ∎

The function $T(x)$ is easy to describe and has a good estimate given by Stirling's formula.[3]

Proposition 16.2.7 (Stirling's Formula). *The following inequalities are true for $n \geq 3$:*

$$\log(n!) < n \log n - n + \frac{1}{2} \log n + 1$$
$$\log(n!) > n \log n - n + \frac{1}{2} \log n.$$

Corollary 16.2.8. *For any $x \geq 3$,*

(a) $\log(\lfloor x \rfloor !) < x \log x - x + \frac{1}{2} \log x + 2$

(b) $\log(\lfloor x \rfloor !) > x \log x - x - \frac{1}{2} \log x + 1.$

PROOF. (a) This follows from Proposition 16.2.7 by using $x - 1 < \lfloor x \rfloor \leq x$.

(b) Using the second inequality in Proposition 16.2.7, we obtain

$$\log(\lfloor x \rfloor !) > (x - 1) \log(x - 1) - x + \frac{1}{2} \log(x - 1)$$
$$= x \log(x - 1) - x - \frac{1}{2} \log(x - 1)$$
$$> x \log(x - 1) - x - \frac{1}{2} \log x.$$

---

[3]For a proof of Stirling's formula, see A. Blanchard. *Initiation à la Théorie Analytique des Nombres Premiers*. Paris: Dunod, 1969.

We have to prove that $x \log(x - 1) \geq x \log s + 1$, or, equivalently, $x \log((x - 1)/x) \geq 1$.

$$x \log \left(1 - \frac{1}{x} \right) = x \left(\frac{1}{x} + \frac{1}{2x^2} + \frac{1}{3x^3} + \cdots \right)$$

$$= 1 + \frac{1}{2x} + \frac{1}{3x^2} + \cdots \geq 1.$$

This completes the proof.   ∎

Lemma 16.2.6 gives $T(x)$, which can be estimated using Stirling's formula, as a series in $\psi(x)$. We are interested in $\psi(x)$, so we must express ψ in terms of the simpler function $T(x)$. This is accomplished by the Möbius inversion formula.

Proposition 16.2.9. *The function $\psi(x)$ has the following formula:*

$$\psi(x) = \sum_{n=1}^{\infty} \mu(n) T(x/n).$$

PROOF. Since $\lfloor x/n \rfloor = \lfloor \lfloor x \rfloor /n \rfloor$, we have $T\left(\lfloor x/n \rfloor\right) = T\left(\lfloor \lfloor x \rfloor /n \rfloor\right)$. We can effectively assume that x is an integer. We prove the result by induction on x. The equation is clearly true for $x = 2$ and $x = 3$. For larger x, we may assume that the formula is valid for $x/2, x/3, \ldots$. Then using Lemma 16.2.6, we obtain

$$\psi(x) = T(x) - \psi(x/2) - \psi(x/3) - \cdots$$

$$= T(x) - \sum_{k=2}^{\infty} \psi(x/k).$$

Using the induction hypothesis, we can write the terms involving ψ on the right in terms of T:

$$T(x) - \sum_{k=2}^{\infty} \sum_{d=1}^{\infty} \mu(d) T(x/kd).$$

We can simplify the sum by grouping all the terms with $T(x/n)$ together for a given n. Then $kd = n$ and $k \geq 2$ imply that d is a divisor of d but not equal to n. Therefore, the sum can be simplified as

$$\psi(x) = T(x) - \sum_{n=2}^{\infty} T(x/n) \sum_{d|n, d<n} \mu(d),$$

and from the identity $\sum_{d|n} \mu(d) = 0$ for $n \geq 2$, the inner sum is equal to $-\mu(n)$; this yields

$$\psi(x) = T(x) - \sum_{n=2}^{\infty} T(x/n)(-\mu(n))$$

$$= \sum_{n=1}^{\infty} \mu(n)T(x/n),$$

which is the desired formula for $\psi(x)$. ∎

───────── **Exercises for Section 16.2** ─────────

1. Show that $\pi(x) = \sum_{2 \leq n \leq x} \dfrac{\Theta(n) - \Theta(n-1)}{\log n}$.

2. Show that $\psi(x) - \psi(\sqrt{x}) = \Theta(x) + \Theta(x^{1/3}) + \Theta(x^{1/5}) + \cdots$.

3. Show that $T(\pi(x)) < \Theta(x)$.

4. Show that $\Theta(x) = \sum_{2^n \leq x} \mu(n)\psi(x^{1/n})$.

16.3 Chebyshev's Theorem

We use Proposition 16.2.9 to prove Chebyshev's Theorem. The main obstacle to this analysis is the Möbius μ function, which is highly irregular. Chebyshev avoided the analysis of the behavior of the μ function by considering only a small part of the infinite series.[4] Chebyshev took the expression

$$T(x) - T(x/2) - T(x/3) - T(x/5) + T(x/30)$$
$$= \log\left(\frac{\lfloor x\rfloor! \lfloor x/30\rfloor!}{\lfloor x/2\rfloor! \lfloor x/3\rfloor! \lfloor x/5\rfloor!}\right).$$

This expression is simpler to manipulate than the one involving the Möbius function. Chebyshev did not give any reason for choosing this expression, and we can only guess that it may have been after some trial and error to see

[4]P. Chebyshev. "Mémoire sur les Nombres Premiers." Journal de Mathématique, 17 (1852), 366-390.

which form gave a simple result. Analysis of the full expression in Proposition 16.2.9 was carried out by Erdös and Selberg in the 1940s to give an elementary proof of the prime number theorem, elementary in the sense that it doesn't use any complex function theory.[5] We reproduce Chebyshev's proof.

Theorem 16.3.1 (Chebyshev). *Given any* $A' < A < A''$, *the function* ψ *satisfies*

$$A'x < \psi(x) < \frac{6A''}{5}x \tag{1}$$

for all x *sufficiently large, where* $A = 0.92129209$. *The inequality*

$$0.9142867x < \psi(x) < 0.9781263x \tag{2}$$

holds for all $x \geq 3000$.

PROOF. We use the expression for $T(x)$ from Lemma 16.2.6 to write $T(x) - T(x/2) - T(x/3) - T(x/5) + T(x/30)$ in terms of ψ. We obtain

$$
\begin{aligned}
T(x)&-T(x/2) - T(x/3) - T(x/5) + T(x/30) \\
&= \psi(x) + \psi(x/2) + \psi(x/3) + \psi(x/4) + \cdots \\
&\quad - \psi(x/2) - \psi(x/4) - \psi(x/6) - \psi(x/8) - \cdots \\
&\quad - \psi(x/3) - \psi(x/6) - \psi(x/9) - \psi(x/12) - \cdots \\
&\quad - \psi(x/5) - \psi(x/10) - \psi(x/15) - \psi(x/20) - \cdots \\
&\quad + \psi(x/30) + \psi(x/60) + \cdots .
\end{aligned}
$$

We can combine terms and write this as $\sum_n a_n\psi(x/n)$, where a_n is determined by $n \bmod 30$. There is a lot of cancellation among the terms. It is not hard to verify that $a_n = 1$ if n is relatively prime to 2, 3, and 5; $a_n = 0$ if n is divisible by only one of the factors 2, 3, or 5, and $a_n = -1$ in all other cases. This analysis yields the following series

$$
\begin{aligned}
T(x)&-T(x/2) - T(x/3) - T(x/5) + T(x/30) \\
&= \psi(x) - \psi(x/6) + \psi(x/7) - \psi(x/10) + \psi(x/11) + \cdots ,
\end{aligned}
\tag{3}
$$

where the sign of the $\psi(x/k)$ term is 1 if $k \equiv 1, 7, 11, 13, 17, 19, 23, 29$ (mod 30) and is -1 if $k \equiv 6, 10, 12, 15, 18, 20, 24, 30$ (mod 30). The other terms do not occur in the series. Equation (3) has the remarkable property that it is alternating (with decreasing terms); hence we have

$$\psi(x) - \psi(x/6) < \log\left(\frac{\lfloor x\rfloor!\lfloor x/30\rfloor!}{\lfloor x/2\rfloor!\lfloor x/3\rfloor!\lfloor x/5\rfloor!}\right) < \psi(x)$$

---

[5]See G. H. Hardy and E. M. Wright. *An Introduction to the Theory of Numbers*, 5th ed. Oxford: Oxford University Press, 1979.

by the properties of alternating series with decreasing terms. To estimate the middle term, we use Stirling's formula for the asymptotic behavior of $\lfloor x \rfloor!$. Using the inequalities in Corollary 16.2.8, for $x \geq 90$, we get

$$(x - 1/2) \log x - x + 1 < \log\lfloor x \rfloor! < (x + 1/2) \log x - x + 2$$
$$-(x/2 + 1/2) \log(x/2) + x/2 - 2 < -\log\lfloor x/2 \rfloor!$$
$$< -(x/2 - 1/2) \log(x/2) + x/2 - 1$$
$$-(x/3 + 1/2) \log(x/3) + x/3 - 2 < -\log\lfloor x/3 \rfloor!$$
$$< -(x/3 - 1/2) \log(x/3) + x/3 - 1$$
$$-(x/5 + 1/2) \log(x/5) + x/5 - 2 < -\log\lfloor x/5 \rfloor!$$
$$< -(x/5 - 1/2) \log(x/5) + x/5 - 1$$
$$(x/30 - 1/2) \log(x/30) - x/30 + 1 < \log\lfloor x/30 \rfloor!$$
$$< (x/30 + 1/2) \log(x/30) - x/30 + 2.$$

Adding all the inequalities yields

$$Ax - \frac{5}{2} \log x + \log 30 - 4 < \log\left(\frac{\lfloor x \rfloor! \lfloor x/30 \rfloor!}{\lfloor x/2 \rfloor! \lfloor x/3 \rfloor! \lfloor x/5 \rfloor!} \right)$$
$$< Ax + \frac{5}{2} \log x - \log 30 + 1,$$

where $A = \log\left(\frac{2^{1/2} 3^{1/3} 5^{1/5}}{30^{1/30}} \right) = 0.92129202$. These inequalities follow from the observation that $x - x/2 - x/3 - x/5 + x/30 = 0$, and

$$x \log x - x/2 \log(x/2) - x/3 \log(x/3) - x/5 \log(x/5) + x/30 \log(x/30)$$
$$= (x - x/2 - x/3 - x/5 + x/30) \log x$$
$$+ x(1/2 \log 2 + 1/3 \log 3 + 1/5 \log 5 - 1/30 \log 30)$$
$$= Ax.$$

This proves that

$$\psi(x) > Ax - \frac{5}{2} \log x - 8 \qquad (4)$$
$$\psi(x) - \psi(x/6) < Ax + 5/2 \log(x) + 7. \qquad (5)$$

Then observing that $(5/2 \log x + 1)/x$ approaches 0 as $x \to \infty$ and is a decreasing function of x, we obtain Equation (1). Evaluating at $x = 3000$ gives the lower bound in Equation (2).

For the upper bound, we iterate the inequality (5).

$$\psi(x) - \psi\left(\frac{x}{6}\right) < Ax + \frac{5}{2}\log x - 2$$

$$\psi\left(\frac{x}{6}\right) - \psi\left(\frac{x}{6^2}\right) < A\frac{x}{6} + \frac{5}{2}\log(x/6) - 2$$

$$\vdots$$

$$\psi\left(\frac{x}{6^{n-2}}\right) - \psi\left(\frac{x}{6^{n-1}}\right) < A\frac{x}{6^2} + \frac{5}{2}\log(x/6^{n-1}) - 2.$$

We proceed until $x/6^{n-2} \geq 90$ but $x/6^{n-1} < 90$ and write

$$\psi\left(\frac{x}{6^{n-1}}\right) - \psi\left(\frac{x}{6^n}\right) \leq \psi(90) = 89.47 < 90,$$

in particular,

$$\psi\left(\frac{x}{6^{n-1}}\right) - \psi\left(\frac{x}{6^n}\right) < A\frac{x}{6^{n-1}} + \frac{5}{2}\log\frac{x}{6^{n-1}} - 2 + 90.$$

Adding all the inequalities, we obtain

$$\psi(x) - \psi\left(\frac{x}{6^n}\right) < Ax\left(1 + \frac{1}{6} + \frac{1}{6^2} + \cdots \frac{1}{6^{n-1}}\right) + \frac{5}{2}n\log x - 2n + 90.$$

Since $x/6^{n-1} < 90$, we have that $n > \log x / \log 6$, and $\psi(x/6^n) = 0$.
Therefore,

$$\psi(x) < \frac{6A}{5}x + \left(\frac{5}{2}\log x - 2\right)\frac{\log x}{\log 6} + 90.$$

Observe that

$$\lim_{x \to \infty} \frac{(5/2\log x - 2)\log x/\log 6 + 90}{x} = 0$$

and it is a decreasing function. This proves that $\psi(x) < (6A''/5)x$ for any $A'' < A$ and x sufficiently large. Evaluating at $x = 3000$, we obtain $\psi(x) < 0.9781263x$. ∎

Proposition 16.3.2. *The function $\Theta(x)$ satisfies*

$$\Theta(x) > Ax - \frac{12}{5}A\sqrt{x} - \frac{5}{4}\frac{\log^2 x}{\log 6} + \left(\frac{2}{\log 6} - \frac{5}{2}\right) - 181$$

$$\Theta(x) < 6/5Ax - A\sqrt{x} + \frac{5}{2}\frac{\log^2 x}{\log 6} + \left(\frac{5}{4} - \frac{2}{\log 6}\right)\log x + 91.$$

PROOF. These follow from the inequalities for $\psi(x)$ in the proof of Chebyshef's Theorem and Lemma 16.2.5

$$\psi(x) - 2\psi(\sqrt{x}) \le \Theta(x) \le \psi(x) - \psi(\sqrt{x}).$$

■

Again, the remainder term is such that the ratio with x approaches 0. In the limit, we obtain

$$Ax \le \Theta(x) \le \frac{6A}{5}x.$$

To get a reasonable estimate, we compute the ratio of the remainder terms for $x = 3000$ and obtain

Corollary 16.3.3. *For $x \ge 3000$,*

$$0.80199x \le \Theta(x) \le 0.9649752x.$$

It is more difficult to deduce an upper bound for $\pi(x)$ from $\Theta(x)$. Our analysis gives an upper bound of $2.13x/\log x$. This is quite weak and a much better result can be obtained.

Corollary 16.3.4. *For $x \ge 10^5$,*

$$0.9142867\frac{x}{\log x} \le \pi(x) \le 2.1315451\frac{x}{\log x}.$$

PROOF. Lemma 16.2.4 says that $\pi(x)\log x \ge \psi(x)$, which implies $\pi(x) \ge \psi(x)/\log x > 0.9142867x/\log x$ for $x \ge 3000$ by Chebyshev's Theorem.

To extract an upper bound for $\pi(x)$ from that of $\Theta(x)$, we use the Abel summation method. We write $\Theta(x) < Bx$ ($B \approx 0.965$) for $x \ge 3000$ and take $\Theta(x) \le 3000$ for $x \le 3000$. It is easy to verify using the definition of Θ that $\pi(x) = \sum_{2 \le n \le x} \frac{\Theta(n)-\Theta(n-1)}{\log n}$. Hence we can write

$$\pi(x) = \sum_{2 \le n \le x} \frac{\Theta(n) - \Theta(n-1)}{\log n}$$

$$= \Theta(2)\left(\frac{1}{\log 2} - \frac{1}{\log 3}\right) + \Theta(3)\left(\frac{1}{\log 3} - \frac{1}{\log 4}\right) + \cdots.$$

Grouping the terms with $\Theta(n)$ together,

$$\pi(x) = \sum_{2 \le n < \lfloor x \rfloor} \Theta(n)\left(\frac{1}{\log n} - \frac{1}{\log(n+1)}\right) + \frac{\Theta(\lfloor x \rfloor)}{\log(\lfloor x \rfloor)}. \qquad (6)$$

To simplify the sum, we note that

$$\frac{1}{\log n} - \frac{1}{\log(n+1)} = \frac{\log(n+1) - \log n}{\log n \log(n+1)} = \frac{\log(1+1/n)}{\log n \log(n+1)} \leq \frac{1}{n(\log n)^2}.$$

We have used the inequality $\log(1+1/n) \leq 1/n$ in the last step. This yields

$$\sum_{2 \leq n \leq \lfloor x \rfloor} \Theta(n) \left(\frac{1}{\log n} - \frac{1}{\log(n+1)} \right)$$

$$\leq \sum_{2 \leq n < 3000} \Theta(n) \frac{1}{n(\log n)^2} + \sum_{3000 \leq n < \lfloor x \rfloor} \frac{\Theta(n)}{n(\log n)^2}$$

$$\leq 3000 \sum_{2 \leq n < 3000} \frac{1}{n(\log n)^2} + B \sum_{3000 \leq n < \lfloor x \rfloor} \frac{1}{(\log n)^2}$$

$$\leq 6000 + B \sum_{2 \leq n < \lfloor x \rfloor} \frac{1}{(\log n)^2}.$$

To estimate the second sum,

$$\sum_{2 \leq n \leq x} \frac{1}{(\log n)^2} = \sum_{2 \leq n \leq \sqrt{x}} \frac{1}{(\log n)^2} + \sum_{\sqrt{x} < n \leq x} \frac{1}{(\log n)^2}$$

$$\leq \frac{\sqrt{x}}{(\log 2)^2} + \frac{1}{(\log \sqrt{x})^2} x$$

$$< 4\sqrt{x} + \frac{4x}{(\log x)^2}.$$

The last term in (6) is

$$\frac{\Theta(\lfloor x \rfloor)}{\log \lfloor x \rfloor} \leq \frac{\Theta(x)}{\log x} \leq B \frac{x}{\log x},$$

and hence we have obtained

$$\pi(x) \leq \frac{Bx}{\log x} + 4B\sqrt{x} + \frac{4Bx}{(\log x)^2} + 6000.$$

The main term is $Bx/\log x$, and we can divide by x to have the remainder term approach 0. The remainder is a decreasing term, and for $x \geq 10^5$, we can easily verify that $\pi(x) \leq 2.24x/\log x$. ∎

The estimates used in the proof above are by no means the best, and they can be improved significantly.

Chebyshev used his theorem to prove a conjecture of Bertrand on the existence of a prime between n and $2n$. Chebyshev's Theorem can be used to show that there is always a prime in the interval $[n, n+n/5]$. The prime number theorem is needed to show that, for every $\epsilon > 0$, there is always a prime between n and $(1 + \epsilon)n$ for n sufficiently large. We will prove Bertrand's conjecture and ask for improvements in the exercises.

Corollary 16.3.5. *For every positive integer n there is a prime between n and $2n$.*

PROOF. To prove this, we need to show that $\Theta(2n) - \Theta(n) > 0$ for all n. We can verify directly for $n < 3000$ that there is a prime between n and $2n$. Using the estimates on $\Theta(x)$, we obtain

$$0.8222(2n) < \Theta(2n) < 1.149(2n)$$
$$0.8222n < \Theta(n) < 1.149n.$$

Subtracting the two equations, we obtain

$$\Theta(2n) - \Theta(n) \geq 1.6444n - 1.149n$$
$$= 0.4954n.$$

This implies that $\Theta(2n) - \Theta(n) = \sum_{n < p \leq 2n} \log p > 0$ for $n \geq 3000$. Therefore, there is a prime between n and $2n$ for all n. ■

───────────── **Exercises for Section 16.3** ─────────────

1. Verify that the series obtained in Equation (3) is alternating and the terms are as described.

2. Complete the proof of Proposition 16.3.2.

3. A simpler proof of Chebyshev's Theorem about $\psi(x)$ can be given using the expression $T(x) - 2T(x/2)$ instead of the expression used by Chebyshev. Use this expression and work through the proof of Chebyshev's Theorem. What constants do you obtain in the following expression?

$$cx < \psi(x) < Cx$$

4. Use Chebyshev's Theorem to show that there is always a prime between n and $3n/2$ for n sufficiently large. How much more can the interval be shortened using the estimates proved on $\Theta(x)$?

5. Show that $1 + \frac{1}{2} + \frac{1}{3} + \cdots + \frac{1}{n}$ is not an integer for any $n > 1$.

6. Show that the product $1 \cdot 3 \cdot 5 \cdot \ldots (2n - 1)$ of consecutive odd integers can never be a square or a higher power.

7. Show that 30 is the largest integer n with the following property: for $1 < k \leq n$, $(k, n) = 1$ implies that k is prime.

16.4   The Riemann Zeta Function

The Riemann zeta-function, $\zeta(s)$ is defined as

$$\zeta(s) = \sum_{n=1}^{\infty} \frac{1}{n^s}.$$

The properties of the ζ-function are intricately linked with the properties of prime numbers. The reason for this is the Euler factorization of the ζ-function,

$$\zeta(s) = \prod_p \frac{1}{1 - p^{-s}},$$

where the product is over all the primes. (See Section 15.3 for a discussion of Euler products.) The ζ-function is viewed as a function of a complex variable s. Recall that if $s = \sigma + i\tau$, with σ and τ real, then

$$n^s = e^{s \log n} = e^{\sigma \log n} e^{i\tau \log n} = e^{\sigma \log n} \left(\cos(\tau \log n) + i \sin(\tau \log n) \right).$$

Also, from the above equation, the ζ-function is absolutely convergent for $\mathrm{Re}(s) = \sigma > 1$. Also $\lim_{s \to 1+} \zeta(s) = +\infty$, which leads to the proof that there are infinitely many primes.

The deeper relations between the ζ-function and arithmetic come from an extension of the domain of definition. The ζ-function can be defined on the entire complex plane, excluding the point $s = 1$, such that $\zeta(s) = \sum_{n=1}^{\infty} \frac{1}{n^s}$ for $\mathrm{Re}(s) > 1$.

Using the following proposition, we can extend the ζ-function to the strip $0 < \mathrm{Re}(s) < 1$ (also known as the critical strip).

Proposition 16.4.1. *For* $\mathrm{Re}(s) > 1$,

$$\zeta(s) = \left(1 - \frac{2}{2^s} \right) = \sum \frac{(-1)^{n-1}}{n^s}.$$

PROOF. We write

$$\frac{1}{2^s} \zeta(s) = \frac{1}{2^s} + \frac{1}{4^s} + \frac{1}{6^s} + \cdots.$$

Therefore

$$\zeta(s) - \frac{2}{2^s}\zeta(s) = 1 - \frac{1}{2^s} + \frac{1}{3^s} - \frac{1}{4^s} + \cdots .$$

All the rearrangements of the infinite series above are valid in the region of absolute convergence. ∎

Now, since the series $\sum(-1)^{n-1}/n^s$ is absolutely convergent for $\mathrm{Re}(s) > 1$, we can define $\zeta(s)$ on the strip $0 < \mathrm{Re}(s) < 1$ to be

$$\zeta(s) = \frac{1}{(1 - 2/2^s)} \sum_{n=1}^{\infty} \frac{(-1)^{n-1}}{n^s}$$

which converges because it has alternate decreasing terms.

The Riemann ζ-function can be defined for $\mathrm{Re}(s) \leq 0$ and satisfies the following functional equation relating $\zeta(s)$ to $\zeta(1-s)$,

$$\pi^{-s/2}\Gamma(s/2)\zeta(s) = \pi^{-(1-s)/2}\Gamma((1-s)/2)\zeta(1-s),$$

where $\Gamma(s)$ is the Gamma function defined for $s > 0$ by $\Gamma(s) = \int_0^{\infty} t^{s-1}e^t dt$. The Gamma function can be defined on the entire complex plane excluding the negative integers. The Gamma function extends the factorial function to the complex numbers, since $\Gamma(n) = (n-1)!$ if n is a positive integer.[6]

Among the numerous properties of the ζ-function are its special value results. These were first obtained by Euler who showed that

$$\zeta(2) = \sum_{n=1}^{\infty} \frac{1}{n^2} = \frac{\pi^2}{6},$$

and more generally

$$\zeta(2k) = (-1)^k \frac{(2\pi)^{2k+2}}{2} B_{2k},$$

where B_{2k} are the Bernoulli numbers.

Here we will give the proof of the result for $\zeta(2)$, and the proof for $\zeta(2k)$, $k > 1$ will be left as a project at the end of the chapter.

Theorem 16.4.2.

$$\zeta(2) = \sum_{n=1}^{\infty} \frac{1}{n^2} = \frac{\pi^2}{6}.$$

Bernhard Riemann (1826-1866)

Riemann

Bernhard Riemann received his early education from his father. Riemann showed remarkable skill in arithmetic from an early age. Riemann went to study in Göttingen, where the legendary Gauss was a professor of mathematics. The mathematical atmosphere in Göttingen was not conductive to research, with even Gauss teaching only elementary courses. Riemann transferred to Berlin and studied with Jacobi, Dirichlet, and Eisenstein. Riemann obtained his doctorate in 1851 for his thesis on "Riemann surfaces."

To qualify for a lectureship in Göttingen, Riemann had to submit an essay and give a lecture. In the essay he introduced the "Riemann integral" and for the lecture he had to submit three topics of which one was selected by the examiners. Riemann had only prepared the first two, but Gauss chose the third topic, the foundations of geometry. This lecture is one of the classics of mathematics, in which Riemann introduced the main ideas of "Riemannian Geometry," including metrics, curvature, and the geometry of n-dimensional spaces. These are the concepts used later by Einstein in his general theory of relativity.

In 1857, Riemann proved a fundamental theorem on algebraic functions, which is one of the cornerstones of algebraic geometry. His only work in number theory is his 1859 paper, "Über die Anzahl der Primzahlen," in which he proved the analytic continuation of the ζ-function and stated his formulas for the number of primes. It is in this paper that he made his observation regarding the zeros of the ζ-function, now known as the "Riemann hypothesis."

Riemann died an untimely death due to tuberculosis in 1866.

PROOF. We give here the same proof that Euler gave. Euler did not prove the convergence of the various infinite products and series involved, but for the proof to be totally rigorous it is necessary to do so.

---

[6]See J. B. Conway. *Functions of One Complex Variable.* New York: Springer–Verlag, 1980.

Recall that a polynomial with roots $\alpha_1, \alpha_2, \ldots, \alpha_k$ can be written as

$$f(x) = c\left(1 - \frac{x}{\alpha_1}\right)\left(1 - \frac{x}{\alpha_2}\right)\cdots\left(1 - \frac{x}{\alpha_k}\right),$$

where $c = f(0)$. Since $\sin \pi x = 0$ for $x = 0, \pm 1, \pm 2, \ldots$, we write

$$\frac{\sin \pi x}{\pi x} = \prod_{k=1}^{\infty}\left[\left(1 - \frac{x}{k}\right)\left(1 + \frac{x}{k}\right)\right]$$

$$= \prod_{k=1}^{\infty}\left(1 - \frac{x^2}{k^2}\right).$$

On the other hand, $\sin \pi x$ has the series expansion

$$\sin \pi x = \pi x - \frac{(\pi x)^3}{3!} + \frac{(\pi x)^5}{5!} - \cdots$$

$$\frac{\sin \pi x}{\pi x} = 1 - \frac{\pi^2}{6}x^2 + \frac{\pi^4}{5!}x^4 - \cdots.$$

The coefficient of x^2 in the expansion of $\prod_{k=1}^{\infty}(1 - x^2/k^2)$ is $-\sum_{k=1}^{\infty}\frac{1}{k^2} = -\zeta(2)$, and by comparison of the coefficient of x^2 on both sides, we obtain the desired result. ∎

Next we give a brief outline of the connection between the zeros of the ζ-function and prime number theory. The exposition is incomplete, since we omit most of the details regarding convergence.[7] Complete details of the relation between the ζ-function and prime numbers is beyond the scope of this book. We hope that the reader will be able to see the source of this relationship and consult the literature for further study.

We write with the Euler product for the ζ-function,

$$\zeta(s) = \prod_{p}\left(1 - \frac{1}{p^s}\right)^{-1},$$

therefore,

$$\log \zeta(s) = -\sum_{p}\log\left(1 - \frac{1}{p^s}\right),$$

---

[7]For complete details, consult T. M. Apostol. *Introduction to Analytic Number Theory.* New York: Springer–Verlag, 1975.

and differentiating we obtain

$$\frac{\zeta'(s)}{\zeta(s)} = \sum_p \frac{1}{1 - 1/p^s} \log p$$

$$= \sum_p \log p \sum k p^{-sk}$$

$$= \sum_n n \frac{\Lambda(n)}{n^s},$$

where $\Lambda(n) = \log p$ if $n = p^m$ for some m and prime p and zero otherwise.
    It can be shown that

$$\frac{1}{2\pi i} \int_{c-i\infty}^{c+i\infty} \frac{y^s}{s} \, dy = \begin{cases} 0 & \text{if } 0 < y < 1 \\ \frac{1}{2} & \text{if } y = 1 \\ 1 & \text{if } y > 1 \end{cases},$$

where the integral is defined over the vertical line $\sigma = c$ if $s = \sigma + i\tau$. The
integral is an improper integral and it is viewed as the principal value

$$\lim_{T \to \infty} \frac{1}{2\pi i} \int_{c-iT}^{c+iT} \frac{y^s}{s} \, ds.$$

Then we obtain

$$\frac{1}{2\pi i} \int_{c-i\infty}^{c+i\infty} \frac{x^s}{s} \left(\frac{-\zeta'(s)}{\zeta(s)} \right) ds = \sum_n \frac{1}{2\pi i} \int_{c-i\infty}^{c+i\infty} \frac{x^s \Lambda(n)}{sn^s} \, ds$$

$$= \sum_n \frac{1}{2\pi i} \Lambda(n) \int_{c-i\infty}^{c+i\infty} \frac{1}{s} \left(\frac{x}{n} \right)^s ds.$$

The last integral is 1 if $x > n$, 0 if $x < n$ and $1/2$ if $x = n$. Then

$$\frac{1}{2\pi i} \int_{c-i\infty}^{c+i\infty} \frac{x^s}{s} \left(\frac{-\zeta'(s)}{\zeta(s)} \right) ds = \sum_{n < x} \Lambda(n) + \frac{1}{2} \Lambda(x)$$

$$= \sum_{n \le x} \Lambda(n) - \frac{1}{2} \Lambda(x)$$

$$= \psi(x) - \frac{1}{2} \Lambda(x).$$

Here $\psi(x)$ is Chebyshev's psi function defined in Section 16.2. The integral
involving $\zeta'(s)/\zeta(s)$ can be evaluated using Cauchy's theorem if we know

the poles of the integrand. These are the poles of $\zeta'(s)/\zeta(s)$, which has poles at the poles of the ζ-function and its zeros.

It can be shown that $\zeta(s)$ is 0 at the negative even numbers; these are called the trivial zeros. All the other zeros have a nonzero imaginary part. Also, there are no zeros for $\mathrm{Re}(s) > 1$ because of the Euler product formula. Hence, the nontrivial zeros lie in the critical strip because of the functional equation relating $\zeta(s)$ and $\zeta(1-s)$. It can be shown that

$$\frac{\zeta'(s)}{\zeta(s)} = \frac{\zeta'(0)}{\zeta(0)} - \frac{1}{s} + \sum_\rho \left(\frac{1}{\rho} - \frac{1}{s-\rho} \right) + \sum_k \left(\frac{-1}{2k} - \frac{1}{s+2k} \right),$$

where the sum over ρ is the sum over the non-trivial zeros of the ζ-function.

Using this in the formula for $\psi(x)$ and applying Cauchy's integral formula, we see that if $\Lambda(x) = 0$, then

$$\psi(x) = x - \frac{\zeta'(0)}{\zeta(0)} - \sum_\rho \frac{x^\rho}{\rho} - \sum_{k=1}^\infty \frac{x^{-2k}}{2k}.$$

The term $\zeta'(0)/\zeta(0)$ can be computed to be $\log(2\pi)$, the first sum is over the nontrivial zeros of the ζ-function and the contribution of the trivial zeros is $\sum_k x^{-2k}/2k = \log(1 - 1/x^2)/2$.

The Riemann Hypothesis asserts that all the nontrivial zeros of $\zeta(s)$ lie on the line $\mathrm{Re}(s) = 1/2$. If $\mathrm{Re}(\rho) = 1/2$, then $|x^\rho| = |e^{\rho \log x}| = e^{\log x/2} = x^{1/2}$.

If the Riemann Hypothesis is true, the error between x and $\psi(x)$ is of the order of $x^{1/2}$, that is

$$|\psi(x) - x| \le c x^{1/2} \log^2 x.$$

--- **Projects for Chapter 16** ---

1. The Bernoulli numbers are defined by

$$\frac{x}{e^z - 1} = \sum_{k=0}^\infty B_k \frac{z^k}{k!}.$$

Recall that $z/(e^z - 1)$ is defined at $z = 0$ by $\lim_{z \to 0} z/(e^z - 1) = 1$.

The above is the Taylor expansion of the function, and there is a simple recurrence formula to compute the Bernoulli numbers.

(a) Using the power series expansions of $z/(e^z - 1)$ and $e^z - 1$, write z as the product of two infinite series. Deduce that

$$\sum_{i=0}^{m-1} B_i/(i!(m-i)!) = 0.$$

Therefore

$$\sum_{i=0}^{m-1} \binom{m}{i} B_i = 0.$$

(b) Prove that the Bernoulli numbers are rational.

(c) Justify the formula

$$\sin x = x \prod_{k=1}^{\infty} \left(1 - \frac{x^2}{\pi^2 k^2}\right).$$

(d) By taking logarithms and differentiating, prove that

$$\frac{\cos x}{\sin x} = \frac{1}{x} + \sum_{k=1}^{\infty} \frac{-2x}{\pi^2 k^2 - x^2}.$$

(e) Prove that

$$\frac{\cos x}{\sin x} = \frac{e^{-ix}(e^{2ix} + 1)}{e^{-ix}(e^{2ix} - 1)}$$

and use the change of variable $u = 2ix$ to obtain

$$\frac{u}{e^u - 1} = 1 - \frac{u}{2} + \sum_{k=1}^{\infty} \frac{2u^2}{4\pi^2 k^2 + u^2}.$$

(f) Using the power series for $(1 + u/2\pi k)^2$, write

$$\frac{u}{e^u - 1} = 1 - \frac{u}{2} + 2\sum_{k=1}^{\infty}(-1)^r \frac{u^{2r+2}}{(2\pi)^{2r+2}} \zeta(2r).$$

Deduce that

$$\zeta(2k) = (-1)^k \frac{(2\pi)^{2k+2}}{2} B_{2k}.$$

Chapter 17

Quadratic Reciprocity Law

17.1 Quadratic Reciprocity

*Engaged in other work, I chanced on an extraordinary arithmetic truth.
Since I considered it so beautiful in itself and since I suspected
its connection with even more profound results, I concentrated on it
all my efforts in order to understand the principles on which
it depended and to obtain a rigorous proof.*

— C. F. GAUSS

In Section 9.1, the evaluation of the Legendre symbol $\left(\dfrac{a}{p}\right)$ was reduced to the study of the symbols $\left(\dfrac{-1}{p}\right)$, $\left(\dfrac{2}{p}\right)$, and $\left(\dfrac{q}{p}\right)$ for an odd prime q. Euler's criterion showed that $\left(\dfrac{-1}{p}\right)$ depends only on the remainder p mod 4. We continue with the study of $\left(\dfrac{2}{p}\right)$, $\left(\dfrac{3}{p}\right)$, ..., with the goal of discovering general principles underlying these symbols. The nature of the symbols $\left(\dfrac{2}{p}\right)$ and $\left(\dfrac{3}{p}\right)$ were known to Fermat and Euler. Euler was able to give complete proofs for $\left(\dfrac{3}{p}\right)$, but the case $\left(\dfrac{2}{p}\right)$ eluded him. The first complete proof of the properties of the "quadratic character" of 2 was given by Legendre in 1775. We have attempted to give direct proofs, similar to those that might have been given by Fermat or Euler.

From the knowledge of $\left(\dfrac{-1}{p}\right)$, we obtain a relation between $\left(\dfrac{2}{p}\right)$ and $\left(\dfrac{-2}{p}\right)$, because

$$\left(\frac{-2}{p}\right) = \left(\frac{-1}{p}\right)\left(\frac{2}{p}\right).$$

Hence we can study either $\left(\dfrac{2}{p}\right)$ or $\left(\dfrac{-2}{p}\right)$. Recall that $\left(\dfrac{-1}{p}\right) = 1$ if and only if $p \equiv 1 \pmod 4$.

Lemma 17.1.1. (a) *If $p \equiv 1 \pmod 4$, then 2 is a quadratic residue modulo p if and only if -2 is a quadratic residue modulo p.*

  (b) *If $p \equiv 3 \pmod 4$, then 2 is a quadratic residue modulo p if and only if -2 is a quadratic nonresidue modulo p.*

PROOF. We have $\left(\dfrac{-2}{p}\right) = \left(\dfrac{-1}{p}\right)\left(\dfrac{2}{p}\right)$ and

$$\left(\frac{-1}{p}\right) = \begin{cases} 1 & \text{if } p \equiv 1 \pmod 4; \\ -1 & \text{if } p \equiv 3 \pmod 4. \end{cases}$$

If $p \equiv 1 \pmod 4$, then $\left(\dfrac{2}{p}\right)$ and $\left(\dfrac{-2}{p}\right)$ are the same sign, and if $p \equiv 3 \pmod 4$, then $\left(\dfrac{2}{p}\right)$ and $\left(\dfrac{-2}{p}\right)$ have opposite signs.   ∎

We would like to give a condition on p so that $\left(\dfrac{2}{p}\right) = 1$. To see what such a condition might be, we consider a few primes for which 2 is a quadratic residue. The simplest way to do this is to factor numbers of the form $x^2 - 2$. If $p \mid x^2 - 2$ for some x, then $x^2 \equiv 2 \pmod p$, hence $\left(\dfrac{2}{p}\right) = 1$.

The odd primes less than 100 that occur as factors of $x^2 - 2$ are:

$$7, 17, 23, 31, 41, 47, 71, 73, 79, 89, 97.$$

Working modulo 8, we observe that these are all congruent to 1 or 7, so we may be led to conjecture (as Fermat and Euler did) that $\left(\dfrac{2}{p}\right) = 1$ if and only if $p \equiv 1, 7 \pmod 8$. This is the content of the following proposition. By Lemma 17.1.1, we can work with either 2 or -2. The reader should verify that the different cases do contain a complete description of $\left(\dfrac{2}{p}\right)$.

Proposition 17.1.2. *Let p be an odd prime.*

(a) *If $p \equiv 1 \pmod 8$, then -2 is a quadratic residue modulo p.*

(b) *If $p \equiv 3, 5 \pmod 8$, then 2 is a quadratic nonresidue modulo p.*

(c) *If $p \equiv 7 \pmod 8$, then -2 is a quadratic nonresidue modulo p.*

PROOF.   (a) If $p \equiv 1 \pmod 8$, write $p = 8k + 1$. Consider the identity

$$x^{4k} + 1 = \left(x^{2k} - 1\right)^2 + 2x^{4k}$$
$$= \left(x^{2k} - 1\right)^2 + 2\left(x^k\right)^2.$$

By Lagrange's theorem, $x^{4k} + 1 \equiv 0 \pmod p$ has exactly $4k$ solutions because $x^{4k} + 1$ is a factor of $x^{p-1} - 1 = x^{8k} - 1$. Hence there exists an x, $x \not\equiv 0 \pmod p$ such that

$$(x^{2k} - 1)^2 + 2(x^k)^2 \equiv 0 \pmod p,$$

or $(x^{2k} - 1)^2 (x^k)^{-2} \equiv -2 \pmod p$; that is, -2 is a quadratic residue modulo p.

(b) The proof will be by contradiction. Suppose p is the smallest prime congruent to 3 or 5 such that 2 is a quadratic residue modulo p. Let $0 < x < p$, x odd, be a solution to $x^2 \equiv 2 \pmod p$. (We can take x odd because one of x or $p - x$ must be odd, and both give the same quadratic residue.)

Let $x^2 - 2 = pk$ with k necessarily odd. Since $x < p$, $0 < pk = x^2 - 2 < x^2 < p^2$, and this implies that $pk < p^2$ or $k < p$. For any odd prime q dividing k, $x^2 \equiv 2 \pmod q$. We must have $q \equiv 1, 7 \pmod 8$ because of our hypothesis that p is the smallest prime congruent to 3 or 5 for which 2 is a quadratic residue. This implies that $k \equiv 1, 7 \equiv \pm 1 \pmod 8$ because k is a product of primes congruent to 1 or 7 modulo 8. Hence $x^2 - 2 \equiv pk \equiv \pm p \pmod 8 \equiv \pm 3 \pmod 8$. This implies that $x^2 \equiv -1, 5 \pmod 8$, a contradiction to the fact that $x^2 \equiv 1 \pmod 8$ for x odd. Thus, 2 is not a quadratic residue for any prime congruent to 3 or 5 modulo 8.

(c) This proof is very similar to the case $p \equiv 3, 5 \pmod 8$. For the sake of contradiction, assume that p is the smallest prime congruent to 7 mod 8 such that -2 is a quadratic residue modulo p. If x is a solution to $x^2 \equiv -2 \pmod p$, then we can take $0 < x < p - 1$ and x odd. (Why?)

Suppose $x^2 + 2 = pk$; then $0 < pk < (p-1)^2 + 2 < p^2$, hence $k < p$. If q is any odd prime dividing k, then $x^2 \equiv -2 \pmod{q}$, so $q \not\equiv 7 \pmod 8$ because of the choice of p. We must have $q \equiv 1 \pmod 8$ [by (a)] or $q \equiv 3 \pmod 8$ (by (b) and Lemma 17.1.1). Hence $k \equiv 1, 3 \pmod 8$, and $x^2 + 2 = pk \equiv 7 \cdot 1, 7 \cdot 3 \pmod 8 \equiv 7, 5 \pmod 8$. This implies that $x^2 \equiv 5, 3 \pmod 8$, a contradiction. ∎

We can state the result of the above proposition in terms of the Legendre symbol as follows using Lemma 17.1.1.

Corollary 17.1.3. *Let p be an odd prime; then,*

$$\left(\frac{-2}{p}\right) = 1 \text{ if and only if } p \equiv 1, 3 \pmod 8 \qquad (1)$$

$$\left(\frac{2}{p}\right) = 1 \text{ if and only if } p \equiv 1, 7 \pmod 8. \qquad (2)$$

See Exercise 2 for a compact formula to express $\left(\dfrac{2}{p}\right)$.

Example 17.1.4.   1. $\left(\dfrac{2}{37}\right) = -1$ as $37 \equiv 5 \pmod 8$.

2. $\left(\dfrac{-2}{41}\right) = 1$ as $41 \equiv 1 \pmod 8$.

Exercise. Compute $\left(\dfrac{2}{59}\right)$ and $\left(\dfrac{-2}{911}\right)$.

Example 17.1.5. We can use Equation (1) to prove the infinitude of primes of the form $8k + 3$.

Suppose there exists only a finite number of primes of the form $8k+3$, say $p_1, \ldots, p_k$. Let $N = (p_1 \cdots p_k)^2 + 2$. If $q \mid N$, then -2 is a square modulo q, that is, $q \equiv 1, 3 \pmod 8$. If every prime divisor of N were congruent to 1 modulo 8, then $N \equiv 1 \pmod 8$, but we know that $N \equiv 3 \pmod 8$; thus, it has a prime divisor congruent to 3 modulo 8. This prime cannot be one of the p_i's, and the desired contradiction is obtained.

Example 17.1.6. Another application of the determination of $\left(\dfrac{2}{p}\right)$ is to find prime factors of some Mersenne numbers. If p is prime, $p \equiv 3 \pmod 4$, then $2p + 1$ divides $M_p = 2^p - 1$ if and only if $2p + 1$ is prime; this implies that if $p > 3$ is a prime such that $2p + 1$ is prime, then M_p is composite. For example, $2^{83} - 1$ and $2^{11939} - 1$ are composite.

To see this, let $q = 2p + 1$ be prime. Since $p \equiv 3 \pmod 4$, we have $q \equiv 7 \pmod 8$, so 2 is a quadratic residue modulo q. By Euler's criterion,

$$2^{(q-1)/2} \equiv 1 \pmod q,$$

which implies that q divides $M_p = 2^{(q-1)/2} - 1$.

Conversely, let $q = 2p + 1$ be a divisor of $2^p - 1$. Since $2^{(q-1)/2} \equiv 1 \pmod q$, $\mathrm{ord}_q(2) \mid (q-1)/2$. But $(q-1)/2$ is prime, so $\mathrm{ord}_q(2) = p$. But $\mathrm{ord}_q(2)$ also divides $\phi(q)$. Hence $\phi(q) \geq p$. On the other hand, $\phi(q) \leq q - 1 = 2p$. Therefore, $\phi(q) = p$ or $\phi(q) = 2p$. It is not possible to have $\phi(q) = p$, as p is an odd prime and the ϕ function never takes an odd value for $q > 2$. Hence $\phi(q) = 2p = q - 1$ and q is prime.

A prime p such that $2p + 1$ is prime is called a Germain prime, after the nineteenth century mathematician Sophie Germain, who first considered such primes in the context of Fermat's Last Theorem. (See Section 14.6.) It is an unsolved problem to determine if there are infinitely many Germain primes. An affirmative answer will prove that there are infinitely many composite Mersenne numbers, which is another famous unsolved question.

Exercise. Is $2^{117959} - 1$ composite?

To study the symbol $\left(\dfrac{3}{p} \right)$, we first study the Legendre symbol of -3. From $\left(\dfrac{-3}{p} \right)$ and $\left(\dfrac{-1}{p} \right)$, we can determine $\left(\dfrac{3}{p} \right)$. Again, the goal is to obtain congruence conditions on p that will guarantee that $\left(\dfrac{-3}{p} \right) = 1$. To determine this condition, we compute a few prime factors of numbers of the form $x^2 + 3$, because $p \mid x^2 + 3$ if and only if $\left(\dfrac{-3}{p} \right) = 1$.

$$2, 3, 7, 13, 19, 37, 43, 61, 67, 73, 79, 97, \dots.$$

Excluding 2 and 3, all these primes are congruent to 1 mod 3, and we can quickly check that every prime $p \equiv 1 \pmod 3$, $p < 100$, appears in this list. Based on this, we are led to expect the following.

Proposition 17.1.7 (Fermat, Euler). *If $p \geq 5$ is a prime, then -3 is a quadratic residue modulo p if and only if $p \equiv 1 \pmod 3$.*

PROOF. Suppose $p \equiv 1 \pmod 3$. Write $p = 3k + 1$ and use the identity

$$4(x^{3k} - 1) = 4(x^k - 1)(x^{2k} + x^k + 1)$$
$$= (x^k - 1)(4x^{2k} + 4x^k + 4)$$
$$= (x^k - 1)\left((2x^k + 1)^2 + 3 \right).$$

Now $x^{3k} - 1 \equiv 0 \pmod{p}$ has $3k$ solutions, while $x^k - 1 \equiv 0 \pmod{p}$ has k solutions, so $(2x^k + 1)^2 + 3 \equiv 0 \pmod{p}$ has $2k$ solutions. There exists an x such that $2x^k + 1 \not\equiv 0 \pmod{p}$, as the congruence $2x^k + 1 \equiv 0 \pmod{p}$ can have at most k solutions. For such an x, let $y \equiv 2x^k + 1 \not\equiv 0 \pmod{p}$, hence $y^2 + 3 \equiv 0 \pmod{3}$, that is, -3 is a quadratic residue modulo p.

Conversely, suppose -3 is a quadratic residue modulo p, that is, $y^2 + 3 \equiv 0 \pmod{p}$ for some y odd, say $y = 2r+1$. Clearly, $r \neq 1$ and $(2r+1)^2 + 3 \equiv 0 \pmod{p}$ implies that $4r^2 + 4r + 1 + 3 \equiv 0 \pmod{p}$, or $4(r^2 + r + 1) \equiv 0 \pmod{p}$. Now $(x^3 - 1) = (x-1)(x^2 + x + 1)$, so $r \neq 1$ is a solution to the equation $x^3 - 1 \equiv 0 \pmod{p}$; that is, $\mathrm{ord}_p(r) = 3$, and this implies $3 \mid p - 1$ or $p \equiv 1 \pmod{3}$.   ∎

Corollary 17.1.8. $\left(\dfrac{3}{p}\right) = 1$ *if and only if* $p \equiv \pm 1 \pmod{12}$.

PROOF. As $\left(\dfrac{3}{p}\right) = \left(\dfrac{-1}{p}\right)\left(\dfrac{-3}{p}\right)$, the symbol $\left(\dfrac{3}{p}\right)$ is 1 if both $\left(\dfrac{-1}{p}\right)$ and $\left(\dfrac{-3}{p}\right)$ are 1 or both are -1.

If the first case, $\left(\dfrac{-1}{p}\right) = 1$ and $\left(\dfrac{-3}{p}\right) = 1$ if and only if $p \equiv 1 \pmod{4}$ and $p \equiv 1 \pmod{3}$, respectively. The two congruences $p \equiv 1 \pmod{4}$ and $p \equiv 1 \pmod{12}$ are equivalent to the single congruence $p \equiv 1 \pmod{12}$.

In the second case, $\left(\dfrac{-1}{p}\right) = -1$ and $\left(\dfrac{-3}{p}\right) = -1$ if and only if $p \equiv 3 \pmod{4}$ and $p \equiv 2 \pmod{3}$, respectively; that is, both symbols are -1 if and only if $p \equiv 11 \pmod{12}$.

(Note that we use the Chinese Remainder Theorem to combine the two congruences on p into a single congruence.)   ∎

Example 17.1.9. We can determine $\left(\dfrac{6}{p}\right)$ using the information obtained so far.

We have $\left(\dfrac{6}{p}\right) = \left(\dfrac{2}{p}\right)\left(\dfrac{3}{p}\right) = 1$ if both $\left(\dfrac{2}{p}\right)$ and $\left(\dfrac{3}{p}\right)$ are 1 or if both are -1.

If the first case, $\left(\dfrac{2}{p}\right) = 1$ and $\left(\dfrac{3}{p}\right) = 1$ if and only if $p \equiv 1, 7 \pmod{8}$, and $p \equiv \pm 1 \pmod{12}$, respectively. Using the Chinese Remainder Theorem, the congruences $p \equiv \pm 1 \pmod{8}$ and $p \equiv \pm 1 \pmod{8}$ can be combined into the single congruence $p \equiv \pm 1 \pmod{24}$.

In the second case, $\left(\dfrac{2}{p}\right) = -1$ and $\left(\dfrac{3}{p}\right) = -1$ if and only if $p \equiv 3, 5$ (mod 8) and $p \equiv 5, 7$ (mod 12), respectively. We combine these congruence to see that both the symbols are -1 if and only if $p \equiv 5, 19$ (mod 24).

This implies that $\left(\dfrac{6}{p}\right) = 1$ if and only if $p \equiv \pm 1, \pm 5$ (mod 24).

The reader should observe that in all the cases considered so far in the determination of $\left(\dfrac{a}{p}\right)$ for $a = -1, 2, 3, 6$, the congruence condition on p has been modulo $4a$. We consider more examples to see if this pattern holds.

The following primes occur as factors of numbers of the form $x^2 - 5$:

$$2, 5, 11, 19, 29, 31, 41, 59, 61, 71, 79, 89, 101, 109, \ldots .$$

We expect that these primes satisfy some condition modulo $20 = 4 \cdot 5$. Excluding 2 and 5, the sequence modulo 20 is

$$11, 19, 9, 11, 1, 19, 1, 11, 19, 9, 9, 11, 19, 19, 1, \ldots ,$$

and every prime less than 100 satisfying $p \equiv 1, 9, 11, 19$ (mod 20) appears in the list of primes for which 5 is a quadratic residue. Based on this, we can conjecture that

$$\left(\frac{5}{p}\right) = 1 \text{ if and only if } p \equiv 1, 9, 11, 19 \quad (\text{mod } 20),$$

or working with the absolute residue system,

$$\left(\frac{5}{p}\right) = 1 \text{ if and only if } p \equiv \pm 1, \pm 9 \quad (\text{mod } 20).$$

Similarly, we compute primes dividing numbers of the form $x^2 - 7$ and their classes modulo 28.

$$2, 3, 7, 19, 29, 31, 37, 47, 53, 59, 83, 103, 109, 113, \ldots .$$

Excluding 2 and 7, the sequence modulo 28 is

$$3, 19, 1, 3, 9, 19, 25, 3, 27, 19, 25, 1, 19, \ldots .$$

Based on this, we can conjecture

$$\left(\frac{7}{p}\right) = 1 \text{ if and only if } p \equiv 1, 3, 9, 19, 25, 27 \quad (\text{mod } 28),$$

or, equivalently,

$$\left(\frac{7}{p}\right) = 1 \text{ if and only if } p \equiv \pm 1, \pm 9, \pm 25 \pmod{28}.$$

Euler noticed a pattern in the rules given above.[1] He observed that the numbers appearing on the right hand side of the congruence are plus or minus odd squares.

Quadratic Reciprocity Law (Euler's version). *If p and q are distinct odd primes, then $\left(\dfrac{q}{p}\right) = 1$ if and only if $p \equiv \pm b^2 \pmod{4q}$ for some odd integer b.*

Remark. We require q to be a prime, as the analysis of $\left(\dfrac{6}{p}\right)$ above shows that the assertion is not true for composite moduli.

Example 17.1.10. If $q = 13$, then Euler's conjecture states that if p an odd prime, then $\left(\dfrac{13}{p}\right) = 1$ if and only if $p \equiv \pm 1^2, \pm 3^2, \pm 5^2, \pm 7^2, \ldots$ (mod 52), that is, $p \equiv \pm 1, \pm 9, \pm 25, \pm 49, \pm 29, \pm 17 \pmod{52}$. It is enough to take b up to 11; after that, the values modulo 52 repeat.

A proof eluded Euler in spite of considerable effort. In fact, after stating these and other related conjectures in a letter to Goldbach, Euler remarked:

> ...I am convinced that I have not exhausted this material, rather, that there are countless wonderful properties of numbers to be discovered here, by means of which the theory of divisors could be brought to much greater perfection; and I am convinced that if Your Excellency were to consider this subject worthy of some attention, he would make very important discoveries in it. The greatest advantage would show itself, however, when one could find proofs of these theorems.

Legendre, in his 1785 memoir *Recherches d'Analyse Indeterminée*, offered a proof of quadratic reciprocity. Though he established some cases [for $p \equiv 1 \pmod 4$ and $q \equiv 1 \pmod 4$], the proof was flawed in other cases due to circular reasoning. He also assumed the existence of primes in arithmetic progressions $a + kb$, $(a, b) = 1$, a result proved much later by Dirichlet.

Gauss rediscovered quadratic reciprocity in 1795 and succeeded in proving the theorem by induction. He offered eight different proofs during his

---

[1] H. M. Edwards. "Euler and Quadratic Reciprocity." *Mathematics Magazine,* 56(1983), 285–291.

lifetime in search of techniques that would generalize to higher degree congruences. We prove the reciprocity law in the next section based on a result discovered by Gauss in 1808. Here is Gauss's formulation of quadratic reciprocity.

Quadratic Reciprocity Law (Gauss's version). *Let p and q be odd primes.*

(a) *If p is of the form $4n + 1$, then q is a square modulo p if and only if p is a square modulo q.*

(b) *If p is of the form $4n + 3$, then q is a square modulo p if and only if $-p$ is a square modulo q.*

We show that the versions of Euler and Gauss are equivalent.

PROOF.   (a) Suppose Euler's conjecture is true. If $\left(\dfrac{q}{p}\right) = 1$, then $p \equiv \pm b^2 \pmod{4q}$ for some odd b, that is, $p \equiv \pm b^2 \pmod 4$ and $p \equiv \pm b^2 \pmod q$. But $b^2 \equiv 1 \pmod 4$, that is, $p \equiv \pm 1 \pmod 4$.

If $p \equiv 1 \pmod 4$, then $p \equiv b^2 \pmod{4q}$ implies $p \equiv b^2 \pmod q$, that is, $\left(\dfrac{q}{p}\right) = 1$ if and only if $\left(\dfrac{p}{q}\right) = 1$.

(b) If $p \equiv 3 \pmod 4$, then $p \equiv -b^2 \pmod{4q}$ implies $-p$ is a square modulo q, that is, $\left(\dfrac{q}{p}\right) = 1$ if and only if $\left(\dfrac{-p}{q}\right) = 1$. ∎

This relation can also be expressed in the following form:

Quadratic Reciprocity Law (Third version).   (a) *If one of p or q is congruent to 1 modulo 4, then* $\left(\dfrac{q}{p}\right) = \left(\dfrac{p}{q}\right)$.

(b) *If both p and q are congruent to 3 modulo 4, then* $\left(\dfrac{q}{p}\right) = -\left(\dfrac{p}{q}\right)$.

This version, together with Euler's formulation, is the most useful in practice. The law is also expressed in the following compact formula due to Legendre.

Quadratic Reciprocity Law (Legendre's version). *Let p and q be odd primes. Then*

$$\left(\frac{p}{q}\right)\left(\frac{q}{p}\right) = (-1)^{\frac{p-1}{2}\frac{q-1}{2}}.$$

The Quadratic Reciprocity Law is proved in the next section.

———————————— **Exercises for Section 17.1** ————————————

1. Evaluate the following Legendre symbols.

   (a) $\left(\dfrac{3}{43}\right)$ (b) $\left(\dfrac{2}{43}\right)$ (c) $\left(\dfrac{3}{67}\right)$

   (d) $\left(\dfrac{2}{47}\right)$ (e) $\left(\dfrac{2}{41}\right)$ (f) $\left(\dfrac{6}{53}\right)$

2. Show that the value of $\left(\dfrac{2}{p}\right)$ can be written as $\left(\dfrac{2}{p}\right) = (-1)^{\frac{p^2-1}{8}}$.

3. Show that $\left(\dfrac{3}{p}\right)\left(\dfrac{p}{3}\right) = (-1)^{\frac{p-1}{2}}$.

4. Use a computer to verify the assertion about prime factors of the polynomials $x^2 - 5$ and $x^2 - 7$. Extend your calculations to the prime factors of $x^2 - 11$ and $x^2 - 13$ to test Euler's conjecture.

5. Prove that there are infinitely many primes in the sequence $3n + 1$ using the properties of $\left(\dfrac{3}{p}\right)$.

6. Prove the infinitude of primes of the form $8n + 5$ and $8n + 7$.

7. Suppose $p \equiv 1 \pmod{4}$ and $q = 2p + 1$ are both prime. Show that 2 is a primitive root for q.

8. Let $p = 8n + 5$ be prime. Show that if $a^{\frac{p-1}{4}} \equiv -1 \pmod{p}$, then $x = \pm 2^{2n+1}a^{n+1}$ satisfies $x^2 \equiv a \pmod{p}$.

9. Show that the different formulations of the Quadratic Reciprocity Law are equivalent.

10. **[Pepin's test]**

    (a) If $F_n = 2^{2^n} + 1$ is the nth Fermat number, then show that $F_n \equiv 5 \pmod{12}$ for $n \geq 1$. Hence conclude that if F_n is prime, $\left(\dfrac{3}{F_n}\right) = -1$. Conversely, using the primality test of Section 7.4, show that $3^{(F_n-1)/2} \equiv -1 \pmod{F_n}$ implies that F_n is prime. (Then, by Euler's Criterion, F_n is prime if and only if 3 is not a quadratic residue modulo F_n.) This is the method by which Fermat numbers are verified to be composite.

    (b) Verify that F_6 and F_7 are composite using Pepin's test.

11. Suppose p is a prime divisor of the nth Fermat number, show that $\operatorname{ord}_p(2) = 2^{n+1}$. Show that if $n \geq 2$, $p \equiv 1 \pmod{8}$, hence 2 is a quadratic residue modulo p. Use Euler's criterion to conclude that $2^{n+2} \mid p - 1$. What does this say about the prime divisors of F_5? Show that a factor of F_5 can be found in four divisions.

12. Use the properties of $\left(\dfrac{-3}{p}\right)$ and Fermat's method of descent (see Section 14.4) to determine the primes that can be written in the form $x^2 + 3y^2$.

13. If p divides a number of the form $x^2 + ny^2$, then $-n$ is a quadratic residue modulo p. It is an interesting problem to determine when p can be expressed in the form $x^2 + ny^2$, that is, when is $p = x^2 + ny^2$? For example, we showed, using the method of descent, that for $n = 1$, $p = x^2 + y^2$ if and only if -1 is a quadratic residue modulo p. The method of descent also applies to $x^2 + 2y^2$ and $x^2 + 3y^2$, but fails for $x^2 + 5y^2$.

(a) Make a table of primes that are of the form $x^2 + 5y^2$. These are a subset of the primes $p \equiv 1, 3, 7, 9 \pmod{20}$. Determine the subset.

(b) Repeat the exercise to determine which primes are of the form $x^2 + 6y^2$, $x^2 + 7y^2$, and so on.

(c) It turns out that for larger values of n, the conditions for representability of primes in the form $x^2 + ny^2$ are more complicated. In this regard, Euler conjectured the following amazing result.
$p = x^2 + 27y^2$ *if and only if* $p \equiv 1 \pmod 3$ *and 2 is a cubic residue modulo* p.
Investigate the truth of this conjecture.

(d) Euler conjectured that
$p = x^2 + 64y^2$ *if and only if* $p \equiv 1 \pmod 4$ *and 2 is a biquadratic residue modulo* p.
Investigate the truth of this conjecture.

17.2   Proof of the Quadratic Reciprocity Law

> *Mathematical proofs, like diamonds, are hard as well as clear, and will be touched by nothing but strict reasoning.*
>
> — JOHN LOCKE

Let p be an odd prime and a an integer. By Euler's criterion, $\left(\dfrac{a}{p}\right) = a^{(p-1)/2} \pmod p$. To characterize the primes for which $\left(\dfrac{a}{p}\right)$ is 1, we need another way to compute $a^{(p-1)/2} \bmod p$. Recall that in the proof of Fermat's theorem, to evaluate $a^{p-1} \bmod p$, we took the elements $a, 2a, \ldots, (p-1)a$ and then formed their product. We had to show that these elements were distinct and hence formed a complete residue system. Then the product of these terms is equal to the product of $1, 2, \ldots, p-1$ modulo p, and cancellation allows us to compute the value of $a^{p-1} \bmod p$.

The evaluation of $a^{(p-1)/2} \bmod p$ is based on the same idea, but instead of the residue system $\{0, 1, \ldots, p-1\}$, we use the absolutely least residue system

$$\{-(p-1)/2, \ldots, -1, 0, 1, \ldots, (p-1)/2\}.$$

We will consider the product of the $(p-1)/2$ elements, $a, 2a, \ldots, a(p-1)/2$. Let's see an example.

Example 17.2.1. Modulo 11, the absolutely least residue system, is

$$\{-5, -4, -3, -2, -1, 0, 1, 2, 3, 4, 5\}.$$

Consider $S = \{3 \cdot 1, 3 \cdot 2, 3 \cdot 3, 3 \cdot 4, 3 \cdot 5\} = \{3, 6, 9, 12, 15\}$. The corresponding representatives in the absolutely least residue system are $S' = \{3, -5, -2, 1, 4\}$. Observe that if we neglect the sign of the terms in S', then the elements are $1, 2, 3, 4, 5$.

Exercise. Write $\{7 \cdot 1, 7 \cdot 2, 7 \cdot 3, 7 \cdot 4, 7 \cdot 5, 7 \cdot 6, 7 \cdot 7, 7 \cdot 8, 7 \cdot 9\}$ in the absolutely least system modulo 19. What are the elements if you drop the sign of the negative terms?

For any set of integers S, we denote by S' the corresponding set of representatives in the absolutely least residue system modulo p.

Proposition 17.2.2 (Gauss's Lemma). *Let p be an odd prime and a an integer such that $p \nmid a$. Let*

$$S = \left\{ a \cdot 1, a \cdot 2, \ldots, a\frac{p-1}{2} \right\}$$

and S' the corresponding set of representatives in the absolutely least residue system. If the number of negative terms in S' is n, then

$$\left(\frac{a}{p}\right) = (-1)^n.$$

PROOF. We first show that the elements of S are distinct modulo p, and up to sign they are $\{1, 2, 3, \ldots, (p-1)/2\}$ in some order.

If i and j are distinct elements in the absolutely least residue system and $ai \equiv aj \pmod{p}$, then $i \equiv j \pmod{p}$; this implies that $i = j$, so S consists of distinct elements modulo p. Now, $ai \equiv -aj \pmod{p}$ implies that $i \equiv -j$ $\pmod{p}$, but this too is not possible if $1 \le i \ne j \le (p-1)/2$. Since there are $(p-1)/2$ elements in S', in absolute value they must be $\{1, 2, \ldots, (p-1)/2\}$. Therefore,

$$(a \cdot 1)(a \cdot 2) \cdots \left(a\frac{p-1}{2}\right) \equiv (-1)^n 1 \cdot 2 \cdot 3 \cdots \frac{p-1}{2} \pmod{p}$$

(the factor $(-1)^n$ is due to the n negative terms in S'); hence

$$a^{(p-1)/2} 1 \cdot 2 \cdot 3 \cdots \frac{p-1}{2} \equiv (-1)^n 1 \cdot 2 \cdot 3 \cdots \frac{p-1}{2} \pmod{p}.$$

As $1 \cdot 2 \cdot 3 \cdots (p-1)/2$ is invertible modulo p, we obtain $a^{(p-1)/2} \equiv (-1)^n$ $\pmod{p}$ or $\left(\dfrac{a}{p}\right) = (-1)^n$.   ∎

Example 17.2.3.   1. To compute $\left(\dfrac{3}{11}\right)$, let $S = \{3 \cdot 1, 3 \cdot 2, 3 \cdot 3, 3 \cdot 4, 3 \cdot 5\}$,

then $S' = \{3, -5, -2, 1, 4\}$. Hence $\left(\dfrac{3}{11}\right) = (-1)^2 = 1$. As $11 \equiv -1$ $\pmod{12}$, and this agrees with Corollary 17.1.8.

2. For $\left(\dfrac{5}{41}\right)$, we compute $S = \{5 \cdot 1, \ldots, 5 \cdot 20\}$, and verify that in S'

there are eight negative terms; hence $\left(\dfrac{5}{41}\right) = (-1)^8 = 1$.

Exercise. Compute $\left(\dfrac{7}{13}\right)$ using Gauss's Lemma.

We study $\left(\dfrac{a}{p}\right)$ for an integer a. For this we need a procedure to count negative elements in the set S'. Which integers are equivalent to negative elements in the absolutely least residue system modulo p? It is not hard to see that integers in $(p/2, p), (3p/2, 2p), (5p/2, 3p), \ldots, (jp/2, (j+1)p/2)$, j odd are all negative in S'.

We will also need the fact that in the interval $[x, y]$, there are $\lfloor y \rfloor - \lfloor x \rfloor$ integers if x is not an integer. If x is an integer, then there are $\lfloor y \rfloor - \lfloor x \rfloor + 1$ integers in $[x, y]$.

The definition of the Legendre symbol implies that if $a \equiv b \pmod{p}$, then $\left(\dfrac{a}{p}\right) = \left(\dfrac{b}{p}\right)$; that is, replacing a "numerator" in the Legendre symbol by its remainder modulo p doesn't change the value of the symbol. The

Carl Friedrich Gauss (1777–1855)

Gauss

Gauss was born into a poor family and owed his success to his own extraordinary abilities. He learned to calculate before he could talk. He was guided by teachers who recognized his unusual abilities and encouraged him in his studies. At the age of 14, Gauss was freed of financial burdens by the Duke of Brunswick, who generously supported him for many years. Gauss was a student at Göttingen and later became professor of mathematics and director of the Göttingen observatory.

Gauss's first significant mathematical achievements were: the solution of the problem of constructability of regular polygons by ruler and compass; the first proof of the quadratic reciprocity law, and the determination of the orbit of the asteroid Ceres. These three achievements, together with the publication of the *Disquisitiones Arithmeticae*, established Gauss as the leading mathematician of the day. The *Disquisitiones* was the first systematic treatment of the theory of numbers. Gauss summarized all the previous work, introduced the notion of congruences, proved the quadratic reciprocity law. One of the remarkable features of the *Disquisitiones* was a deep study of binary quadratic forms, including a composition law for such forms, which makes them into a group. Gauss proved the Fundamental Theorem of Algebra in his doctoral dissertation.

Among his other discoveries include the method of least squares and non-Euclidean geometry. Unfortunately, Gauss was reluctant to publish many of his results, and these results became known only after his death.

Gauss was also an astronomer and a physicist. He made significant contributions to electromagnetism and invented the first telegraph.

fact that a similar statement is true for the "denominator" is at the heart of quadratic reciprocity.

Theorem 17.2.4. *Let p be an odd prime and $a > 0$ an integer; then $\left(\dfrac{a}{p}\right)$ depends only on $p \bmod 4a$; that is, if p and q are two odd primes with $p \equiv q$*

$\pmod{4a}$, *then* $\left(\dfrac{a}{p}\right) = \left(\dfrac{a}{q}\right)$.

PROOF. Let n be the number of negative terms in S where

$$S = \{a, 2a, 3a, \ldots, \frac{p-1}{2}a\}.$$

Using Gauss's Lemma, we have to determine $(-1)^n$. This value depends only on $n \bmod 2$, the parity of n. We are going to show that the parity of n in S' depends only on $p \bmod 4a$.

If a multiple of a lies in the interval $(jp/2, (j+1)p/2)$ for j odd, then we write $jp/2 < sa < (j+1)p/2$ for some s satisfying $0 < s \le (p-1)/2$. This implies that $jp/2a < s < (j+1)p/2a$. Also, $(j+1)p/2 \le ap/2$, that is, $j+1 \le a$. Let A_j be the number of multiples of a in $(jp/2, (j+1)p/2)$. Then A_j is the number of integers in $(jp/2a, (j+1)p/2a)$, hence

$$A_j = \lfloor (j+1)p/2a \rfloor - \lfloor jp/2a \rfloor.$$

(As j and p are odd, $jp/2a$ is not an integer.)

Now $n = \sum_{j=1, j\text{odd}}^{a-1} A_j$, so

$$n \bmod 2 = (\sum_{j=1, j\text{odd}}^{a-1} A_j \bmod 2) \bmod 2.$$

Writing $p = 4ak + r$, with $r = p \bmod 4a$, we obtain

$$A_j = \lfloor 2kj + 2k + (j+1)r/2a \rfloor - \lfloor 2kj + jr/2a \rfloor;$$

that is,

$$A_j \equiv \lfloor (j+1)r/2a \rfloor - \lfloor jr/2a \rfloor \pmod 2.$$

This shows that modulo 2, $n = \sum_j A_j$ depends only on r, hence $(-1)^n$ depends only on $r = p \bmod 4a$.

Therefore, if two odd primes p, q satisfy $p \equiv q \pmod{4a}$, then $p \bmod 4a = q \bmod 4a$, hence $\left(\dfrac{a}{p}\right) = \left(\dfrac{a}{q}\right)$. ∎

Using this result, we can derive the following.

Proposition 17.2.5. *Suppose p and q are odd primes with $p \equiv q \pmod 4$, then $\left(\dfrac{p}{q}\right) = \left(\dfrac{-q}{p}\right)$.*

PROOF. Suppose $p - q = 4k$; then $p \equiv 4k \pmod{q}$, and the properties of the Legendre symbol imply that

$$\left(\frac{p}{q}\right) = \left(\frac{4k}{q}\right). \tag{1}$$

Because $p \equiv q \pmod{4k}$, Theorem 17.2.4 implies

$$\left(\frac{4k}{q}\right) = \left(\frac{4k}{p}\right), \tag{2}$$

and $4k \equiv -q \pmod{p}$ implies

$$\left(\frac{4k}{p}\right) = \left(\frac{-q}{p}\right). \tag{3}$$

Equations (1), (2), and (3) imply that $\left(\dfrac{p}{q}\right) = \left(\dfrac{-q}{p}\right)$. ∎

Corollary 17.2.6.   (a) *If $p \equiv q \equiv 1 \pmod{4}$, then* $\left(\dfrac{p}{q}\right) = \left(\dfrac{q}{p}\right)$.

(b) *If $p \equiv q \equiv 3 \pmod{4}$, then* $\left(\dfrac{p}{q}\right) = -\left(\dfrac{q}{p}\right)$.

PROOF. Both assertions follow from the proposition and the fact that

$$\left(\frac{-1}{p}\right) = \begin{cases} 1 & \text{if } p \equiv 1 \pmod{4}; \\ -1 & \text{if } p \equiv 3 \pmod{4}. \end{cases}$$ ∎

To complete the proof of the quadratic reciprocity law, we must deal with the case $p \not\equiv q \pmod{4}$.

Proposition 17.2.7. *If $p \equiv -q \pmod{4a}$, then* $\left(\dfrac{a}{p}\right) = \left(\dfrac{a}{q}\right)$.

PROOF. In this case, we have $p + q = 4ak$ for some k, and we can write

$$p = 4as + r \qquad\qquad 0 < r < 4a$$
$$q = 4as' + r' \qquad\qquad 0 < r' < 4a.$$

Since $4a \mid p + q$, we must have $r + r' = 4a$. Using the same notation as in the proof of Theorem 17.2.4, let A_j be the number of multiples of a in the interval $(jp/2, (j+1)p/2)$; then $A_j = \lfloor (j+1)r/2a \rfloor - \lfloor jr/2a \rfloor$.

Let A'_j be the number of multiples of a in the interval $(jq/2, (j+1)q/2)$. Then $A'_j = \lfloor (j+1)r'/2a \rfloor - \lfloor jr'/2a \rfloor$. But $r' = 4a - r$; therefore,

$$A'_j = \lfloor 2(j+1) - (j+1)r/2a \rfloor - \lfloor 2j - jr/2a \rfloor$$
$$= 2(j+1) - 1 - \lfloor (j+1)r/2a \rfloor - 2j + 1 + \lfloor jr/2a \rfloor.$$

Since $A_j \equiv A'_j \pmod 2$, we have $\left(\dfrac{a}{p}\right) = \left(\dfrac{a}{q}\right)$. ∎

Corollary 17.2.8. *If $p \not\equiv q \pmod 4$, then $\left(\dfrac{p}{q}\right) = \left(\dfrac{q}{p}\right)$.*

PROOF. If $p \not\equiv q \pmod 4$, then $p \equiv -q \pmod 4$, so $p + q = 4a$ for some a. Then $q \equiv 4a \pmod p$ implies that

$$\left(\frac{q}{p}\right) = \left(\frac{4a}{p}\right),$$

and the multiplicativity of the Legendre symbol gives

$$\left(\frac{q}{p}\right) = \left(\frac{4}{p}\right)\left(\frac{a}{p}\right) = \left(\frac{a}{p}\right).$$

Also, since $p \equiv 4a \pmod q$, we have

$$\left(\frac{p}{q}\right) = \left(\frac{4a}{q}\right) = \left(\frac{a}{q}\right).$$

Hence $\left(\dfrac{p}{q}\right) = \left(\dfrac{q}{p}\right)$. ∎

Quadratic Reciprocity Law. (a) *If one of p or q is congruent to 1 modulo 4, then $\left(\dfrac{q}{p}\right) = \left(\dfrac{p}{q}\right)$.*

(b) *If both p and q are congruent to 3 modulo 4, then $\left(\dfrac{q}{p}\right) = -\left(\dfrac{p}{q}\right)$.*

PROOF. This follows immediately from Corollaries 17.2.6 and 17.2.8. ∎

An immediate application of the reciprocity law is to give another method to compute the Legendre symbol. To evaluate $\left(\dfrac{a}{p}\right)$, we first take the remainder of a upon division by p, factor the remainder, and use the reciprocity law to simplify each term that arises. Other applications are to determine prime divisors of binary quadratic forms. Some of these applications are explored in the next chapter.

Example 17.2.9. Compute $\left(\dfrac{33}{67}\right)$.

$$\left(\frac{33}{67}\right) = \left(\frac{3}{67}\right)\left(\frac{11}{67.}\right)$$

$$\left(\frac{3}{67}\right) = -1 \quad \text{as } 67 \equiv 7 \pmod{12}.$$

and as both 67 and 11 are congruent to 3 modulo 4, we have

$$\left(\frac{11}{67}\right) = -\left(\frac{67}{11}\right).$$

Since $67 \equiv 1 \pmod{11}$, we get

$$\left(\frac{67}{11}\right) = \left(\frac{1}{11}\right) = 1.$$

Therefore, $\left(\dfrac{33}{67}\right) = (-1)(-1) = 1$, that is, 33 is a quadratic residue modulo 67.

Example 17.2.10. Determine all possible prime divisors of numbers of the form $x^2 - 10$ for integers x.

First, $p \mid x^2 - 10$ for some x if and only if $x^2 \equiv 10 \pmod{p}$, that is, if and only if $\left(\dfrac{10}{p}\right) = 1$. Now, $\left(\dfrac{10}{p}\right) = \left(\dfrac{2}{p}\right)\left(\dfrac{5}{p}\right)$, and we have to determine $\left(\dfrac{5}{p}\right)$ using the reciprocity law.

From Corollary 17.1.3, we obtain

$$\left(\frac{2}{p}\right) = 1 \Leftrightarrow p \equiv 1, 7 \pmod{8}. \tag{4}$$

Using the reciprocity law, we have $\left(\dfrac{5}{p}\right) = \left(\dfrac{p}{5}\right)$ [as $5 \equiv 1 \pmod{4}$]. Next, $\left(\dfrac{p}{5}\right)$ is determined by the value $p \bmod 5$. We have $\left(\dfrac{1}{5}\right) = \left(\dfrac{4}{5}\right) = 1$, and $\left(\dfrac{2}{5}\right) = \left(\dfrac{3}{5}\right) = -1$; therefore,

$$\left(\frac{5}{p}\right) = 1 \Leftrightarrow p \equiv 1, 4 \pmod{5}. \tag{5}$$

Combining Equations (4) and (5), we see that if $p \equiv 1, 9, 31, 39 \pmod{40}$, then $\left(\dfrac{10}{p}\right) = 1$.

Also, $\left(\dfrac{10}{p}\right) = 1$ if $\left(\dfrac{2}{p}\right) = -1$ and $\left(\dfrac{5}{p}\right) = -1$. We have seen that

$$\left(\frac{2}{p}\right) = -1 \Leftrightarrow p \equiv 3, 5 \pmod{8} \tag{6}$$

$$\left(\frac{5}{p}\right) = -1 \Leftrightarrow p \equiv 2, 3 \pmod{5}. \tag{7}$$

Combining these two, we obtain $p \equiv 3, 13, 27, 37 \pmod{40}$, that is,

$$\left(\frac{10}{p}\right) = 1 \Leftrightarrow p \equiv \pm 2, \pm 3, \pm 9, \pm 13 \pmod{40}.$$

Exercise. Use the quadratic reciprocity law to compute $\left(\dfrac{65}{101}\right)$.

The existence of a nice solution to the quadratic case raises similar questions for the cubic and higher degree congruences. First, we have to define the cubic residue symbol. Let $p \equiv 1 \pmod{3}$, and suppose we write $\left(\dfrac{a}{p}\right)_3 = 1$ if a is a cubic residue modulo p. What should $\left(\dfrac{a}{p}\right)_3$ be if a is not a cubic residue? Is it -1? We would like the symbol to have the same properties as the Legendre symbol; that is,

$$\left(\frac{a^3}{p}\right)_3 = 1$$

$$\left(\frac{a}{p}\right)_3 = \left(\frac{b}{p}\right)_3 \quad \text{if } a \equiv b \pmod{p}$$

$$\left(\frac{ab}{p}\right)_3 = \left(\frac{a}{p}\right)_3 \left(\frac{b}{p}\right)_3.$$

Suppose a is not a cubic residue; then

$$\left(\frac{a^3}{p}\right)_3 = \left(\frac{a}{p}\right)_3^3 = 1.$$

This implies that $\left(\dfrac{a}{p}\right)_3$ must be a cubic root of 1, so it cannot be -1.

Exercise. What are the three cube roots of 1 in the complex numbers?

The above shows that a study of the cubic residue symbol requires an extension of the number system by introducing a cubic root of unity. Similarly, for a biquadratic residue symbol $\left(\dfrac{a}{p}\right)_4$ we would require a fourth root of unity (a square root of -1).

Let ω satisfy $\omega \neq 1$, $\omega^3 = 1$ and i satisfy $i^4 = 1$, $i^2 = -1$. Results like Euler's criterion for the cubic and biquadratic symbol come from a study of the number system $\mathbb{Z}[\omega]$ and $\mathbb{Z}[i]$, where

$$\mathbb{Z}[\omega] = \{a + b\omega \ : \ a, b \text{ are integers}\}$$
$$\mathbb{Z}[i] = \{a + bi \ : \ a, b \text{ are integers}\}$$

(The set $\mathbb{Z}[i]$ is also known as the set of Gaussian integers.) Gauss introduced these number systems to give a cogent description of cubic and biquadratic reciprocity. He remarked,

> The theorem on biquadratic residues gleam with the greatest simplicity and genuine beauty only when the field of arithmetic is extended to *imaginary* numbers so that without restriction, the numbers of the form $a + bi$ constitute the object, i denotes $\sqrt{-1}$ and the indeterminates a, b denote integral real numbers between $-\infty$ and $+\infty$. We will call such numbers *integral complex numbers*...,

The study of higher reciprocity laws was one of the milestones of nineteenth century mathematics, culminating in the Abelian Reciprocity Law of Artin in 1922.[2]

———————————— **Exercises for Section 17.2** ————————————

1. Evaluate the following Legendre symbols using the Quadratic Reciprocity Law.

(a) $\left(\dfrac{19}{37}\right)$ (b) $\left(\dfrac{205}{349}\right)$ (c) $\left(\dfrac{60}{79}\right)$

(d) $\left(\dfrac{777}{911}\right)$ (e) $\left(\dfrac{95}{101}\right)$ (f) $\left(\dfrac{220}{997}\right)$

2. Use Gauss's Lemma to evaluate $\left(\dfrac{5}{37}\right)$.

---

[2]For an introduction to higher reciprocity laws, see D. Cox. *Primes of the Form $x^2 + ny^2$. Fermat, Class Field Theory and Complex Multiplication.* New York: Wiley, 1989.

3. This exercise evaluates $\left(\dfrac{2}{p}\right)$, the "quadratic character" of 2, using Gauss's Lemma.

   (a) Show that the number of multiples of 2 in the interval $(p/2, p)$ is $n = \lfloor p/2 \rfloor - \lfloor p/4 \rfloor$.

   (b) Show that n is even for $p \equiv 1, 7 \pmod 8$ and odd for $p \equiv 3, 5 \pmod 8$.

4. For the computation of $\left(\dfrac{3}{p}\right)$, find the number of multiples of 3 in the intervals $(p/2, p)$ and $(3p/2, 2p)$. By considering values of $p \bmod 12$, determine when this number is even.

5. Find congruences characterizing prime numbers p for which the following integers are quadratic residues.

   (a) 15   (b) 11
   (c) 7   (d) 33

6. Characterize prime numbers that can occur as factors of numbers of the form $x^2 - 13$.

7. Characterize all primes such that every quadratic nonresidue is also a primitive root.

8. Use the properties of $\left(\dfrac{5}{p}\right)$ to show that there are infinitely many primes of the form $5n + 1$.

17.3   The Jacobi Symbol

The Legendre symbol $\left(\dfrac{a}{p}\right)$ is extended by the Jacobi symbol to include composite "denominators." The extension also satisfies the same nice properties of the Legendre symbol, including the Quadratic Reciprocity Law. The Jacobi symbol can be computed in about the same time as the GCD (without requiring any prime factorization), hence gives another way to compute the Legendre symbol.

Definition 17.3.1. Let $m > 0$ be an odd integer and a an integer. Suppose m has the factorization $m = p_1 \cdots p_k$. The **Jacobi symbol** $\left(\dfrac{a}{m}\right)$ is defined as

$$\left(\frac{a}{m}\right) = \left(\frac{a}{p_1}\right) \cdots \left(\frac{a}{p_k}\right),$$

where the symbols on the right are the Legendre symbols.

Remarks.   1. If m is prime, then the Jacobi symbol is the same as the Legendre symbol; hence no confusion can arise by using the same notation.

2. If $(a, m) \neq 1$, then there exists $p \mid m$ such that $p \mid a$, and $\left(\dfrac{a}{p} \right) = 0$, so $\left(\dfrac{a}{m} \right) = 0$.

3. In writing $m = p_1 \cdots p_k$, we do not need the primes to be distinct. It will not be important to know the exponents, so we drop them for simplicity. This is because the Legendre symbol takes values ± 1 and 0, so only the values of the exponents modulo 2 is needed for the computation.

The Jacobi symbol is defined this way so that it has many nice properties. From the definition, it is clear that the symbol is multiplicative in the denominator, that is, $\left(\dfrac{a}{mn} \right) = \left(\dfrac{a}{m} \right) \left(\dfrac{a}{n} \right)$.

Example 17.3.2.   1. The Jacobi symbol $\left(\dfrac{a}{m} \right)$ is not directly related to a being a quadratic residue or not modulo m. For example, $\left(\dfrac{2}{15} \right) = \left(\dfrac{2}{3} \right) \left(\dfrac{2}{5} \right) = (-1)(-1) = 1$, but 2 is not a quadratic residue modulo 15. (Only 1 and 4 are quadratic residues modulo 15.) Unlike the Legendre symbol, $\left(\dfrac{a}{m} \right) = 1$ does not imply that a is a quadratic residue modulo m.

2. If a is a quadratic residue modulo m, then a is a quadratic residue modulo every prime divisor of m, hence $\left(\dfrac{a}{m} \right) = 1$. This shows that $\left(\dfrac{a}{m} \right) = -1$ implies that a is not a quadratic residue modulo m.

3. If $\left(\dfrac{a}{m} \right) = 1$, then a is a quadratic residue modulo m if and only if all the symbols $\left(\dfrac{a}{p} \right)$ occurring in the definition are 1.

4. If the Jacobi symbol were defined by the condition $\left(\dfrac{a}{m} \right) = 1$ when a is a quadratic residue modulo m and -1 otherwise, then it fails to

be multiplicative. It is easy to find an example of a composite number m where the product of two quadratic nonresidues modulo m is also a quadratic nonresidue modulo m.

Lemma 17.3.3 (Properties of the Jacobi symbol). *Let a, b, m, and n be integers, $m > 0$, $n > 0$.*

(a) $\left(\dfrac{ab}{m}\right) = \left(\dfrac{a}{m}\right)\left(\dfrac{b}{m}\right)$.

(b) $\left(\dfrac{a}{mn}\right) = \left(\dfrac{a}{m}\right)\left(\dfrac{a}{n}\right)$.

(c) *If $a \equiv b \pmod{m}$, then* $\left(\dfrac{a}{m}\right) = \left(\dfrac{b}{m}\right)$.

PROOF. All of these follow from the definition and from the properties of the Legendre symbol. ∎

The following lemma will be useful in proving the reciprocity law of the Jacobi symbol.

Lemma 17.3.4. *Let m be an odd, positive integer with the factorization $m = p_1 \cdots p_k$ (not necessarily distinct primes). Then*

(a) $m \equiv 1 \pmod{4}$ *if and only if an even number of the $p_1, \ldots, p_k$ are congruent to 1 modulo 4.*

(b) $m \equiv 1, 7 \pmod{8}$ *if and only if an even number of the $p_1, \ldots, p_k$ are congruent to 3 or 5 modulo 8.*

The lemma is easy to prove by looking at multiplication modulo 4 and modulo 8.

Theorem 17.3.5. *Let $m, n > 0$ be odd integers such that $(m, n) = 1$. Then*

(a) $\left(\dfrac{-1}{m}\right) = 1$ *if and only if $m \equiv 1 \pmod{4}$.*

(b) $\left(\dfrac{2}{m}\right) = 1$ *if and only if $m \equiv 1, 7 \pmod{8}$.*

(c) *(Reciprocity)* $\left(\dfrac{m}{n}\right)\left(\dfrac{n}{m}\right) = (-1)^{\frac{n-1}{2}\frac{m-1}{2}}$.

PROOF. We prove (b) and (c). The proof of (a) is left as an exercise.

(b) Let $m = p_1 \cdots p_k$, then $\left(\dfrac{2}{m}\right) = \left(\dfrac{2}{p_1}\right) \cdots \left(\dfrac{2}{p_k}\right)$. Now, $\left(\dfrac{2}{p_i}\right) = -1$

if and only if $p_i \equiv 3, 5 \pmod 8$, so $\left(\dfrac{2}{m}\right) = 1$ if and only if there are

an even number of -1's on the right. This happens if and only if there
is an even number of $p_i \equiv 3, 5 \pmod 8$, that is, if and only if $m \equiv 1, 7$
$\pmod 8$ (by Lemma 17.3.4 above).

(c) Let $m = p_1 \cdots p_k$ and $n = q_1 \cdots q_l$. Let r be the number of p_i's,
$1 \leq i \leq k$, which are congruent to 3 modulo 4, and s the number of
q_j's, $1 \leq j \leq l$, which are congruent to 3 modulo 4. Then

$$\left(\frac{m}{n}\right) = \prod_{i,j}\left(\frac{p_i}{q_j}\right).$$

Now, $\left(\dfrac{p_i}{q_j}\right) = \left(\dfrac{q_j}{p_i}\right)$, unless both are congruent to 3 modulo 4, and
there are rs factors in the product, where both primes are congruent to
3 modulo 4; hence

$$\left(\frac{m}{n}\right) = (-1)^{rs}\prod_{i,j}\left(\frac{q_i}{p_j}\right)$$

$$= (-1)^{rs}\left(\frac{n}{m}\right).$$

If one of r or s is even, that is, either n or m is congruent to 1 mod 4,
then $\left(\dfrac{m}{n}\right) = \left(\dfrac{n}{m}\right)$. If both r and s are odd, then both n and m are

congruent to 3 modulo 4 and $\left(\dfrac{m}{n}\right) = -\left(\dfrac{n}{m}\right)$.   ∎

The reciprocity law for the Jacobi symbol gives another method to compute the Legendre symbol. Consider the following example.

Example 17.3.6. Compute $\left(\dfrac{11}{21}\right)$.

Using reciprocity of the Jacobi symbol, we have

$$\left(\frac{11}{21}\right) = \left(\frac{21}{11}\right).$$

As $21 \equiv 10 \pmod{11}$, we get

$$= \left(\frac{10}{11}\right) = \left(\frac{2}{11}\right)\left(\frac{5}{11}\right)$$

Using reciprocity and $11 \equiv 3 \pmod 8$, we have

$$= -\left(\frac{5}{11}\right) = -\left(\frac{11}{5}\right)$$

$$= -\left(\frac{1}{5}\right) = -1.$$

Exercise. Decide if 57 is a quadratic residue modulo 97 using the Jacobi symbol.

─────────────────── **Exercises for Section 17.3** ───────────────────

1. Evaluate the following Jacobi symbols

   (a) $\left(\dfrac{45}{93}\right)$

   (b) $\left(\dfrac{703}{1553}\right)$

   (c) $\left(\dfrac{1054}{1069}\right)$

   (d) $\left(\dfrac{1069}{1995}\right)$

2. Prove that $\left(\dfrac{-1}{m}\right) = 1$ if and only if $m \equiv 1 \pmod 4$.

3. Write a computer program to evaluate the Jacobi Symbol using the reciprocity law for the Jacobi symbol. Unlike the evaluation of the Legendre symbol, the evaluation of the Jacobi symbol does not require any integer factorizations. Your algorithm should be very similar to the Euclidean Algorithm for the greatest common divisor.

4. The **Kronecker symbol** extends the Jacobi symbol to include all "denominators" as follows. Define

$$\left(\frac{a}{2}\right) = \begin{cases} 0 & \text{if } a \text{ is even} \\ (-1)^{a^2-1/8} & \text{if } a \text{ is odd.} \end{cases}$$

$$\left(\frac{a}{-1}\right) = \begin{cases} 1 & \text{if } a \geq 0 \\ -1 & \text{if } a < 0. \end{cases}$$

   (a) Show that for all a, b, and c,

$$\left(\frac{ab}{c}\right) = \left(\frac{a}{c}\right)\left(\frac{b}{c}\right)$$

and, if $bc \neq 0$

$$\left(\frac{a}{bc}\right) = \left(\frac{a}{b}\right)\left(\frac{a}{c}\right).$$

(b) Use the reciprocity law for the Jacobi symbol to show that for $a \neq 0$,
$$\left(\frac{a}{b}\right) = \left(\frac{a}{b'}\right) \text{ if } b \equiv b' \pmod{4|a|}.$$

(c) If $a \equiv 0, 1 \pmod 4$, show that $\left(\frac{a}{b}\right) = \left(\frac{a}{b'}\right)$ if $b \equiv b' \pmod{|a|}$.

5. Euler's criterion states that for an odd prime p,

$$a^{(p-1)/2} \equiv \left(\frac{a}{p}\right) \pmod p.$$

The converse of this assertion is not true. The difference in the properties of $\left(\dfrac{a}{n}\right)$ depending on whether n is prime or composite can be exploited to yield a probabilistic compositeness test. If n is prime, then $a^{(n-1)/2} \equiv \left(\dfrac{a}{n}\right) \pmod n$. If we don't know that n is prime, then a priori, $\left(\dfrac{a}{n}\right)$ is the Jacobi symbol and can be computed using the reciprocity laws. We can also compute $a^{(n-1)/2} \bmod n$ directly. This gives us a probabilistic compositeness test, as $a^{(n-1)/2} \not\equiv \left(\dfrac{a}{n}\right) \pmod n$ implies that n is composite. Just as with Fermat's theorem and the related concept of pseudoprime, we define an **Euler pseudoprime to base** a (epsp(a)) to be a composite number that satisfies $a^{(n-1)/2} \equiv \left(\dfrac{a}{n}\right) \pmod n$.

(a) Find the smallest Euler pseudoprime to base 2.

(b) Is a pseudoprime to base a necessarily an Euler pseudoprime to base a?

(c) Show that if n is epsp(a) and $\left(\dfrac{a}{n}\right) = -1$, then n is a strong pseudoprime to base a.

─────────────── **Projects for Chapter 17** ───────────────

1. **Lucas–Lehmer test for Mersenne primes.**   The Fibonacci numbers and the associated sequence of Lucas numbers satisfy many wonderful divisibility properties. An interesting application of these divisibility properties

is a test due to Lucas (modified by Lehmer) for the primality of numbers of the form $M_p = 2^p - 1$, where p is prime. Lucas used this method to show that $2^{127} - 1$ is prime and Lehmer showed that $2^{257} - 1$ is composite. Derive the test for $p \equiv 3 \pmod 4$ by answering the following questions. Recall that the Fibonacci numbers are defined by

$$f_n = f_{n-1} + f_{n-2}$$
$$f_1 = f_2 = 1.$$

The Lucas numbers v_n are defined as

$$v_n = v_{n-1} + v_{n-2}$$
$$v_1 = 1, \quad v_2 = 3.$$

(a) Make a table of Fibonacci and Lucas numbers.

(b) Prove the following properties:
   i. $v_n = f_{n+1} + f_{n-1}$.
   ii. $f_{n+m} = f_n f_{m+1} + f_m f_{n-1}$.
   iii. $f_{2n-1}^2 = f_n^2 + f_{n-1}^2$.
   iv. $v_n^2 - 5f_n^2 = 4(-1)^n$.
   v. $v_{2n} = v_n^2 - 2(-1)^n$.

(c) Show that $(f_n, f_{n-1}) = 1$ and $(f_n, v_n) = 1$ for all n.

(d) Recall (see Exercise 2.5.24) that

$$f_n = \frac{1}{\sqrt{5}} \left\{ \left(\frac{1 + \sqrt{5}}{2} \right)^n - \left(\frac{1 - \sqrt{5}}{2} \right)^n \right\}.$$

Use the Binomial Theorem to show that

$$f_n = \frac{1}{2^{n-1}} \sum_{k=0}^{\lfloor \frac{n-1}{2} \rfloor} \binom{n}{2k+1} 5^k$$

or

$$2^{n-1} f_n = \begin{cases} \binom{n}{1} + 5\binom{n}{3} + 5^2\binom{n}{5} + \cdots 5^{\frac{n-1}{2}}\binom{n}{n} & \text{for } n \text{ odd} \\ \binom{n}{1} + 5\binom{n}{3} + 5^2\binom{n}{5} + \cdots 5^{\frac{n-2}{2}}\binom{n}{n} & \text{for } n \text{ even.} \end{cases}$$

(e) Show, using part (d) and Euler's criterion, that if p is an odd prime,

$$f_p \equiv \left(\frac{5}{p} \right) \pmod p$$

and

$$2f_{p+1} \equiv 1 + \left(\frac{5}{p} \right) \pmod p.$$

Conclude that if $\left(\dfrac{5}{p}\right) = 1$, then $f_{p+1} \equiv 0 \pmod{p}$ and if $\left(\dfrac{5}{p}\right) = -1$, then $f_{p-1} \equiv 0 \pmod{p}$.

(f) Consider the case $q = 2^p - 1$, where p and q are prime. Show that if $p \equiv 3 \pmod 4$, then $\left(\dfrac{5}{q}\right) = -1$. Show that $f_{2n} = f_n v_n$. Use the fact that $f_{q+1} \equiv 0 \pmod{q}$ to show that $q \mid v_{\frac{q+1}{2}}$.

(g) Conversely, suppose p is prime $q \mid v_{\frac{q+1}{2}}$ with $q = 2^p - 1$. If q is not prime, then let t be prime factor of q such that $t \le \sqrt{q}$. Show that $f_{q+1} \equiv 0 \pmod t$ and $(f_{\frac{q+1}{2}}, t) = 1$.

Show that the smallest integer r such that $f_r \equiv 0 \pmod t$ is a power of 2. Part (e) implies that $r \le t + 1$, $r \le t + 1 \le \sqrt{q} + 1 < q + 1$. Hence $f_{\frac{q+1}{2}}$ is divisible by t, a contradiction. This implies that q is prime.

(h) Explain how v_{2^p-1} can be computed using the identities of part (b).

(i) Apply the test to show that $2^{127} - 1$ is prime.

The test given here does not include the case $p \equiv 1 \pmod 4$.[3] To include these primes, it is necessary to consider a generalized Fibonacci sequence given by

$$u_n = 0, 1, 1, 2, 3, \ldots$$
$$v_n = 2, 2, 4, 6, 10, 16, \ldots.$$

2. **Fibonacci numbers modulo primes:** Consider the Fibonacci sequence modulo 13:

$$1, 1, 2, 3, 5, 8, 0, \quad 8, 8, 3, 11, 1, 12, 0,$$
$$12, 12, 11, 10, 8, 5, 0, \quad 5, 5, 10, 2, 12, 1, 0, \quad 1, 1, 2, \ldots.$$

Note that the sequence begins to repeat itself after 28 terms. In addition, the second block of seven numbers is 8 times the first block, the third block is 8^2 times the first, and so on. The goal of the project is to study the period length and explain this phenomena.

(a) Write a computer program that outputs the Fibonacci sequence modulo p. Does the sequence always repeat by returning to the starting values?

(b) Make a table of the period length of the sequence for primes p, $2 < p < 100$.

(c) The sequence can be better understood by writing the recurrence for Fibonacci numbers in terms of matrices. The equations,

$$f_{n+1} \equiv f_n + f_{n-1} \pmod p$$
$$f_n \equiv f_n \pmod p$$

---

[3]For the general test, consult P. Ribenboim. *The Little Book of Big Primes*. New York: Springer–Verlag, 1991.

are the same as the single matrix equation,

$$\begin{pmatrix} f_{n+1} \\ f_n \end{pmatrix} \equiv \begin{pmatrix} 1 & 1 \\ 1 & 0 \end{pmatrix} \begin{pmatrix} f_n \\ f_{n-1} \end{pmatrix} \quad (\text{mod } p).$$

Let $A = \begin{pmatrix} 1 & 1 \\ 1 & 0 \end{pmatrix}$ and $\text{ord}_p(A)$ be the smallest integer r such that $A^r \equiv I \pmod p$, where I is the identity matrix. Show that the period length is $\text{ord}_p(A)$, and that, if $A^s \equiv I \pmod p$, then $\text{ord}_p(A) \mid S$.

(d) Use the formula for Fibonacci numbers (as given in the project above) and the binomial theorem to show that if $\left(\dfrac{5}{p}\right) = 1$, then $f_{p-1} \equiv 0$ $(\text{mod } p)$ and $f_p \equiv 1 \pmod p$. Conclude that $\text{ord}_p(A) \mid p - 1$.

(e) Show that if $\left(\dfrac{5}{p}\right) = -1$, then $f_p \equiv -1 \pmod p$ and $f_{p+1} \equiv 0$ $(\text{mod } p)$. Show that $\text{ord}_p(A) \mid 2p + 2$.

(f) Can you explain the pattern mentioned at the beginning of the project?

(g) Show that if we take a more general Fibonacci sequence,

$$f_n = f_{n-1} - f_{n-2}$$
$$f_1 = a, \ f_2 = b,$$

the period of the sequence modulo p is still the same.

Chapter 18

Binary Quadratic Forms

18.1 Basic Notions

The theory of quadratic forms arose from the study of representation of numbers as sums of squares. The beginning of the theory can be seen in the works of Fermat, Euler, and Legendre. They defined some of the basic concepts and proved some of the first results. The theory of forms was put on a solid footing by Gauss in his *Disquisitiones Arithmeticae*. He discovered many of the deeper properties of binary quadratic forms. Many of his results are now easier to understand through the study of the ideal class group of quadratic fields. The theory of binary forms itself is very concrete and provides examples for the more general theory of ideals. In this chapter, we study the most elementary aspect of the theory of indefinite binary quadratic forms.

Definition 18.1.1. A **quadratic form** in n variables is a function,

$$f(x_1, \ldots, x_n) = \sum_{i=1}^{n} q_{ii} x_i^2 + \sum_{1 \leq i < j \leq n} q_{ij} x_i x_j,$$

where the numbers q_{ij} are assumed to integers.

Example 18.1.2.   1. The simplest quadratic forms are the sums of squares,

$$f(x_1, \ldots, x_n) = x_1^2 + \ldots x_n^2.$$

2. Other examples of quadratic forms include $f(x, y) = x^2 + ny^2$ and $g(x, y, z) = x^2 + y^2 + z^2 + xy + yz + xz$.

A binary quadratic form is a quadratic form in two variables. We usually write a binary quadratic form f as $f(x, y) = ax^2 + bxy + cy^2$. This can be written in matrix notation as

$$f(x, y) = (x, y) \begin{pmatrix} a & b/2 \\ b/2 & c \end{pmatrix} \begin{pmatrix} x \\ y \end{pmatrix}.$$

In this representation, we say that $f(x,y) = ax^2 + bxy + cy^2$ is associated with the matrix $\begin{pmatrix} a & b/2 \\ b/2 & c \end{pmatrix}$. We also denote a binary quadratic form $f(x,y) = ax^2 + bxy + cy^2$ as $f(x,y) = \{a,b,c\}$. Most of our study is restricted to binary quadratic forms. Some properties of forms in more variables are explored in the exercises.

Definition 18.1.3. A binary form $f(x,y) = ax^2 + bxy + cy^2$ is said to be **primitive** if the coefficients a, b, and c are relatively prime.

Definition 18.1.4. A number m is **represented** by a form if $f(a,b) = m$ for some a and b. The form **properly represents** m if $f(a,b) = m$ and $\gcd(a,b) = 1$.

The fundamental question in the theory of quadratic forms is to determine the numbers represented by a form, and the number of such representations. For example, the problem for sums of two squares was solved in Section 14.4. The method of descent was used to show that every prime $p \equiv 1$ (mod 4) was representable as a sum of two squares. We also showed that the representation was essentially unique, and from this fact, we can obtain a formula for the number of representations of an integer as a sum of two squares. Similar techniques can be used to show that any positive number can be represented as a sum of four squares. For other forms, the problem is more difficult. The goal of our study is to see to what extent such questions can be answered for binary quadratic forms using elementary techniques.

To understand the notion of representation of a number by a binary quadratic form, we define the concept of equivalence. It is clear that the numbers represented by the form $x^2 + 2y^2$ are the same as the one of $2x^2 + y^2$. It is not so obvious that $f(x,y) = x^2 + y^2$ and $g(x,y) = x^2 + 2xy + 2y^2$ represent the same numbers. This is because $g(x,y) = f(x+y,y)$ or $f(x,y) = g(x-y,y)$. If $f(a,b) = m$, then g also represents m as $g(a-b,b) = a^2 + b^2 = m$. Conversely, if $g(a,b) = m$, then $f(a+b,b) = m$.

We have two simple transformations of one form into another: one in which $g(x,y) = f(y,x)$ and another in which $g(x,y) = (x-ky,y)$. These transformations can be written using 2×2 matrices.

$$(y,x) = (x,y) \begin{pmatrix} 0 & 1 \\ 1 & 0 \end{pmatrix}$$

$$(x - ky, y) = (x,y) \begin{pmatrix} 1 & 0 \\ -k & 1 \end{pmatrix}.$$

Combinations of the two matrices in the transformations above gives rise to all integral matrices of determinant ± 1. (See Project 4 of Chapter 11.) Hence we make the following definition.

Definition 18.1.5. Two forms f and g are said to be **equivalent** if there exist integers p, q, r, and s, such that

$$f(x,y) = g(px + qy, rx + sy), \quad ps - qr = \pm 1.$$

The equivalence is **proper** if $ps - qr = 1$.

If f and g are equivalent, then we write $f \sim g$.

Example 18.1.6.   1. $f(x,y) = x^2 + 3xy + 3y^2$ and $g(x,y) = x^2 - xy + y^2$ are equivalent, as it is easy to verify that $f(x,y) = g(x + y, x + 2y)$. Moreover, they are properly equivalent since $1 \cdot 2 - 1 \cdot 1 = 1$.

2. The form $ax^2 + bxy + cy^2$ is equivalent to $ay^2 + bxy + cx^2$ by interchanging x and y.

The transformation from g to f can be written as

$$(px + qy, rx + sy) = (x, y) \begin{pmatrix} p & r \\ q & s \end{pmatrix},$$

and the condition $ps - qr = \pm 1$ is equivalent to the fact that the matrix $T = \begin{pmatrix} p & r \\ q & s \end{pmatrix}$ has an inverse with integer entries. We also use the notation $T \cdot f = f(px + qy, rx + sy)$. Then f and g are equivalent if there exists a matrix T with integer entries and determinant ± 1 such that $T \cdot f = g$.

If f is associated to the matrix A and g is associated to B, and f and g are equivalent, then

$$f(x,y) = g(px + qy, rx + sy)$$

$$(x, y)A \begin{pmatrix} x \\ y \end{pmatrix} = (px + qy, rx + sy)B \begin{pmatrix} px + qy \\ rx + sy \end{pmatrix}$$

$$= (x, y) \begin{pmatrix} p & r \\ q & s \end{pmatrix} B \begin{pmatrix} p & q \\ r & s \end{pmatrix} \begin{pmatrix} x \\ y \end{pmatrix}$$

$$= (x, y)TBT^t \begin{pmatrix} x \\ y \end{pmatrix},$$

hence $A = TBT^t$, where T^t is the transpose of T.

Lemma 18.1.7. *The notion of equivalence of binary quadratic forms is an equivalence relation.*

PROOF. Suppose the binary forms, f, g, and h are associated to the matrices A, B, and C, respectively. If $f \sim g$, then there exists T such that $B = TAT^t$. The condition on T is that it has integer entries and its determinant is ± 1.

Hence T^{-1} also has integer entries. We can write $A = T^{-1}B\left(T^{-1}\right)^t$, hence $g \sim f$. If $f \sim g$ and $g \sim h$, then there exist matrices T and U such that $B = TAT^t$ and $C = UBU^t$. Then $C = (UT)A(UT)^t$, hence $f \sim h$. This proves the transitivity of $\sim$, because UT is also an integral matrix of determinant ± 1. ∎

Lemma 18.1.8. *If f is equivalent to g, then f and g represent the same numbers.*

PROOF. Suppose there exist p, q, r, and s such that $ps - qr = \pm 1$ and $f(x,y) = g(px+qy, rx+sy)$. If $f(a,b) = m$, then $g(pa+qb, ra+sb) = m$. Conversely, if g represents m, $g(a,b) = m$, then $f(sa - qb, -ra + pb) = (ps - qr)^2 g(a,b) = m$. Hence f and g represent the same numbers. ∎

We can also verify quite easily that equivalent forms preserve proper representations. (See Exercise 1.)

It is clear that if f and g are properly equivalent, then they are equivalent, but equivalence does not imply proper equivalence. Under the transformation $x \mapsto x, y \mapsto -y$ (of determinant -1) $ax^2+bxy+cy^2$ is improperly equivalent to $ax^2 - bxy + cy^2$. These two are properly equivalent in a few cases. For example, $2x^2 + 2xy + 3y^2$ is properly equivalent to $2x^2 - 2xy + 3y^2$ as

$$\begin{pmatrix} 2 & 1 \\ 1 & 3 \end{pmatrix} = \begin{pmatrix} 1 & 0 \\ 1 & 1 \end{pmatrix} \begin{pmatrix} 2 & -1 \\ -1 & 3 \end{pmatrix} \begin{pmatrix} 1 & 1 \\ 0 & 1 \end{pmatrix},$$ but $3x^2 + xy + 5y^2$ and $3x^2 - xy + 5y^2$ are not properly equivalent. This will be clear in Section 18.3 after we study the reduction theory for these forms.

The connection between proper equivalence and proper representations is given by the following proposition.

Proposition 18.1.9. *A form $f(x,y)$ properly represents an integer m if and only if it is properly equivalent to $g(x,y) = mx^2 + b'xy + c'y^2$ for some b' and c'.*

PROOF. If f and g are properly equivalent, then $g(x,y) = f(px + qy, rx + sy)$ with $ps - qr = 1$. Since $g(1,0) = m$, we have $f(p,r) = m$, and $(p,r) = 1$, as a linear combination of them equals 1. This shows that f properly represents m.

Conversely, if $f(p,r) = m$, with $(p,r) = 1$, then there exist s and q such that $ps - qr = 1$. To compute the transformation, we use the matrix notation.

$$f(px + qy, rx + sy) = (px + qy, rx + sy)A \begin{pmatrix} px + qy \\ rx + sy \end{pmatrix} p$$

$$= ((px, rx) + (qy, sy)) A \left[\begin{pmatrix} px \\ rx \end{pmatrix} + \begin{pmatrix} qy \\ sy \end{pmatrix} \right]$$

$$= (px, rx)A \begin{pmatrix} px \\ rx \end{pmatrix} p + (qy, sy)Ap \begin{pmatrix} qy \\ sy \end{pmatrix}$$

$$+ (px, rx)A \begin{pmatrix} qy \\ sy \end{pmatrix} + (qy, sy)A \begin{pmatrix} px \\ rx \end{pmatrix}$$

$$= f(p, r)x^2 + (2apq + bps + bqr + 2crs)xy$$
$$+ f(r, s)y^2$$
$$= mx^2 + b'xy + c'y^2$$

for some b' and c'.   ∎

A fundamental quantity that determines the behavior of a form is its discriminant. The **discriminant** of $f(x, y) = ax^2 + bxy + cy^2$ is defined to be $\Delta(f) = b^2 - 4ac$. The determinant of the associated matrix $A = \begin{pmatrix} a & b/2 \\ b/2 & c \end{pmatrix}$ is $ac - b^2/4$. Therefore, $\Delta(f) = -4\det(A)$. The discriminant is preserved under equivalence (see Exercise 7).

To understand the effect of the discriminant on the form, we complete the square.

$$4af(x, y) = 4a^2x^2 + 4abxy + 4acy^2$$
$$= (2ax + by)^2 - b^2y^2 + 4acy^2$$
$$= (2ax + by)^2 - \Delta y^2.$$

If Δ is a perfect square, then the quadratic form factors into linear forms. Then the numbers represented by f can be determined by solving linear equations.

Example 18.1.10. Consider the representations of 15 by the form $f(x, y) = x^2 + 3xy + 2y^2$. The discriminant is 1, hence f factors into a product of linear forms. It is easy to see that $f(x, y) = (x + 2y)(x + y)$. There are eight ways to write 15 as a product of two integers: $1 \cdot 15$, $15 \cdot 1$, $3 \cdot 5$, $5 \cdot 3$, $-1 \cdot -15$, $-15 \cdot -1$, $-3 \cdot -5$, and $-5 \cdot -3$. Each one of these gives a set of two equations for x and y that can be solved for x and y. For example, $x + 2y = 1$ and $x + y = 15$ imply that $x = 29$ and $y = -14$. Continuing this way, we can obtain all the solutions to the equation $x^2 + 3xy + 2y^2 = 15$.

We restrict attention to the case when Δ is not a perfect square. If Δ is negative, then $4af$ is always nonnegative, hence f is always of the same sign, depending on the sign of a. If Δ is positive, then $4af$ can take both positive and negative values, and so does f. There is significant difference in the behavior of these two types of forms. In the first case, $f(x,y) = m$ has only finitely many solutions, whereas in the second case, there can be infinitely many solutions. For example, $x^2 - 2y^2$ is form with discriminant 8, and the equation $x^2 - 2y^2 = 1$ has infinitely many solutions.

Definition 18.1.11. A form f is **definite** if it represents only positive or negative values; otherwise, it is **indefinite**. It is **positive definite** if it represents only positive values and **negative definite** if it represents only negative values.

Binary quadratic forms with negative discriminants are definite, while those with positive discriminants are indefinite. We will restrict our study to definite forms. Indefinite forms are more complicated to study. A detailed treatment of indefinite forms is given by Buell.[1]

──────────────── **Exercises for Section 18.1** ────────────────

1. Show that if f and g are properly equivalent, then they properly represent the same numbers.

2. Show that proper equivalence is an equivalence relation, but improper equivalence is not.

3. Show that if there exist matrices T and S such that $T \cdot f = g$ and $S \cdot g = h$, then $ST \cdot f = h$.

4. Suppose f represents m. Show that we can write m as $d^2 m'$, where f properly represents m'.

5. Show that the discriminant Δ satisfies $\Delta \equiv 0, 1 \pmod 4$.

6. (a) Find all representations of 17 by the form $2x^2 + 9xy - 5y^2$.

    (b) Find all representations of 21 by the form $x^2 + 6xy - 16y^2$.

    (c) Solve the equation $2x^2 + 9xy + 4y^2 = 120$.

7. Show that if f and g are equivalent, then $\Delta(f) = \Delta(g)$. Can $13x^2 + 4xy + 21y^2$ and $14x^2 + 3xy + 20y^2$ be equivalent?

8. Determine which of the following forms are definite.

    (a) $2x^2 - 2xy + 3y^2$ (b) $3x^2 + 7xy + 19y^2$
    (c) $3x^2 + 13xy + 7y^2$ (d) $17x^2 + 19xy + 23y^2$

[1]D. Buell. *Binary Quadratic Forms*. New York: Springer–Verlag, 1980.

9. The form $3x^2 + 8y^2$ properly represents 35. Find a transformation T such that $T \cdot (3x^2 + 8y^2) = 35x^2 + bxy + cy^2$ for some b and c.

10. Suppose f is a quadratic form in n-variables, $f(x_1, \ldots, x_n) = \sum_{i=1}^{n} a_{ii}x_i^2 + \sum_{1 \le i < j \le n} a_{ij}x_i x_j$. The determinant of f is defined to be the determinant of the matrix A, where the ij entry of A is a_{ij}. We say that the matrix A is associated with f. If A and B are associated with forms f and g in n variables, we say that $f \sim g$ (f is equivalent to g) if there exists an integral matrix T, such that $B = TAT^t$ and $\det T = \pm 1$. Show that equivalence of forms is an equivalence relation and equivalent forms represent the same numbers.

11. Show that $f = \{a, b, c\}$ is positive definite if and only if $a > 0$ and $\Delta < 0$.

12. For quadratic forms in more than two variables, the determinant is not sufficient to determine if a form is definite. To obtain a condition on definiteness for a ternary quadratic form $f = \sum_{i=1}^{3} \sim_{j=1}^{3} a_{ij}x_i x_j$, show that

$$a_{11}f = (a_{11}x_1 + a_{12}x_2 + a_{13}x_3)^2 + g(x_2, x_3),$$

where g is the binary form $g = \{a_{11}a_{22} - a_{12}^2, a_{11}a_{23} - a_{12}a_{13}, a_{11}a_{33} - a_{13}^2\}$. Suppose δ is the determinant of f. Show that f is positive definite if and only if $a_{11} > 0$, $a_{11}a_{22} - a_{12}^2 > 0$ and $\delta > 0$.

18.2 Reduction of Definite Forms-I

> The number of a class is the class of all classes
> similar to the given class.
>
> — BERTRAND RUSSELL

We now turn our attention to the main problem in the theory. Given a form $f(x, y)$ and an integer m, we want to decide if m is represented by the form. For example, we know that $n = x^2 + y^2$ if and only if n does not have a prime factor of the form $4k + 3$ occurring to an odd exponent. All primes of the form $4k + 1$ are representable as $x^2 + y^2$. For other forms, it is not possible to give such a simple characterization, like a congruence condition modulo the discriminant (-4 in this case). In later sections, we will explore the issue of congruence conditions on m, which can settle the problem of representation. We will start with a partial answer to the problem of representation, which shows that proper equivalence of forms is the key to the solution.

Proposition 18.2.1. *Let Δ be an integer satisfying $\Delta \equiv 0, 1 \pmod 4$ and let m be an integer such that $(m, \Delta) = 1$; then m is properly represented by a primitive form of discriminant Δ if and only if Δ is a quadratic residue modulo $4m$.*

PROOF. If m is properly represented by a primitive form f of discriminant Δ, then by Proposition 18.1.9, f is properly equivalent to $mx^2 + bxy + cy^2$. Hence $\Delta = b^2 - 4mc$, and $\Delta \equiv b^2 \pmod{4m}$.

Conversely, if $\Delta \equiv b^2 \pmod{4m}$, there exists c such that $\Delta = b^2 - 4mc$. The form $mx^2 + bxy + cy^2$ properly represents m and is primitive since $(m, \Delta) = 1$ implies $(m, b) = 1$. ∎

Remark. The condition $\Delta \equiv 0, 1 \pmod{4}$ is necessary since $\Delta = b^2 - 4ac$ and $b^2 \equiv 0, 1 \pmod{4}$.

The proposition gives a condition on when m is represented by some form of discriminant Δ, but it doesn't answer the question whether m is represented by a particular form. Since equivalent forms represent the same numbers, if all the forms of a given discriminant Δ are equivalent to one another, then m is represented by any form of discriminant Δ if Δ is a quadratic residue modulo $4m$, and, in this case, the problem is fully solved. In the general case, we consider equivalence classes of forms by using Lagrange's theory of reduced forms, which allows us to determine the equivalence classes of forms of a given discriminant.

To study forms by equivalence, it is enough to find a form in each equivalence class. For this, Lagrange defined the notion of a reduced form. It turns out that these form a complete set of representatives for equivalence classes of forms; that is, every equivalence class contains one, and only one, reduced form. We restrict our study to positive definite forms, hence $\Delta < 0$ and $a > 0$.

Definition 18.2.2. A primitive positive definite binary form $ax^2 + bxy + cy^2$ is said to be **reduced** if $|b| \le a \le c$, and $b \ge 0$ if either $|b| = a$ or $a = c$.

Example 18.2.3.   1. $x^2 + y^2$ is a reduced form since $|0| < 1 \le 1$ and $b \ge 0$. The form $x^2 + ny^2$ of discriminant $-4n$ is reduced for every $n > 0$.

2. The form $x^2 + xy + y^2$ is reduced, but $x^2 - xy + y^2$ is not reduced. In the latter form, the middle coefficient is -1, but it should be positive when $a = c$.

3. The form $2x^2 + 2xy + 3y^2$ is reduced, but $2x^2 - 2xy + 3y^2$ is not. This is because $|-2| = 2$, that is, $|b| = a$, so we must have $b > 0$.

4. If a is given, the possible values of b in a reduced form $ax^2 + bxy + cy^2$ are $-a + 1, -a + 2, \ldots, 0, 1, \ldots, a$.

If f and g are properly equivalent, then there is a unimodular transformation, (that is, an integer matrix of determinant 1) that transforms f to g. The proof of the reduction of positive definite forms is based on the fact that such a matrix can be written as products of two elementary transformations, U and W, which are given by

$$U = \begin{pmatrix} 1 & 0 \\ -1 & 1 \end{pmatrix}, \qquad W = \begin{pmatrix} 0 & 1 \\ -1 & 0 \end{pmatrix}.$$

U has the effect $x \mapsto x - y$ and $y \mapsto y$, and W, the effect of interchanging x and y with a sign change. The transformation $U^k = \begin{pmatrix} 1 & 0 \\ -k & 1 \end{pmatrix}$ sends $x \mapsto x - ky$ and $y \mapsto y$. If we allow k to be negative, then $U^{-1} = \begin{pmatrix} 1 & 0 \\ 1 & 1 \end{pmatrix}$ is, of course, the inverse of U, and $U^{-k} = \left(U^k\right)^{-1}$.

Starting from a form $f = ax^2 + bxy + cy^2$, we repeatedly apply these transformations until the form is reduced. First, if $c < a$, then we can interchange x and y using W and make $c \geq a$. If $|b| > a$, then we replace x by $x - ky$ to obtain

$$a(x - ky)^2 + b(x - ky)y + cy^2$$
$$= ax^2 + (-2ak + b)xy + (ak^2 - bk + c)y^2.$$

Let $b' = -2ak + b$ and $c' = ak^2 - bk + c$. We want to choose k so that $|b'| \leq a$, that is, $-a < -2ak + b < a$. Solving for k, we obtain the inequalities,

$$\frac{b}{2a} - \frac{1}{2} < k < \frac{b}{2a} + \frac{1}{2}.$$

This implies that k is the nearest integer to $b/2a$. For this choice of k, it is possible that $c' < a$, in which case, we repeat the procedure. The algorithm eventually stops because we are reducing the size of the middle coefficient.

We illustrate the technique with an example.

Example 18.2.4. Consider the form $70x^2 + 230xy + 189y^2$. This has discriminant -20 and is not reduced as $230 > 70$. We apply U^k for a suitable k to reduce the middle coefficient. This gives

$$70(x - ky)^2 + 230(x - ky)y + 189y^2 = 70x^2 - 140kxy + 230xy$$
$$+ 70k^2y^2 - 230ky^2 + 189y^2.$$

It is clear that only $k = 2$ will reduce the coefficient of xy so that it is less than 70 in absolute value, and we obtain

$$70x^2 - 50xy + 9y^2.$$

The coefficient of y^2 is less than that of x^2, so we switch x and y to get the form

$$9x^2 + 50xy + 70y^2.$$

Again applying U^k with $k = 3$ yields

$$9x^2 - 4xy + y^2.$$

We switch the terms by applying W to obtain

$$x^2 + 4xy + 9y^2.$$

Applying U^k with $k = 2$ yields

$$x^2 + 5y^2,$$

which is a reduced form.

Theorem 18.2.5. *The number of reduced forms of a given discriminant is finite.*

PROOF. If $ax^2 + bxy + cy^2$ is reduced, then $\Delta = b^2 - 4ac$ and $|b| \le a \le c$. This implies that

$$b^2 \le a^2 \le c^2$$

and

$$a^2 \le ac \le c^2.$$

Hence

$$b^2 - 4ac \le a^2 - 4a^2$$
$$\le -3a^2.$$

Hence we obtain

$$a^2 \le -\Delta/3$$

and

$$c^2 \ge -\Delta/4.$$

There are only finitely many choices for a and hence for b. Once a and b are known, then c is determined by the equation $b^2 - 4ac = \Delta$. ∎

The number of reduced forms of discriminant Δ is called the **class number** of Δ and denoted by $h(\Delta)$. We give a few examples of determining the number of reduced forms of a given discriminant.

Example 18.2.6.   1. For $\Delta = -3$, since $a^2 \le -\Delta/3$, we must have $a \le 1$. We cannot have $a = 0$; otherwise, Δ would be perfect square, so $a = 1$. Also, b cannot be 0 since Δ is not a multiple of 4, hence $b = 1$. Therefore, $c = 1$, and the only reduced form of discriminant -3 is $x^2 + xy + y^2$, so $h(-3) = 1$.

Joseph Louis Lagrange (1736–1813)

Lagrange

Lagrange was born in Turín and studied at the University of Turín. He stayed in Turín until 1766 when he was invited to Berlin by Frederick the Great. In 1786, Lagrange moved to Paris and stayed there until his death.

Apart from Euler, Lagrange was the most productive mathematician of the eighteenth century. Lagrange discovered the method of Calculus of Variations in 1754. Problems of the path of quickest descent or isoperimetry are examples of the calculus of variations. Lagrange considered the possibility of showing the unsolvability of polynomial equations by considering the permutations of roots, an idea that led to the foundation of group theory in the works of Abel and Galois. In number theory, Lagrange showed that every positive integer is a sum of four squares. He was the first to give complete proofs regarding the solutions of the equation $x^2 - dy^2 = 1$. Lagrange made important contributions to the theory of binary quadratic forms, work that paved the way for the deep results of Gauss.

In 1793, he was appointed to the committee charged with standardizing the weights and measures, and his work led to the creation of the metric system.

2. For $\Delta = -4$, $a \leq 1$. If $b = 0$, then $c = 1$ is a solution. If $b = 1$, then there is no solution for c in $1^2 - 4c = -4$. Hence $x^2 + y^2$ is the only reduced form of discriminant -4, and the class number is $h(-4) = 1$.

3. For $\Delta = -20$, we must have $a \leq 2$. If $a = 1$, then $b = 0$ has a solution $c = 5$. $a = 1$ and $b = 1$ has no solution for c. If $a = 2$, then $b = 0$ is not possible. $b = \pm 1$ is not possible either. If $b = 2$, then $c = 3$ is a solution. Hence the reduced forms of discriminant -20 are $x^2 + 5y^2$ and $2x^2 + 2xy + 3y^2$ and $h(-20) = 2$.

In the next section, we prove that any binary quadratic form is equivalent to a unique reduced form, and no two reduced forms are equivalent. The methods given in this section can be used to compute all the reduced forms of a given discriminant. The problem of representability of integers is solved using Proposition 18.2.1. Using this, the theorem for representation of integers, when the class number is one, is direct, but for other discriminants, we need the genus theory.

─────────────────── **Exercises for Section 18.2** ───────────────────

1. Determine which of the following forms are reduced.

   (a) $x^2 - xy + 2y^2$

   (b) $x^2 + xy + 2y^2$

   (c) $2x^2 - xy + 2y^2$

   (d) $2x^2 - xy + y^2$

2. Reduce the following quadratic forms using the technique outlined in Example 18.2.4.

   (a) $651x^2 + 2718xy + 2837y^2$

   (b) $178x^2 - 3520xy + 14835y^2$

   (c) $26x^2 - 381xy + 1396y^2$

3. Write a program to reduce a definite form using the method given in this section.

4. Show that if $\Delta \equiv 1 \pmod 4$, then $x^2 + xy + \dfrac{1-\Delta}{4}y^2$ is a reduced form, and if $\Delta \equiv 0 \pmod 4$, then $x^2 - \dfrac{\Delta}{4}y^2$ is always a reduced form.

5. Determine all reduced forms of discriminant -15, -23, and -96.

6. Write a computer program to determine all reduced primitive forms of a discriminant $\Delta < 0$.

7. If f is a reduced form, show that $f(x,y) \geq (a - |b| + c) \min(x^2, y^2)$. Show that a is the smallest nonzero value of f, and c is the next smallest number properly represented by f.

8. There is a remarkable formula due to Dirichlet for computing the class number $h(\Delta)$. Let p be an odd prime, $p \equiv 3 \pmod 4$. Let A be the sum of all the quadratic residues modulo p and B the sum of all the quadratic non-residues modulo p. (Here the residues are taken in the standard residue system.) Then Dirichlet's formula is

$$h(-p) = \frac{B - A}{p}.$$

Verify Dirichlet's formula for $p = 19$, 23. Write a computer program to compute the class number for $\Delta = -p$ by this method.

9. A binary quadratic form $ax^2 + bxy + cy^2$ of discriminant $\Delta > 0$ is said to be reduced if $0 < b < \sqrt{\Delta}$ and $\sqrt{\Delta} - b < |2a| < \sqrt{\Delta} + b$. Show that there are a finite number of reduced forms for a given discriminant.

18.3 Reduction of Definite Forms-II

The next step in the theory of binary quadratic forms is to show that each equivalence class of forms has only one reduced form.

A positive definite form $f(x, y) = ax^2 + bxy + cy^2$ can be factored as

$$a(x - \tau y)(x - \bar{\tau} y),$$

where τ is a complex number. Here τ is the root of the quadratic equation $ax^2 + bx + c = 0$, with discriminant $\Delta < 0$. Either τ or $\bar{\tau}$ has positive imaginary part, and we select τ such that $\operatorname{Im} \tau > 0$. We say that τ is **associated** to f and write $f \to \tau$.

Now we see which complex numbers correspond to positive definite forms. Let $\tau = r + is$; then

$$\frac{1}{a} f(x, y) = (x - (r + is)y)(x - (r - is)y)$$
$$= x^2 - 2rxy + (r^2 + s^2)y^2.$$

Since $f(x, y)$ has integer coefficients, we need r to be a rational number, and this implies that s^2 must also be a rational number. Therefore, we have a one-to-one correspondence between primitive positive definite forms and complex numbers with rational real parts, and imaginary parts that are square roots of rational numbers.

Lemma 18.3.1. *Let f and g be two positive definite forms that are properly equivalent, that is, $g(x, y) = f(px + qy, rx + sy)$ with $ps - qr = 1$. If $f \to \tau$ and $g \to \tau'$, then $\tau' = \dfrac{s\tau - q}{-r\tau + p}$.*

PROOF. Since $f(x, y) = a(x - \tau y)(x - \bar{\tau} y)$, we have

$$g(x, y) = f(px + qy, rx + sy)$$
$$= a(px + qy - \tau(rx + sy))(px + qy - \bar{\tau}(rx + sy))$$
$$= a(x(p - r\tau) - y(s\tau - q))(x(p - r\bar{\tau}) - y(s\bar{\tau} - q))$$
$$= a|p - r\tau|^2 \left(x - \left(\frac{s\tau - q}{-r\tau + p} \right) y \right) \left(x - \left(\frac{s\bar{\tau} - q}{-r\bar{\tau} + p} \right) \right).$$

Then either $\tau' = \dfrac{s\tau - q}{-r\tau + p}$ or $\tau' = \dfrac{s\bar{\tau} - q}{-r\bar{\tau} + p}$. It is easy to check that

$\mathrm{Im}\left(\dfrac{s\tau - q}{-r\tau + p}\right) = \dfrac{1}{|p - r\tau|^2}\,\mathrm{Im}\,\tau$. Since we selected the root with positive imaginary part as corresponding to a form, we have completed the proof of the lemma. ∎

Definition 18.3.2. We say that τ and τ' are **equivalent** if there exists a matrix $A = \begin{pmatrix} a & b \\ c & d \end{pmatrix}$ of determinant 1 such that $A\tau := \tau' = \dfrac{a\tau + b}{c\tau + d}$.

The above lemma shows that if f transforms to g under T, then $\tau' = (T^t)^{-1}\tau$.

Theorem 18.3.3 (Lagrange). *Every primitive positive definite form is properly equivalent to a unique reduced form.*

PROOF. If $f(x, y) = ax^2 + bxy + cy^2 = a(x - \tau y)(x - \bar{\tau} y)$, then by comparing both sides we obtain

$$a(\tau + \bar{\tau}) = 2a\,\mathrm{Re}\,\tau = -b$$

and

$$a(\tau\bar{\tau}) = a|\tau|^2 = c.$$

We repeatedly use the transformations U^k and W on τ where $U^k = \begin{pmatrix} 1 & 0 \\ -k & 1 \end{pmatrix}$, $k \in \mathbb{Z}$, and $W = \begin{pmatrix} 0 & 1 \\ -1 & 0 \end{pmatrix}$.

These correspond to

$$
\begin{array}{ccc}
\begin{aligned}
x &\to x - ky \\
y &\to y, \\
\tau &\to \tau - k
\end{aligned}
& \text{and} &
\begin{aligned}
x &\to y \\
y &\to -x \\
\tau &\to -1/\tau.
\end{aligned}
\end{array}
$$

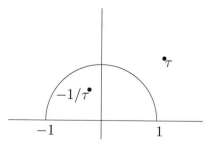

Figure 18.3.1 $\tau' = W\tau$

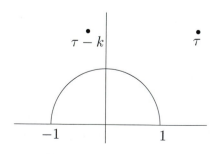

Figure 18.3.2 $\tau' = U^k\tau$

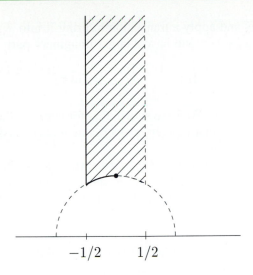

Figure 18.3.3 The domain $\mathcal{F}$

Consider the domain $\mathcal{F} = \{-1/2 \le \mathrm{Re}\,\tau < 1/2, |\tau| > 1\} \cup \{-1/2 \le \mathrm{Re}\,\tau \le 0, |\tau| = 1\}$ in the upper half plane. (See Figure 18.3.3.)

Suppose τ lies in $\mathcal{F}$. This implies that $|\tau + \bar{\tau}| = |2\,\mathrm{Re}\,\tau| \le 1$, and $|\tau| \ge 1$, which imply that $|b| \le a$ and $c \ge a$. Moreover, if $|b| = a$, then $|\mathrm{Re}\,\tau = 1/2$, so $\mathrm{Re}\,\tau = -1/2$, which implies that $b \ge 0$; and, if $c = a$, then $|\tau| = 1$, so $\mathrm{Re}\,\tau \le 0$, and again, $b \ge 0$. We have proved that if τ lies in the region $\mathcal{F}$, then the corresponding form is reduced.

Given any other τ, we apply the transformations U^k and W to bring it into the region $\mathcal{F}$. These transformations are the same as the ones used in the reduction algorithm of the previous section.

For a given τ, apply a suitable U^k to make $\mathrm{Re}\,\tau$ between $-1/2$ and $1/2$. Now, if τ is not in $\mathcal{F}$, then either $|\tau| < 1$ or $|\tau| = 1$ and $0 < \mathrm{Re}\,\tau \le 1/2$. In the latter case, if we apply W, then $|\tau| = 1$ and $-1/2 \le \mathrm{Re}\,\tau \le 0$ so it is in $\mathcal{F}$. If $|\tau| < 1$, we can apply W to make the absolute value larger than 1. If τ is not in $\mathcal{F}$, then we apply U^k again, and so on. The process will eventually stop because in each step we are reducing the values of a, b, and c; when it stops, then τ will be in $\mathcal{F}$.

To see this in another way, recall that

$$\mathrm{Im}\,A\tau = \frac{\mathrm{Im}\,\tau}{|r\tau + s|^2}, \qquad A = \begin{pmatrix} p & q \\ r & s \end{pmatrix}.$$

If $\mathrm{Im}\,A\tau \ge \mathrm{Im}\,\tau$, then $|r\tau + s|^2 \le 1$. This inequality has only finitely many solutions for r and s, because, for a fixed τ, $|r\tau + s|$ describes an ellipse in the (r, s) plane; hence, it contains a finite number of points with integer coordinates. Therefore, among all $A\tau$ there exists one with the largest imaginary part. We pick this and apply a translation to bring it into $\mathcal{F}$. Clearly, it must be in $\mathcal{F}$; otherwise, $-1/\tau$ will have a larger imaginary part:

$$\operatorname{Im} \frac{-1}{\tau} = \frac{\operatorname{Im}\tau}{|\tau|^2} > \operatorname{Im}\tau,$$

contradicting the choice. We have shown that for every τ' there is a τ equivalent to it that lies in $\mathcal{F}$, that is, every proper equivalence class of binary forms contains a reduced form.

Now, we need to see that two reduced forms cannot be properly equivalent. Suppose τ and τ' are both in $\mathcal{F}$ and are equivalent. Then $\operatorname{Im}\tau = \operatorname{Im}\tau'/|r\tau + s|^2$. We may assume (by taking the inverse if necessary) that $|r\tau + s|^2 \le 1$.

If $\tau = x + iy$, then $|r\tau + s|^2 = (rx + s)^2 + (ry)^2 \le 1$. Since $y \ge \sqrt{3}/2$, we have that $1 \ge (rx + s)^2 + (ry)^2 \ge (rx + s)^2 + 3/4r^2$, which implies that $|r| \le 1$.

If $r = 0$, then $s = \pm 1$; hence $p = \pm 1$ and $\tau = \tau' + q$. But q can only be 0 since a translation by one unit or more sends τ' outside $\mathcal{F}$.

If $r = 1$, then $|\tau + s|^2 \le 1$ or $(x + s)^2 + y^2 \le 1$, since $|x| < 1/2$, and $|s| \le 1$. If $s = 0$, then $|\tau|^2 = 1$, so τ lies in the portion of the circle that belongs to $\mathcal{F}$, that is, $-1/2 < \operatorname{Re}\tau \le 0$, and we have $A = \begin{pmatrix} p & -1 \\ 1 & 0 \end{pmatrix}$ so that $\tau' = (p\tau - 1)/\tau$ or $\tau' = p - 1/\tau$. But $-1/\tau$ has absolute value equal to 1 and the real part between 0 and $1/2$, so if $\tau' \in \mathcal{F}$, we must have $p = 0$ and $\operatorname{Re}\tau = 0$, in which case $\tau' = \tau$. We cannot have $s = \pm 1$ because that sends τ outside $\mathcal{F}$.

The study of $r = -1$ is similar to the one for $r = 1$.

We have proved that it is not possible to have different τ and τ' in $\mathcal{F}$ that are equivalent, and this implies that each equivalence class of forms can only have one reduced form. ∎

This proof can be written completely without the use of complex numbers translating each operation into the language of forms, but it is more difficult to visualize the extra conditions in the definition of reduced forms. The equivalence of complex numbers in the upper half plane under the linear fractional transformations with integer matrices of determinant 1 is a fundamental object in mathematics. It has important applications in number theory via the connection between elliptic curves and modular forms.

We give a table of reduced forms for small discriminants in Table 18.3.1. These include only the primitive reduced forms. The problem of representation of a number by a form can be solved in a few cases where the class number is 1. We give examples of the use of Proposition 18.2.1.

Example 18.3.4. The form $x^2 + y^2$ has discriminant -4, and all forms of discriminant -4 are properly equivalent. An odd integer m is properly repre-

Table 18.3.1 Class numbers and reduced forms for small discriminants

Δ	$h(\Delta)$	Reduced forms
-3	1	$x^2 + xy + y^2$
-4	1	$x^2 + y^2$
-7	1	$x^2 + xy + 2y^2$
-8	1	$x^2 + 2y^2$
-11	1	$x^2 + xy + 3y^2$
-12	1	$x^2 + 3y^2$
-15	2	$x^2 + xy + 4y^2$
		$2x^2 + xy + 2y^2$
-16	1	$x^2 + 4y^2$
-19	1	$x^2 + xy + 5y^2$
-20	2	$x^2 + 5y^2$
		$2x^2 + 2xy + 3y^2$
-23	3	$x^2 + xy + 6y^2$
		$2x^2 - xy + 3y^2$
		$2x^2 + xy + 3y^2$

sented by some form of discriminant -4 (and hence by $x^2 + y^2$) if and only if -4 is a quadratic residue modulo $4m$, that is, $b^2 \equiv -4 \pmod{4m}$ has a solution. In this case, b must be even, and, dividing by 4, we obtain $b'^2 \equiv -1 \pmod{m}$. By the Chinese Remainder Theorem, we have to solve this congruence for each prime power occurring in the factorization of m. We have already solved this congruence in Sections 3.4 and 9.3, and we give a brief outline of the steps for the reader's benefit.

Now, for an odd prime p, $b^2 \equiv -1 \pmod{p^r}$ implies that $b^2 \equiv -1 \pmod{p}$, which is not solvable for $p \equiv 3 \pmod{4}$. The congruence $b^2 \equiv -1 \pmod{p}$ has a solution for $p \equiv 1 \pmod{4}$ and we show by induction on r that the congruence $b^2 \equiv -1 \pmod{p^r}$ also has a solution. Suppose $b^2 \equiv -1 \bmod p$. Then $b^2 + 1 = kp$ for some k. We look for a solution to $x^2 \equiv -1 \pmod{p^2}$ of the form $x = b + tp$ for some t. The condition is

$$x^2 + 1 \equiv (b + tp)^2 + 1 \pmod{p^2}$$
$$\equiv b^2 + 1 + 2tbp + t^2 p^2 \pmod{p^2}$$
$$\equiv kp + 2tbp \pmod{p^2}$$
$$\equiv p(k + 2tb) \pmod{p^2}$$

and the equation $2tb + k \equiv 0 \pmod{p}$ is solvable as $2b$ is relatively prime to

p. For this value of t, $b + tp$ is a solution to $x^2 \equiv -1 \pmod{p^2}$. Similarly, we can continue this for higher powers of p.

For $p = 2$, -1 is a square. But there is no solution for higher exponents.

Therefore, we can conclude that m is properly represented by the form $x^2 + y^2$, or by any positive definite form of discriminant -4 if and only if it has no prime factor congruent to 3 mod 4 and is not divisible by 4.

Example 18.3.5. A number m relatively prime to 20 is represented by some form of discriminant -20 if and only if $b^2 \equiv -20 \pmod{4m}$ has a solution. Then b must be even, and, dividing by 4, we obtain $b'^2 \equiv -5 \pmod{m}$. For simplicity, we consider the case when m is an odd prime not equal to 5.

We want to determine when $\left(\dfrac{-5}{p}\right) = 1$. By quadratic reciprocity,

$$\left(\frac{5}{p}\right)\left(\frac{p}{5}\right) = (-1)^{\frac{p-1}{2}\frac{4}{2}} = 1.$$

Hence

$$\left(\frac{-5}{p}\right) = \left(\frac{p}{5}\right)(-1)^{(p-1)/2}.$$

Then, using the Chinese Remainder Theorem, we see that -5 is a quadratic residue modulo p if and only if $p \equiv 1, 3, 7, 9 \pmod{20}$. This implies that these p are represented by some form of discriminant -20, hence by some reduced form of discriminant -20. There are two reduced forms of discriminant -20: $x^2 + 5y^2$ and $2x^2 + 2xy + 3y^2$, and these represent primes $p \equiv 1, 3, 7, 9 \pmod{20}$. We can check that the primes $p = x^2 + 5y^2$ are congruent to 1 and 9 modulo 20, and the primes represented by $2x^2 + 2xy + 3y^2$ are congruent to 3 and 7 modulo 20.

─────────────────────── **Exercises for Section 18.3** ───────────────────────

1. If $f = \{a, b, c\}$ is a nonprimitive form with $(a, b, c) = d$, show that $d^2 \mid \Delta$ and $\{a/d/b/d/c/d\}$ is a primitive form of discriminant Δ/d^2. Show that the number of equivalence classes of forms of discriminant Δ is

$$\sum_{d^2 \mid \Delta, d > 0} h(\Delta/d^2).$$

2. Show that $\mathrm{Im}(A\tau) = \dfrac{\mathrm{Im}\,\tau}{|c\tau + d|^2}$.

3. This exercise concerns representability by forms of discriminant -3.

      (a) Show that there is only one equivalence class of forms of discriminant -3 with a reduced representative $x^2 + xy + y^2$.

      (b) Show that if m is properly represented by $x^2 + xy + y^2$, then all prime factors of m are congruent to 1 modulo 3.

      (c) Show that if $b^2 \equiv -3 \pmod{p}$ has a solution, then $b^2 \equiv -3 \pmod{p^r}$ has a solution for all $r > 1$.

      (d) Conclude that m is properly represented by a form of discriminant -3 if and only if it has no prime factor congruent to 2 modulo 3 and is not divisible by 9.

4. Which primes are represented by the unique reduced form of discriminant -19?

5. Find all discriminants Δ, $-1000 < \Delta < 0$, such that $h(\Delta) = 1$. Determine which primes are represented by the unique reduced form for these discriminants.

18.4   Genus Theory

To answer our question about which numbers are represented by a quadratic form, Proposition 18.2.1 provides a partial answer. This lemma is most useful when there is only one equivalence class of forms for a given discriminant, that is, $h(-\Delta) = 1$. The problem of representation is more difficult in general, and much more information can be obtained by studying the values represented by a form modulo m for different m. We consider a few examples first to see what this theory is about and then develop the general techniques in the next section.

    To solve problems over the integers, we need to put together information obtained by considering the problem modulo p^r for different primes p.

Example 18.4.1. Consider forms of discriminant $\Delta = -20$. There are two equivalence classes of forms with representatives $x^2 + 5y^2$ and $2x^2 + 2xy + 3y^2$. Proposition 18.2.1 shows that for representability of primes, we obtain

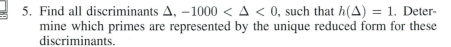

$$\left. \begin{array}{c} x^2 + 5y^2 = p \\ 2x^2 + 2xy + 3y^2 = p \end{array} \right\} \Longleftrightarrow \left\{ \begin{array}{l} -20 \equiv b^2 \pmod{4p} \text{ has a solution,} \\ \text{or, equivalently,} \\ p \equiv 1, 3, 7, 9 \pmod{20} \end{array} \right.$$

The equivalence on the right is a consequence of the quadratic reciprocity law. A computation shows that these forms represent different primes. At first, it is not clear how to separate the two forms to make a more precise statement about the primes represented by each. To distinguish between these, we look at the values represented by the forms in $(\mathbb{Z}/20\mathbb{Z})^{\times}$, the invertible elements modulo 20.

We can compute (as there are not too many possibilities) that $x^2 + 5y^2$ represents 1 and 9 in $(\mathbb{Z}/20\mathbb{Z})^\times$, and $2x^2 + 2xy + 3y^2$ represents 3 and 7 in $(\mathbb{Z}/20\mathbb{Z})^\times$.

If $p \equiv 1,\ 3,\ 7,\ 9 \pmod{20}$, one of these two forms represents p, and if it is represented in the integers, the same equation must be true modulo 20, so we must have

$$x^2 + 5y^2 = p \Leftrightarrow p \equiv 1,\ 9 \pmod{20}$$
$$2x^2 + 2xy + 3y^2 = p \Leftrightarrow p \equiv 3,\ 7 \pmod{20}.$$

The above excludes $p = 2$ and $p = 5$, which are represented by the second and first forms, respectively. This completely solves the problem of representing primes by each form. We can then extend this to look at representation of other integers.

Example 18.4.2. Now consider the forms of discriminant $\Delta = -84$. There are four reduced forms (hence equivalence classes) of discriminant -84, these are $x^2 + 21y^2$, $2x^2 + 2xy + 11y^2$, $3x^2 + 7y^2$, and $5x^2 + 4xy + 5y^2$. As an application of Proposition 18.2.1, we obtain

$$\left.\begin{array}{r} x^2 + 21y^2 = p \\ 2x^2 + 2xy + 11y^2 = p \\ 3x^2 + 7y^2 = p \\ 5x^2 + 4xy + 5y^2 = p \end{array}\right\} \Longleftrightarrow \left\{\begin{array}{l} -84 \equiv b^2 \pmod{4p} \\ \text{or, equivalently,} \\ p \equiv 1, 5, 11, 17, 19, 23, 25, \\ 31, 37, 41, 55, 71 \pmod{84}. \end{array}\right.$$

To distinguish among these forms, we look at the values represented modulo 84. There are not many values to be computed, and the task can easily be performed on a computer. This yields the following representations.

$$x^2 + 21y^2 \text{ represents } 1, 25, \text{ and } 37 \text{ in } (\mathbb{Z}/84\mathbb{Z})^\times.$$
$$2x^2 + 2xy + 11y^2 \text{ represents } 11, 23, \text{ and } 71 \text{ in } (\mathbb{Z}/84\mathbb{Z})^\times.$$
$$3x^2 + 7y^2 \text{ represents } 19, 31, \text{ and } 55 \text{ in } (\mathbb{Z}/84\mathbb{Z})^\times.$$
$$5x^2 + 4xy + 5y^2 \text{ represents } 5, 17, \text{ and } 41 \text{ in } (\mathbb{Z}/84\mathbb{Z})^\times.$$

Using the previous result, we can state the following theorems about representation of primes by forms of discriminant -84.

$$x^2 + 21y^2 = p \Leftrightarrow p \equiv 1,\ 25,\ 37 \pmod{84}$$
$$2x^2 + 2xy + 11y^2 = p \Leftrightarrow p \equiv 11,\ 23,\ 71 \pmod{84}$$
$$3x^2 + 7y^2 = p \Leftrightarrow p \equiv 19,\ 31,\ 55 \pmod{84}$$
$$5x^2 + 4xy + 5y^2 = p \Leftrightarrow p \equiv 5, 17, 41 \pmod{84}$$

The above two examples illustrate the basic idea of genus theory, that is, looking at the values represented by the form modulo the discriminant. This can be used to separate the forms and obtain theorems on representation of primes. To fix these ideas, we make the following definition.

Definition 18.4.3. Two forms f_1 and f_2 of discriminant Δ are said to be in the same **genus** if f_1 and f_2 represent the same values in $(\mathbb{Z}/\Delta\mathbb{Z})^\times$.

The set of equivalence classes of binary quadratic forms is divided into **genera**, and each genus consists of a finite number of classes.

We will study quadratic forms modulo m. Suppose $m = p^{e_1} \cdots p_k^{e_k}$.

Lemma 18.4.4. *Two forms f and g represent the same values in $(\mathbb{Z}/m\mathbb{Z})^\times$ if and only if they represent the same values in each $(\mathbb{Z}/p_i^{e_i}\mathbb{Z})^\times$.*

PROOF. This is an easy consequence of the Chinese Remainder Theorem. ∎

In the above examples, each genus consisted of only one form, and we were able to state the theorems about representation. This is a powerful theory, as one can obtain many more theorems by considering the forms modulo the discriminant.

This theory has its limitations. It will not always happen that each genus will consist of a single form. Consider the following example.

Example 18.4.5. Let $\Delta = -56$. There are four reduced forms of discriminant -56. Using Proposition 18.2.1, we obtain

$$
\left.
\begin{array}{r}
x^2 + 14y^2 = p \\
3x^2 + 2xy + 5y^2 = p \\
2x^2 + 7y^2 = p \\
3x^2 - 2xy + 5y^2 = p
\end{array}
\right\}
\Longleftrightarrow
\left\{
\begin{array}{l}
-56 \equiv b^2 \pmod{4p} \text{ has a solution} \\
\text{or, equivalently,} \\
p \equiv 1, 3, 5, 9, 13, 15, 19, 23 \\
\quad 25, 27, 39, 45 \pmod{56}
\end{array}
\right.
$$

Looking at the values modulo 56, we notice that $x^2 + 14y^2$ and $2x^2 + 7y^2$ both represent the same values $1, 9, 15, 23, 25, 25, 39$ in $(\mathbb{Z}/56\mathbb{Z})^\times$. This does not allow us to distinguish between these forms at the level of the integers. For example, $x^2 + 14y^2$ represents 23, but $2x^2 + 7y^2$ does not. The forms $3x^2 + 2xy + 5y^2$ and $3x^2 - 2xy + 5y^2$ both represent $3, 5, 13, 19, 27, 45$ in $(\mathbb{Z}/56\mathbb{Z})^\times$, but they represent different primes in the integers. Hence we cannot distinguish between the values represented by these two forms by a condition modulo the discriminant.

The number of discriminants in which each genus has only one form is finite. Thus the general problem of deciding when a number is represented by a form cannot be answered by this method. There is much more one can say by considering higher reciprocity laws.[2]

---

[2]See D. Cox. *Primes of the Form $x^2 + ny^2$. Fermat, Class Field Theory, and Complex Multiplication.* New York: Wiley, 1989.

1. Prove Lemma 18.4.4.

2. Determine the genera for discriminants -15, -23 and -24. What assertions can you prove regarding representability of primes?

3. Let p be an odd prime. Show that $x^2 + 10y^2 = p$ if and only if $p \equiv 1, 9, 11, 19 \pmod{40}$.

4. Characterize the primes represented by the form $x^2 + 22y^2$.

18.5   Generic Characters and Genera

To begin our study of quadratic forms modulo m for different m, we first define the notion of equivalence of forms modulo m.

Definition 18.5.1. Given two forms f_1 and f_2, we say that f_1 **is equivalent to f_2 modulo m** ($f_1 \sim f_2 \pmod m$) if there exists a matrix $T = \begin{pmatrix} a & b \\ c & d \end{pmatrix}$ such that for all x and y,

$$f_1(x,y) \equiv T \cdot f_2(x,y) = f_2(ax + by, cx + dy) \pmod m$$

and $(\det T, m) = 1$.

It is clear that equivalent forms represent the same values in $(\mathbb{Z}/m\mathbb{Z})^\times$, and if f represents k, with $(k, m) = 1$, then $f \sim kx^2 + bxy + cy^2 \pmod m$, for some b and c. (This is analogous to Proposition 18.1.9.)

Example 18.5.2. The forms $x^2 + 5y^2$ and $2x^2 + 2xy + 3y^2$ are not equivalent modulo 5 as they represent different values. The first form represents 1 and 4, while the second represents 2 and 3. These are equivalent for any other odd prime $p \neq 5$. Consider the case when -5 is a quadratic residue modulo p. We can complete the square and obtain

$$2(2x^2 + 2xy + 3y^2) = 4x^2 + 4xy + 6y^2$$
$$= (2x + y)^2 + 5y^2.$$

By the transformation $2x + y \mapsto x$ and $y \mapsto y$ of determinant 2, we obtain

$$2(2x^2 + 2xy + 3y^2) \sim x^2 + 5y^2 \pmod p;$$

hence

$$2x^2 + 2xy + 3y^2 \sim 2^{-1}x^2 + 2^{-1}5y^2 \pmod p$$
$$\sim 2x^2 + 2 \cdot 5y^2 \pmod p.$$

In the last step, we have used the transformation $2^{-1}x \mapsto x$ and $2^{-1}y \mapsto y$.

If -5 is a quadratic residue modulo p, then we can use the transformation $x \mapsto x$ and $cy \mapsto y$, where $c^2 \equiv -5 \pmod{p}$, and get

$$2x^2 + 2xy + 3y^2 \sim 2x^2 - 2y^2 \pmod{p}$$
$$\sim 2x - yx + y \pmod{p}$$
$$\sim 2xy \pmod{p}$$
$$\sim xy \pmod{p}$$

and $x^2 + 5y^2 \sim xy \pmod{p}$ by the same process. We will see in Proposition 18.5.5 that the forms are equivalent even when -5 is not quadratic residue modulo p.

Definition 18.5.3. A form f **represents** $0 \pmod{m}$ if there exist relatively prime integers x and y, such that $f(x, y) \equiv 0 \pmod{m}$.

The following lemma is the key to many of the reductions we make with quadratic forms for odd primes.

Lemma 18.5.4. *For an odd prime p, f represents $0 \pmod{p^r}$ if and only if Δ is a quadratic residue modulo p.*

PROOF. Write $f(x, y) = ax^2 + 2bxy + cy^2$ where b can be half integral; then

$$af(x, y) = a^2x^2 + 2abxy + acy^2 =$$
$$= (ax + by)^2 - (b^2 - ac)y^2.$$

If $p \nmid a$, we can use the transformation $ax + by \mapsto x$, $y \mapsto x$ to obtain

$$af(x, y) \sim x^2 - \Delta/4y^2 \pmod{p^r},$$

and it is clear that it is 0 for nontrivial x and y if and only if Δ is a quadratic residue modulo p^r. The assertion of the proposition follows from the fact that for an odd prime p, Δ is a quadratic residue modulo p^r if and only if Δ is a quadratic residue modulo p.

If $p \mid a$, then the above transformation is not valid as its determinant is a and $(a, p^r) \neq 1$. In this case $\Delta \equiv 4b^2 - 4ac \equiv 4b^2 \pmod{p}$ is clearly a quadratic residue modulo p, and hence modulo p^r. Let $a = kp^\alpha$, $p \nmid k$. If $\beta = \lceil (r - \alpha)/2 \rceil$, then

$$f(-cp^\beta, 2bp^\beta) \equiv kc^2p^{\alpha+2\beta} - 4b^2cp^{2\beta} + 4b^2cp^{2\beta} \equiv 0 \pmod{p^r}.$$

Since $(a, b, c) = 1$, one of $-cp^\beta$ or $2bp^\beta$ must be nonzero modulo p^r ∎

Our goal is to classify the forms modulo p^r for each p and then use this classification to separate forms over the integers.

Proposition 18.5.5. *Let p be an odd prime, $p \nmid \Delta$; then*

$$f(x,y) \sim x^2 - y^2 \quad (\text{mod } p^r) \text{ if } \left(\frac{\Delta}{p}\right) = 1 \tag{1}$$

$$f(x,y) \sim x^2 - ny^2 \quad (\text{mod } p^r) \text{ if } \left(\frac{\Delta}{p}\right) = -1, \tag{2}$$

where n is any quadratic nonresidue.

PROOF. If Δ is a quadratic residue modulo p, then f represents 0 by Proposition 18.5.4. This implies that by a suitable transformation, we can make the first coefficient 0, that is,

$$f(x,y) \sim bxy - cy^2 \quad (\text{mod } p^r)$$
$$\sim (bx - cy)y \quad (\text{mod } p^r)$$

and $p \nmid b$, as $p \nmid \Delta$. By the transformation $bx - cy \mapsto x$, $y \mapsto y$, we obtain

$$f(x,y) \sim xy \quad (\text{mod } p^r),$$

and, replacing x by $x - y$ and y by $x + y$,

$$f(x,y) \sim x^2 - y^2 \quad (\text{mod } p^r).$$

If $\left(\frac{\Delta}{p}\right) = -1$, then $p \nmid ac$, and we can complete the square as in the proof of Lemma 18.5.4 to obtain

$$af(x,y) = (ax + by)^2 - \frac{\Delta}{4}y^2;$$

therefore,

$$f(x,y) \sim a^{-1}x^2 - a^{-1}\Delta/4y^2 \quad (\text{mod } p^r).$$

In the last step we used the transformation $ax + by \mapsto x$ and $y \mapsto y$. Since $\Delta/4$ is not a quadratic residue, we can change y to ky for some k such that $\Delta/4k^2 \equiv n$ for any nonresidue n.

$$f(x,y) \sim ax^2 - any^2 \quad (\text{mod } p^r)$$

If a is a residue, we obtain $f(x,y) \sim x^2 - ny^2$ for some nonresidue n, and if a is not a residue we obtain $f(x,y) \sim nx^2 - y^2$. Here we make use of the fact that Δa^{-1} is a quadratic residue if a and Δ are not residues. To see that $x^2 - ny^2$ and $nx^2 - y^2$ are equivalent modulo p^r, observe that if $nx^2 - y^2$ represents a quadratic residue α, then $nx^2 - y^2 \sim \alpha x^2 + \beta xy + \gamma y^2$ for some β, γ, and since α is a quadratic residue, $\alpha x^2 + \beta xy + \gamma y^2 \sim x^2 - ny^2$, proving the equivalence of $nx^2 - y^2$ and $x^2 - ny^2$. Let α be a quadratic residue such that $\alpha + 1$ is a nonresidue, then $nx^2 - y^2 \sim (\alpha+1)x^2 - y^2$, and this last form clearly represents α. ∎

Lemma 18.5.6. *If f has an odd discriminant, then $f \sim xy$ (mod 2) if $\Delta \equiv 1$ (mod 8). If $\Delta \not\equiv 1$ (mod 8), then $f \sim x^2 + xy + y^2$ (mod 2).*

PROOF. Since $\Delta = b^2 - 4ac$, b must be odd for Δ to be odd. If $\Delta \equiv 1$ (mod 8), then a or c must be even. If both a and c are even, then $f(x,y) \equiv xy$ (mod 2). If only one of them is even,

$$f(x,y) \sim x^2 + xy \quad (\text{mod } 2)$$
$$x(x+y) \quad (\text{mod } 2)$$
$$xy \quad (\text{mod } 2).$$

If $\Delta \not\equiv 1$ (mod 8), then a and c are odd and $f(x,y) \equiv x^2 + xy + y^2$ (mod 2). ∎

The two forms in the above lemma are inequivalent. This can be easily checked by the reader. (See Exercise 4.)

Theorem 18.5.7. *If $\Delta \equiv 1$ (mod 4), then for every $p \nmid \Delta$, all forms of discriminant Δ are equivalent modulo p.*

PROOF. This follows from Lemma 18.5.5 for odd primes and Lemma 18.5.6 for $p = 2$. ∎

Corollary 18.5.8. *If $\Delta \equiv 1$ (mod 4) and $(m, \Delta) = 1$, then all forms of discriminant Δ represent the same values in $(\mathbb{Z}/m\mathbb{Z})^{\times}$.*

We now consider the primes dividing the discriminant and $p = 2$ for the study of even discriminants. For this, we need the following lemma.

Lemma 18.5.9. *Let $f(x,y) = ax^2 + bxy + cy^2$ be a primitive form. For all m there exist integers x and y such that $(x,y) = 1$ and $(f(x,y), m) = 1$, that is, f properly represents a number prime to m.*

PROOF. Let m have the prime factorization $p_1^{e_1} \cdots p_k^{e_k}$. Since a, b, and c are coprime, for each i, there exist x_i and y_i such that $f(x_i, y_i) \not\equiv 0$ (mod p_i). By solving $x \equiv x_i$ (mod $p_i^{e_i}$) and $y \equiv y_i$ (mod $p_i^{e_i}$) for $i = 1, \ldots, k$, we see that $f(x,y)$ is not divisible by any of the primes $p_1, \ldots, p_k$. Hence f properly represents a number prime to m. ∎

Corollary 18.5.10. *Let f be a primitive form; then there exists an integer k with $(k, \Delta) = 1$ such that f properly represents k.*

Theorem 18.5.11. *Let f_1 and f_2 be two primitive forms of discriminant Δ representing k_1 and k_2, respectively, where k_1 and k_2 are relatively prime to Δ. Let p be an odd prime such that $p \mid \Delta$; then $f_1 \sim f_2 \pmod{p^r}$ if and only if*

$$\left(\frac{k_1}{p}\right) = \left(\frac{k_2}{p}\right).$$

PROOF. Let $f_1(x, y) = a_1 x^2 + 2b_1 xy + c_1 y^2$. If $p \mid \Delta$, we can assume that $p \nmid a_1$, and by completing the square, we obtain

$$a_1 f(x, y) = (a_1 x + b_1 y)^2 - \frac{\Delta}{4} y^2$$
$$\sim x^2 - C y^2 \pmod{p^r}$$

or

$$f_1(x, y) \sim \alpha_1 x^2 - \gamma_1 y^2.$$

Since $p \nmid \alpha_1$, then $p \mid \gamma_1$ so $f_1 \sim \alpha_1 x^2 \pmod{p}$; therefore, since $k_1 \equiv \alpha_1 m^2$ (mod p) for some m, we must have $\left(\dfrac{k_1}{p}\right) = \left(\dfrac{\alpha_1}{p}\right)$. Similarly, we get $\left(\dfrac{k_1}{p}\right) = \left(\dfrac{\alpha_2}{p}\right)$.

If f_1 and f_2 are equivalent modulo p^r, then they are equivalent modulo p, and $\left(\dfrac{k_1}{p}\right) = \left(\dfrac{k_2}{p}\right)$. Conversely, if $\left(\dfrac{k_1}{p}\right) = \left(\dfrac{k_2}{p}\right)$, then $\left(\dfrac{\alpha_1}{p}\right) = \left(\dfrac{\alpha_2}{p}\right)$, so either both α_1 and α_2 are residues or both are not, and $f_1 \sim f_2 \pmod{p}$ in either case. This implies that $f_1(x, y) \sim \alpha x^2 - c_1 y^2 \pmod{p^r}$ and $f_2 \sim \alpha x^2 - c_2 y^2 \pmod{p^r}$. Since $\Delta \equiv 4 a c_1 \equiv 4 a c_2 \pmod{p^r}$, we have $c_1 \equiv c_2 \pmod{p^r}$ so $f_1 \sim f_2 \pmod{p^r}$. ∎

Definition 18.5.12. Let f be a primitive positive definite form of discriminant Δ, and let k be a number prime to Δ represented by f. If $p_1, \ldots, p_r$ are the primes dividing Δ, the **generic characters** χ_i are defined by $\chi_i = \left(\dfrac{k}{p_i}\right)$.

The characters do not depend on the number k used to compute them. (See Exercise 6.)

The generic characters are ± 1, and we fix an ordering on the set of primes and consider the characters as an ordered set.

Now, we consider the case Δ even. Here we have to look at the discriminant modulo 8.

We define two characters,

$$\alpha(k) = \left(\frac{-1}{k}\right) = (-1)^{(k-1)/2}$$

$$\beta(k) = \left(\frac{2}{k}\right) = (-1)^{(k^2-1)/8}.$$

It is easy to verify that $\alpha(k) = 1$ if $k \equiv 1,5 \pmod 8$, and $\alpha(k) = -1$ if $k \equiv 3,7 \pmod 8$; and $\beta(k) = 1$ if $k \equiv 1,7 \pmod 8$, and $\beta(k) = -1$ if $k \equiv 3,5 \pmod 8$. We use a combination of these characters to distinguish between the values represented by forms modulo 8. Let k_1 and k_2 be odd numbers represented by the forms f_1 and f_2 of discriminant $\Delta = 4\Delta^*$. The following theorem gives a description of equivalence classes of forms modulo 8 and uses the characters α and β to distinguish between them. There are several cases depending on Δ^*.

Theorem 18.5.13. *Let f_1 and f_2 be two positive definite forms of even discriminant $\Delta = 4\Delta^*$. Let k_1 and k_2 be two odd numbers represented by f_1 and f_2, respectively. Then*

(a) *If $\Delta^* \equiv 1, 5 \pmod 8$, then $f_1 \sim f_2 \pmod 8$.*

(b) *If $\Delta^* \equiv 2 \pmod 8$, then $f_1 \sim f_2$ if and only if $\beta(k_1) = \beta(k_2)$.*

(c) *If $\Delta^* \equiv 3, 4, 7 \pmod 8$, then $f_1 \sim f_2$ if and only if $\alpha(k_1) = \alpha(k_2)$.*

(d) *If $\Delta^* \equiv 6 \pmod 8$, then $f_1 \sim f_2$ if and only if $\alpha(k_1)\beta(k_1) = \alpha(k_2)\beta(k_2)$.*

(e) *If $\Delta^* \equiv 0 \pmod 8$, then $f_1 \sim f_2$ if and only if $\alpha(k_1) = \alpha(k_2)$ and $\beta(k_1) = \beta(k_2)$.*

PROOF. Since $2 \mid \Delta$, then $2 \mid b$ in $\Delta = b^2 - 4ac$. Thus we can write the form as $f(x) = ax^2 + 2bxy + cy^2$ and $\Delta = 4b^2 - 4ac$. We can assume that a is odd, and since the middle term is even, we can complete the square.

$$af(x,y) \equiv a^2x^2 + 2abxy + acy^2 \pmod{2^r}$$
$$\equiv (ax + by)^2 - (b^2 - ac)y^2 \pmod{2^r}$$
$$\equiv (ax + by)^2 - \Delta^*y^2 \pmod{2^r}$$
$$\sim x^2 - \Delta^*y^2 \pmod{2^r};$$

therefore, $f(x,y) \sim a^{-1}x^2 - \Delta^*a^{-1}y^2 \pmod{2^r}$, which we can write as $ax^2 - a\Delta^*y^2 \pmod{2^r}$.

We can modify x by cx and y by cy, which implies that a can be adjusted by quadratic residues:

$$a = \begin{cases} 1 & \text{for } r = 1 \\ 1,\ 3 & \text{for } r = 2 \\ 1,\ 3,\ 5,\ 7 & \text{for } r = 3; \end{cases}$$

for $r \geq 4$, there are nontrivial quadratic residues modulo 2^r. For example, modulo 16, the residues are 1 and 9, and the invertible elements can be written as $1, 3, 5, 7, 9 \cdot 1, 9 \cdot 3, 9 \cdot 5$, and $9 \cdot 7$. This means that we can take 1, 3, 5, 7 as representatives for the set of invertible elements modulo the quadratic residues.

Similarly, for higher values of r, we can take $a = 1,\ 3,\ 5, 7$ as the possibilities for a. It suffices to do the calculation modulo 8, as it will be the same modulo 2^r for $r \geq 3$, that is,

$$f_1 \sim f_2 \pmod{2^3} \Leftrightarrow f_1 \sim f_2 \pmod{2^r} \qquad \text{for } r \geq 3.$$

Now we study the different cases in the theorem.

(a) Let $\Delta^* \equiv 1 \pmod 8$; then by interchanging x and y, we have $x^2 - y^2 \sim 7x^2 - 7y^2 \pmod 8$ and $3x^2 - 3y^2 \sim 5x^2 - 5y^2 \pmod 8$. These two classes also transform to one another $x^2 - y^2 \sim 3x^2 - 3y^2 \pmod 8$. The last equation is true because of the transformation

$$(2x + y)^2 - (x + 2y)^2 \equiv 3x^2 - 3y^2 \pmod 8$$

and $\det \begin{pmatrix} 2 & 1 \\ 1 & 2 \end{pmatrix} = 3$.

If $\Delta^* \equiv 5 \pmod 8$, then we can have the forms $x^2 - 5y^2, 3x^2 - 7x^2$, $5x^2 - y^2$, and $7x^2 - 3y^2$. $x^2 - 5y^2 \sim 3x^2 - 7y^2$ by switching x and y and $x^2 - 5y^2 \sim 5x^2 - y^2$ by the transformation $\begin{pmatrix} 1 & 2 \\ 2 & 1 \end{pmatrix}$. Since $(x + 2y)^2 - 5(2x + y)^2 \equiv 5x^2 - y^2 \pmod 8$, hence all the forms are equivalent modulo 8. This finishes the proof of the first case.

(b) Now let $\Delta^* \equiv 2 \pmod 8$. Then $f(x, y) \sim ax^2 - 2ay^2$. We have four forms, $x^2 - 2y^2, 3x^2 - 6y^2, 5x^2 - 2y^2$, and $7x^2 - 6y^2$, and we want to consider equivalences among them. It is not hard to compute that $x^2 - 2y^2$ represents 1 and 7 in $(\mathbb{Z}/8\mathbb{Z})^\times$, $3x^2 - 6y^2$ represents 3 and 5, $5x^2 - 2y^2$ represents 3 and 5, and $7x^2 - 6y^2$ represents 1 and 7. So we get at least two equivalence classes of forms.

Now, $x^2 - 2y^2 \sim 7x^2 - 6y^2$ via the transformation $\begin{pmatrix} 1 & 2 \\ 1 & 1 \end{pmatrix}$, since $(x + 2y)^2 - 2(x + y)^2 \sim 7x^2 - 6y^2 \pmod 8$.

Similarly, $3x^2 - 6y^2$ and $5x^2 - 2y^2$ are equivalent by the transformation
$\begin{pmatrix} 1 & 2 \\ 1 & 1 \end{pmatrix}$ or its inverse.

This shows that $f_1 \sim f_2$ if and only if $\beta(k_1) = \beta(k_2)$. The proofs of all the other cases are left as exercises. ∎

The following is a table of the generic characters: for $\Delta \equiv 1 \pmod 4$. If $p_1, \ldots, p_r | \Delta$ then $\chi_1(k) = \left(\dfrac{k}{p_1} \right), \ldots, \chi_r(k) = \left(\dfrac{k}{p_r} \right)$ are the characters.

For $\Delta \equiv 0 \pmod 2$, let $\Delta^* = \Delta/4$.

Δ^* mod 8	Characters
1, 5	$\chi_1, \ldots, \chi_r$
3, 4, 7	$\chi_1, \ldots, \chi_r, \alpha$
2	$\chi_1, \ldots, \chi_r, \beta$
6	$\chi_1, \ldots, \chi_r, \alpha\beta$
0	$\chi_1, \ldots, \chi_r, \alpha, \beta$

We give a few examples of the determination of the character set and use them to find the values represented in $(\mathbb{Z}/\Delta\mathbb{Z})^\times$. Our algorithm is as follows.

Algorithm 18.5.14. (1) Determine the reduced forms of discriminant Δ.

(2) Find the primes dividing Δ, and, if Δ is even, find its character set based on the above table.

(3) For each reduced form, find an odd number prime to the discriminant represented by the form and compute the character set.

(4) If the sets of characters are different, then each genus has only one class.

(5) Use the Chinese Remainder Theorem and the character sets to determine the primes represented by each form.

Example 18.5.15. Let $\Delta = -20$, $p = 5$, $\Delta^* = -5 \equiv 3 \mod 8$; then $\left(\dfrac{k}{5} \right)$ and $\alpha(k)$ are the characters. The reduced forms are $x^2 + 5y^2$ and $2x^2 + 2xy + 3y^2$.

$x^2 + 5y^2$ represents 29 and $\left(\dfrac{29}{5} \right) = 1$ and $\left(\dfrac{-1}{29} \right) = 1$.

$2x^2 + 2xy + 3y^2$ represents 7 and $\left(\dfrac{7}{5} \right) = -1$ and $\left(\dfrac{-1}{7} \right) = 1$. The two

sets of characters are different, so the forms are in different genera. To use the characters to identify the primes represented by the forms, we proceed as follows. From the characters of $x^2 + 5y^2$, we find that it represents the squares in $(\mathbb{Z}/5\mathbb{Z})^\times$ and represents 1 in $(\mathbb{Z}/4\mathbb{Z})^\times$. that is, $x^2 + 5y^2 = p$ if and only if $p \equiv 1, 4 \pmod 5$ and $p \equiv 1 \pmod 4$. Using the Chinese Remainder Theorem, this is equivalent to $p \equiv 1, 9 \pmod{20}$. The form $2x^2 + 2xy + 3y^2$ represents nonsquares in $(\mathbb{Z}/5\mathbb{Z})^\times$ and 3 in $(\mathbb{Z}/4\mathbb{Z})^\times$; that is, $p = 2x^2 + 2xy + 3y^2$ if and only if $p \equiv 2, 3 \pmod 5$ and $p \equiv 3 \pmod 4$. By the Chinese Remainder Theorem, this is equivalent to $p \equiv 3, 7 \pmod{20}$.

Example 18.5.16. Let $\Delta = -56$, $\Delta^* = -14 \equiv 2 \pmod 8$, so we have two characters, $\chi_1(k) = \left(\dfrac{k}{7}\right)$ and $\beta(k)$. The reduced forms are $x^2 + 14y^2$, $2x^2 + 7y^2$, $3x^2 + 2xy + 5y^2$, and $3x^2 - 2xy + 5y^2$. We compute an odd number represented by the form and the value of the characters.

form	k	$\chi_1(k)$	$\beta(k)$
$x^2 + 14y^2$	15	1	1
$2x^2 + 7y^2$	9	1	1
$3x^2 + 2xy + 5y^2$	5	-1	-1
$3x^2 - 2xy + 5y^2$	5	-1	-1

Here there are four forms and two genera with each genus consisting of two equivalence classes of forms. The first two forms have the same character set, and hence represent the same numbers in $(\mathbb{Z}/56\mathbb{Z})^\times$. The same is true for the last two forms. Here it would not be possible by this method to identify the primes represented by each form.

Example 18.5.17. Let $\Delta = -96$. Since $-96 = 4 \cdot -24$, we have $\Delta^* \equiv 0 \pmod 8$. From the table, we have three characters. $\chi_1(k) = \left(\dfrac{k}{3}\right)$, $\alpha(k)$, and $\beta(k)$. There are four reduced forms of this discriminant. We compute an odd number prime to 96 represented by each form and evaluate the characters.

form	k	$\chi_1(k)$	$\alpha(k)$	$\beta(k)$
$x^2 + 24y^2$	25	1	1	1
$3x^2 + 8y^2$	11	-1	-1	-1
$4x^2 + 4xy + 7y^2$	7	1	-1	-1
$5x^2 + 2xy + 5y^2$	5	-1	1	-1

Hence these forms are in different genera, and we can determine the primes represented by these forms. For example, the character set for the second form implies that $3x^2 + 8y^2$ represents nonsquares modulo 3, and 3 modulo 8. This implies that $3x^2 + 8y^2 = p$ if and only if $p \equiv 2 \pmod 3$ and $p \equiv 3 \pmod 8$. By the Chinese Remainder Theorem, this is equivalent to $p \equiv 11 \mod 24$. We can extend this modulo the discriminant and obtain $p \equiv 11, 35, 59, 83 \pmod{96}$ is represented by $3x^2 + 8y^2$. We can obtain similar theorems about the representation of prime numbers by other forms.

─────────────────── **Exercises for Section 18.5** ───────────────────

1. Show that $p = x^2 + 7y^2$ for some x and y if and only if $p \equiv 1, 9, 11, 15, 23, 25$ (mod 28).

2. Show that $x^2 + 15y^2 = p$ for some x and y if and only if $p \equiv 1, 19, 31, 49$ (mod 60).

3. Show that when $\Delta \equiv 1$ (mod 4), the form $x^2 + xy + \dfrac{1 - \Delta}{4}y^2$ represents the squares in $(\mathbb{Z}/\Delta\mathbb{Z})^\times$.

4. Show that $x^2 + xy + y^2$ and xy are inequivalent modulo 2.

5. Complete the proofs of the remaining cases in Theorem 18.5.13.

6. Prove that if $k_1 \neq k_2$, $(k_1, \Delta) = (k_2, \Delta) = 1$, and p is an odd prime dividing Δ, then $\left(\dfrac{k_1}{p}\right) = \left(\dfrac{k_2}{p}\right)$. Then the generic characters do not depend on the choice of k.

7. Consider the binary form $f(x, y) = ax^2 + bxy + cy^2$. Suppose that $f(x, y) = m$ and $f(u, v) = m'$, with m and m' relatively prime to the discriminant Δ of f. Verify that
$$4mm' = A^2 - \Delta B^2,$$
where $A = 2axu + b(xv + yu) + 2cyv$ and $B = xv - yu$. Use this to establish the following results.

   (a) Show that mm' is congruent to a square modulo the discriminant, hence to a square modulo each divisor of Δ.

   (b) Show that for every odd prime $p \mid \Delta$, f either represents only squares or only nonsquares modulo p^r. Hence all representable integers (for the form f) have the same quadratic character $\left(\dfrac{m}{p}\right)$.

   (c) For even discriminants, $\Delta = 4\Delta^*$, consider several cases. If $\Delta^* \equiv 0$ (mod 8), show that $mm' \equiv 1$ (mod 8), and hence f only represents 1 or only 3 or only 5 or only 7 modulo 8. If $\Delta^* \equiv 2$ (mod 8), show that $mm' \equiv 1, 7$ (mod 8), and hence f either represents 1 or 7 or only 3 or 5 modulo 8. Similarly, consider the other cases for Δ^*.

18.6 The Number of Representations by Forms

The number of representations of m by a quadratic form can be determined in a few cases by elementary techniques. More general theorems are possible by analytical methods beyond the scope of this book. Our starting point is Proposition 18.2.1. Suppose $(m, \Delta) = 1$, and m is represented by some form of discriminant Δ. Then Δ is a quadratic residue modulo $4m$. We will use this to show that solutions to the congruence $b^2 \equiv \Delta$ (mod $4m$) determine representations of m by equivalence classes of forms. In the case when each genus consists of a single class, we can completely determine the number of representations.

We start with the number of solutions to $b^2 \equiv \Delta \pmod{4m}$.

Lemma 18.6.1. *Let m be an odd integer with $(m, \Delta) = 1$. The number of solutions to the equation $x^2 \equiv \Delta \pmod{4m}$ in a complete residue system is*

$$2 \prod_{p \mid m} \left(1 + \left(\frac{\Delta}{p} \right) \right).$$

PROOF. Using the Chinese Remainder Theorem, we must solve

$$x^2 \equiv \Delta \pmod{4}$$
$$x^2 \equiv \Delta \pmod{p^r}$$

for each p^r occurring in the factorization of m.

Since $\Delta \equiv 0, 1 \pmod{4}$, the congruence $x^2 \equiv \Delta \pmod{4}$ has two solutions. For p odd, $x^2 \equiv \Delta \pmod{p^r}$ has two solutions or none depending on if Δ is a quadratic residue modulo p or not. Hence, the number of solutions to $x^2 \equiv \Delta \pmod{p^r}$ is $1 + \left(\dfrac{\Delta}{p} \right)$. Combining these yields the formula. ∎

Corollary 18.6.2. *The number of solutions to $b^2 \equiv \Delta \pmod{4m}$ in the absolutely least residue system modulo $4m$ that satisfy $|b| < m$ is*

$$\prod_{p \mid m} \left(1 + \left(\frac{\Delta}{P} \right) \right).$$

PROOF. If $b^2 \equiv \Delta \pmod{4m}$, then $-b$ is also a solution, and so are $b \pm 2m$. Hence half the solutions in the least absolute residue system satisfy $-m < b < m$. ∎

Example 18.6.3. Let $\Delta = -20$ and $m = 3$; then $x^2 \equiv -2 \pmod{12}$ has

$$2 \left(1 + \left(\frac{-2}{3} \right) \right) = 2 \left(1 + \left(\frac{1}{3} \right) \right) = 4 \text{ solutions.}$$

The solutions are $x = \pm 2$ and $x = \pm 4$. Only $x = \pm 2$ satisfy $|x| < 3$.

Lemma 18.6.4. *Every proper representation of m by a form f of discriminant Δ determines a solution b of $x^2 \equiv \Delta \pmod{4m}$ satisfying $|b| < m$.*

PROOF. By Proposition 18.1.9, if f properly represents m, then f is equivalent to $mx^2 + bxy + cy^2$ for some b and c. By using the technique of reduction theory, we can reduce this form and assume that $|b| \leq m$. Since $\Delta = b^2 - 4mc$, we have that $b^2 \equiv \Delta \pmod{4m}$. ∎

Recall that the transformation from f to $mx^2 + bxy + cy^2$ depends on integers p and q such that $f(p,q) = m$. To emphasize this dependence, we write the solution to the congruence $b^2 \equiv \Delta \pmod{4m}$ arising in the proof as $b_{f,(p,q)} \in (\mathbb{Z}/4m\mathbb{Z})^{\times}$. If there is no ambiguity about the representation, we can denote this by b_f to indicate the dependence of b on f.

Lemma 18.6.5. *If f and g are inequivalent forms of discriminant Δ that properly represent m, then $b_f \neq b_g$.*

PROOF. Suppose that $b_f = b_g = b$. Since f and g properly represent m, there exist c and c' such that

$$f \sim mx^2 + bxy + cy^2$$
$$g \sim mx^2 + bxy + c'y^2.$$

But $c = c'$ because the discriminant is preserved under the equivalence. The transitivity of $\sim$ implies that f and g are equivalent. ∎

We see that proper representations of m by inequivalent forms give different solutions to the congruence $x^2 \equiv \Delta \pmod{4m}$. The function

$$f(p,q) = m \mapsto b_{f,(p,q)},$$

from proper representations of m to solutions of $x^2 \equiv \Delta \pmod{4m}$, $|x| < m$, separates equivalence classes. To count the total number of proper representations by equivalence classes, we must determine when two different proper representations by a form f give the same b.

Suppose $f(u,v) = m$ and $f(t,w) = m$ are two distinct proper representations and $b_{f,(u,v)}$ and $b_{f,(t,w)}$ are the same, denoted by b. Then there exists u', v', such that $uv' - vu' = 1$ and $T \cdot f = mx^2 + bxy + cy^2$, where $T = \begin{pmatrix} u & v \\ u' & v' \end{pmatrix}$. Similarly, there exists a matrix $S = \begin{pmatrix} t & w \\ t' & w' \end{pmatrix}$ with $tw' - wt' = 1$ such that $S \cdot f = mx^2 + bxy + cy^2$. Since the two representations are distinct, $S \neq T$, that is, $S^{-1}T \neq I$. By the transitivity, $S^{-1}T$ transforms f to f. Then we need to determine the number of equivalences from f to f.

Proposition 18.6.6. *Suppose $f(x,y)$ is a definite binary quadratic form. If $\Delta \neq -4, -3$ and $T \cdot f = f$, then $T = \pm \begin{pmatrix} 1 & 0 \\ 0 & 1 \end{pmatrix}$.*

If $\Delta = -4$, there are two additional transformations of f onto itself given by

$$x \rightarrow y \qquad\qquad\qquad\qquad\qquad x \rightarrow -y$$
$$y \rightarrow -x, \qquad\qquad\qquad\qquad\qquad y \rightarrow x.$$

If $\Delta = -3$, there are four other transformations (in addition to those in the first case) given by

$$x \rightarrow x + y \qquad x \rightarrow y \qquad\qquad x \rightarrow -x - y \qquad x \rightarrow -y$$
$$y \rightarrow -x, \qquad y \rightarrow -x - y, \qquad y \rightarrow x, \qquad\quad y \rightarrow x + y.$$

PROOF. Suppose $f(x, y) = ax^2 + bxy + cy^2$ with $(a, b, c) = 1$. We can assume that f is reduced. (Why?) Suppose that $T = \begin{pmatrix} p & r \\ q & s \end{pmatrix}$ is such that $f(px + qy, rx + sy) = f(x, y) = ax^2 + bxy + cy^2$. Then $f(p, r) = f(1, 0) = a$ and $f(q, s) = f(0, 1) = c$.

Now, $bxy \geq -|b| \min(x^2, y^2)$, hence

$$f(x, y) \geq ax^2 - |b| \min(x^2, y^2) + cy^2$$
$$\geq (a - |b| + c) \min(x^2, y^2).$$

Suppose $pr \neq 0$. Since $f(p, r) = a$, then $a \geq (a - |b| + c) \min(p^2, r^2) \geq a - |b| + c$. As f is reduced and $(a, b, c) = 1$, this is only possible for $a = b = c = 1$. Then $f(x, y) = x^2 + xy + y^2$ and $\Delta = -3$. Similarly, if $qs \neq 0$, $f(q, s) = c$, then $f(x, y) = x^2 + xy + y^2$.

From $p^2 + pr + r^2 = 1$, we can deduce that $pr = -1$. Of the four transformations that satisfy $ps - qr = 1$ and $pr = -1$, only $\pm \begin{pmatrix} 1 & -1 \\ 1 & 0 \end{pmatrix}$ transform f into itself. Similarly, from $q^2 + qs + s^2 = 1$, we obtain the transformations $\pm \begin{pmatrix} 0 & -1 \\ 1 & -1 \end{pmatrix}$.

If $p = 0$, then $f(0, r) = a = cr^2$; since $a \leq c$, we must have that $a = c$ and $r = \pm 1$ and $c \geq a - |b| + c$. Again, this is only possible if $a = c = 1$ and $b = 0, 1$. If $b = 0$, then $f(x, y) = x^2 + y^2$, $\Delta = -4$, and we obtain the transformations $\pm \begin{pmatrix} 0 & 1 \\ -1 & 0 \end{pmatrix}$. If $b = 1$, then $f(x, y) = x^2 + xy + y^2$, and if $p = 0$ and $r = \pm 1$, we obtain $\pm \begin{pmatrix} 0 & -1 \\ 1 & -1 \end{pmatrix}$, which was obtained before. If $\Delta \neq -3, -4$, we must have $r = 0, p = \pm 1$ and the theorem is proved. ∎

Remark. The additional transformations obtained for $\Delta = -3, -4$ can also be obtained by observing that there are four representations of 1 by $x^2 + y^2$ and six representations by $x^2 + xy + y^2$. (See Exercise 2.)

Theorem 18.6.7. *Let $\Delta \neq -4, -3$, and m such that $(m, \Delta) = 1$; then the number of proper representations of an odd integer m by all the primitive reduced forms of discriminant Δ is*

$$2 \prod_{p \mid m} \left(1 + \left(\frac{\Delta}{p} \right) \right).$$

PROOF. If $\Delta \neq -4, -3$, and $f(p, q) = m$ for a form of discriminant Δ, then $f(-p, -q) = m$ is the only other proper representation of m by f giving the same solution b_f to $b^2 \equiv \Delta \pmod{4m}$. Hence the map from proper representations to solutions $b^2 \equiv \Delta \pmod{4m}$, $|b| \leq m$ is 2 to 1. Using Corollary 18.6.2, we obtain that the number of proper representations by equivalence classes of primitive forms of discriminant Δ is $2 \prod_{p \mid m} \left(1 + \left(\frac{\Delta}{p} \right) \right)$. ∎

Corollary 18.6.8. *If each genus consists of only one equivalence class, and f represents m, then the number of proper representations of m by f is*

$$2 \prod_{p \mid m} \left(1 + \left(\frac{\Delta}{p} \right) \right).$$

Example 18.6.9. Let $\Delta = -56$ and $m = 15$; then

$$2 \left(1 + \left(\frac{-56}{3} \right) \right) \left(1 + \left(\frac{-56}{5} \right) \right) = 8.$$

There are four equivalence classes with two genera. The first genus consisting of the forms $x^2 + 14y^2$ and $2x^2 + 7y^2$ can represent 15, and each represents 15 in four ways. (If only positive representations are counted, then there is only one each.)

Example 18.6.10. Let $\Delta = -96$ and $m = 35$; then

$$2 \left(1 + \left(\frac{-96}{5} \right) \right) \left(1 + \left(\frac{-96}{7} \right) \right) = 8.$$

Hence there are eight representations of 35 by forms of discriminant -96. There are three reduced forms, each in a different genus, so only one class represents 35. We see that $3x^2 + 8y^2$ can represent 15, and if only positive representations are counted, then there are two representations. These are

$$3 \cdot 3^2 + 8 \cdot 1^2 = 35$$
$$3 \cdot 1^2 + 8 \cdot 2^2 = 35.$$

─────────────────── **Exercises for Section 18.6** ───────────────────

1. Show that if p is represented by the form $x^2 + ny^2$, then the equation $p = x^2 + ny^2$ has a unique solution in the positive integers.

2.   (a) Show that the six representations of 1 by $x^2 + xy + y^2$ give the six transformations stated in Theorem 18.6.6.

    (b) Show that the four representations of 1 by $x^2 + y^2$ give the four transformations of f for $\Delta = -4$ in Theorem 18.6.6.

3.   (a) Use the multiplicativity of the Jacobi symbol to show that the number of solutions to $b^2 \equiv \Delta \pmod{4m}$, $|b| < m$ for $(m, \Delta) = 1$ are

$$\sum_{q|m,\, q \text{ squarefree}} \left(\frac{\Delta}{q}\right)$$

    (b) Let

$$w = \begin{cases} 4 & \text{if } \Delta = -4; \\ 6 & \text{if } \Delta = -3; \\ 2 & \text{otherwise.} \end{cases}$$

Show that the number of proper representations by primitive reduced forms of discriminant $\Delta < 0$ is $w \sum_{q|m,\, q \text{ squarefree}} \left(\frac{\Delta}{q}\right)$, and the total number of representations of m by all reduced forms of discriminant Δ is

$$f(m) = w \sum_{d|m} \left(\frac{\Delta}{q}\right).$$

    (c) Show that $\dfrac{1}{w} f(m)$ is a multiplicative function.

4. This exercise will show Fermat's claim that $y^2 = x^3 - 2$ has $(3, \pm 5)$ as the only integral solutions.

    (a) If m is odd, show that m and m^3 have the same number of proper representations by the form $x^2 + 2y^2$.

    (b) If $m = a^2 + 2b^2$ is a proper representation of m, show that

$$m^3 = (a^3 - 6ab^2)^2 + 2(3a^2b - 2b^3)^2$$

is a proper representation of m^3. (This is obtained by taking the cube of $(a + \sqrt{-2}b)$ and expressing it as $x + \sqrt{-2}y$.)

    (c) Show that two different proper representations of m by $x^2 + 2y^2$ give rise to two different representations of m^3 in part (b). (The map $(a, b) \mapsto (a^3 - 6ab^2, 3a^2b - 2b^3)$ is an injection from proper representations of m to proper representations of m^3). Conclude that every proper representation of m^3 is obtained this way.

(d) Consider the representation $x^3 = y^2 + 2 \cdot 1^2$. Show that x must be odd. Conclude using parts (a)-(c) that there exist a and b so that $y = a^3 - 6ab^2$ and $1 = 3a^2b - 2b^3$. Show that the latter equation has only $a = \pm 1$ and $b = \pm 1$ as solutions. Use this to prove Fermat's claim.

5. Follow the technique of the previous exercise to find all integral solutions of $y^2 = x^3 - 4$.

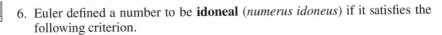

6. Euler defined a number to be **idoneal** (*numerus idoneus*) if it satisfies the following criterion.

> Let m be an odd number relatively prime to n, which is properly represented by $x^2 + ny^2$. If the equation $x^2 + ny^2 = m$ has only one solution with $x, y \geq 0$, then m is a prime number.

Euler used these forms to construct some large primes. Gauss made the following remarkable observation.

> A positive integer n is idoneal if and only if for forms of discriminant $-4n$, every genus consists of a single class.

(a) Prove Gauss's observation regarding idoneal numbers.

(b) Show that for n to be idoneal, its class number must be a power of 2 because the number of generic characters is always a power of 2.

(c) Determine all n, $n \leq 200$, for which $h(-4n) = 2$. Which of these n are idoneal?

(d) Determine all $n \leq 200$, for which $h(-4n) = 4$. Which of these are idoneal?

(e) Find a few examples of discriminants $\Delta = -4n$, where the class number is 8 and 16, mostly for $n \leq 2000$. Find all that are idoneal.

Chapter 19

Elliptic Curves

19.1 Introduction

Elliptic curves are plane curves that are the locus of points satisfying a cubic equation in two variables. The first instances of elliptic curves occur in the works of Diophantus and Fermat. Classically, the problem of computing the arc length of an ellipse gave rise to elliptic functions that satisfy cubic equations; hence plane cubic curves are called elliptic curves. In number theory, cubic equations arise naturally in many Diophantine problems, and many problems can be converted into a problem about elliptic curves.

In this section, we introduce some of these classical problems. In later sections, we discuss a few elementary properties of elliptic curves, particularly the fact that there is a way to add points on the curve so that the set of points becomes a group. This is the beginning of a deep and beautiful theory, which is also one of the most active areas of research in number theory. We will not be able to explore the recent applications of the theory of elliptic curves to Fermat's Last Theorem. The interested reader is invited to continue the study in the numerous texts that have recently appeared.[1]

Definition 19.1.1. A plane algebraic curve is defined as the locus of points (x, y) satisfying $f(x, y) = 0$, where f is a nonconstant polynomial in two variables.

Example 19.1.2. 1. If $f(x, y) = ax + by + c$, then the associated algebraic curve is a straight line. The integer points on this curve give the solutions to the linear Diophantine equation $ax + by = -c$.

2. If $f(x, y) = x^2 + y^2 - 1$, then the algebraic curve is the unit circle. The solutions (x, y) for which x and y are rational correspond to the Pythagorean triples.

---

[1]J. H. Silverman and J. Tate. *Rational Points on Elliptic Curves*. New York: Springer–Verlag, 1992.

3. In general, a polynomial $f(x,y)$ of degree 2 gives either an ellipse, a parabola, or a hyperbola.

If $f(x,y)$ is a cubic polynomial in (x,y), then the locus of points satisfying $f(x,y) = 0$ is a cubic curve. A general cubic equation in two variables is of the form

$$ax^3 + bx^2y + cxy^2 + dy^3 + ex^2 + fxy + gy^2 + hx + iy + j = 0.$$

Using the classification of cubic curves,[2] we find that a general cubic can be transformed into an equation of the form

$$y^2 + axy + by = x^3 + cx^2 + dx + c.$$

Further, we can complete the square on the left and make a linear change of variables to transform this into an equation of the form

$$y^2 = x^3 + ax^2 + bx + c.$$

We will take this to be the standard form of an elliptic curve.

Definition 19.1.3. An **elliptic curve** is the set of points (x,y) satisfying the equation

$$y^2 = x^3 + ax^2 + bx + c.$$

Example 19.1.4. A classical problem in number theory is to determine when a product of two consecutive integers is a product of three consecutive integers. Equivalently, we want integer solutions to the equation

$$y(y+1) = x(x+1)(x+2).$$

This is an example of an elliptic curve. A solution is $x = 1$, $y = 2$. Can you find more solutions of this equation? The equation can be written as

$$y^2 + y = (x-1)x(x+1) = x^3 - x.$$

Further, we can complete the square on the left to get

$$(y + 1/2)^2 = x^3 + x + 1/4.$$

Replacing $y + 1/2$ by y and clearing the denominators, we get

$$4y^2 = 4x^3 + 4x + 1.$$

Replacing y by $y/8$ and x by $x/4$, and clearing denominators, we get an equation in standard form

$$y^2 = x^3 + 32x + 8.$$

---

[2]A. W. Knapp. *Elliptic Curves.* Princeton, NJ: Princeton University Press, 1992.

Example 19.1.5. Another example of an elliptic curve arises in solving Lucas's problem of the square pyramid. The pyramidal numbers are of the form $x(x + 1)(x + 2)/6$, and the problem is to determine which of these numbers is a perfect square. We want to know the integer solutions to the equation

$$y^2 = \frac{x(x + 1)(x + 2)}{6}.$$

Example 19.1.6. The congruent number problem is a classical problem related to right triangles with rational sides. We say that an integer n is **congruent** if it is the area of right triangle with rational sides. If a and b are the legs of the triangle and c is the hypotenuse, then we have

$$a^2 + b^2 = c^2$$
$$\frac{1}{2}ab = n.$$

For example, 6 is congruent because it is the area of the triangle with sides $(3, 4, 5)$. The numbers $1, 2, 3$ are not congruent, but 5 is a congruent number.

Let $x^2 = c^2/4$; then $x \pm n = ((a \pm b)/2)^2$ are perfect squares. If n is congruent, then we have three rational squares, $x - n, x, x - n$, in arithmetic progression with a common difference of n.

Conversely, if we have three rational squares in an arithmetic progression with common difference n, then we can find a rational right triangle of area n. Denote the terms in the arithmetic progression by $x - n$, x, and $x + n$. Since $x, x - n$ and $x + n$ are perfect squares, a, b, and c can be found from

$$a = \sqrt{x + n} + \sqrt{x - n}$$
$$b = \sqrt{x + n} - \sqrt{x - n}$$
$$c = 2\sqrt{x}.$$

The relation with elliptic curves is obtained by observing that $x(x - n)(x + n)$ is a perfect square. Hence we obtain a rational solution to $y^2 = x^3 - n^2 x$. If we can show the converse, that is, a rational solution (x, y) with $y \neq 0$ means that $x - n, x$, and $x + n$ are perfect squares, then the congruent number problem will be equivalent to finding rational points on the elliptic curve $y^2 = x^3 - n^2 x$.[3]

The problem of finding integer points on an elliptic curve $y^2 = f(x)$ is quite difficult, and it turns out that it is simpler to find new rational points. Recall that in the determination of Pythagorean triples in Section 14.3, we considered the intersection of a straight line of rational slope through $(-1, 0)$

---

[3]For further details, consult N. Koblitz. *Introduction to Elliptic Curves and Modular Forms.* New York: Springer–Verlag, 1984.

with the unit circle. Because the resulting equation is quadratic, knowing one root allowed us to compute the other root (which was necessarily rational). In the case of a cubic curve, if we take a line of rational slope through a point, then the line can meet the curve at two other points, which need not have rational slopes. For example, consider the intersection of the line $y = x$ with $y^2 = x^3 - x$. This gives $x^2 = x^3 - x$. This has one rational root $x = 0$. The other two roots satisfy $x^2 - x - 1 = 0$.

To succeed with the geometric construction, what we need is a guarantee that the line through a rational point will meet the curve at one other rational point. To accomplish this, instead of using an arbitrary line, Diophantus used the tangent line at a point.

Example 19.1.7. Consider the equation of Example 19.1.4, (with x replaced by $x - 1$ for a simpler equation)

$$y(y + 1) = (x - 1)x(x + 1).$$

The slope of the tangent line at a point can be obtained by implicit differentiation:

$$2y\,\frac{dy}{dx} + \frac{dy}{dx} = 3x^2 - 1$$

or

$$\frac{dy}{dx} = \frac{3x^2 - 1}{2y + 1}.$$

The point $P = (1, 0)$ is on the curve. The slope of the tangent line at P is $m = (3 - 1)/(2 \cdot 0 + 1) = 2$. We find the points of intersection of the line $y = 2(x - 1)$ with the curve by substituting this equation into the equation of the curve. We obtain

$$2(x - 1)^2 + (x - 1) = x^3 - x.$$

Simplifying, and writing all terms on one side, we obtain

$$x^3 - 4x^2 + 2x - 1 = 0.$$

This equation has a double root at $x = 1$ and factors as

$$(x - 1)^2(x - 2) = 0.$$

The third root is $x = 2$, which gives $y = 2(2 - 1) = 2$. Therefore, $(2, 2)$ is another rational solution.

Exercise. Determine the intersection of the tangent line at $(0,0)$ with the curve $y^2 + y = x^3 - x$.

Newton observed that another method of obtaining rational solutions is to determine the intersection of a line joining two rational points P_1 and P_2 with the curve. If there is a third point of intersection, then this is necessarily rational. This is because if a polynomial of degree 3 has two rational roots, then all its roots are rational. An example will make this clear.

Example 19.1.8. Consider the points $P = (1,0)$ and $Q = (-1,-1)$ on the elliptic curve $y^2 + y = x^3 - x$.

The line joining P and Q has slope $m = 1/2$, and equation $y = (x-1)/2$. Substituting into the equation for the curve, we get

$$\frac{1}{4}x^2 - \frac{1}{2}x + \frac{1}{4} + \frac{1}{2}x - \frac{1}{2} = x^3 - x$$

or

$$x^3 - \frac{1}{4}x^2 - x - \frac{1}{4} = 0.$$

This is a cubic equation with three roots x_1, x_2, and x_3 and can be factored as

$$(x - x_1)(x - x_2)(x - x_3)$$
$$= x^3 - (x_1 + x_2 + x_3)x^2 + (x_1 x_2 + x_2 x_3 + x_1 x_2)x - x_1 x_2 x_3. \quad (1)$$

By comparing the coefficients of x^2, we see that $x_1 + x_2 + x_3 = 1/4$. From the two known roots, $x_1 = 1$ and $x_2 = -1$, of the equation, we obtain the third root $x_3 = 1/4$. From this we compute $y_3 = (1/4 - 1)/2 = -3/8$ using the equation of the straight line. We have obtained another rational point $(1/4, -3/8)$.

The example shows that even if we start with points having integer coordinates, the points of intersection can have rational coordinates. Hence it is convenient to study rational solutions first, and then determine the integer solutions.

The construction above fails when the tangent line or a line joining two points is vertical. In this case, the line only meets the curve at two points. In the next section, we will introduce a point at infinity, which is a common point of all the vertical lines. Adding this point at infinity to the elliptic curve, we will conclude that even a vertical line meets the curve at three points.

To apply Diophantus's technique, we have to be able to define the tangent line. This is not always possible. For example, for $y^2 = x^3$, the tangent line at $(0,0)$ cannot be defined. (See Figure 19.1.1.)

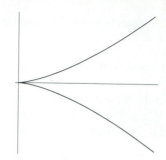

Figure 19.1.1 Graph of $y^2 = x^3$

For the curve $y^2 = f(x)$, we obtain $f'(x) = 2y \, dy/dx$. The tangent line is undefined at a point (x_0, y_0), where both y_0 and $f'(x_0)$ are zero. This happens when $f(x_0)$ and $f'(x_0)$ are zero, that is, $f(x_0)$ has a double root at x_0. The discriminant of a polynomial is used to determine if an equation has a multiple root.

If $f(x) = ax^2 + bc + x$, then the discriminant of f, $\Delta(f)$ is $(r_1 - r_2)^2$ where r_1, r_2 are the roots of $f(x) = 0$. In this case, the discriminant is given by $b^2 - 4ac$.

If $f(x) = ax^2 + bx^2 + cx + d$, then $\Delta(f) = (r_1 - r_2)^2 (r_2 - r_3)^2 (r_1 - r_3)^2$, where r_1, r_2, and r_3 are the three roots of $f(x) = 0$. The formula for the discriminant in terms of the coefficients of $f(x)$ is $\Delta(f) = 27c^2 + 4a^3c + 4b^3 - a^2b^2 + 8abc$.[4]

We will consider elliptic curves of the form $y^2 = f(x)$, where $\Delta(f) \neq 0$. The discriminant will also be useful in computing certain integer points on elliptic curves. An elliptic curve for which the discriminant in nonzero is called **nonsingular**.

The real points of $y^2 = f(x)$ are sketched by first drawing a graph of $y = f(x)$ and then taking the two square roots of the positive part of the graph. The three possibilities for the graph of $f(x)$ and the corresponding graphs of $y^2 = f(x)$ are shown on the following page.

———————————— **Exercises for Section 19.1** ————————————

1. Write a computer program to find integer solutions to the equation

$$y(y + 1) = (x - 1)x(x + 1).$$

Use the chord–tangent construction to find some rational points.

---

[4]See A. W. Knapp. *Elliptic Curves*. Princeton, NJ: Princeton University Press, 1992, page 59.

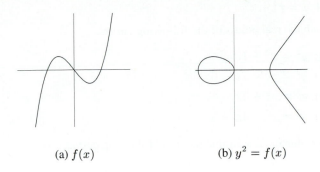

(a) $f(x)$ (b) $y^2 = f(x)$

Figure 19.1.2 $y = f(x)$ and the corresponding elliptic curve

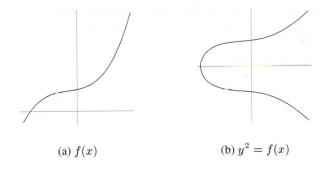

(a) $f(x)$ (b) $y^2 = f(x)$

Figure 19.1.3 $y = f(x)$ and the corresponding elliptic curve

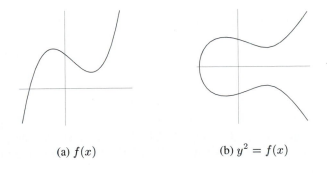

(a) $f(x)$ (b) $y^2 = f(x)$

Figure 19.1.4 $y = f(x)$ and the corresponding elliptic curve

2. Determine the point of intersection of the tangent line at $(0,0)$ on the curve $y^2 + y = x^3 + x^2$.

3. Sketch the real points of the following curves.

   (a) $y^2 = x^3 + 1$.

   (b) $y^2 = x^3 - 7x - 6$.

   (c) $y^2 = x^3 + 4x$.

   (d) $y^2 = x^3 - 4x - 3$.

4. Determine the discriminants of the following curves.

   (a) $y^2 + y = x^3 - x$.

   (b) $y^2 + y = x^3 + x^2$.

   (c) $y^2 = x^3 + x^2 - 16x - 32$.

   (d) $y^2 = x^3 - 81x - 243$.

5. The equation $a^3 + b^3 = c^3$ has only trivial integral solutions. If $c \neq 0$, we can write this as $u^3 + v^3 = 1$, where $u = a/c$ and $v = b/c$.

   Verify that if $x = 3/(u + v)$ and $y = 9(u - v)/(u + v) + 1/2$, then $y^2 - y = x^3 - 7$. Hence a rational point on $u^3 + v^3 = 1$ gives a rational point on $y^2 - y = x^3 - 7$. Conversely, if we are given a rational point (x, y) on $y^2 - y = x^3 - 7$, show that we can find a rational solution to $u^3 + v^3 = 1$. Determine all rational solutions to $y^2 - y = x^3 - 7$.

19.2 Projective Spaces

We give a brief description of projective spaces in order to introduce an operation of addition on the points of an elliptic curve. The addition law makes the points on an elliptic curve into an abelian group. The definition of the addition law requires the introduction of the point at infinity, a point which lies on every vertical line. It is also necessary to study of intersection of cubic curves in order to understand the associative law in the abelian group. This section treats only a few basic facts that are necessary for our treatment of elliptic curves. Even this is not complete, as we ignore multiplicities of intersection.[5]

Projective geometry studies the properties of geometric objects invariant under projection. For example, perspective projection is a tool employed by artists to give a realistic perception of depth. Parallel lines on a square tiled floor can be projected as in Figure 19.2.1. The figure shows the point at infinity where all the parallel lines meet. Similarly, the horizon in a photograph represents a line at infinity.

---

[5]For a more detailed introduction, consult M. Reid. *Undergraduate Algebraic Geometry.* Cambridge: Cambridge University Press, 1988.

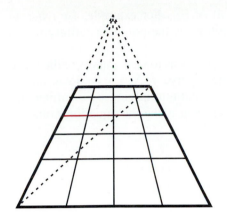

Figure 19.2.1 Perspective projection

Figure 19.2.2 Projective line

To treat points at infinity on an equal footing with ordinary points, we introduce projective coordinates. A point P in $\mathbb{R}^2$ with coordinates (a, b) is represented as a 3-tuple (x, y, z) with $x = a$, $y = b$, $z = 1$. We consider (tx, ty, tz), with $t \neq 0$ to be the same point as (x, y, z). Projective spaces can be defined for any field k. For our study, k will denote $\mathbb{Q}$, $\mathbb{R}$, $\mathbb{C}$, or the finite field $\mathbb{F}_p$ of integers module a prime p.

Definition 19.2.1. The **projective space** $\mathbb{P}^n(k)$ is defined as the set of points $\mathbb{P}^n(k) = \{(x_0, x_1, \ldots, x_n)\} - (0, 0, \ldots, 0)$ with $(tx_0, tx_1, \ldots, tx_n)$ identified with $(x_0, x_1, \ldots, x_n)$. We denote a point in projective space as $(x_0 : x_1 : \cdots : x_n)$.

Example 19.2.2. The projective line, $\mathbb{P}^1(\mathbb{R})$, is the set of points (x, y) excluding $(0, 0)$ with the point (tx, ty) identified with (x, y). If we select $P = (x, y)$, then all the points (tx, ty) are on the line joining P to the origin, and these are identified with P. We can take a representative for the point on the unit circle. The projective line $\mathbb{P}^1(\mathbb{R})$ is then represented by the unit circle with diagonally opposite points identified together. A point on the line $y = ax$ corresponds to the point $(a : 1)$ in $\mathbb{P}^1(\mathbb{R})$, and the vertical line $x = 0$ corresponds to $(1 : 0)$ in the projective line. This point can be considered as the point at infinity.

The space k^n is called the affine space and viewed as a subspace of $\mathbb{P}^n(k)$ by the identification $(x_0 : x_1 : \cdots : x_{n-1} : 1)$.

Projective spaces arise very naturally when we look at the set of all lines in a plane. Consider a line $a + bx + cy = 0$. Here $(a, b, c) \neq (0, 0, 0)$, and (a, b, c) and (ta, tb, tc) represent the same line. The coefficients of the line can be considered as a point $(a : b : c)$ of $\mathbb{P}^2$. For example, $(a : b : 1)$ corresponds to a line of slope $-b$. In this view, the points at infinity in $\mathbb{P}^2$, $(a : b : 0)$, correspond to vertical lines.

The use of homogeneous coordinates allows us to view curves in the plane as curves in projective space. An algebraic curve $f(X, Y) = 0$ can be homogenized by letting $X = x/z$, $Y = y/z$ and multiplying by a sufficiently high power of z to clear the denominators and make the equation homogeneous.

Example 19.2.3. The line

$$aX + bY + c = 0$$

gets converted to

$$a\frac{x}{z} + b\frac{y}{z} + c = 0,$$

and multiplying by z we get

$$ax + by + cz = 0.$$

The projective line $ax + by + cz = 0$ is the set of points $(x : y : z)$ in $\mathbb{P}^2$ satisfying this equation. The affine points (x, y) satisfying $ax + by + c = 0$ correspond to the projective point $(x : y : 1)$ on the projective line. The projective line has additional points, the points at infinity given by $z = 0$. On $ax + by + cz = 0$, the points at infinity are $(-b, a, 0)$. If we fix a and b, then the family of lines $ax + by + c = 0$ are parallel, and they have a point in common, the point at infinity $(-b, a, 0)$. Two nonparallel lines meet at a point in k^2, and parallel lines meet at their common point at infinity; hence two lines always meet at a point in the projective space $\mathbb{P}^2$.

The use of projective coordinates leads to a beautiful generalization of this result. Bezout's Theorem states that a curve of degree n and a curve of degree m meet in at most nm points, and by careful interpretation of points at infinity and multiplicities of points of intersection, and the use of complex coordinates, it is possible to show that the number of points of intersection is exactly nm.[6]

Example 19.2.4. The unit circle $X^2 + Y^2 = 1$ can be homogenized by writing $(x/z)^2 + (y/z)^2 = 1$, or $x^2 + y^2 = z^2$.

The points at infinity correspond to $z = 0$. In $\mathbb{P}^2(\mathbb{R})$, $x^2 + y^2 = 0$ has no solutions, so there are no points at infinity on the circle. Similarly, one

---

[6]See M. Reid. *Undergraduate Algebraic Geometry.* Cambridge: Cambridge University Press, 1988.

can show that an ellipse has no points at infinity, a parabola has one point at infinity, and a hyperbola has two points at infinity.

We showed that all solutions to $x^2 + y^2 = z^2$ are given by the formula $x = r^2 - s^2$, $y = 2rs$, and $z = r^2 + s^2$. This gives a rational parameterization of the projective curve $x^2 + y^2 = z^2$. We can use the same technique to give parameterizations for any conic

$$ax^2 + bxy + cy^2 + dxz + eyz + fz^2 = 0.$$

(Recall the examples of Section 14.3.) This is useful to show that two conic curves meet in at most four points. (See the exercises.)

The cubic curve $y^2 = x^3 + ax^2 + bx + c$ becomes $y^2 z = x^3 + ax^2 z + bxz^2 + cz^3$ in homogeneous coordinates. The points at infinity correspond to $z = 0$. This implies that $x = 0$; hence there is only one point at infinity, $(0 : 1 : 0)$, on the elliptic curve. This point is also the common point on the family of vertical lines $x = c$.

A general cubic curve in projective space has the equation

$$\begin{aligned} a_0 x^3 &+ a_1 x^2 y + a_2 xy^2 \\ &+ a_3 y^3 + a_4 x^2 z + a_5 xyz \\ &+ a_6 y^2 z + a_7 xz^2 + a_8 yz^2 + a_9 z^3 = 0. \end{aligned} \quad (1)$$

The ten coefficients $(a_0, \dots, a_9)$ determine a point in projective space $\mathbb{P}^9$. Conversely, a point in $\mathbb{P}^9$ gives a cubic curve in P^2.

Lemma 19.2.5. *Suppose $f(x, y)$ is a homogeneous polynomial of degree d in two variables. Then $f(x, y) = 0$ has at most d solutions in $\mathbb{P}^2$.*

PROOF. Suppose $f(x, y) = a_0 x^d + a_1 x^{d-1} y \cdots + a_d y^d$. If the point at infinity $(1 : 0)$ is a solution, then $a_0 = 0$. Let r be the smallest integer such that $a \ne 0$; then $f(x, y) = y^r (a_r x^{d-r} + \cdots + a_d y^{d-r})$. We say that $(1 : 0)$ is a root with multiplicity r because of the y^r term. The other solutions of $f(x, y) = 0$ are of the form $(x : 1)$ with $a_r x^{d-r} + \cdots + a_d = 0$. This is a polynomial of degree $d - r$, hence has at most $d - r$ roots, counting multiplicities. Thus, the total number of solutions to $f(x, y) = 0$ is d. ∎

Example 19.2.6. Consider the intersection of a line $ax + by + cz = 0$ with the cubic curve given by Equation (1). We can assume that one of the coefficients a, b, or c is nonzero, say c. Then substituting $z = a'x + b'y$ into (1), we get a homogeneous equation of degree three in two variables. By the previous lemma, the equation has at most three solutions. Hence there are at most three points of intersection between a line and a cubic curve.

Example 19.2.7. Two cubic curves C_1 and C_2 have at most nine points of intersection. Suppose $f(x, y, z)$ and $g(x, y, z)$ are two homogeneous polynomials of degree 3 such that C_1 and C_2 are the locus of f and g, respectively. We can write f and g as polynomials in z.

$$f(x, y, z) = F(z) = a_0(x, y)z^3 + a_1(x, y)z^2 + a_2(x, y)z + a_3(x, y)$$
$$g(x, y, z) = G(z) = b_0(x, y)z^3 + b_1(x, y)z^2 + b_2(x, y)z + b_3(x, y),$$

where $a_i(x, y)$ (respectively b_i) is a homogeneous polynomial in two variables of degree i. The polynomials $z^2 F(z)$, $zF(z)$, $F(z)$, $z^2 G(z)$, $zG(z)$, $G(z)$ are in the space of polynomials of degree at most 6. Since they have a common zero, they do not form a basis of this space. Hence there exist nonzero constants $c_1, \ldots, c_6$ such that

$$c_1 z^2 F(z) + c_2 zF(z) + c_3 F(z) + c_4 z^2 G(z) + c_5 zG(z) + c_6 G(z) = 0.$$

Writing out the equation for F and G, we get a system of six linear equations with a nonzero solutions. This implies that the determinant

$$R(f, g) = \det \begin{pmatrix} a_0 & a_1 & a_2 & a_3 & 0 & 0 \\ 0 & a_0 & a_1 & a_2 & a_3 & 0 \\ 0 & 0 & a_0 & a_1 & a_2 & a_3 \\ b_0 & b_1 & b_2 & b_3 & 0 & 0 \\ 0 & b_0 & b_1 & b_2 & b_3 & 0 \\ 0 & 0 & b_0 & b_1 & b_2 & b_3 \end{pmatrix}$$

must be zero. Now, $R(f, g)$ is a homogeneous polynomial of degree 9 in x and y, so the equation has at most nine solutions.

If we use complex coordinates, then the intersection consists of exactly nine points in $\mathbb{P}^2(\mathbb{C})$. The proof requires a careful treatment of intersection multiplicities, which is beyond the scope of this book.

We need one more result about intersection of cubics to describe the group law on rational points of elliptic curves.

Proposition 19.2.8. *Let C_1 and C_2 be two cubic curves whose intersection consists of nine distinct points, $C_1 \cap C_2 = \{P_1, \ldots, P_9\}$. Then any cubic passing through eight points $\{P_1, \ldots, P_8\}$ also passes through the ninth point, P_9.*

PROOF. An equation for a cubic determines a point in 9-dimensional projective space $\mathbb{P}^9$, and conversely, every point in $\mathbb{P}^9$ gives a cubic curve. Let $f(P) = 0$ be the equation of the cubic curve C corresponding to a point f in $\mathbb{P}^9$. Assume that C passes through the eight points $\{P_1, \ldots, P_8\}$. Since P_1 is on the curve, $f(P_1) = 0$ gives a linear equation in 10 variables, whose solution consists of a set of points of dimension 8. Each additional point on the curve gives an additional condition on f, and reduces the dimension by 1. This implies that the cubic curves passing through the eight points $\{P_1, \ldots, P_8\}$ correspond to a straight line in $\mathbb{P}^9$. If C_1 has equation $f_1(P) = 0$ and C_2 has equation $f_2(P) = 0$, then the line joining these two in $\mathbb{P}^9$ has equation $c_1 f_1 + c_2 f_2 = 0$. Hence $f = c_1 f_1 + c_2 f_2$ for some c_1 and c_2. But P_9 is also on the curves f_1 and f_2, so $f(P_9) = c_1 f_1(P_9) + c_2 f_2(P_9) = 0$; hence P_9 is also on the curve C. ∎

Exercises for Section 19.2

---

1. Show that in projective coordinates there are no points at infinity on an ellipse, and only one point at infinity on a parabola.

2. Show that a line and a conic have at most two points of intersection in projective coordinates. Show that if we use complex coordinates and count multiplicities, then the number of points of intersection is always two.

3. Show that two conics have at most four points of intersection.

19.3 The Group Law on Elliptic Curves

The group law on the rational points on an elliptic curve is defined using the intersection of straight lines with the cubic curve. Recall that a line that meets the cubic curve at two rational points also meets it at a third rational point. In the case of a vertical line, this third point is the point at infinity on the elliptic curve

$$y^2 = x^3 + ax^2 + bx + c. \tag{1}$$

The point at infinity on this curve in projective coordinates is $(0 : 1 : 0)$, which we denote by $\mathcal{O}$.

The structure of an abelian group on the curve will be given by defining an addition operation (denoted by $+$) satisfying the commutative and associative laws. Let $E(\mathbb{Q})$ denote the set of rational points (x, y) satisfying (1) together with the point at infinity. We choose this point to be the identity element of the group. For every point $P = (x, y)$, we define $-P$ to be the third point on the line joining P and $\mathcal{O}$, that is, the vertical line through P. Hence $-P = (x, -y)$.

The addition rule for rational points is defined as follows. If $P = (x_1, y_1)$ and $Q = (x_2, y_2)$ are distinct points with $x_2 \neq x_1$, then $-(P+Q)$ is the third point of intersection of the line joining P and Q with E. The equation of the line is $y - y_1 = m(x - x_1)$, where the slope m is $\dfrac{y_2 - y_1}{x_2 - x_1}$. We can find the coordinates of the third point of intersection (x_3, y_3) by substituting this in (1) to obtain

$$m^2(x - x_1)^2 + 2m(x - x_1)y_1 + y_1^2 = x^3 + ax^2 + bx + c.$$

Expanding the left side and combining the terms, we obtain

$$x^3 + (a - m^2)x^2 + (b')x + (c') = 0,$$

where b' and c' are expressions involving b, c, m, x_1, and y_1. This is a cubic equation with roots x_1, x_2 and x_3. The coefficient of x^2 is $-(x_1 + x_2 + x_3)$, hence

$$x_1 + x_2 + x_3 = -(a - m^2)$$

or

$$x_3 = m^2 - a - x_1 - x_2.$$

The sum $P + Q$ has coordinates $(x_3, -y_3)$, where $y_3 = m(x_3 - x_1) + y_1$.

The rule for addition when $P = Q$ is a little different. To get a third point of intersection with the elliptic curve, we have to take the tangent line to the curve at $P = (x_1, y_1)$. The slope of the tangent line can be found by implicit differentiation:

$$2y\frac{dy}{dx} = 3x^2 + 2ax + b;$$

hence the slope m of the tangent line at P is

$$m = \frac{3x_1^2 + 2ax_1 + b}{2y_1}.$$

Here we assume that $y_1 \neq 0$. If $y_1 = 0$, then the tangent line is vertical, in which case, the third point of intersection is $\mathcal{O}$. Substituting the equation of the line $(y - y_1) = m(x - x_1)$ into (1), we obtain

$$x^3 + (a - m^2)x^2 + b'x + c' = 0.$$

This has a double root at x_1. If the third point has coordinates (x_3, y_3), then the coefficient of x^2 must be $-(2x_1 + x_3)$, hence

$$2x_1 + x_3 = -(a - m^2)$$

or

$$x_3 = m^2 - a - 2x_1.$$

Then $P + P = 2P = (x_3, -y_3)$, where $y_3 = m(x_3 - x_1) + y_1$.

The different rules for addition can be concisely expressed by the following rule. If P, Q, R are three points on the curve that lie on a line, then $P + Q + R = 0$. Here P, Q, and R need not be distinct points and can include the point at infinity.

Example 19.3.1. Consider the curve $y^2 = x^3 + 1$. The tangent line at $P = (-1, 0)$ is vertical, hence $P = -P$ or $2P = \mathcal{O}$. The line L_1 through $(-1, 0)$ and $(0, 1)$ has slope $m = 1$. Hence the third point of intersection has coordinates

$$x_3 = 1^2 - 0 - (-1 - 0)$$
$$= 1 + 1 = 2$$
$$y_3 = (2 + 1) = 3.$$

Hence $(-1, 0) + (0, 1) = -(2, 3) = (2, -3)$.

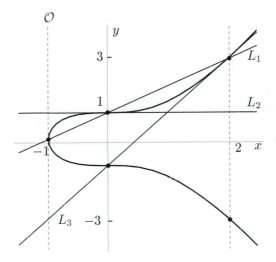

Figure 19.3.1 Group law on $y^2 = x^3 + 1$

The tangent line L_2 at $(0, 1)$ is horizontal. If we substitute $y = 1$ into $y^2 = x^3 + 1$, then we get $x^3 = 0$. This has a triple root at 0, hence the tangent line meets the curve thrice at $(0, 1)$. By definition, $(0, 1) + (0, 1) = -(0, 1) = (0, -1)$.

Exercise. The tangent line L_3 at $(2, 3)$ on $y^2 = x^3 + 1$ is shown in Figure 19.3.1. Determine $(2, 3) + (2, 3)$ by inspection. Compute $(0, 1) + (2, 3)$ on $y^2 = x^3 + 1$.

Example 19.3.2. The real points on $y^2 = x^3 + x^2 + 4x + 4$ are shown in Figure 19.3.2 . The tangent line at $(0, 2)$ has slope

$$m = \frac{3 \cdot 0^2 + 2 \cdot 0 + 4}{2 \cdot 3} = 1.$$

The equation of the tangent is $y - 2 = x$. Substituting this into the equation for the curve, we get

$$x^2 + 4x + 4 = x^3 + x^2 + 4x + 4$$

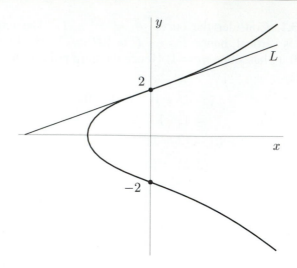

Figure 19.3.2 Group law on $y^2 = x^3 + x^2 + 4x + 4$

or $x^3 = 0$. Therefore, the tangent line meets the curve thrice at $(0, 2)$. We can write

$$(0, 2) + (0, 2) = -(0, 2) = (0, -2)$$

or $3 \cdot (0, 2) = \mathcal{O}$.

Here we use the notation $3P$ to denote $P + P + P$. This is justified by the associative law, $P + (Q + R) = (P + Q) + R$. A sketch of the proof of the associative law is given at the end of this section. For any positive integer k, we can write $k \cdot P = P + P \cdots + P$, where P is added to itself k times.

Exercise. Consider $P = (4, 10)$ on the curve $y^2 = x^3 + x^2 + 4x + 4$. Determine kP for $k = 2, 3, 4, 5, 6$.

Example 19.3.3. Consider $y^2 = x^3 - x^2 - 4x + 8$ given in Figure 19.3.3. The figure shows that $(1, 2) + (-2, 2) = (2, -2)$ and $2 \cdot (2, 2) = (-2, 2)$.

We derived the addition law for the standard form of the elliptic curve given in Equation (1). The same formula for x_3 can be used for curves of the form $y^2 + dy = x^3 + ax^2 + bx + c$ because the coefficient of x^2 is unchanged by the introduction of the addition y term on the left. In this case, the coordinates of $-(x, y)$ are given by $-(x, y) = (x, -y - d)$.

Example 19.3.4. Recall the elliptic curve $y^2 + y = x^3 - x$ from Example 19.1.4 . We are interested in the integer points on this curve to solve the problem of determining when a product of two consecutive integers is equal to a product of three consecutive integers. If (x, y) is on $y^2 + y = x^3 - x$, then $-(x, y) = (x, -y - 1)$.

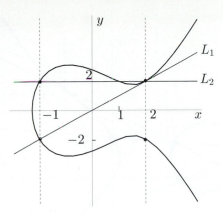

Figure 19.3.3 Group law on $y^2 = x^3 - x^2 - 4x + 8$

The line L_1 joining $(0,0)$ and $(-1,-1)$ has equation $y = x$, which meets the curve at a third point with x-coordinates 2. Hence

$$(0,0) + (-1,-1) = (2,-3).$$

We also see from the line L_2 gives $2 \cdot (0,-1) = (1,-1)$. Here is a table of

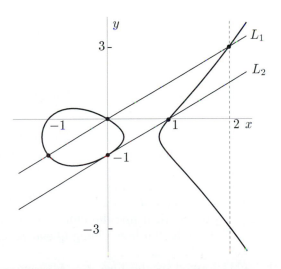

Figure 19.3.4 Group law on $y^2 + y = x^3 - x$

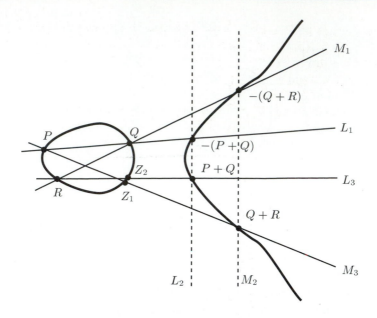

Figure 19.3.5 Associative law on elliptic curves

some multiples of $P = (0, 0)$.

P	$(0, 0)$	$-P$	$(0, -1)$
$2P$	$(1, 0)$	$-2P$	$(1, -1)$
$3P$	$(-1, -1)$	$-3P$	$(-1, 0)$
$4P$	$(2, -3)$	$-4P$	$(2, 2)$
$5P$	$(1/4, -5/8)$	$-5P$	$(1/4, -3/8)$
$6P$	$(6, 14)$	$-6P$	$(6, -15)$

We see that $6P$ gives another integer solution to the problem $14 \cdot 15 = 5 \cdot 6 \cdot 7$.

Exercise. If $P = (0, 0)$ is on $y^2 + y = x^3 - x$, determine $8P$ by doubling $4P$ and by computing $5P + 3P$.

The construction given above makes the rational points $E(\mathbb{Q})$ into a group. To see this, observe that the line joining P and $\mathcal{O}$ is the vertical line through P, hence $P + \mathcal{O} = -(-P) = P$. Similarly, it is easy to see that $P + Q = Q + P$. The proof of the associative is more difficult. Consider a curve C and three distinct points P, Q, and R. (See Figure 19.3.5.) Let L_1 be the line joining P and Q, L_2 the vertical line through $P + Q$, and L_3 the line joining $P + Q$ and R. Let M_1 be the line joining Q and R, M_2 the vertical line through $Q + R$, and M_3 the line joining P and $Q + R$.

Let C_1 be the cubic formed by the lines L_1, M_2, and L_3, and C_2 the cubic formed by the lines M_1, L_2, and M_3. We make the assumption that the points P, Q, $-(P+Q)$, $P+Q$, R, $Q+R$, $-(Q+R)$, $\mathcal{O}$ are distinct. Let $Z_1 = -((P+Q)+R)$ and $Z_2 = -(P+(Q+R))$. By construction, the curves C_1 and C have nine points in common and C_2 has eight points in common with C_1 and C. This implies, by Proposition 19.2.8, that C_2 passes through the ninth point of intersection of C and C_1. Hence C_2 contains Z_2. This implies that Z_1 and Z_2 are the same; otherwise, C and C_2 would have ten points in common, which is not possible because two cubic curves can have at most nine points of intersection.

The proof of associativity assumed that all the points were distinct. The general case requires a study of intersection multiplicities.

The rational points on an elliptic curve can be classified into two types. If P is a rational point such that $kP = \mathcal{O}$ for some integer k, then P is called a point of finite order. The smallest positive integer k such that $kP = \mathcal{O}$ is called the order of P. In Example 19.3.1, the point $(-1,0)$ has order 2 and $(0,1)$ has order 3. In Example 19.3.2, the point $(0,2)$ has order 3, and it can be verified that $(4,10)$ has order 6.

A point of finite order is also called a **torsion point** and the set of points of finite order is called the torsion subgroup of $E(\mathbb{Q})$. We will see that it is not difficult to determine all the torsion points on an elliptic curve. If P is not of finite order, then we say that P is a point of infinite order. We can show that the point $(0,0)$ on $y^2 + y = x^3 - x$ in Example 19.3.4 has infinite order. This will follow from the determination of all points of finite order. In 1923, Mordell proved that the set of points of infinite order is finitely generated; that is, there exists points $P_1, \ldots, P_r$ such that every point of infinite order is an integral linear combination

$$P = c_1 P_1 + \cdots + c_r P_r.$$

The number r is called the **rank** of $E(\mathbb{Q})$. The study of the rank of an elliptic curve is one of the most active areas of research in number theory.

—————————— **Exercises for Section 19.3** ——————————

1. Consider the points $P = (0,0)$ and $Q = (1,2)$ on the curve $y^2 = x^3 + x^2 + 2x$. Determine $2P$, $P+Q$, and $2Q$.

2. Write a computer program to add points on an elliptic curve $y^2 = x^3 + ax^2 + bx + c$. (Be careful to check when the lines become vertical.)

3. Derive addition formulas for rational points on the curve

$$y^2 + a_1 xy + a_2 y = x^3 + a_3 x^2 + a_4 x + a_5.$$

4. On the curve, $y^2 = x^3 - x^2 + 3x + 9$, determine the sum $(0,3) + (-1,2)$ and $(-1,2)$ and $(3,6)$.

5. Determine $2 \cdot (1, 2)$ and $(1, 2) + (2, 4)$ on the curve $y^2 = x^3 - x^2 + 8x - 4$.

6. On $y^2 + y = x^3 - x^2$, compute $(0, 0) + (1, -1)$.

7. Write a program to compute multiples of a point P on an elliptic curve E. Write the program so that if $kP = (a/c, b/d)$, then your program should give you the prime factorization of c and d. Compute the multiples kP, $1 \leq k \leq 12$ in the following cases.

   (a) $E : y^2 = x^3 + x^2 + 2x$, $P = (1, 2)$.

   (b) $E : y^2 = x^3 - x^2 + 3x + 9$, $P = (0, 3)$.

   (c) $E : y^2 = x^3 - x^2 - 4x + 8$, $P = (1, 2)$.

   What relationship do you observe between b and d? Does it hold for any curve? Prove your conjecture.

8. Fix a prime p (say 2, 3, or 5), and determine the values of k such that the coordinates of kP have a factor of p in the denominator. Do this for $k \leq 50$ for the curves given in the previous exercise. What can you say about the set of points that have a factor of p in the denominator?

9. A point of order 2 in $E(\mathbb{Q})$ is such that the tangent line at that point is vertical. Give a geometric interpretation for a point of order 3.

19.4   Elliptic Curves Modulo Primes

The further study of elliptic curves requires an analysis of elliptic curves defined over fields different from the field of rational numbers, particularly, the finite field of integers modulo a prime. This is similar to the study of binary quadratic forms in Section 18.5. Using that study, we were able to obtain results about representation of integers by binary quadratic forms. For elliptic curves, we will be able to determine some integer points by studying solutions modulo primes. In the next section, we will see how to obtain some information about rational points on a elliptic curve from the knowledge of elliptic curves modulo p. Additional applications of this study include methods of factoring, primality testing, and construction of public–key cryptosystems.[7]

Let E be an elliptic curve given by the equation $y^2 = x^3 + ax^2 + bx + c$. Let p be a prime number. The elliptic curve modulo p, E_p, is the set of solutions (x, y) satisfying

$$y^2 \equiv x^3 + ax^2 + bx + c \pmod{p}. \tag{1}$$

Here we can take x and y to be in a fixed complete residue system modulo p, so E_p is a finite set. The group law on an elliptic curve is defined when

---

[7]A. J. Menezes. *Elliptic Curve Cryptosystems*. Boston, MA: Kluwer, 1993.

the discriminant is nonzero. The discriminant of the curve in (1) is $\Delta(f) = (27c^2 + 4a^3c + 4b^3 - a^2b^2 + 8abc) \bmod p$.

The integers modulo a prime p form a field, which is denoted by $\mathbb{F}_p$. Recall that the integers modulo p have the two operations of addition and multiplication defined, and every nonzero element has a multiplicative inverse. This is all that we need to define the group law on E_p. Again, it will be convenient to use projective coordinates to denote the point at infinity. In homogeneous coordinates, the elliptic curve has the equation

$$y^2 z = x^3 + ax^2 z + bxz^2 + cz^3,$$

and the point at infinity is $\mathcal{O} = (0 : 1 : 0)$.

The rules for addition of points on $E(\mathbb{Q})$ also apply to E_p with the interpretation that the reciprocal is the inverse modulo p. When the inverse modulo p does not exist, then the corresponding line is "vertical" modulo p. The rules are the following.

If $P_1 = (x_1, y_1)$ and $P_2 = (x_2, y_2)$, then $P_1 + P_2 = \mathcal{O}$ when $x_1 \equiv x_2$ (mod p). If $y_1 \equiv 0$ (mod p), then $P_1 = -P_1$ and $2P_1 = \mathcal{O}$. In all other cases, the sum $P_1 + P_2$ is obtained by computing

$$m = \begin{cases} (y_2 - y_1)(x_2 - x_1)^{-1} \bmod p & \text{if } P_1 \neq P_2 \\ (3x_1^2 + 2ax_1 + b)(2y_1)^{-1} \bmod p & \text{if } P_1 = P_2, \end{cases} \tag{2}$$

and then let

$$x_3 \equiv m^2 - a - (x_1 + x_2) \pmod{p}.$$

Then $P_1 + P_2 = (x_3, y_3)$, where $y_3 \equiv -m(x_3 - x_1) - y_1 \pmod{p}$.

Example 19.4.1. Consider the curve E defined by $y^2 = x^3 + 1$. The curve has discriminant 27, so if $p \neq 3$, then the group law is defined on E_p. If $p = 5$, then E_5 consists of six points. This can be seen by substituting values of x and then solving the resulting congruence for y. If $x = 1$ or $x = 3$, then the congruence $y^2 \equiv x^3 + 1$ (mod 5) has no solutions. If $x = 0$ or $x = 2$, then the congruence $y^2 = x^3 + 1$ (mod p) has two solutions. If $x = 4$, then there is only one solution. Therefore, the points are $(0, 1)$, $(0, -1)$, $(2, 3)$, $(2, -3)$, $(4, 0)$, and $\mathcal{O}$.

We see that $(4, 0) = -(4, 0)$ or $2 \cdot (4, 0) = \mathcal{O}$. To compute $(2, 3) + (4, 0)$, the slope m is $(0 - 3)(4 - 2)^{-1} \bmod 5$, which is $-3 \cdot 3 \bmod 5$. Hence $m = 1$. Since $a = 0$, we have $x_3 = 1 - (2 + 4) \bmod 5$. This gives $(2, 3) + (4, 0) = (0, -1)$.

Similarly, we can compute $(2, 3) + (2, 3) = (0, 1)$ and $(2, 3) + (0, 1) = (4, 0)$.

If we denote $(2, 3)$ by P, then the multiples of P are

$$
\begin{array}{ll}
P & (2, 3) \\
2P & (0, 1) \\
3P & (4, 0) \\
4P & (0, -1) \\
5P & (2, -3) \\
6P & \mathcal{O}.
\end{array}
$$

So E_5 is a cyclic group generated by P.

Exercise. Compute the number of points in E_7. Compute $(5, 0) + (2, 3)$ on E_7.

Example 19.4.2. Consider the curve E defined the equation $y^2 = x^3 - 4x - 3$. The discriminant of E is 52. If $p \neq 2, 13$, then the curve is nonsingular and the group law is defined. The number of points N_p for some small primes is given below.

p	N_p
3	4
5	4
7	10
11	14

For $p = 7$, the 10 points on the curve are $\mathcal{O}$, $(-1, 0)$, $(0, 2)$, $(0, -2)$, $(1, 1)$, $(1, -1)$, $(2, 2)$, $(2, -2)$, $(-2, 2)$, $(-2, -2)$. It is easy to verify that the point $P = (-2, 2)$ has order 10; hence all the points in E_7 are obtained by computing multiples of P.

We give a formula for the number of points on an elliptic curve E_p using the Legendre symbol. Let $y^2 = f(x)$ be the equation of the curve. For x_0 in a residue system modulo p, if $\left(\dfrac{f(x_0)}{p} \right) = 1$, then there exists a solution $y = y_0$ to $y^2 \equiv f(x_0) \pmod{p}$, so the two points (x_0, y_0) and $(x_0, -y_0)$ are on the curve. If $\left(\dfrac{f(x_0)}{p} \right) = -1$, then there is no point on the curve with first coordinate x_0. If $\left(\dfrac{f(x_0)}{p} \right) = 0$, then $p \mid f(x_0)$ and $(x_0, 0)$ is a point on the curve. By combining the three cases we see that the number of points of the form (x_0, y_0) is

$$
1 + \left(\frac{f(x_0)}{p} \right).
$$

We have to also include the point at infinity in E_p. Hence the total number of points N_p on E_p is

$$N_p = 1 + \sum_{x \bmod p} \left(1 + \left(\frac{f(x)}{p} \right) \right) = p + 1 + \sum_{x \bmod p} \left(\frac{f(x)}{p} \right). \qquad (3)$$

Here the sum $\sum_{x \bmod p}$ is a sum over a complete residue system modulo p. This formula for computation of N_p is suitable for small primes, say primes less than 10^5. For larger primes, there are more efficient procedures are available.[8]

Example 19.4.3. We can use (3) to compute the number of points on $y^2 = x^3 + 1 \pmod{p}$ for different p. The values of N_p for a few small primes are given below.

p	5	7	11	13	19	23	29	31	37
N_p	6	12	12	12	18	12	24	30	36

The quantity that plays an important role in the applications of elliptic curves is $a_p = p + 1 - N_p$. The values of a_p are

p	5	7	11	13	19	23	29	31	37
a_p	0	-4	0	-2	0	-12	-6	0	0

We see that $a_p = 0$ whenever $p \equiv 2 \pmod{3}$. This can be verified for more primes. (See Exercise 3.)

Since the points on an elliptic curve modulo p form a finite group, many of the results from group theory can be used. In particular, if the number of points on E_p is N_p, then for every point P on E_p, we have $N_p P = \mathcal{O}$. The analogous statement for the group of invertible elements modulo p is Fermat's Theorem. The general statement for any finite group is known as Lagrange's Theorem.

Proposition 19.4.4. *Let N_p be the number of elements on the curve E_p. Then for any point P on E,*

$$N_p P = \mathcal{O}.$$

PROOF. The proof is similar to the proof of Fermat's Theorem. Denote N_p by m and let $P_1, P_2, \ldots, P_m$ be the distinct points on E. Then the points $P_1 + P, P_2 + P, \ldots, P_m + P$ are also distinct. Hence

$$P_1 + P_2 + \cdots + P_m = (P_1 + P) + (P_2 + P) + \cdots + (P_m + P).$$

Canceling the sum of $P_1 + P_2 + \cdots + P_m$ on both sides, we obtain that $mP = \mathcal{O}$. ∎

---

[8] R. Schoof. "Elliptic Curves over Finite Fields and the Computation of Square Roots Mod p." *Mathematics of Computation*, 43(1985), 483–494.

Another advantage of using projective coordinates is that we can extend the $\mod p$ function $a \mapsto a \mod p$ to elliptic curves $E(\mathbb{Q}) \to E_p$ so that it respects the group law.

The map from $E(\mathbb{Q})$ to E_p is defined by first defining the reduction between $\mathbb{P}^2(\mathbb{Q})$ and $\mathbb{P}^2(\mathbb{F}_p)$. If $(a : b : c) \in \mathbb{P}^2(\mathbb{Q})$, then we can multiply the greatest common denominator of a, b, and c to assume that the coordinates are integers. Further, we divide by any common divisors, so we can assume that $\gcd(a, b, c) = 1$. This form of a point is called reduced. The reduced form of a point is unique up to multiplication by -1.

Suppose $(a : b : c)$ is the reduced form of a point P in $\mathbb{P}^2(\mathbb{Q})$. For a prime p, we define the reduction $r_p(P)$ to be $r_p(P) = (a \mod p : b \mod p : c \mod p)$. The reduction $r_p(P)$ is in $\mathbb{P}^2(\mathbb{F}_p)$ because all three numbers a, b, and c cannot be divisible by p. If there is no ambiguity about the prime p, then we will also denote $r_p(P)$ by $\overline{P}$.

Let L be the line given by the equation $\alpha x + \beta y + \gamma z = 0$. By clearing denominators, we can assume that $(\alpha : \beta : \gamma)$ is a reduced point. The reduction of L, $\overline{L}$ is defined to be the line

$$\alpha x + \beta y + \gamma z \equiv 0 \quad (\text{mod } p).$$

Similarly, if $y^2 = x^3 + ax^2 + bx + c$ is an elliptic curve such that a, b, c are integers, then its reduction $\overline{E} \,(= E_p)$ is given by

$$y^2 \equiv x^3 + ax^2 + bx + c \quad (\text{mod } p).$$

The important property of the reduction $r_p(E) \to E_p$ is that it preserves addition. If P_1 and P_2 are in $E(\mathbb{Q})$ such that $P_1 + P_2 = Q$, then $\overline{P_1} + \overline{P_2} = \overline{Q}$.

This can be proved by studying the reduction of the intersection of the line joining P_1 and P_2 with the elliptic curve. If the line is given by the equation $\alpha x + \beta y + \gamma z = 0$, then we can assume that one of α, β, or γ is not divisible by p, say $p \nmid \gamma$. Solve for z and substitute into the equation of the elliptic curve to obtain a homogeneous equation of degree three in x and y. Since the solutions are rational, the equation can be factored into the form

$$(b_1 x - a_1 y)(b_2 x - a_2 y)(b_3 x - a_3 y) = 0.$$

Thus we can write the points on the curve in reduced form. If we let $c_1 = -(\alpha a_1 + \beta b_1)$ and $c_2 = -(\alpha a_2 + \beta b_2)$, then $P_1 = (\gamma a_1, \gamma b_1, c_1)$ and $P_2 = (\gamma a_2, \gamma b_2, c_2)$. If P_3 is the third point of intersection, then $P_3 = (a_3, b_3, c_3)$. If $p \nmid \gamma$), then these are in reduced form. Hence $\overline{P_1}$, $\overline{P_2}$, and $\overline{P_3}$ are in the intersection of $\overline{L}$ and $\overline{E}$. Since the group law is defined using the intersection of lines with the elliptic curve, we see that reduction modulo p respects the group law.

─────────────── **Exercises for Section 19.4** ───────────────

1. Let E be the elliptic curve $y^2 = x^3 + 1$. Find a point P on E_7 so that every point in E_7 is a multiple of P.

2. Write a program to add points on an elliptic curve modulo p. This will be useful in the elliptic curve factorization method of Section 19.6.

3. Let N_p be the number of points on the elliptic curve $y^2 \equiv x^3 + 1 \pmod{p}$. If $p \equiv 2 \pmod 3$. then show that the map $x \mapsto x^3$ is a one-to-one function modulo p. Conclude that $N_p = p + 1$.

4. Compute the group of points modulo 3 and 7 on the elliptic curve $y^2 = x^3 + x + 1$.

5. Write a computer program to compute N_p and a_p on an elliptic curve modulo p.

6. Determine the number of points of the curve $y^2 = x^3 - 1$ for all odd primes less than 50.

19.5 Rational Points on Elliptic Curves

The rational points on an elliptic curve can be divided into two classes. If P is a point on an elliptic curve E, and there is an integer k such that $kP = \mathcal{O}$, then P is called a point of finite order on E. A set of points of finite order is called the **torsion subgroup** of E and denoted by E_{tor}. In this section, we study a theorem of Lutz and Nagell by which it is possible to determine all the torsion points on an elliptic curve. It is much more difficult to determine points of infinite order, and we will not be able to give a treatment of this topic in this book.[9]

Recall that the reduction map r_p for a prime p gives a homomorphism of elliptic curves $E(\mathbb{Q}) \mapsto E_p$. Let E_0 be the set of points P in $E(\mathbb{Q})$ such that $r_p(P) = \overline{\mathcal{O}}$. Since r_p preserves addition, it follows that if P_1 and P_2 are in E_0, then so does $P_1 + P_2$, that is, E_0 is a subgroup of $E(\mathbb{Q})$. Suppose that a rational point in E_0 has the reduced form $P = (x : y : z)$; then $\overline{P} = (\overline{x} : \overline{y} : \overline{z}) = (0 : 1 : 0)$. Hence $y \neq 0$, and x and z are divisible by p. If $P \neq \mathcal{O}$, then $z \neq 0$, and the point $(x : y : z)$ corresponds to the point $(x/z, y/z) = (x', y')$. Since z is divisible by p, but not y, the denominator of y' is a multiple of p. Conversely, it is easy to see that if $P = (x, y)$ is a rational point such that the denominator of y is a multiple of p, then $r_p(P) = \overline{\mathcal{O}}$.

To classify rational points, we will show that for any prime, E_0 consists only of points of infinite order. Here p is a fixed prime. This will imply that

─────────────────────────
[9]For a more in-depth discussion, consult J. H.Silverman and J. Tate. *Rational Points on Elliptic Curves*. New York: Springer–Verlag, 1992.

a point of finite order has integer coordinates, and using this fact, it is easy to give a criterion that will determine all points of finite order. To show that E_0 has no points of finite order, we need to use the valuation function v_p. This function measures the order of divisibility of a rational number by the powers of p.

Definition 19.5.1. Suppose $x = p^n a/b$ is a rational number such that a and b are relatively prime to p. Then define $v_p(x) = n$.

For any rational number, v_p is well defined because of the unique factorization into primes. The function v_p is known as the p-**adic valuation** on the rational numbers.

Lemma 19.5.2. *The valuation v_p satisfies the following properties.*

(a) $v_p(xy) = v_p(x) + v_p(y)$.

(b) $v_p(x + y) \geq \min(v_p(x), v_p(y))$ *with equality occurring when* $v_p(x) \neq v_p(y)$.

PROOF. (a) Let $x = p^n a/b$ and $y = p^m a'/b'$ where a, b, a' and b' are coprime to p. Then $xy = p^{n+m}(aa')/(bb')$ implies that $v_p(xy) = n + m$, hence the assertion follows. (b) Using the same notation as before, suppose that $n \leq m$. Then

$$x + y = p^n \left(\frac{a}{b} + p^{m-n} \frac{a'}{b'} \right)$$
$$= p^n \left(\frac{ab' + p^{m-n} a'b}{bb'} \right).$$

If $n < m$, then $ab' + p^{m-n} ba'$ is not divisible by p, hence $v_p(x + y) = n$. If $n = m$, then it is possible that $ab' + ba'$ is a multiple of p, so the valuation of $x + y$ can be greater than n. ∎

Example 19.5.3. It is possible that if $v_p(x) = v_p(y)$, then $x + y$ has a much larger valuation. For example, $v_5(119/5) = -1$ and $v_5(6/5) = -1$, but $v_5(119/5 + 6/5) = v_5(25) = 2$.

Proposition 19.5.4. *Let $P = (x, y)$ be a rational point on the elliptic curve $y^2 = x^3 + ax^2 + bx + c$, where a, b, and c are integers. If $v_p(y) < 0$, then $3v_p(x) = 2v_p(y)$.*

PROOF. The proof follows by computing the valuation of both sides of the equation

$$y^2 = x^3 + ax^2 + bx + c.$$

We have $v_p(y^2) = 2v_p(y) < 0$ and

$$v_p(x^3 + ax^2 + bx + c) \geq \min(v_p(x^3), v_p(ax^2), v_p(bx), v_p(c)).$$

We must have $v_p(x) < 0$; otherwise, we obtain $v_p(x^3 + ax^2 + bx + c) \geq 0$, but $v_p(x^3 + ax^2 + bx + c) = v_p(y^2) = 2v_p(y) < 0$. Since a, b, and c are integers, $v_p(x^3) < v_p(ax^2)$, and $v_p(x^3) < v_p(bx)$, so we have

$$v_p(x^3 + ax^2 + bx + c) = v_p(x^3).$$

This implies that $2v_p(y) = 3v_p(x)$.   ∎

The proposition shows that if a point $(a/c, b/d)$ is on an elliptic curve with integer coefficients, then $c^3 = d^2$, so the rational points are of the form $\left(\dfrac{a}{c^2}, \dfrac{b}{c^3} \right).$

We define subsets of $E(\mathbb{Q})$ using the powers of p dividing the coordinates of the points. Define

$$E^{(n)} = \{(x,y) \in E \mid v_p(x) \leq 2n, v_p(y) \leq 3n.\} \cup \{\mathcal{O}\}.$$

It is clear from the definition that $E_0 = E^{(1)}$ and that

$$E^{(1)} \supseteq E^{(2)} \supseteq E^{(3)} \cdots .$$

It is possible to show that each $E^{(r)}$ is a subgroup of $E(\mathbb{Q})$, the proof of which is implicit in the proof of the following theorem.

Proposition 19.5.5. *There are no nontrivial torsion points in $E^{(1)}$.*

PROOF. We will use projective coordinates in the proof. If $P = (x : y : 1)$ with $v_p(y) < 0$, then by multiplying by $1/y$, we get $P = (x' : 1 : z')$ with $v_p(x'), v_p(z') \geq 0$. If $P \in E^{(n)}$ and $P \notin E^{(n+1)}$, then $v_p(y) = -3n$. Since $v_p(x) = -2n$, $P = (x' : 1 : z')$ satisfies $v_p(x') = n$ and $v_p(z') = 3n$.

For simplicity, we take the equation of the curve to be $y^2 = x^3 + Ax + B$. Let $P_1 = (x_1 : 1 : z_1)$ and $P_2 = (x_2 : 1 : z_2)$ be two points in $E^{(n)}$. We will show that if $P_3 = P_1 + P_2 = (x_3 : 1 : z_3)$, then $x_3 \equiv x_1 + x_2 \pmod{p^{5n}}$.

Using the homogeneous form of the curve, we obtain

$$z_1 = x_1^3 + Ax_1 z_1^2 + Bz_1^3$$
$$z_2 = x_2^3 + Ax_2 z_2^2 + Bz_2^3.$$

Subtracting the equations, we get

$$z_1 - z_2 = (x_1 - x_2)(x_1^2 + x_1 x_2 + x_2^2)$$
$$+ A(x_1 z_1^2 - x_2 z_2^2) + B(z_1 - z_2)(z_1^2 + z_1 z_2 + z_2^2).$$

Write $x_1 z_1^2 - x_2 z_2^2 = (x_1 - x_2)z_2^2 + x_1(z_1^2 - z_2^2)$, and simplify the expression to obtain

$$(z_1 - z_2)[1 - B(z_1^2 + z_1 z_2 + z_2^2) - Ax_1(z_1 + z_2)]$$
$$= (x_1 - x_2)[x_1^2 + x_1 x_2 + x_2^2 + Az_2^2].$$

Let $u = B(z_1^2 + z_1 z_2 + z_2^2) - Ax_1(z_1 + z_2)$, and let the line joining P_1 and P_2 have the equation $z = mx + c$; then the slope m is given by

$$m = \begin{cases} \dfrac{z_1 - z_2}{x_1 - x_2} = \dfrac{x_1^2 + x_1 x_2^2 + x_2^2 + Az_2^2}{1 - u} & \text{if } x_1 \neq x_2 \\[3ex] \dfrac{dz}{dx}(P_1) = \dfrac{Ax_1^2 + 3z_1^2}{1 - 2Ax_1 z_1 - 3Bz_1^2} & \text{if } P_1 = P_2, \end{cases}$$

The equations above show that if $v_p(x) \geq n$, then $v_p(m) \geq 2m$ because the denominator in the expression for m is an expression of the form $1 + u$, where u is a multiple of p. Thus, the denominator is not a multiple of p. Also $z = mx + c$ implies that $c = -mx_1 + z_1$, hence $v_p(c) \geq 3n$. To find the third point of intersection P_3, we intersect the line $z = mx + c$ with the curve $z = x^3 + Axz^2 + Bz^3$, and compare the coefficient of x^2 to find

$$x_1 + x_2 + x_3 = -\frac{2Amc + 3Bm^2 c}{1 + Am^2 + Bm^3}.$$

The denominator in the expression on the right has valuation 0 because it is of the form $1 + u$ where u is a multiple of p. Hence

$$v_p(x_1 + x_2 + x_3) \geq v_p(mc) \geq 5n,$$

and

$$x_1 + x_2 + x_3 \equiv 0 \pmod{p^{5n}}.$$

Let $g(P) = x$ if $P = (x : 1 : z)$. We have shown that $g(P_1 + P_2) \equiv -(g(P_1) + g(P_2)) \pmod{p^{5n}}$.

Suppose $P_1 = (x_1 : 1 : z_1)$ is a point of finite order in $E^{(1)}$. Let n be the smallest integer such that P_1 is in $E^{(n)}$ but not in $E^{(n+1)}$. Then $kP_1 = \mathcal{O}$ implies that $g(kP_1) \equiv -kg(P_1) \equiv 0 \pmod{p^{5n}}$. If $p \nmid k$, then $g(P_1) \equiv 0 \pmod{p^{5n}}$, hence $v_p(x) \geq 5n$. Since $5n > (n+1)$, this implies that $P_1 \in E^{(n+1)}$, a contradiction.

If p divides the order of P_1, then write $k = pk_1$ and let $P_2 = k_1 P_1$. Then the order of P_2 is p. Choose n such that P_2 is in $E^{(n)}$ but not in $E^{(n+1)}$. The $g(pP_2) \equiv -pg(P_2) \equiv 0 \pmod{p^{5n}}$ implies that $g(P_2) \equiv 0 \pmod{p^{5n-1}}$. Again, $5n - 1 > n + 1$ for $n \geq 2$, and we obtain a contradiction. This completes the proof of the proposition. ∎

The proof also applies to the curve $y^2 = x^3 + ax^2 + bx + c$. We leave the verification of this fact to the reader.

Corollary 19.5.6. *A torsion point on the elliptic curve* $y^2 = x^3 + Ax + B$ *with A and B integers has integer coordinates.*

PROOF. If $P = (x, y)$ is a torsion point, then the proposition shows that the denominators of the coordinates are not divisible by any prime. Hence x and y are integers. ∎

Theorem 19.5.7 (Lutz–Nagell). *If* $P = (x_1, y_1)$ *is a torsion point on* $y^2 = x^3 + Ax + B$, *then either* $y_1 = 0$ *or* $y_1^2 \mid 4A^3 + 27B^2$.

PROOF. Let $f(x) = x^3 + Ax + B$. If $P = (x_1, y_1)$ has order 2, then $y_1 = 0$. If P has order greater than 2, then $Q = 2P = (x_2, y_2)$ is also a point of finite order, hence its coordinates are integers. The slope of the tangent line at P is $m = f'(x_1)/(2y_1)$ and $2x_1 + x_2 = m^2$. Since x_1 and x_2 are integers, m must be an integer, that is, $2y_1 \mid (3x_1^2 + A)$. But $y_1^2 = f(x_1)$, hence $y_1^2 \mid f(x_1)$. This implies that y_1^2 divides any linear combination of $f(x_1)$ and $f'(x_1)$. In particular, we can express $4A^3 + 27B^2$ as

$$4A^3 + 27B^2 = (-27f(x) + 54B)f(x) + (f'(x) + 3A)f'(x)^2.$$

This shows that $y_1^2 \mid 4A^3 + 27B^2$. ∎

The proposition actually determines all integer points P such that $2P$ is also an integer point. It is easy to find all the points satisfying the condition. From this, we can determine if a point is a torsion point or a point of infinite order by computing its multiples. A torsion point can be quickly found because the orders of torsion points are small. If a multiple of an integer point is not an integer point, then the point has infinite order.

Example 19.5.8. Let E be the curve $y^2 = x^3 + 1$. By the proposition, a torsion point $P = (x_1, y_1)$ satisfies $y_1 = 0$ or $y_1^2 \mid 27$. If $y_1 = 0$, then $x_1 = -1$. If $y_1^2 \mid 27$, then $y_1 = \pm 1$ or ± 3. This gives four points $(0, 1)$, $(0, -1)$, $(2, 3)$, and $(2, -3)$. Together with $(-1, 0)$ and $\mathcal{O}$, we obtain a total of six torsion points. It is possible to show that the curve has no other rational points. (See Exercise 9.)

Example 19.5.9. Consider the curve $y^2 = x^3 - 2$. If $p = (x_1, y_1)$ is of finite order, then $y_1 = 0$ or $y_1^2 \mid 27 \cdot 4$. We see that there is no point on the curve with $y = 0$. If $y_1^2 = 4$, then $x_1^3 - 2 = 4$, hence x_1 is not an integer. Similarly, there are no solutions with $y_1^2 = 9$ or $y_1^2 = 36$. This implies that there no nontrivial torsion points. It is shown in Exercise 18.6.4 that $(3, \pm 5)$ are the only integer points on the curve. Hence $2 \cdot (3, \pm 5)$ does not have integer coordinates, and these points are of infinite order.

Example 19.5.10. If $P = (x_1, y_1)$ is a torsion point on $y^2 = x^3 - x$, then $y_1 = 0$ ior $y_1^2 \mid -4$. From this, it is easy to see that $(0,0)$, $(1,0)$, $(-1,0)$, and $\mathcal{O}$ are the only points of finite order on the curve.

The results we have proved allow us to determine if a point has infinite order. It is much more difficult to determine all the points of infinite order. In 1923, L. J. Mordell proved the remarkable theorem that the group of rational points is finitely generated. Hence there exists a finite set $P_1, \ldots, P_r$ such that every point P of infinite order on the curve can be written as

$$P = k_1 P_1 + \cdots + k_r P_r.$$

The smallest integer r satisfying this property is called the **rank** of the elliptic curve. The determination of the rank is a difficult problem and is an active area of research in number theory.

────────────────────── **Exercises for Section 19.5** ──────────────────────

1. Construct examples such that $v_5(x) = -1$ and $v_5(y) = -1$, but $v_5(x + y)$ can be made as large as we wish. Generalize to any prime.

2. Show that if $p \neq 2$ and $v_p(x + y) > \min(v_p(x), v_p(y))$, then $v_p(x - y) = \min(v_p(x), v_p(y))$. What can you say when $p = 2$?

3. Show that if $v_p(x) \geq 0$ for all p, then x is an integer.

4. Use the proof of Proposition 19.5.5 to show that $E^{(n)}$ is a subgroup of E for every $n \geq 1$.

5. Find all torsion points on the curves $y^2 = x^3 - 21x + 37$ and $y^2 = x^3 - 43x + 166$.

6. Modify the proof of Proposition 19.5.5 to show that any point of finite order on $y^2 = x^3 + ax^2 + bx + c$ has integer coordinates. Let $f(x) = x^3 + ax^2 + bx + c$ be a point of finite order on the curve, and let $\Delta_0 = 27c^2 + 4a^3c + 4b^3 - a^2b^2 - 18abc$. Verify the identity

$$\Delta_0 = (-27f(x) + 54c + 4a^3 - 18ab)f(x) + (f'(x) + 3b - a^2)f'(x)^2,$$

   and conclude that if $P = (x_1, y_1)$ is a torsion point, then either $y_1 = 0$ or $y_1^2 \mid \Delta_0$.

7. Determine the points of finite order on the curves $y^2 = x^3 - x^2 - 4x + 4$, $y^2 = x^3 + x^2 + 7x$, and $y^2 = x^3 + x^2 - 160x + 308$.

8. Find points of infinite order on the curves $y^2 = x^3 - 2x$, $y^2 = x^3 - 5x$, and $y^2 = x^3 + 5$.

9. Consider the curve $y^2 = x^3 + 1$. Write $(y - 1)(y + 1) = x^3$ and show that if $\gcd(y - 1, y + 1) = 1$, then $y - 1 = r^3$, and $y + 1 = s^3$ for some r and s. Conclude that there are no integer solutions. If $\gcd(y - 1, y + 1) = 2$,

then show that there exist r and s satisfying $r^3 - 2s^3 = \pm 1$. Using the fact that the only solutions to this equation are $r = \pm 1$ and $s = \pm 1$, determine all the integer solutions to $y^2 = x^3 + 1$.

19.6   Elliptic Curve Factorization Method

The elliptic curve factorization method, due to H. W. Lenstra, uses ideas from Pollard's $(p-1)$-method.[10] The $(p-1)$-method is based on the structure of the multiplicative group of invertible elements modulo p, and the elliptic curve method uses the group of points on an elliptic curve modulo p. The advantage of using these groups is that the group operation in these two groups can be computed very rapidly. There are other factorization methods that use similar ideas; for example, the class group method of Shanks uses computations in the class group of quadratic forms.

Let us first recall the idea of the $(p-1)$-method. Suppose n is a composite number with a prime factor p. Suppose further that $p - 1$ has small prime factors so that $p - 1 \mid k!$ for a reasonably small integer k. Then $a^{k!} \equiv 1 \pmod{p}$, hence $p \mid (a^{k!} - 1, n)$. If we are lucky, then this greatest common divisor gives a proper factor of n. The $(p-1)$-method is successful when n has a prime factor p such that $p - 1$ has small prime factors. Here $p - 1$ is the number of invertible elements modulo p. In the elliptic curve method, the number of elements in an elliptic curve modulo p will play a similar role.

Suppose E is an elliptic curve given by the equation $y^2 = x^3 + ax + b$. If n is composite, then the equations (2) given in Section 19.4 do not define a group law. This is because nonzero numbers can fail to have inverses modulo n. Recall that we have to compute the slopes $m = (y_2 - y_1)(x_2 - x_1)^{-1} \bmod n$. If $(x_2 - x_1)$ is not invertible, then $\gcd(x_2 - x_1, n) > 1$. There are two possibilities. If the greatest common divisor is a proper divisor of n, then we have found a factor of n. If the greatest common divisor is n, then we can try another point on the curve or a different curve.

The reader may wonder what elliptic curves have to do with this. We can randomly select a number x and try to compute (x, n) to find a factor of n. This is unlikely to succeed. The reason elliptic curves are helpful is the following. Suppose p is a prime divisor of n. The group law is defined on E_p. The formulas for the group law use arithmetic modulo p. In this case, we can first do the computation modulo n and then reduce modulo p to obtain the same result. Suppose we compute kP modulo n for a point P on E and $\overline{P}$ is the reduction modulo p. Then $\overline{kP} = k\overline{P}$. If N_p is the number of points on E_p, then $k\overline{P} = \mathcal{O}$ when $N_p \mid k$. In computing kP, we will have $x_2 - x_1 \equiv 0 \pmod{p}$, and so $(x_2 - x_1, n) > 1$.

---

[10]H. W. Lenstra. "Factoring Integers with Elliptic Curves." *Annals of Mathematics*, 126 (1987), 649–673.

The utility of elliptic curves comes from our ability to compute kP very rapidly for large values of k and the fact that there are many elliptic curves such that N_p has small prime factors. This is the reason the elliptic curve method is a lot more powerful than the $(p-1)$-method. When the $(p-1)$-method fails, there is nothing we can do. When the elliptic curve method fails on a particular curve, we can always try another elliptic curve.

The basis of the elliptic curve method is the addition rule using the following formulas. If $P_1 = (x_1, y_1)$ and $P_2 = (x_2, y_2)$, and if $(x_1 - x_2, n) = 1$ and $(y_1, n) = 1$, then

$$
m = \begin{cases} (y_2 - y_1)(x_2 - x_1)^{-1} \bmod n & \text{if } P_1 \neq P_2 \\ (3x_1^2 + 2ax_1 + b)(2y_1)^{-1} \bmod n & \text{if } P_1 = P_2, \end{cases} \tag{1}
$$

and

$$
x_3 \equiv m^2 - a - (x_1 + x_2) \pmod{n}.
$$

Then $P_1 + P_2 = (x_3, y_3)$, where $y_3 \equiv -m(x_3 - x_1) - y_1 \pmod{n}$. When $(x_1 - x_2, n) \neq 1$ or $(y_1, n) \neq 1$, then these formulas do not apply. But it does not matter, because in that case, we will have a factor of n. Remember that these formulas do not define a group law on the points on E modulo n.

The computation of kP for large k can be done by addition and duplication laws. This is similar to the computation of $a^m \bmod n$ for large values of n. To compute kP, we can write $k = k_0 + 2k_1 + 2^2 k_2 + \cdots + 2^r k_r$, where each k_i is 0 or 1. Then

$$
kP = k_0 P + k_1(2P) + k_2(2^2 P) + \cdots + k_r(2^r P), \tag{2}
$$

where $2^i P$ can be computed from $2^{i-1} P$ using the doubling formula.

Let us look at some examples of factorization using elliptic curves. Additional details of the algorithm are discussed after the examples.

Example 19.6.1. Let $n = 137703491$. We take the point $P = (2, 1)$ on the elliptic curve $y^2 = x^3 + x - 9$. (These are chosen by first selecting P and a and solving for b in $y^2 = x^3 + ax + b$.) We compute $kP \bmod n$ using Equation 2 for increasing values of k. The best values of k are $r!$ for increasing r. For example,

$$
20!P = (38765800, 102761480)
$$
$$
40!P = (73059078, 50101112).
$$

Next, the computation fails when we try to compute $60!P$. In this computation, we come to the addition $(98622427, 37062796) + (25032179, 18303780)$, and we see that $\gcd(98622427 - 25032179, n) = 17389$, a proper divisor of n.

Example 19.6.2. Let $n = 271811237833$. As in the previous example, we take $P = (2, 1)$ on the curve $y^2 = x^3 + ax + b$. Start with $a = 1$ so that $b = -9$. We compute $20!P$, $40!P$, $60!P$, and $80!P$ without any trouble. At this stage, we can continue with higher multiples of P or use another curve. Taking $a = 2$ gives the curve $y^2 = x^3 + 2x - 11$. We can compute $20!P$, $40!P$, $60!P$, and $80!P$ without any trouble. So we try another curve $y^2 = x^3 + 3x - 13$. We are able to compute $60!P$ successfully, but computation of $80!P$ fails at the addition

$$(16755565661, 260664116207) + (229018473606, 100653888225).$$

We compute the greatest common divisor

$$\gcd(16755565661 - 229018473606, n) = 2595377,$$

to obtain a proper factor of n.

There are two factors that affect the implementation of the algorithm. The first is the choice of elliptic curves, and the second is the choice of k. For simplicity, the elliptic curves can be chosen to be of the form $y^2 = x^3 + ax + 1$, which contain the point $(0, 1)$ for all values of a. We try these curves in succession for increasing values of a starting at $a = 1$. The simplest choice for k in computing kP is to compute $2P$, $3!P$, $4!P$, and so on until $r!P$ for some prespecified bound r. A better choice is to take k to be products of primes and prime powers smaller than an upper bound B. Suppose $q_1, \ldots, q_r$ are the primes and prime powers less than B. Then we compute $P_1 = q_1 P$, $P_2 = q_2 P_1$, and so on until the computation fails, or we exhaust our supply of primes. In the latter case, we have to try another elliptic curve.

The algorithm for computing kP is similar to Algorithm 4.3.3. The only difference is that we have to check if the addition law is valid.

Algorithm 19.6.3 (Multiples of P on an elliptic curve). This algorithm computes kP_1 on an elliptic curve modulo n. The computation can fail, in which case, the algorithm terminates by returning the two points where the failure occurred. Let $P_1 = (x_1, y_1)$ and $P_2 = (x_2, y_2)$ in the course of the algorithm.

1. [Initialize] Let $P_2 = \mathcal{O}$.

2. [Check if done] If $k = 0$, return P_2, and terminate.

3. [k is odd] If $k \bmod 2 = 1$, then compute $d = (x_2 - x_1, n)$, and set $k = k - 1$. If $d = 1$, then compute $P_2 = P_2 + P_1$, and go to step 2; otherwise, terminate algorithm, and return P_1 and P_2.

4. [k is even] If k is even, then compute $d = (2y_1, n)$, and let $k = k/2$. If $d = 1$, then compute $P_1 = 2P_1$, and go to step 2; otherwise, terminate, and return P_1.

The complete algorithm for the elliptic curve factorization method follows. The algorithm uses a table of the first m primes.

Algorithm 19.6.4. The algorithm attempts to factor n using the elliptic curve method.

1. [Table of Primes] Let T be the table of the first m primes.

2. [Initialize] $a = 1$.

3. [Choose E] $P = (0, 1)$, $i = 1$ and $E : y^2 = x^3 + ax + 1$.

4. If $i <= m$, then let $k = T[i]$; otherwise, $a = a + 1$, and go to step 3.

5. [Compute multiple] Compute $P = kP$. If the computation fails, go to step 6; otherwise, $i = i + 1$, and repeat step 4.

6. [Is Factor Found?] Check if a noninvertible element modulo n is found. If a factor is obtained, then return the factor, and terminate algorithm; otherwise, increase a, and return to step 3.

Example 19.6.5. We use the previous algorithm to factor the number $n = 357564082969$ using the first 100 primes. The algorithm fails to yield a factor for the curves $y^2 = x^3 + x + 1$ and $y^2 = x^3 + 2x + 1$. On $y^2 = x^3 + 3x + 1$, the computation of $(149 \cdot 139 \cdots 5 \cdot 3 \cdot 2)P$ (with $P = (0, 1)$) fails in the doubling of $(269927330530, 348260288613)$. Here the second coordinate 348260288613 is not invertible modulo n, and we find $(348260288613, n)$ to obtain the factor 430811.

 **Exercises for Section 19.6**

1. Write a program to implement the elliptic curve factoring method.

2. Factor the following integers using the elliptic curve factoring method.

   (a) 387234344593757.
   (b) 51776736578406607.

Appendix A

Mathematical Induction

An important principle in proving mathematical assertions is the Principle of Mathematical Induction, which applies to the class of assertions of the form "$P(n)$ is true for all natural numbers," where $P(n)$ is some property of a natural number n. For example, $P(n)$ can be the statement that $n^3 - n$ is a multiple of 6. In many cases, a formula or theorem is guessed by experiment, and the only way to establish it rigorously is to use the Principle of Mathematical Induction.

The set of natural numbers $\{1, 2, 3, \dots\}$ is denoted by $\mathbb{N}$. The Principle of Mathematical Induction is the following property of natural numbers.

Principle of Mathematical Induction (PMI). *Let S be a subset of the natural numbers with the properties:*

(a) $1 \in S$.

(b) *If $n \in S$, then $n + 1 \in S$.*

Then $S = \mathbb{N}$.

We will say that a proof is by induction whenever we use the Principle of Mathematical Induction or its variants. A typical proof by induction is the following.

Example A.1. Prove that $1 + 2 + \cdots + n = n(n+1)/2$ for every natural number n.

PROOF. Let S be the set of natural numbers for which the formula holds, that is,

$$S = \left\{ n \in \mathbb{N} : 1 + 2 + \cdots + n = \frac{n(n+1)}{2} \right\}.$$

We want to prove that $S = \mathbb{N}$.

First we check that $1 \in S$. Indeed $1 = 1(1+1)/2$.

Next, we prove that if $k \in S$, then $k + 1 \in S$. Then the Principle of Mathematical Induction will imply that $S = \mathbb{N}$.

Assume that $k \in S$, that is,

$$1 + 2 + \cdots + k = \frac{k(k+1)}{2}. \tag{3}$$

Using this, we want to show that

$$1 + 2 + \cdots + (k+1) = \frac{(k+1)(k+2)}{2}. \tag{4}$$

Consider

$$1 + 2 + \cdots + (k+1) = \underbrace{1 + 2 + \cdots + k}_{k(k+1)/2} + k + 1$$

$$= \frac{k(k+1)}{2} + k + 1$$

$$= \frac{k(k+1) + 2(k+1)}{2} = \frac{(k+1)(k+2)}{2}.$$

We have been able to prove (4) using (3), so it follows that $S = \mathbb{N}$, and the assertion is true for all natural numbers. ∎

A proof by induction of a property $P(n)$ of the natural numbers consists of the following steps.

1. Define a set of natural numbers S to be the set of n such that $P(n)$ holds.

2. Show that $1 \in S$, that is, $P(1)$ is true.

3. Show that if $k \in S$, then $k + 1 \in S$, that is, the truth of $P(k)$ implies that of $P(k+1)$.

Example A.2. Prove by induction that for every $n \in \mathbb{N}$, $n + 3 < 5n^2$.

PROOF. Let $S = \{n \in \mathbb{N} : n + 3 < 5n^2\}$. We verify that $1 \in S$: $1 + 3 < 5 \cdot 1^2$ is true. Assume that $k \in S$, that is,

$$k + 3 < 5k^2 \text{ (this is the induction hypothesis).}$$

We want to show that $k + 1 \in S$, that is,

$$(k+1) + 3 < 5(k+1)^2.$$

We use the induction hypothesis on the expression $k+3$, which occurs on the left-hand side as follows.

$$(k+1)+3 = \underbrace{k+3}_{<5k^2}+1,$$
$$< 5k^2 + 1,$$
$$< 5k^2 + 10k + 5,$$
$$= 5(k+1)^2.$$

This shows that $k + 1 \in S$, and the statement follows by the Principle of Mathematical Induction. ∎

It is not necessary to write down the set S explicitly when it is clear what is meant. The following is a good model on how to write a proof by induction.

Example A.3. Prove by induction that $2^{2n-1} + 3^{2n-1}$ is divisible by 5 for every natural number n.

PROOF. The result is true for $n = 1$ because $2^{2\cdot 1 - 1} + 3^{2\cdot 1 - 1} = 5$. Assume the result is true for $n = k$, that is, $2^{2k-1} + 3^{2k-1} = 5c$ for some integer c. We want to prove that $2^{2(k+1)-1} + 3^{2(k+1)-1} = 5d$ for some d.

$$2^{2(k+1)-1} + 3^{2(k+1)-1} = 2^{2k+1} + 3^{2k+1}$$
$$= 4 \cdot 2^{2k-1} + 9 \cdot 3^{2k-1}$$
$$= 4(\underbrace{2^{2k-1} + 3^{2k-1}}_{5c}) + 5 \cdot 3^{2k-1}$$
$$= 5(4c + 3^{2k-1}).$$

Then $2^{2(k+1)-1} + 3^{2(k+1)-1} = 5d$ with $d = 4c + 3^{2k-1}$, so by induction, the desired result is true. ∎

In some cases, we do not want to prove that a result is true for all natural numbers, but rather for all numbers greater than a fixed one. In such cases, we use the following generalization of the PMI.

Principle of Mathematical Induction (Extension). *Let S be a subset of $\mathbb{N}$ such that*

1. $n_0 \in S$ for some natural number n_0.

2. If $k \geq n_0$ and $k \in S$, then $k + 1 \in S$.

Then $\{n \in \mathbb{N} : n \geq n_0\} \subseteq S$.

In other words, if we show that $P(n_0)$ is true, and the truth of $P(k)$ implies that $P(k+1)$ is true for $k \geq n_0$, then the property P holds for every $n \geq n_0$.

Example A.4. Prove that $n^2 < 2^n$ for all $n \geq 5$.

Note that the result is true for $n = 1$, but false for $n = 2, 3, 4$, so we cannot use the first PMI.

PROOF. The result is true for $n_0 = 5$ because $5^2 < 2^5$. Assume that $k^2 < 2^k$ for $k \geq 5$; we want to prove that $(k+1)^2 < 2^{k+1}$. Now

$$(k+1)^2 = \underbrace{k^2}_{<2^k} + 2k + 1 < 2^k + 2k + 1.$$

If we prove that $2k + 1 < k^2$, then the proof will be complete, as we have

$$(k+1)^2 < 2^k + k^2 < 2^k + 2^k = 2^{k+1}.$$

But $2k + 1 < k^2$ is equivalent to $k^2 - 2k = k(k-2) > 1$, which is true when $k \geq 3$, and in particular when $k \geq 5$. ∎

In some cases, we cannot prove directly that if $n \in S$, then $n + 1 \in S$. It is useful to make use of the the seemingly stronger hypothesis that if $k \in S$ for every $k \leq n$, then $n+1 \in S$. This variation is the called Strong Induction or the Principle of Complete Induction (PCI). As it turns out, strong induction is equivalent to ordinary induction.

Principle of Complete Induction. *Let S be a subset of the natural numbers such that*

(a) $1 \in S$.

(b) $\{1, 2, \ldots, n\} \subset S$ *implies that* $n + 1 \in S$.

Then $S = \mathbb{N}$.

The principle of complete induction can also be extended to prove results that are true for natural numbers larger than some n_0.

Example A.5. Every natural number $n > 1$ has a prime factor.

PROOF. Clearly, 2 has a prime factor as it is prime. Assume that every number less than $k + 1$ has a prime factor larger than 1. If $k + 1$ does not have a prime factor, then $k + 1$ is not prime and can be written as $k + 1 = ab$ with $a, b \geq 2$. It follows that a and b are less that $k+1$, so they have prime factors; therefore, $k + 1$ has a prime factor. ∎

The principle of induction is equivalent to another important property of the natural numbers, the Well-Ordering Principle.

The Well-Ordering Principle. *If S is a nonempty set of positive integers, then S contains a smallest element.*

We use the well-ordering principle very frequently.

———————————— **Exercises for Appendix A** ————————————

1. Show that $1^2 + 2^2 + \cdots + n^2 = \dfrac{n(n+1)(2n+1)}{6}$ for all natural numbers n.

2. Find a closed form expression for each of the following expressions, and use induction to prove the formulas.

   (a) $1 + 3 + 5 + \cdots + (2n+1)$.
   (b) $1^3 + 2^3 + \cdots + n^3$.
   (c) $\dfrac{1}{1 \cdot 2} + \dfrac{1}{2 \cdot 3} + \cdots + \dfrac{1}{n(n+1)}$.

3. **(The Towers of Hanoi)** Suppose that n disks of different radii are stacked on peg A, with the largest at the bottom, and with each disk having a smaller radius than the one below it (See Figure A.1.) The problem is to transfer all the disks to peg B, one at a time, to form an identical tower. The transfer is performed so that at no stage is a larger disk placed on a smaller disk (hence we require the use of the auxiliary peg C). Let M_n be the number of moves required to transfer a tower of n disks. Explain why

$$M_n = 2M_{n-1} + 1.$$

Find a closed form for M_n and prove it by induction.

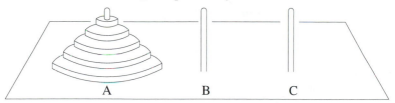

Figure A.1 The three pegs in the Towers of Hanoi

4. Show that $a + ar + ar^2 + \cdots + ar^{n-1} = \dfrac{a(r^n - 1)}{r - 1}$ for $r \neq 1$ and any natural number n.

5. For which natural numbers does the assertion $n! > n^2$ hold? Prove your claim. Similarly, study the inequality $n! > n^3$.

6. Show that $5^{2n} - 4^n$ is divisible by 7 for all natural numbers.

7. Show that every positive integer is a sum of distinct Fibonacci numbers.

8. Bertrand's postulate states that there is a prime between n and $2n$ for every natural number n. Use this to show that for every $n > 1$, one of n or $n - 1$ can be written as a sum of distinct primes.

9. Show that $(n/3)^n < n! < (n/2)^n$ for all $n \geq 5$.

Appendix B

Binomial Theorem

The Binomial Theorem generalizes the identities for computing $(a+b)^2$ and $(a+b)^3$. These are:

$$(a+b)^2 = a^2 + 2ab + b^2$$
$$(a+b)^3 = a^3 + 3a^2b + 3ab^2 + b^3.$$

Similarly, it is not difficult to write down the expansion of $(a+b)^4$. The Binomial Theorem gives a method to compute $(a+b)^n$ for any positive integer n. In order to evaluate $(a+b)^n$, we have to know the coefficient of each term. The expression $(a+b)^n$ is a product of n terms $(a+b)(a+b)\cdots(a+b)$, and a term in the product is formed by selecting a term from each factor. We can select an a or a b from each factor; suppose we choose r a's; then the other $n-r$ factors are b, so we obtain $a^r b^{n-r}$. Hence the terms in the expansion of $(a+b)^n$ are a^n, $a^{n-1}b$, $a^{n-2}b^2, \ldots, a^{n-r}b^r, \ldots, b^n$ for r between 0 and n. We denote the coefficient of $a^{n-r}b^r$ in the expansion of $(a+b)^n$ by $C(n,r)$ or $\binom{n}{r}$. This coefficient is called the binomial coefficient, and it is the number of ways of selecting r elements out of n.

By definition we have the expansion

$$(a+b)^n = \binom{n}{0}a^n + \binom{n}{1}a^{n-1}b + \binom{n}{2}a^{n-2}b^2 + \cdots$$

$$= \sum_{r=0}^{n} \binom{n}{r} a^{n-r}b^r.$$

We derive the properties of these coefficients and then give a formula that simplifies their computation.

Proposition B.1. *Let n and r be nonnegative integers such that $1 \leq r \leq n$.*

(a) $\binom{n}{0} = \binom{n}{n} = 1.$

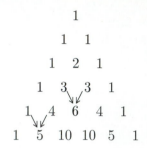

Figure B.2 Pascal's triangle

(b) $\displaystyle\binom{n}{r} = \binom{n-1}{r} + \binom{n-1}{r-1}.$

PROOF. Both these properties follow from the expansion of

$$(a+b)^n = (a+b)(a+b)^{n-1}$$
$$= a(a+b)^{n-1} + b(a+b)^{n-1}.$$

(a) The only way to get a^n in $(a+b)^n$ is to multiply the a^{n-1} term in $(a+b)^{n-1}$ with a. All the powers of a in $b(a+b)^{n-1}$ are smaller than n. Using this fact, we see by induction that $\binom{n}{0} = 1$. Similarly, the coefficient of b^n should be 1.

(b) We use the expansion of $(a+b)^{n-1}$ in terms of the binomial coefficients $\binom{n-1}{r}$. Then

$$(a+b)^n = a(a+b)^{n-1} + b(a+b)^{n-1}$$
$$= a \sum_{r=0}^{n-1} \binom{n-1}{r} a^{n-1-r} b^r + b \sum_{r=0}^{n-1} \binom{n-1}{r} a^{n-1-r} b^r.$$

The term $a^{n-r}b^r$ on the left is the sum of two terms on the right. The contribution is from the term $a^{n-1-r}b^r$ in the first sum and $a^{n-r}b^{r-1}$ in the second sum. The coefficients of these are $\binom{n-1}{r-1}$ and $\binom{n-1}{r}$, respectively. This completes the proof. ■

The recursion relation for binomial coefficients is the basis for the famous "Pascal's triangle" for computing the binomial coefficient. See Figure B.2. The coefficients in the expansion of $(a+b)^5$ are in the sixth row of the triangle. A number in this row is obtained by adding the two numbers above it in the fifth row. Similarly, the coefficients in the expansion of $(a+b)^n$ lie in the $n+1$st row.

Use of Pascal's triangle is impractical for large n, as it requires knowledge of all the smaller binomial coefficients. A better method is given by the following formula using the factorial function. We define $n!$ (pronounced "n factorial") for $n \geq 1$ as the product of the natural numbers less than or equal to n, that is, $n! = n(n-1)(n-2)\cdots 3\cdot 2\cdot 1$. By convention, $0!$ is defined to be 1. For example, $5! = 1\cdot 2\cdot 3\cdot 4\cdot 5 = 120$.

Proposition B.2. *If $n \geq 1$ and $0 \leq r \leq n$, then*

$$\binom{n}{r} = \frac{n!}{r!(n-r)!}.$$

PROOF. The proof is by induction on n. The assertion is clear for $n = 1$, as both $\binom{1}{0}$ and $\binom{1}{1}$ are 1 and so are all the factorials involved. Assume that the assertion is true for $n = k-1$, that is,

$$\binom{k-1}{r} = \frac{(k-1)!}{r!(k-1-r)!} \text{ for } 0 \leq r \leq k-1.$$

From part (b) of the previous proposition, we compute

$$\begin{aligned}
\binom{k}{r} &= \binom{k-1}{r} + \binom{k-1}{r-1} \\
&= \frac{(k-1)!}{r!(k-1-r)!} + \frac{(k-1)!}{(r-1)!(k-r)!} \\
&= \frac{(k-1)!}{(r-1)!(k-r-1)!}\left[\frac{1}{r} + \frac{1}{k-r}\right] \\
&= \frac{(k-1)!}{(r-1)!(k-r-1)!}\frac{k}{r(k-r)} \\
&= \frac{k!}{r!(k-r)!}.
\end{aligned}$$

We used the properties $r! = r(r-1)!$ and $(k-r)! = (k-r)(k-r-1)!$ to group the terms together. This completes the proof. ∎

The binomial coefficient $\binom{n}{r}$ has the combinatorial interpretation that it is the number of ways of selecting r objects from a set of n distinct objects. The selection of objects is as sets, without concern for their order. For example, from a set of 5 objects $\{a, b, c, d, e\}$, we could select two elements in 10 ways: $ab, ac, ad, ae, bc, bd, be, cd, ce, de$. We can prove that $\binom{n}{r} = \frac{n!}{r!(n-r)!}$ using this interpretation. Consider the permutations of r distinct letters chosen from a total of n letters. The first letter can be chosen in n ways, the second in $n-1$ ways, and so on. There are $n-r+1$ choices for the rth letter. Therefore, the total number of permutations of r letters from n is

$$n(n-1)(n-2)\cdots(n-r+1) = \frac{n!}{(n-r)!}.$$

On the other hand, any set of r letters has $r!$ permutations, so if we select r letters in $\binom{n}{r}$ ways, then the total number of such permutations is $r!\binom{n}{r}$. This implies that

$$\binom{n}{r} = \frac{n!}{r!(n-r)!}.$$

The combinatorial interpretation of the binomial coefficients can be related to the definition given above by observing that in the expansion of $(a+b)^n$, to obtain a term of $a^r b^{n-r}$, we have to choose a from r factors and b from the other $n-r$ factors.

Example B.3. Many properties of the binomial coefficients can be proved by using either the Binomial Theorem or combinatorics. For example, to show that

$$\binom{n}{0} + \binom{n}{1} + \binom{n}{2} + \cdots + \binom{n}{n} = 2^n,$$

we can use the Binomial Theorem with $a = b = 1$, or observe that the left-hand side of the equality is the number of ways of selecting any subset of a set of n elements. This is equal to the number of subsets, and it is easy to see that the number of subsets of an n element set is 2^n because an element is either in a subset or it is not. There are two choices for each element, so a total of 2^n subsets.

An important property of the binomial coefficients is that for an odd prime p, all the coefficients $\binom{p}{k}$ for $1 \le k \le p-1$ are multiples of p. This implies that $(a+b)^p - (a^p + b^p)$ is a multiple of p for any a and b. The binomial coefficients also satisfy many other divisibility properties with respect to prime numbers. These are explored in a project at the end of Chapter 3.

———————————— **Exercises for Appendix B** ————————————

1. Compute $\binom{12}{7}$ and $\binom{10}{6}$.

2. Determine the expansion of $(a+b)^7$ using the Binomial Theorem.

3. What is the coefficient of $a^{89} b^{11}$ in the expansion of $(a+b)^{100}$?

4. The triangular numbers are defined in Section 14.5. Show that the nth triangular number T_n is equal to $\binom{n+1}{2}$. Prove the following result of Aryabhata regarding the sum of the first n triangular numbers:

$$\binom{2}{2} + \binom{3}{2} + \cdots + \binom{n+1}{2} = \binom{n+2}{3}.$$

5. Show that

$$\binom{n}{r} = \binom{n-2}{r} + 2\binom{n-2}{r-1} + \binom{n-2}{r-2}.$$

6. Show that

$$\binom{n}{1} + 2\binom{n}{2} + 3\binom{n}{3} + \cdots + n\binom{n}{n} = n2^{n-1}.$$

7. Establish the following identities.

(a) $\binom{k}{k} + \binom{k+1}{k} + \binom{k+2}{k} + \cdots + \binom{n}{k} = \binom{n+1}{k+1}.$

(b) $\binom{n}{0} - \binom{n}{1} + \binom{n}{2} - \binom{n}{3} + \cdots + (-1)^n \binom{n}{n} = 0.$

(c) $\binom{n}{0}^2 + \binom{n}{1}^2 + \binom{n}{2}^2 + \cdots + \binom{n}{n}^2 = \binom{2n}{n}.$

Appendix C

Algorithmic Complexity and O-notation

In the text, we discuss many algorithms for solving number theoretic problems. An algorithm is a method to solve a problem in a finite amount of time. An important quantity to measure the effectiveness of an algorithm and compare different algorithms is the number of steps used by the procedure to solve a problem. The time taken for an algorithm to solve the problem is proportional to the number of steps; hence this is known as the **time complexity** of the algorithm. In comparing different algorithms to solve a problem, it is important to have an estimate for the running time of each procedure.

The time complexity of an algorithm is measured as a function of the input size. In describing the time complexity, it is simpler to give an estimate on the running time than to give the exact number of steps. The O-notation is very useful in describing the running time estimates.

Definition C.1. If $f(x)$ and $g(x)$ are two functions, then we say that $f = O(g)$ if there exist constants C and x_0 such that $|f(x)| \le Cg(x)$ for all $x \ge x_0$.

Example C.2.   1. If $f(n) = n^2 + n + 2$, then we can write $f(n) = O(n^2)$ because $|f(n)| \le 2n^2$ for $n \ge 3$.

2. If $p(n)$ is a polynomial of degree d in n, then $p(n) = O(n^d)$.

3. If $f(x) = \sin x$, then $f(x) = O(1)$. In fact, any bounded function f can be written as $f = O(1)$.

4. The O-notation is used to describe the principal error term in many number theoretic functions. For example, $\pi(x) = x/\log x + O(\sqrt{x})$ means that the error $|\pi(x) - x/\log x|$ is bounded by a multiple of $\sqrt{x}$ for all x sufficiently large.

The time complexity of an algorithm is a function of the input size. If the input size is k, then a method with time complexity $O(k)$ or $O(k^2)$ or $O(k^d)$ for some d is called a polynomial time algorithm. For example, sorting a list of names into an increasing order can be done in time proportional to the square of the number of names; hence it is a polynomial time algorithm. If the time taken by an algorithm is $O(a^k)$ for some a, then it called an exponential time algorithm. An example of an exponential time algorithm is integer factoring by trial division.

In number theoretic methods, we use the number of digits in the input as a measure of its size. For example, the addition of two numbers takes time proportional to the number of digits; and multiplication takes time proportional to the product of the number of digits. The number of digits in the binary representation of an integer n is $\lfloor \log_2 n \rfloor + 1$. Hence an algorithm with integer input n is polynomial time if its time complexity is $O([\log n]^d)$. Examples of polynomial time algorithms are the Euclidean algorithm, the solution of linear, and simultaneous linear congruences. If the time taken by an algorithm with integer input n is $O(a^{\log n})$ or $O(n^k)$, then it is an exponential time algorithm. For example, factoring an integer n by trial division can take up to $\sqrt{n}$ steps, so it is an exponential time algorithm. Loosely speaking, a polynomial time algorithm for a problem is usually considered acceptable, but if all the algorithms to solve a problem are of exponential time, then the problem is considered intractable. It may happen in practice that an exponential time algorithm is better than a polynomial time method for small inputs, but eventually, for large enough input, the polynomial time algorithm will be better. (See Exercise 3.) There are also algorithms that are better than exponential time, but not polynomial time. For example, it is expected that the quadratic sieve method (Section 12.2) has a running time of

$$O\left(e^{C\sqrt{\log(n)\log(\log(n))}}\right).$$

We also study some *probabilistic algorithms* that depend on a source of random numbers. Because of this randomness, it is not possible to give an estimate on the actual running time, but only an estimate on the expected running time. We consider a probabilistic algorithm to be good if it is expected to be polynomial time for most inputs. An example of this is the procedure to compute square roots modulo primes, which is discussed in Section 9.2. This method depends on computing a random quadratic nonresidue. In this case, a probabilistic algorithm is preferable to a deterministic method, which takes exponential time.

——————————— **Exercises for Appendix C** ———————————

1. Show that $\log n = O(n)$ and $(\log n)^d = O(n)$ for any d.

2. Show that for any polynomial p, $p(n) = O(2^n)$.

3. Suppose $p(n)$ is a polynomial. Show that if $a > 1$, then $\dfrac{p(n)}{a^n} \to 0$, as $n \to \infty$. This shows that a polynomial time algorithm is better than an exponential time algorithm for large-enough inputs.

4. Let $L(n) = e^{C\sqrt{\log(n)\log\log n}}$. Show that $p(n) = O(L(n))$ for any polynomial $p(n)$ and $L(n) = O(2^n)$.

Answers and Hints

Introduction

4. Yes. You can see this by setting the difference between the expressions for c and b (of Exercise 1) to 1.

6. The nth pentagonal number satisfies, $P(n) = P(n-1) + 3n - 2$ with $P(1) = 1$. Using this, it is easy to show that $P(n) = n(3n-1)/2$.

2.1 Divisibility

10. It is enough to give an example. Take $x = 14.1$ and $n = -7$.

24. You can try to prove the conjecture by induction on n. For this, you will find the Binomial Theorem and Exercise 21.

25. We guess that every third is even. This can be proved by induction using the formula $f_{3n} = 2f_{3n-2} + f_{3n-3}$.

29. The first part uses the factorization of $x^n - 1$ obtained from the geometric series. In (b), if $n = mk$, use the formula for the geometric series with $r = x^m$ to show that $x^m - 1 | x^n - 1$. For the converse, assume that $m \nmid n$ and derive a contradiction by doing the long division of $\frac{x^n - 1}{x^m - 1}$.

2.2 Primes

10. If $a_0 = 1$, proceed as follows. Suppose $f(a) = b$ for some a and b. Use the Binomial Theorem to show that $f(kb + a)$ is a multiple of b for any k.

2.3 Unique Factorization

4. If $\nu(n) = 6$, then n can have a prime factorization $n = p^5$ or $n = p^3 q^2$. Taking $p = 2$ and $q = 3$, we see that $n = 32$ is the smallest possible value.

6. Compare exponents in the prime factorizations of a^3 and b^2.

15. Determine the exponent of p in the prime factorization of the denominator and compare with that of the numerator.

17. You can prove this by induction on the number of prime factors. It is easy if n has only one prime factor. If $n = p^e m$ where $p \nmid m$, then we can write any divisor of n as $d = p^a d'$, where $d' \mid m$. From this conclude that there are $e(e+1)/2$ factors of p for each divisor of m.

19. In part (e), suppose the sum converges; then we can choose m such that $1/p_m + 1/p_{m+1} + \cdots < 1/2$, where p_j represents the jth prime. Let x be an arbitrary positive integer. Then $x - N(x)$ is the number of integers less than x that are divisible by at least one prime greater than or equal to p_m. This implies that $x - N(x) \leq x/2$, hence we obtain $x/2 \leq N(x) \leq 2^m\sqrt{x}$. Hence $x \leq 2^{2m+2}$, contradicting the choice that x was arbitrary.

2.4 Elementary Factoring Methods

8. Write $n = 16q + r$; then $n^2 = 16q' + r^2$. Hence it suffices to consider $r^2 \bmod 16$.

2.5 GCD and LCM

1. (a) 71, (b) 350, (c) 211, (d) 51, (e) 1, (f) 1.

10. Use unique factorization into primes and the formulas for the GCD and LCM in terms of their prime factorization.

11. The formula is established by using the formula for gcd and lcm in terms of the prime factorization. The formula for the lcm of four numbers is

$$[a,b,c,d] = \frac{|abcd|(a,b,c)(b,c,d)(c,d,a)(d,a,b)}{(a,b)(b,c)(c,d)(d,a)(a,c)(b,d)(a,b,c,d)}.$$

15. Observe that $2^{2^n} = (2^{2^{n-1}})^2$. Use this to simplify $F_n - 2$, using the identity $(a^2 - b^2) = (a-b)(a+b)$.

22. You can take $u = (a,b)$ and $v = 1$.

23. Try long division to obtain an analog of Proposition 2.5.15. More precisely, if $m > n$, then let $m = nq + r$ with $0 \leq r < n$. Show that $(a^m - 1, a^n - 1) = (a^n - 1, a^r - 1)$.

24. In (b), show that the right-hand side satisfies the definition of the Fibonacci numbers.

2.6 Linear Diophantine Equations

4. Since $(11, 29) = 1$, the equation has $11x + 29y = m$ has solutions. Write the general solution, and obtain a condition on m so that the equation has one nonnegative solution.

6. Express the coordinates of an arbitrary red line and a blue line as x/a and y/b. Minimize the difference by solving a linear Diophantine equation.

10. You may need to use the identity $(a, b, c) = (a, (b, c))$.

14. The solutions are $x = 15, y = 82, z = 15$, and $x = 50, y = 40, z = 30$.

3.1 Congruences

2. (a) 8, (b) 21, (c) 89, (d) 5.

3. Any divisor of 56.

8. Show that when n and m are odd and not multiples of 3, then $8 \mid n^2 - m^2$ and $3 \mid n^2 - m^2$.

9. Show that $3^{2n+5} \equiv 5 \cdot 2^n$ (mod 7) and $2^{4n+1} \equiv 2^n \cdot 2$ (mod 7).

11. Determine the values of $0^3, 1^3, \ldots, 8^3$ modulo 9.

16. Study the proof of Proposition 3.1.10.

17. In (c), use the fact that $10 \equiv -1$ (mod 11), and in (d), use $10^3 \equiv -1$ (mod 13) and $10^3 \equiv -1$ (mod 7).

3.2 Inverses Modulo m and Linear Congruences

2. Since $16 \cdot 8 - 1 \cdot 119 = 1$, the inverse of 16 modulo 119 is 8.

4. (a) 16, and (b) $x \equiv 17$ (mod 26), which gives six solutions modulo 156, $x = 17, 43, 69, 95, 121, 147$.

9. Show that $(4k)! \equiv [(2k)!]^2$ (mod p), and use Wilson's Theorem.

15. Use Wilson's Theorem in (a) and (b). In (b), use $n \equiv -2$ (mod $n + 2$) and $n + 1 \equiv -1$ (mod $n + 2$).

3.3 Chinese Remainder Theorem

1. (a) $x \equiv 149$ (mod 210), and (b) $x \equiv 430$ (mod 630).

2. (a) $x \equiv 19$ (mod 120). (b) No solution because $x \equiv 4$ (mod 6) and $x \equiv 12$ (mod 18) are incompatible.

3.4 Polynomial Congruences

5. You can do this directly by showing that if $p^k \mid x^2 - 1$, then p^k divides $x - 1$ or $x + 1$.

7. The four solutions to $x^2 \equiv 1$ (mod 2^k) for $k \geq 3$ are ± 1, and $2^{k-1} \pm 1$.

4.1 Fermat's Theorem

5. Use $2730 = 2 \cdot 3 \cdot 5 \cdot 7 \cdot 13$, Fermat's Theorem and the Chinese Remainder Theorem.

4.2 Euler's Phi Function

8. It seems plausible that if $\phi(n) \mid n - 1$, then n is prime, though no one has been able to prove this yet.

4.3 Euler's Theorem

10. Use $561 = 3 \cdot 11 \cdot 17$, Fermat's Theorem and the Chinese Remainder Theorem.

12. (a) 13^4, (b) 29^8, (c) $n = 486695567$ is not a prime power because $(2^n - 2, n) = 1$.

4.4 Lagrange's Theorem

3. The equation has $3k$ distinct solutions.

5.1 Classical Cryptosystems

7. The permutation used in part (a) is `legrandqstuvwxyzbcfhijkmop`. In part (b), it is `treasurgoldminzyxvwqkjhfcb`

5.3 The RSA Scheme

1. The plaintext is `A goal is a dream with a deadline.`

7. Do the computation modulo p and q, and use Fermat's Theorem together with the Chinese Remainder Theorem.

6.1 Pseudoprimes and Carmichael Numbers

6. If $a^{n-1} \not\equiv 1 \pmod{n}$, and $x^{n-1} \equiv 1 \pmod{n}$, then $(ax)^{n-1} \not\equiv 1 \pmod{n}$.

7.1 The Concept of Order

2. Compute the powers of 2 and 3 modulo 17, and see if $2^n \equiv -3^n \pmod{17}$ is possible for any n.

7.2 The Primitive Root Theorem

5. Assume by induction that it is true for proper divisors. Show that $\sum_{\delta \mid d} f(d) = d$, and then compare with $\sum_{\delta \mid d} \phi(\delta)$, and use the induction hypothesis to establish that $\phi(d) = f(d)$.

6. Use the existence of a primitive root, and write both sides in terms of the primitive root.

10. For (b), assume, by induction, that $3^{2^{k-4}} \not\equiv 1 \pmod{2^{k-1}}$, but $3^{2^{k-3}} \equiv 1 \pmod{2^{k-1}} \Rightarrow 3^{2^{k-4}} \equiv 1 + 2^{k-2}t$ for t odd, and then continue as in the proof of Theorem 7.2.10.

14. Find all integers n such that $\lambda(n) \mid 12$.

7.3 The Discrete Logarithm

5. (a) $x = 206$, (b) $x = 251$, and (c) 329.

9.1 Quadratic Residues

4. Yes, 5 is a quadratic residue, and the smallest positive value of n is 11.

7. Suppose there exist only finitely many such primes $p_1, \ldots, p_k$. Let $n = (p_1 p_2 \ldots p_k)^2 + 1$, and derive a contradiction.

10. Use the fact that 4 and 9 are residues, and combine with the previous exercise. You will need to show that if there are two consecutive residues, then there are two consecutive nonresidues.

16. Use the classification of cubic residues for $p \equiv 1 \pmod{3}$ to show that $a^{(p-1)/3} \equiv 1 \pmod{p}$ if and only if a is a cubic residue modulo p.

17. The point of the question is: What should be the value of $\left(\dfrac{a}{p}\right)_3$ when a is not a cubic residue modulo p?

9.3 Complete Solution of Quadratic Congruences

1. (a) ± 68. (b) ± 435. (c) No solution. (d) $\pm 123, \pm 379$.

2. (b) $\pm 25, \pm 143$. (c) $\pm 423, \pm 617, \pm 1199, \pm 1393$.

3. (a) $\pm 542, \pm 457$. (b) ± 137.

4. (b) $99, 82, 90, 107$. (c) When you multiply the congruence by 24, multiply the modulus by 3. Justify this, and find the two solutions.

11.1 Introduction

1. (a) $[4, 3, 1, 4]$, (d) $[2, 1, 3, 3, 1, 1, 3, 1, 1, 2]$.

3. The simple continued fraction is unique if the last term is greater than 1. Otherwise, there is a continued fraction representation with the last term equal to 1.

6. Write the approximations as p/q, where q is a suitable power of 10, and apply the Euclidean Algorithm to p and q.

11.2 Convergents

1. (a) $-1, -1/2, -4/7, -29/51, -33/58$. (b) $1, 4/3, 25/19, 279/212$.

4. n should be odd.

7. Try induction.

10. It suffices to solve this for $a_0 = 0$. In this case, the numbers are of the form a/b, $a > 0, b > 0$, and $b < 2a$.

11. A direct proof is possible using $p_n = a_n p_{n-1} + p_{n-2}$ and applying the Euclidean algorithm. Alternately, you can transpose both sides of the identity in Exercise 6.

12. Multiply out both sides, and use an appropriate identity from this section.

13. Replace $[a_n, a_{n-1}, \ldots, a_0]$ by p_n/p_{n-1}, using Exercise 11.

14. Since the fraction is symmetric, apply the result of Exercise 11 to determine p_{2n} and use the identity of Proposition 11.2.4.

19. In part (a), use the result of Exercise 11.

11.3 Infinite Continued Fractions

5. Show that the series

$$\frac{p_0}{q_0} + \left(\frac{p_1}{q_1} - \frac{p_0}{q_0} \right) + \cdots + \left(\frac{p_{k+1}}{q_{k+1}} - \frac{p_k}{q_k} \right) + \cdots$$

is convergent and converges to $\lim_{n \to \infty} \frac{p_n}{q_n} = \alpha$.

9. In part (c) solve the equation $Ax - Cy = 1$, and show that x is determined if $0 < x < C$, hence $x = q_{n-1}$. Show that this choice then determines y. In part (d) write $y = -(q_k x_{k+1} + q_{k-1})/(p_k x_{k+1} + p_{k-1})$, and show that we can apply the result of part (c).

11.4 Quadratic Irrationals

4. (a) $(4 + \sqrt{37})/3$, (b) $(6 + \sqrt{15})/7$.

7. Use Exercise 11.3.3, and the fact that the partial quotients of α are bounded.

11.5 Purely Periodic Continued Fractions

3. Use $x_{k-1} = a_{k-1} + 1/x_k$ to conclude that $\lfloor -1/\overline{x_k} \rfloor = a_{k-1}$. Next use $x_k = 2a_0 + 1/x_1$ to compute $-1/\overline{x_k}$.

4. Show that $x = \overline{[2a_0, a_1, a_2, \ldots, a_{k-1}]}$, and $-1/\overline{x}$ is the first complete quotient in the expansion of $\sqrt{n}$. This will prove the palindromic property of the period of $\sqrt{n}$.

5. Prove the first part by induction. For the second part, use the fact that $B_k = 1$ when the period occurs.

Projects for Chapter 11

5. $a_1 = k$ on the interval $[\frac{1}{k+1}, \frac{1}{k}]$. If $a_1 = k$, then $a_2 = 1$ on the interval $[\frac{1}{k+1}, \frac{2}{2k+1}]$, hence the probability that $a_2 = 1$ is $\sum_{k=1}^{\infty} \left(\frac{2}{2k+1} - \frac{1}{k+1} \right)$. The series is telescoping and can be evaluated using the series for $\log(2)$.

13.1 Best Approximations

12. Write the equation as $1 + (z - y)^2 = y(z - y) + y^2$ to show that $(z - y)/y$ is a convergent of $(1 + \sqrt{5})/2$.

13. Use the formula for Fibonacci numbers of Exercise 2.5.24 to show that if n is a Fibonacci number, then $5n^2 \pm 4$ is a square. Conversely, if $5n^2 \pm 4 = m^2$, then take $p = (m + 2)/2$, and show that p/n is a convergent of $(1 + \sqrt{5})/2$.

13.3 Algebraic and Transcendental Numbers

1. (a) $0.880008000000000000000008\ldots$, (b) $0.900009000\ldots$, where the digit in the k!th place is a 9 if k is odd and 0 otherwise.

Projects for Chapter 13

I(b). Another way to think about this is to ask "What are the two closest good approximations of p/q?"

I(f). You will need to show that if $ps - qr = \pm 1$, then p/q and r/s are eventually adjacent in some Farey series. First, determine all the solutions to $px - qy = \pm 1$ from the two fundamental solutions determined at the beginning of the project.

I(g). Show that $\sqrt{(p/q - r/s)^2 + (1/2q^2 - 1/2s^2)^2} \geq 1/2q^2 + 1/2s^2$. This, together with part (f), will prove the result.

II(a) Look at the solution to the equation $py - qx = 1$ with $0 < y < q$. Show that $\frac{p}{q} = \frac{1}{yq} + \frac{x}{y}$, and then use induction. For rational numbers greater than 1, first show that there exists N such that $\sum_{i=1}^{N} \frac{1}{i} < p/q < \sum_{i=1}^{N+1} \frac{1}{i}$, and then use the previous part.

14.1 Introduction

7. A solution satisfies $x^{1/x} = y^{1/y}$. Sketch a graph of this function, and find the maximum. From this, finding the integer solutions is not hard.

14.2 Congruence Methods

5. Work modulo 9.

7. Work modulo 13.

14.4 **Sums of Two Squares**

8. Show that $x = p_n q_n + p_{n-1} q_{n-1}$. After squaring x, use $p_n q_{n-1} - q_n p_{n-1} = \pm 1$ to simplify the expression.

13. First show that the only divisors of n are of the form $p \equiv 1 \pmod 4$ and then show that if n_1 and n_2 are sums of relatively prime squares, then so is $n_1 n_2$.

15.1 **Arithmetical Functions**

11. Since $\sigma(n) = 2n$, look at the power of 2 on both sides using the formula for $\sigma(n)$ in terms of the prime factorization of n.

12. n must be prime.

20. Show that both sides are multiplicative functions.

15.4 **Möbius Inversion Formula**

7. Express each term on the right as a geometric series, and then use the Möbius inversion formula.

16.1 **Counting Primes**

2. Let S_i be the set of multiples of p_i which are $\le n$.

16.3 **Chebyshev's Theorem**

5. Multiply by $n!$ and use Bertrand's postulate.

6. Use Bertrand's postulate to see that at least one prime occurs only to the first power in the product.

17.1 **Quadratic Reciprocity**

6. For $8n + 5$, if $p_1, p_2, \ldots, p_k$ are finitely many primes of the form $8n + 5$, then consider $Q = (p_1 \ldots, p_k)^2 + 4$ to derive a contradiction.

7. Use the fact that $q \equiv 3 \pmod 8$, hence $2^p \equiv -1 \pmod q$ by Euler's Criterion.

17.2 **Proof of the Quadratic Reciprocity Law**

3. Write $p = 8k + 1$ and use $\lfloor n + \alpha \rfloor = n + \lfloor \alpha \rfloor$, for $\alpha > 0$ real.

Appendix B. **Binomial Theorem**

6. Consider the binomial expansion of $(1 + x)^n$ and differentiate both sides.

7. In part (c), expand $(1+x)^{2n}$ and $(1+x)^n (1+x)^n$ using the Binomial Theorem and look at the coefficients of x^n.

Index of Notation

The following notation is used in the book. The table lists the symbols used in the order of first occurrence in the text. In addition to these symbols, we use the standard notation for different number systems.

$\mathbb{N}$	Natural Numbers
$\mathbb{Z}$	Integers
$\mathbb{Q}$	Rational Numbers
$\mathbb{R}$	Real Numbers
$\mathbb{C}$	Complex Numbers
$\mathbb{Z}/m\mathbb{Z}$	Equivalence class of integers modulo m
$(\mathbb{Z}/m\mathbb{Z})^{\times}$	Equivalence classes of invertible numbers modulo m
$\mathbb{F}_p$	Finite field of integers modulo p

Symbol	Meaning	Page
$a \mid b$	a divides b	7
$a \nmid b$	a does not divide b	7
$\lfloor x \rfloor$	Floor Function	11
$\lceil x \rceil$	Ceiling Function	11
$a \bmod b$	Remainder when a is divided by b	11
f_n	nth Fibonacci number	14
$\pi(x)$	Prime Number Function	20
$\nu(n)$	Number of positive divisors of n	28
(a, b)	Greatest Common Divisor	38
$[a, b]$	Least Common Multiple	40
F_n	nth Fermat number	46
$\sigma(n)$	Sum of positive divisors of n	55
$\displaystyle\sum_{d\mid n}$	Sum over the positive divisors of n	55

Symbol	Meaning	Page
$a \equiv b \pmod{m}$	a congruent to b modulo m	60
$a^{-1} \bmod m$	Inverse of a modulo m	69
$\mathbb{F}_p$	Finite field of p elements	70
$\phi(n)$	Euler's Totient Function	108
$\mathrm{psp}(a)$	Pseudoprime to base a	147
$\mathrm{spsp}(a)$	Strong Pseudoprime to base a	152
$\mathrm{ord}_n(a)$	order of a modulo n	169
$\lambda(n)$	Minimal universal exponent	177
$\mathrm{ind}_g(y)$	Index of y to base g	182
$\left(\dfrac{a}{p}\right)$	Legendre Symbol	216
$[a_0, a_1, \ldots, a_n]$	Simple Continued Fraction	246
$[a_0, \ldots, a_{N-1}, \overline{b_0, \ldots, b_T}]$	Periodic Continued Fraction	267
$Z[i]$	Gaussian Integers	332
$\mu(n)$	Möbius function	362
$\displaystyle\prod_p$	Euler product	370
$\Theta(x)$	Chebyshev's theta function	392
$\psi(x)$	Chebyshev's psi function	392
$\zeta(s)$	Riemann zeta function	404
$\left(\dfrac{a}{m}\right)$	Jacobi Symbol for m composite	432
$f \sim g$	Equivalence of quadratic forms	443
$h(\Delta)$	Class number of Δ	450
$\mathbb{P}^n(k)$	Projective n-space	487
$\mathcal{O}$	Point at infinity on elliptic curves	491
$E(\mathbb{Q})$	Rational points on an elliptic curve	491
E_p	Elliptic curve modulo p	498
$v_p(x)$	p-adic valuation of x	504
$n!$	n factorial	521

Index